Radio Frequency Technologies

Holger Heuermann

Radio Frequency Technologies

Components and Measurement, Radio, RFID, and Localization Systems

Springer

Holger Heuermann
IMP
FH Aachen
Aachen, Nordrhein-Westfalen, Germany

ISBN 978-3-658-50169-3 ISBN 978-3-658-50170-9 (eBook)
https://doi.org/10.1007/978-3-658-50170-9

Translation from the 4th German language edition: "Hochfrequenztechnik" by Holger Heuermann, © 2023. Published by Springer Fachmedien Wiesbaden GmbH. All Rights Reserved.

Responsible Editor: Alexander Grün
This Springer imprint is published by the registered company Springer Fachmedien Wiesbaden GmbH, part of Springer Nature.
The registered company address is: Abraham-Lincoln-Str. 46, 65189 Wiesbaden, Germany

Forword

Textbooks and reference books are almost exclusively written by people with the best academic qualifications. This book is different. I grew up far removed from education, in a house without books. When electronics emerged, as AI has now, I completed a two-stage apprenticeship, first as a plumber and then as an electronics technician. This was followed by a degree course at what was then a second-class technical college and finally a degree course at an academic university, followed by a doctorate.

I then learned a lot during my industrial activities at the two exceptional companies Rosenberger Hochfrequenztechnik and Infineon, which I was then able to continue even more intensively in my numerous research activities at the Applied University FH Aachen.

This book is not written by an academic, but by a maker. The aim is to enable students to develop RF circuits, but this is better described by the word "synthesize".

Circuit development in RF technology has undergone a very rapid transformation in recent decades due to the boom in mobile communications and communication technology. Much of what was science fiction 40 years ago is now part of everyday life (for example, bidirectional communication from smartphones to satellites).

The need for low-cost RF mass products has changed the RF development, which in the past was focused on transmitters, radio relay, military and satellite systems, immensely. For a large number of RF engineers, this shift in focus means that components and assemblies have changed.

In addition to the increased use of semiconductor technology, assembly techniques using so-called LTCC technology and, in particular, SMD technology have been added.

As products have changed, the use of computers and modern measurement technology has also changed a great deal in the development environment. Classic development tools such as the so-called Smith chart are now used in a completely different way in development than was previously the case.

The precise measurement technology and the support of circuit simulators allow RF engineers to model their components precisely and thus make excellent predictions in circuit design.

We are getting closer and closer to the wish of all managers: from simulation to production! Thanks to precise modeling, RF technology is losing its reputation as *black magic* science. Instead, it enables the construction of precision electronic components.

With the introduction of Bachelor's and Master's degrees in Germany, a two-stage professional qualification became necessary. Bachelor graduates should already be able to work as qualified engineers in industry. This book is designed precisely for this training.

The contents of this textbook and reference book are intended to do justice to this aspect of qualification for a career as a development engineer with bachelor's degree tasks. The tasks include linear circuits in the GHz range.

This book takes a completely different approach to the other textbooks. The other textbooks explain how something works. These explanations are also included in this book. However, much greater emphasis is placed on the introduction to creative circuit development using various circuit synthesis techniques.

Despite the pronounced textbook character, emphasis has been placed on the fact that the book also offers valuable assistance for experienced engineers in industry and science, not least through a large number of auxiliary sheets on circuit synthesis in the appendix and supplementary chapters, and is a tool as a reference work for carrying out a good initial design of linear circuits, which can then be used to develop an RF circuit with optimum performance using the circuit simulator.

This book presents a model for separating Bachelor's and Master's content. In addition to an unabridged presentation of the most important fundamentals, devices, components and systems of RF technology, complex components and subsystems such as RF mixers, oscillators and synthesegenerators are only presented in their functionality.

This means that Bachelor's graduates can also use these circuits as qualified ICs. The associated basics such as non-linear circuit technology are not covered.

A major focus is clearly on the linear components of highly integrated RF circuits. Furthermore, it was a concern to emphasize the contents of RF electronics and therefore to leave out waveguide theories, which are extensively explained in numerous textbooks.

A very important part of the book are the RF systems contained in the back section, starting with the clear concepts of the measurement systems, through an innovative dual-mode radio that Bachelor's graduates implemented in their theses without the use of ICs, to radar systems and RFID technology, where many tuning measures are introduced that cannot be replaced even by AI.

This RF book is complemented by the much more scientifically elaborated master book "Microwave Technology", [50], so that the master student also receives a complete education with both books to perform a professional circuit synthesis of the much more complex master topics.

Aachen, September 2025 Holger Heuermann

Acknowledgments

I would like to thank all employees and final theses of the RF Technology Department of the FH Aachen and the Institute for Microwave and Plasma Technology IMP, who have made numerous contributions.

Dipl.-Ing. H. Erkens made many contributions to the mixed-mode parameters. In addition, the book contains some illustrations based on work by Prof. Dr.-Ing. R. Knöchel of the University of Kiel, which he made available. Further illustrations and fundamentals of mixed-mode parameters were kindly provided by Prof. Dr.-Ing. M. Möller, University of Saarbrücken.

Thanks also go to Dr.-Ing. A. Sadeghfam, whose section on coils has enhanced the book, especially for LTCC and chip developers. I would like to thank Dr.-Ing. R. Stolle, Infineon Munich, and Prof. Dr.-Ing. W. Bogner, FH Degendorf, for proofreading the first German-language edition.

I would like to thank my teacher Prof. Dr.-Ing. Schiek, Ruhr University Bochum, and also Prof. Dr.-Ing. E. Voges. Both were also passionate RF book authors, along with many more German-speaking RF book authors who were still around at the end of the last century. All of them motivated me greatly with their special works.

The first editor of this book, Prof. Dr.-Ing. Mildenberger and the publisher Vieweg for their excellent cooperation and the high quality of the printing. The third and fourth German-language editions, as well as this English-language edition, were then professionally supervised by the Springer team under the direction of Executive Editor Dr.-Ing. A. Grün, for which I am grateful to everyone and would like to single out Ms. Nörthen.

Aachen, September 2025 Holger Heuermann

Contents

Chapter 1

Introduction

1.1 Contents of Modern Radio Frequency Technology

Following the first German edition of this basic book, many of the then the novel contents of individual chapters of the book have now become radio frequency technology (RF technology for short). This book can therefore can still rightly claim to describe the state of the art in RF technology.

This reference and textbook is intended to explain the mechanisms of the propagation phenomena of electromagnetic waves in electrical circuits. Furthermore, the reader should be able to analyze and optimize a variety of RF frequency technology circuits and optimize them. It also helps electronics engineers to understand the effects of fast circuits and provides a basic framework for future RF frequency engineers.

First and foremost, this book is intended to help engineers in the field of RF technology to develop modern products. Since new innovative products cannot be realized without a development environment and the necessary basic principles, the following topics formed the guideline for the structure of this book:

1. **Based on the RF Products to be Developed**
    (a) Integrated transceiver circuits with small signal components and sub-circuits such as PLL's, VCO's, mixers, LNA's, buffers, AGC's, filters, transistors, diodes
    (b) Digital high-speed switching and transmission technology components: ECL and other high-speed digital components as well as high-speed data lines
    (c) Power components: Amplifiers, switches, signal dividers
    (d) Passive components: RF cables, HQ filters, impedance transformers, couplers, baluns, signal splitters, attenuators, isolators, circulators, MEMS, antennas
    (e) Systems: Concepts for transceivers, radar, sensors, measurement systems
        i. Subsystems and their partitioning: Modulators, synthesizers
        ii. Optimizations: Sensitivity, noise, efficiency, costs, size

H. Heuermann, *Radio Frequency Technologies*,
https://doi.org/10.1007/978-3-658-50170-9_1

2. **Based on the Equipment of an RF Laboratory**
   (a) Circuit simulator: Linear and non-linear. E. g.: ADS, Designer, Microwave Office
   (b) Field simulator: 2D, 2.5D, 3D, moment method, difference method, finite elements method. E. g.: ADS with Momentum, Sonnet, CST Microwave Studio, HFSS
   (c) System simulator, e. g.: ADS, Designer
   (d) Measuring equipment: Vector network analyzer, spectrum analyzer, power meter, contacting technology (prober, test fixture)

3. **Based on the Necessary Fundamentals**
   (a) Propagation of electromagnetic waves in free space (antennas, EMC)
   (b) Wave types, modes and conductor systems. E. g.: Z_ℓ, Z_{even}, Z_{odd}, ϵ_r, γ and many more
   (c) System parameters: E. g.: S, Y, Z, M, k, F, and many more
   (d) Measurement technology
      i. Scattering parameters (small or large signal in coaxial or planar line systems, scalar or complex
      ii. Measurement of power, frequency, current, voltage, material constants
      iii. Noise in circuits (noise figure, noise power, phase noise, amplitude noise)

Many of these basics are covered below. This overview is intended to help students develop a feel for the range of knowledge required to become a qualified RF frequency technician. to start a career in industry.

<u>Software Recommendations</u>

Every student of engineering professions and every active development engineer is the mathematical program `Matlab` is recommended. Matlab (or alternative programs) can be used as a high-level programming language. with the simplest syntax for even the most complex mathematical operations.

For students and developers as well as researchers who have a penchant for solving theoretical problems, the program `Maple` is very interesting. Maple can be used to achieve analytical solutions.

The following two programs are available as free versions for students who want to deepen their knowledge of RF and microwave technology, as well as for RF development engineers working in industrial companies:

1. RF circuit simulator: Old versions of Serenade, RFSim99 and more
2. 2.5D RF field simulator: Sonnet

The Serenade version SV 8.5 was also used for the creation of this work used in many places. Although this version is quite old, it allows to optimize circuits, which other free versions do not offer. This means that it is easy to work from the reader to understand the calculations. Serenade is also available in the download area of the FH Aachen homepage, [70].

Sonnet is an easy-to-use field simulator, which will be discussed less in the following. A good description of field simulators can be found in [137].

The basics of field simulation as well as many other advanced chapters can be found in the book *Microwave Technology* in, [50]. This book contains the master's course in the field of hardware for communications engineers at the FH Aachen and builds directly on this RF frequency technology book.

Circuit Technology in the GHz Range

The aim of this radio frequency book is to coach linear circuit technology in the single-digit GHz range, in which there are still large correlations with electronics, and in the two-digit GHz range, which is well illustrated by the Fig. 1.1.

Fig. 1.1: Illustration of a low-noise converter (LNC) for satellite television recieving

1.2 Structure of RF Frequency Arrangements

Circuit development in RF technology can be structured in a similar be structured hierarchically in a similar manner to the structure of software programs, Fig. 1.2.

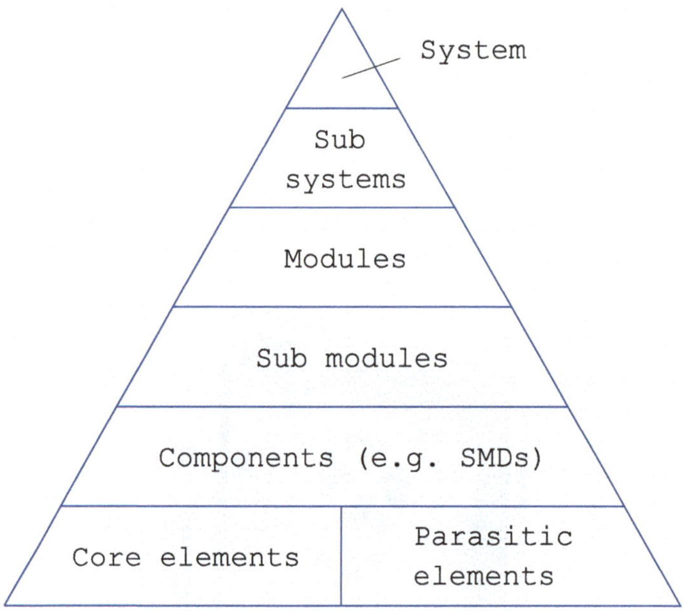

Fig. 1.2: *Pyramid of the development of RF frequency arrangements*

The smallest element is not the component itself, as in electronics itself, but the components used to model a component: Core elements (e.g. inductors) and parasitic elements (e.g. resistors for the losses). The core element that contains the desired properties of the component, usually differs from the absolute value that the component should have. The parasitic elements are generally determined by the finite size and the losses of the component.

From several components you can create so-called sub modules or also modules (e.g. amplifier) can be produced. The frequency dependent properties of linear modules can be excellently specified by means of the scattering parameters, which will be will be discussed in detail below.

In practice, more complex modules (e.g. IQ modulators) are made up of several submodules (e.g. phase shifters, couplers and mixers). For a simple understanding of subsystems (e.g. receiver unit) or systems (e.g. 5G, 6G), this multi-level description is very advantageous. This simplifies the associated block diagrams.

Systems can often still be described using scattering parameters. However, additional measurements are always necessary to evaluate them, as the overall system must have the lowest possible bit error rates. or similar aspects.

This book is guided by this hierarchical structure. In particular is a detailed description of the components down to the elements. With regard to the consideration of systems measurement, localization and RFID systems are dealt with in more detail, from which the foundations for modern communication systems can be derived very well.

1.3 Development of RF Components and Systems via Circuit Synthesis

The aim of this book is to explain and teach the development process of RF components and RF systems.

Circuit Analysis

Often, only circuits are analyzed in universities and books. The ability to analyze circuits is, of course, a necessary foundation for the development of circuits. However, the development of RF hardware is a very creative process.

Circuit analysis is the process of determining how an existing circuit behaves, given its components and connections. It starts with a known circuit diagram and uses methods such as Ohm's law, Kirchhoff's laws, and network theorems to calculate voltages, currents, and power.

The goal of analysis is to understand or verify the performance of a circuit that is already designed. For example, in analysis, you might calculate the output voltage of an amplifier based on its resistor values.

Circuit Synthesis

This book shows that the components can be specified using S-parameters and that circuits can be derived from these using various approaches contained in the book.

This creative process is referred to as "circuit synthesis".

The aim of this book is to highlight several approaches to circuit synthesis. In addition, other important peripheral aspects such as component modeling are presented, since meaningful RF circuit development is not possible without the correct RF models.

Circuit synthesis, on the other hand, is the process of designing a circuit to meet specific functional requirements.

It starts with desired performance criteria, such as a target gain or filter frequency, and determines what components and topology will achieve them.

Synthesis is creative and constructive, often involving trade-offs between cost, size, efficiency, and performance. For example, in synthesis, you might design a low-pass filter that has a 1 kHz cutoff frequency and a specific impedance.

In short: Circuit analysis works from a given circuit to its behavior, while circuit synthesis works from desired behavior to a circuit design. Both skills are essential for engineers-analysis helps validate and troubleshoot, while synthesis enables innovation and creation.

1.4 RF Technology: Between Electronics and Optics

In electronics, the lengths of the connecting cables between different between different components have no significant influence on the electrical behavior of the circuit: The electrical voltages of a source and of a connected load are the same (Fig. 1.3).

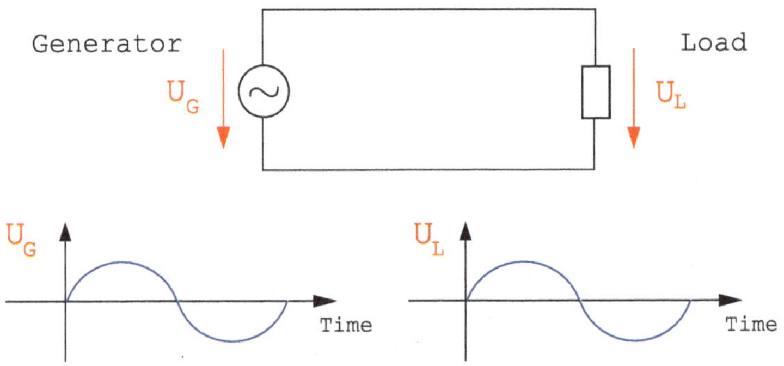

Fig. 1.3: Influence of line conductors in the world of electronics: Representation of the sinusoidal voltages at the generator and at the load with low frequencies

RF technology differs from electronics among other things, in that the electrical voltages at a signal source and a connected load are no longer the same (Fig. 1.4). The length of the interconnected line and the frequency determine the phase difference between the voltages at the source and load.

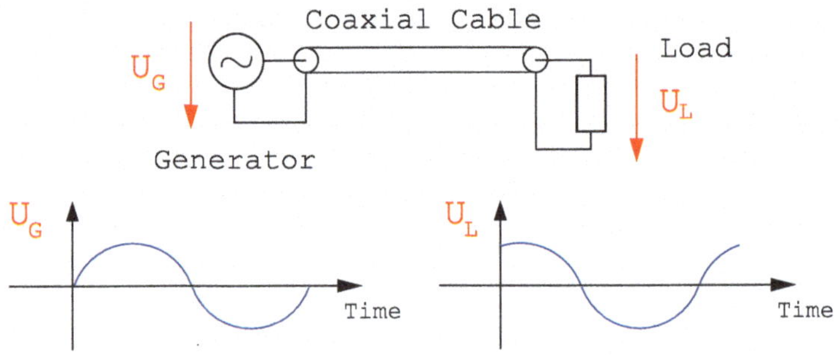

Fig. 1.4: Influence of cable lengths in the world of RF technology: Different phase values of the voltages at the generator and at the load

In optics, there is generally no longer a fixed phase relationship between source and load. For this reason, optics often only works with the two digital states: $1 \equiv$ strong light signal, $0 \equiv$ weak or no light signal at all. However, switching between these two states takes place in the GHz range, so that many Gbit/s can be transmitted.

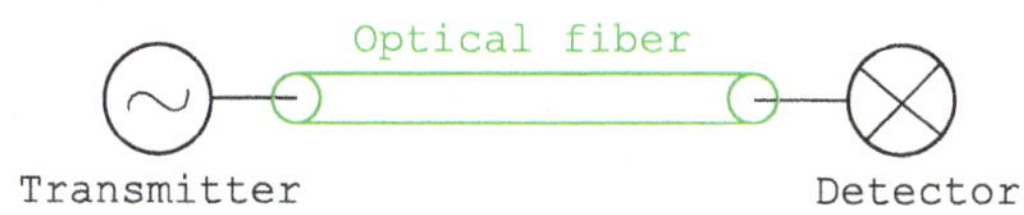

Fig. 1.5: Influence of line lengths in the world of optics: Outside the coherence length: No fixed relationship between the phases of transmitter and detector

If you are able to detect the voltage distribution on a RF transmission line with a snapshot, you can see that it is sinusoidal along the line (Fig. 1.6), provided the generator generates a sinusoidal signal.

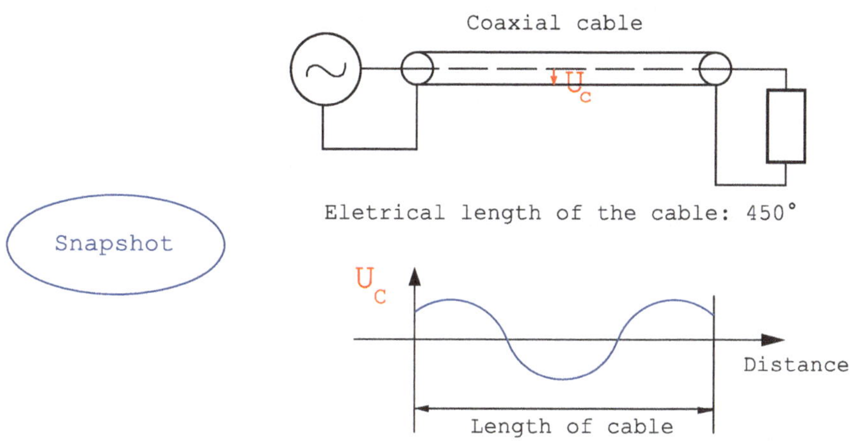

Fig. 1.6: Voltage amplitudes on a line in the world of RF technology: Voltage distribution along the line for a "snapshot"

The fact that, just as in electronics, RF technology has stationary systems allows precise analysis and synthesis techniques to be used.

In contrast to electronics, RF technology deals with wave effects, similar to those known from optics. It is therefore advantageous to describe RF circuits not with currents and voltages at the terminals, but with incident and outgoing wave impulses at the ports. Their ratios result in the so-called scattering parameters, which can be divided into reflection and transmission coefficients.

The connection between the optical and the high-frequency world is illustrated in the Fig. 1.7.

The signals shown in the Fig. 1.7 are referred to as waves in RF technology and optics. These waves and the forward scattering parameters (also S-parameters) are defined as shown below.

a_1 : Incident wave

b_1 : Reflected wave

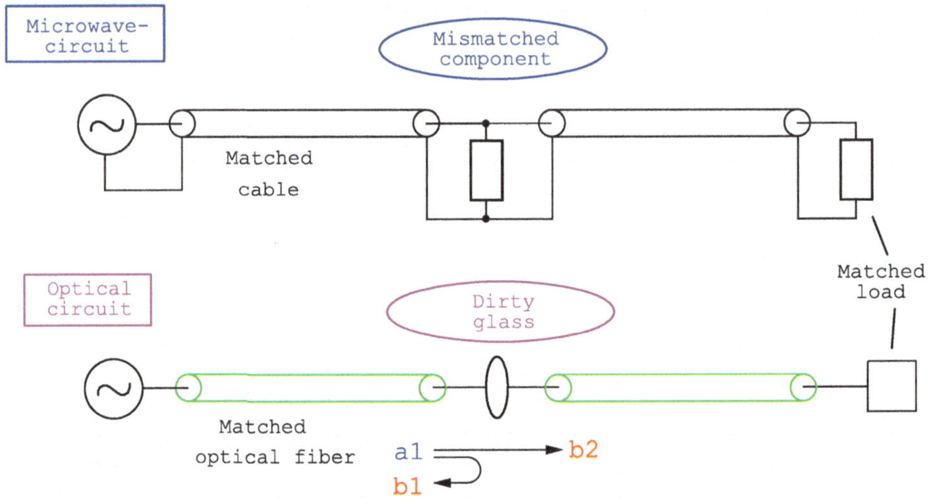

Fig. 1.7: Wave effects in the world of RF technology and optics: Explanation of the scattering matrix

$$b_2 \; : \; \text{Transmitted wave}$$

$$S_{11} = \frac{b_1}{a_1} \; : \; \text{Reflection coefficient}$$

$$S_{21} = \frac{b_2}{a_1} \; : \; \text{Transmission coefficient}$$

The scattering parameters in the form of the reflection coefficient and the transmission coefficient for the forward direction are thus formed from the ratio of the reflected or transmitted waves to the incident wave. The direct conversion of these wave parameters into current and voltage parameters is presented in the following chapter.

Chapter 2

Circuit Theory Basics

For many decades, in the RF technology the use of scattering parameters is the standard for the design of circuit components.

Scattering parameters are used in the design as a result of the circuit and field simulation. After setting up an RF circuit, the scattering parameters are measured using a network analyzer (Fig. 2.1) and compared with the target values of the simulation. For this

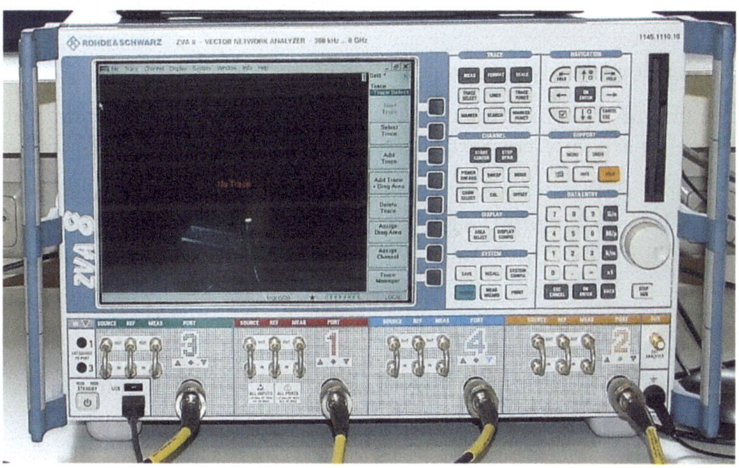

Fig. 2.1: Vector Network Analyzer (VNA) for measuring the vectorial scattering parameters (here a device with 4 measuring ports of type ZVA 8 from Rohde & Schwarz)

reason, this chapter also focuses on the circuit-theoretical description with scattering parameters.

In this chapter of circuit fundamentals, the necessary tools for the development of *source-free*, *linear* and *time-independent* RF circuits will be presented. Therefore, the system requirements are briefly discussed at the beginning.

The scattering matrix and the circuit design of two-ports via the scattering parameters are

H. Heuermann, *Radio Frequency Technologies*,
https://doi.org/10.1007/978-3-658-50170-9_2

then presented for a quick introduction to practice. The scattering parameter properties of two-port and multi-port devices are explained in detail and the usefulness of transmission and chain parameters is presented at the end of the chapter.

2.1 System Requirements

Fig. 2.2 shows a general circuit for further clarification, which has an input signal $x_{(t)}$ and an output signal $y_{(t)}$. Both signals can only be characterized by the associated currents, voltages or a combination of both (this is the RF case).

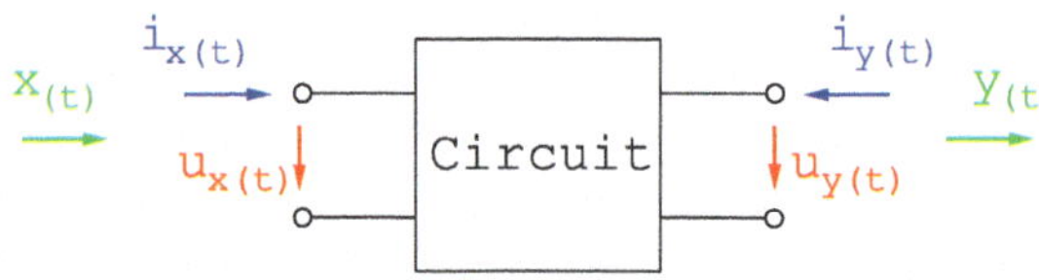

Fig. 2.2: Representation of an electrical circuit from a system perspective with an input signal $x_{(t)}$ and an output signal $y_{(t)}$

RF circuits are predominantly *source-free, linear* and *time-independent*.

Source-free is a circuit if there is no signal source inside. In the following, only circuits in which signal sources are only connected to external terminals or ports are considered.

A circuit is *linear* if it fulfills the two properties of the equations (2.2) and (2.3). The condition for this is that two input signals ($x_1(t)$ and $x_2(t)$) can be applied, which in turn generate two output signals ($y_1(t)$ and $y_2(t)$).

$$x_1(t) \rightarrow y_1(t) \quad , \qquad x_2(t) \rightarrow y_2(t) \tag{2.1}$$

If it is now the case that the addition of the two input signals results in the sum of the output signals,

$$x_1(t) + x_2(t) \quad \rightarrow \quad y_1(t) + y_2(t) \quad , \tag{2.2}$$

and the multiplication of the input signal by a constant A results in an output signal increased by the factor A,

$$A\,x(t) \quad \rightarrow \quad A\,y(t) \quad , \tag{2.3}$$

then a system is *linear*!

A circuit is *time-independent* or *time-invariant*, if a shift in the arrival of the input signal continues to produce the same output signal with the same time shift:

$$x(t - t_0) \quad \rightarrow \quad y(t - t_0) \quad . \tag{2.4}$$

Furthermore, we limit ourselves to analyzing the circuits to purely sinusoidal signals. For the complex voltage U and the complex current I, the following relationship exists to the time-dependent signals $u(t)$ and $i(t)$:

$$u(t) = \hat{u}\cos(\omega t + \alpha) = \sqrt{2}\,\mathrm{Re}\{U\,e^{j\omega t}\} \quad \text{with} \quad U = \frac{\hat{u}}{\sqrt{2}}\,e^{j\alpha} \quad , \tag{2.5}$$

$$i(t) = \hat{i}\cos(\omega t + \beta) = \sqrt{2}\,\mathrm{Re}\{I\,e^{j\omega t}\} \quad \text{with} \quad I = \frac{\hat{i}}{\sqrt{2}}\,e^{j\beta} \quad . \tag{2.6}$$

It can be seen that here, as in most publications, RMS pointers for U and I are used[1]. Some books also mention the use of peak value pointers.

The circuits that are examined below do not have any internal signal sources and are therefore referred to as *source-free*.

2.2 The Scattering Matrix

The transmission line theory is usually presented before the scattering parameters, as shown in chapter 4. In the following, this path was not chosen and therefore some findings of the transmission line theory must be assumed without derivation[2].

However, this structure emphasizes the importance of the scattering parameters and allows everything to be derived from them or the transmission properties of the components in a simple way.

2.2.1 Introduction of Wave Quantities

In both electronics and RF technology, a large number of components such as amplifiers and filters are connected in series to produce systems such as transmission paths. The majority of all RF components therefore have one input and one output.

In electronics, a so-called four-pole is connected via Z, Y or similar parameters[3] described. Two poles are often assigned one input and the other two poles to an output. The schematic representation of a four-pole with the adjacent currents and voltages is

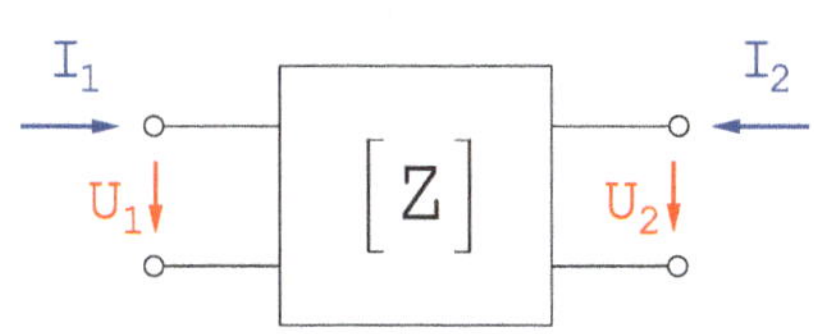

Fig. 2.3: Z-parameter representation of a four-pole

illustrated in Fig. 2.3.

The voltages applied between two terminals (U_i, $i = 1, 2$) and the currents flowing into or out of a terminal pair (I_i) can be calculated in electronics using node and mesh equations

[1] In this book the letter U is used for the voltage, similar to [44], [50].

[2] This non-optimal approach is necessary, to be able to carry out a practical course parallel to the lecture.

[3] These are also presented in detail below.

can be calculated. However, if the connecting lines between individual two-ports have expansions that in the range of a few degrees of the electrical length p, calculated from:

$$p = \frac{\sqrt{\epsilon_r}\,\omega}{c_0}\,\ell \cdot \frac{180^\circ}{\pi} = \frac{\sqrt{\epsilon_r}}{\lambda_0}\,\ell \cdot 360^\circ \quad . \tag{2.7}$$

The letter ℓ corresponds here to the mechanical length. As a rough approximation, every cable and every circuit that has an electrical length of more than 10° must, from the point of view of be considered from the point of view of RF technology.

Example: $\ell = 10\,\text{cm}$ and $\epsilon_r = 10$: $\text{f} = 26\,\text{MHz}$.

For these RF lines, the voltages U_i and the currents I_i are calculated from a leading (U_{pi}) and a trailing voltage wave (U_{ri}), according to Fig. 2.4. The indices r and p stand for the English designations *reflection* and *propagation*.

$$U_1 = U_{p1} + U_{r1}\,, \quad I_1 = \frac{U_{p1}}{Z_{L1}} - \frac{U_{r1}}{Z_L} \tag{2.8}$$

$$U_2 = U_{p2} + U_{r2}\,, \quad I_2 = \frac{U_{p2}}{Z_L} - \frac{U_{r2}}{Z_{L2}} \tag{2.9}$$

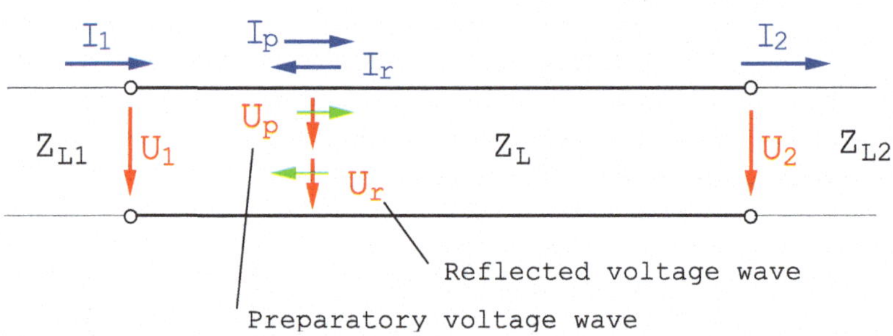

Fig. 2.4: Forward and reverse voltage and current wave along a transmission line

A detailed derivation of Fig. 2.4 is described in the chapter 4.

The so-called characteristic impedances are denoted by Z_{Li}. The characteristic impedance is used to link the voltages and currents propagating in one direction along a currents along a line are linked with each other according to Ohm's law. The characteristic impedance Z_L is often referred to as the reference or standardization impedance.

This reference impedance is often also called the system impedance and is labeled Z_0. It is advantageous to choose a *real* reference impedance. This is defined as follows selected as follows:

General RF technology: 50 Ω,

Antenna technology: 75 Ω,

High-speed digital technology: 100 Ω.

In recent decades, it has proven to be very practical not to use voltages and currents, but with so-called

incident waves (a_i) and
reflected or transmittierted waves (b_i).

These wave quantities are the normalized voltage waves voltage waves U_{pi} and U_{ri}. For normalization, the roots of the respective characteristic impedances of the connecting cables are used for normalization:

$$a_i = \frac{U_{pi}}{\sqrt{\mathrm{Re}\{Z_{Li}\}}} \quad , \qquad b_i = \frac{U_{ri}}{\sqrt{\mathrm{Re}\{Z_{Li}\}}} \quad . \tag{2.10}$$

For real normalization resistors, half of the product of the wave a with the conjugate complex value a^* corresponds to the active power P_{wa} running into the terminal pair:

$$P_{wa} = \frac{1}{2}\, a\, a^* = \frac{U_p\, U_p^*}{2\, Z_L} \quad . \tag{2.11}$$

The same applies to the power wave b and the outgoing active power P_{wb}:

$$P_{wb} = \frac{1}{2}\, b\, b^* = \frac{U_r\, U_r^*}{2\, Z_L} \quad . \tag{2.12}$$

The power waves a and b have the dimension $\sqrt{\text{Watt}}$.

It has proven to be just as practical to no longer work with terminal pairs, but with ports. With this change, a component with one input port and one output port is no longer classified as a four-pole element, but as a two-port (network). (Fig. 2.5). While

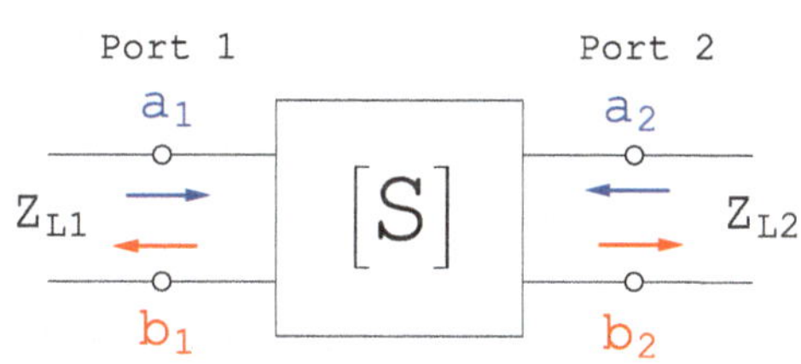

Fig. 2.5: S-parameter representation of a two-port with power waves in two-pole representation

the representation of a two-port in Fig. 2.5 for the representation of the internal electrical circuits of the components is used, the so-called single-pole representation (Fig. 2.6) is used for representations in the system level. Both forms are equivalent in the S-parameter description.

If the Eqs. (2.8) and (2.9) are used in the equations under (2.10), the links between the currents and voltages applied to the ports and the incident and outgoing waves are obtained:

$$a_i = \frac{U_i + Z_{Li} \cdot I_i}{2\sqrt{\mathrm{Re}\{Z_{Li}\}}} \quad , \qquad b_i = \frac{U_i - Z_{Li} \cdot I_i}{2\sqrt{\mathrm{Re}\{Z_{Li}\}}} \quad , \tag{2.13}$$

$$U_i = \frac{Z_{Li} \cdot a_i + Z_{Li} \cdot b_i}{\sqrt{\mathrm{Re}\{Z_{Li}\}}} \quad , \qquad I_i = \frac{a_i - b_i}{\sqrt{\mathrm{Re}\{Z_{Li}\}}} \quad . \tag{2.14}$$

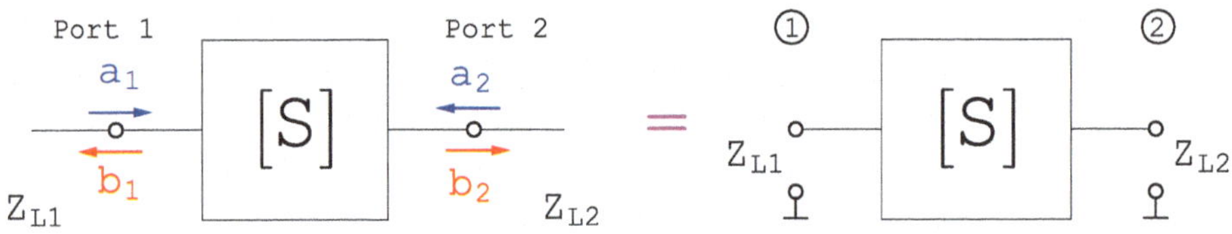

Fig. 2.6: S-parameter representation of a two-port in single-pole representation

2.2.2 Significance of the Scattering Parameters

The scattering matrix links the incoming waves a_i with the outgoing waves outgoing waves b_i. The link takes the form of linear equations

$$\begin{aligned} b_1 &= S_{11}\, a_1 + S_{12}\, a_2 \\ b_2 &= S_{21}\, a_1 + S_{22}\, a_2 \end{aligned} \tag{2.15}$$

or matrix equations:

$$\begin{pmatrix} b_1 \\ b_2 \end{pmatrix} = \begin{pmatrix} S_{11} & S_{12} \\ S_{21} & S_{22} \end{pmatrix} \begin{pmatrix} a_1 \\ a_2 \end{pmatrix} \quad . \tag{2.16}$$

The following applies in abbreviated form

$$\vec{\mathbf{b}} = [\mathbf{S}]\ \vec{\mathbf{a}} \quad . \tag{2.17}$$

The meaning of the scattering parameters S_{ij} (also scattering matrix coefficients) can be illustrated as follows using the example of a two-port:

$$S_{21} = \frac{b_2}{a_1}|_{a_2=0} \quad \text{Forward transmission factor with matched output}$$

$$S_{11} = \frac{b_1}{a_1}|_{a_2=0} \quad \text{Input reflection coefficient with matched output}$$

$$S_{22} = \frac{b_2}{a_2}|_{a_1=0} \quad \text{Output reflection coefficient with matched input}$$

$$S_{12} = \frac{b_1}{a_2}|_{a_1=0} \quad \text{Backward transmission factor with matched input}$$

The scattering parameters are often also referred to as S-parameters. In practice, the logarithmized scattering coefficients are mainly used. These are calculated as follows:

$$S_{ij}^{dB} = 10\ \lg |S_{ij}|^2\ dB = 20\ \lg |S_{ij}|\ dB \quad , \tag{2.18}$$

where lg corresponds to the decadic logarithm on the base 10 ($\lg = \log_{10}$). The dimensionless unit "dB" (decibel) indicates that the numerical value is a logarithmic specification.

The incoming and outgoing power at a two-port are linked via the magnitude squared of the scattering parameters. Thus, the output power P_{wb_2} for a two-port terminated with Z_L is calculated via

$$P_{wb_2} = |S_{21}|^2\ P_{wa_1} \tag{2.19}$$

from the input power P_{wa_1}. Powers are usually referred to 1 mW and are calculated as a logarithmic value as follows:

$$P_w^{dBm} = 10 \lg \frac{P_w}{1\,mW}\ dBm \quad . \tag{2.20}$$

This makes it easy to calculate the reflected and transmitted power of a two-port using the logarithmic conversion formulas selected. For example, the following applies to the transmitted power

$$P_{wb_2}^{dBm} = S_{21}^{dB} + P_{wa_1}^{dBm} \quad . \tag{2.21}$$

Reflection factors are often referred to as return loss, RL:

$$RL = 10 \lg \frac{1}{|S_{ii}|^2}\ dB = 20 \lg \frac{1}{|S_{ii}|}\ dB = -S_{ii}^{dB} \quad . \tag{2.22}$$

Logarithmized transmission factors are also referred to as insertion loss, IL:

$$IL = 20 \lg \frac{1}{|S_{ij}|}\ dB = -S_{ij}^{dB} \quad , \quad i \neq j \quad . \tag{2.23}$$

The following applies to insertion loss,

Attenuation	$\lvert S_{21}\rvert$	Transmitted power	Example
0 dB	1	100 %	Lossless line
3 dB	0.708	50 %	Lossless signal splitter 1 in 2
6 dB	0.501	25 %	Lossless signal splitter 1 in 4
10 dB	0.316	10 %	Attenuator
20 dB	0.1	1 %	Decoupling value for control

In the case of reflection attenuation, the aim is usually to achieve values of at least 15 dB for RF products so that only a negligible amount of power is reflected (radiated back).

The reflection factors (often also called reflection coefficients) can be calculated directly from the input impedance Z_{INi} of the associated port and the characteristic impedance of the line Z_{Li} that is contacted at the port:

$$S_{ii} = \frac{Z_{INi} - Z_{Li}}{Z_{INi} + Z_{Li}} \quad . \tag{2.24}$$

The derivation is carried out in chapter 4. The input resistance is calculated by transformation:

$$Z_{INi} = Z_{Li} \frac{1 + S_{ii}}{1 - S_{ii}} \quad . \tag{2.25}$$

The transmission coefficients (also transmission factors) can be calculated from the characteristic impedances of the lines at the input and output as well as the voltage U_{0i} of a generator at the input and the voltage U_j at the output (see Fig. 2.7).

The following generally applies to the transmission factors:

$$S_{ji} = \frac{2\,U_j}{U_{0i}} \sqrt{\frac{Z_{Li}}{Z_{Lj}}} \quad . \tag{2.26}$$

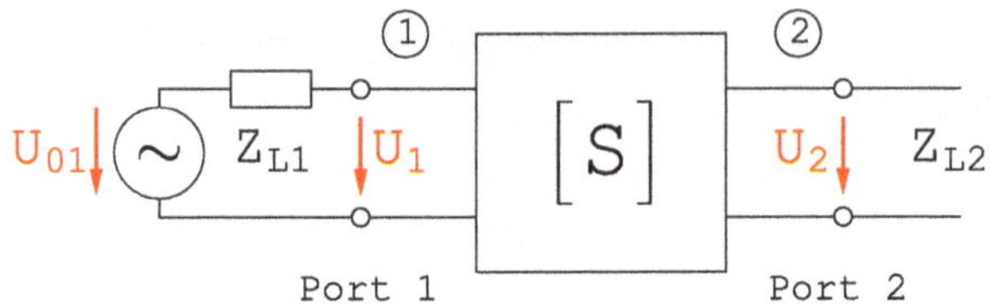

Fig. 2.7: Representation of a two-port for calculating the forward transmission parameters

The forward coefficient of a two-port according to Fig. 2.7 results from:

$$S_{21} = \frac{2\,U_2}{U_{01}} \sqrt{\frac{Z_{L1}}{Z_{L2}}} \quad . \tag{2.27}$$

Two circled numbers can be seen in the Fig. 2.7. These are the short names for the left port 1 and the right port 2. This abbreviation is predominantly used.

The properties shown for a two-port can be extended to a linear n-port. The scatter matrix remains quadratic and therefore has n^2 parameters:

$$\begin{pmatrix} b_1 \\ b_2 \\ b_3 \\ \vdots \\ b_n \end{pmatrix} = \begin{pmatrix} S_{11} & S_{12} & S_{13} & \cdots & S_{1n} \\ S_{21} & S_{22} & S_{23} & & \\ S_{31} & S_{32} & S_{33} & & \\ \vdots & & & \ddots & \\ S_{n1} & & & & S_{nn} \end{pmatrix} \begin{pmatrix} a_1 \\ a_2 \\ a_3 \\ \vdots \\ a_n \end{pmatrix} \quad . \tag{2.28}$$

Fig. 2.1 of the multi-port network analyzer already shows this, that working with multi-port circuits is part of RF technology practice.

2.2.3 The Network Parameters of the Electronics

In historical RF technology and in electronics one does not work with scattering parameters, but with Z-, Y-, G- and H-parameters.

These network parameters are shown in the Tab. 2.1 and the corresponding Fig. 2.8 for the symmetrical counting arrow system and in the Tab. 2.2 and the corresponding Fig. 2.9 for the asymmetrical counting arrow system.

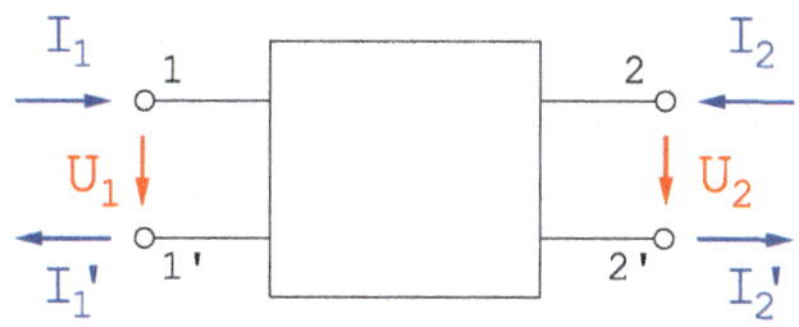

Fig. 2.8: Representation of a four-pole with symmetrical counting arrow system

Tab. 2.1: Definition of two-port matrices, which are preferably used in electronics

	Matrices	Two-port equations
Resistance matrix	$[\mathbf{Z}] = \begin{pmatrix} Z_{11} & Z_{12} \\ Z_{21} & Z_{22} \end{pmatrix}$	$\begin{aligned} U_1 &= Z_{11} I_1 + Z_{12} I_2 \\ U_2 &= Z_{21} I_1 + Z_{22} I_2 \end{aligned}$
Conductance matrix	$[\mathbf{Y}] = \begin{pmatrix} Y_{11} & Y_{12} \\ Y_{21} & Y_{22} \end{pmatrix}$	$\begin{aligned} I_1 &= Y_{11} U_1 + Y_{12} U_2 \\ I_2 &= Y_{21} U_1 + Y_{22} U_2 \end{aligned}$
Series parallel matrix	$[\mathbf{H}] = \begin{pmatrix} H_{11} & H_{12} \\ H_{21} & H_{22} \end{pmatrix}$	$\begin{aligned} U_1 &= H_{11} I_1 + H_{12} U_2 \\ I_2 &= H_{21} I_1 + H_{22} U_2 \end{aligned}$
Parallel series matrix	$[\mathbf{G}] = \begin{pmatrix} G_{11} & G_{12} \\ G_{21} & G_{22} \end{pmatrix}$	$\begin{aligned} I_1 &= G_{11} U_1 + G_{12} I_2 \\ U_2 &= G_{21} U_1 + G_{22} I_2 \end{aligned}$

In the Figs. 2.8 and 2.9, numbers are shown at the four terminals of the four-poles. This nomenclature is found in electronics when describing four-poles as terminal elements. A pair of terminals forms a port. The following must always apply to the currents at the ports: $I_i = I_i'$ $(i = 1, 2)$.

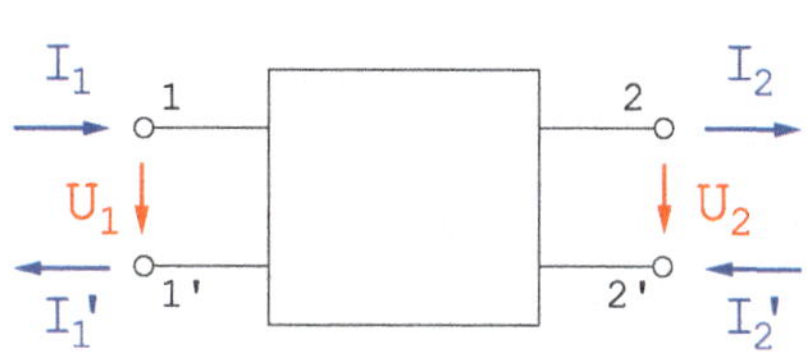

Fig. 2.9: Representation of a four-pole with asymmetrical counting arrow system

Tab. 2.2: Definition of chain matrices, which are preferably used in electronics

	Matrices	Two-port equations
Chain matrix	$[\mathbf{A}] = \begin{pmatrix} A_{11} & A_{12} \\ A_{21} & A_{22} \end{pmatrix}$	$\begin{aligned} U_1 &= A_{11}\,U_2 + A_{12}\,I_2 \\ I_1 &= A_{21}\,U_2 + A_{22}\,I_2 \end{aligned}$
Inverse chain matrix	$[\mathbf{B}] = \begin{pmatrix} B_{11} & B_{12} \\ B_{21} & B_{22} \end{pmatrix}$	$\begin{aligned} U_2 &= B_{11}\,U_1 + B_{12}\,I_1 \\ I_2 &= B_{21}\,U_1 + B_{22}\,I_1 \end{aligned}$

The three Figs. 2.10, 2.11 and 2.12 illustrate the usefulness of these parameters for specially connected two-ports.

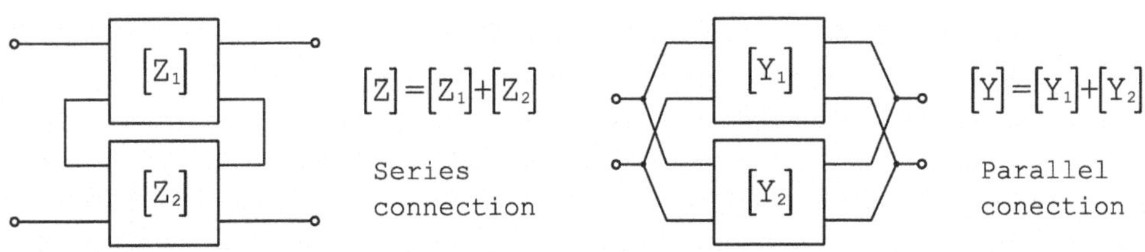

Fig. 2.10: Interconnection of two four-poles networks via Z- and Y-parameters

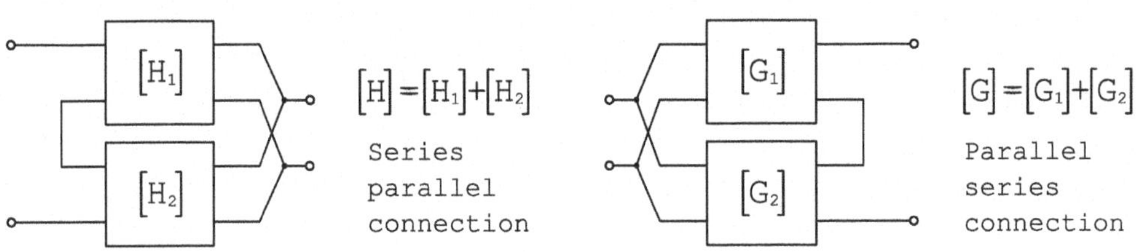

Fig. 2.11: Interconnection of two four-poles via H- and G-parameters

Such interconnections occur more frequently in electronics. In RF circuits, only series or chain circuits with network parameter calculations are solved analytically in practice. Circuit simulators are used for more complex problems.

The auxiliary sheets A.2, A.3, A.4, A.5 and A.6 in the appendix are again very useful for the application of these network parameters.

In RF technology, the Z and Y parameters are used in particular, to obtain the scattering parameters in a simple way.

The first step is to set up the Z or Y matrix. If you have series elements at the ports as in the following example of a T-circuit (Fig. 2.13), then the Z-matrix is more favorable.

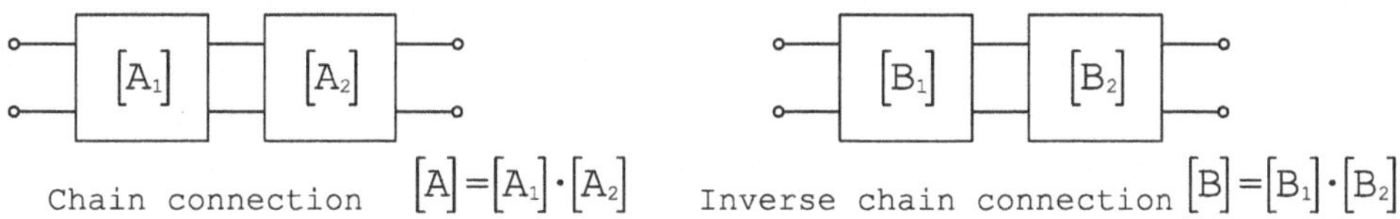

Fig. 2.12: Interconnection of two four poles via A and B parameters

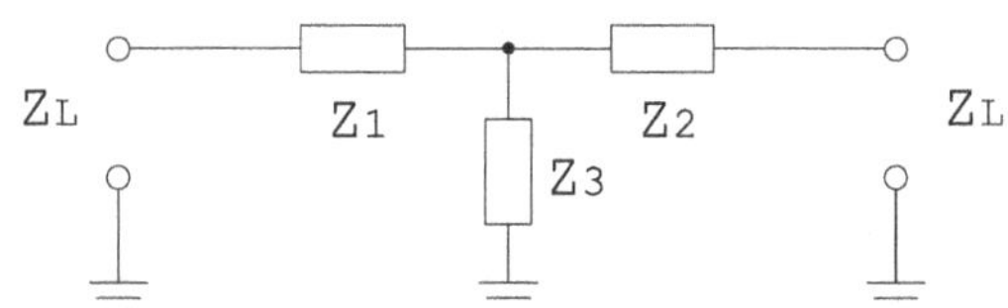

Fig. 2.13: Calculation of the Z-parameters of a T-circuit

To calculate Z_{11} of this T circuit, the current I_2 must be zero. Consequently, there is an open circuit at port 2. Thus, Z_{11} is calculated from the simple addition of Z_1 and Z_3. To determine Z_{22}, I_1 must now be zero. Z_{22} is calculated from the ratio U_2/I_2 and therefore from the input resistance $Z_2 + Z_3$. For the element Z_{12}, the ratio U_1/I_2 must be calculated for the case that $I_1 = 0$ applies. In this case, as well as for Z_{21}, the simple solution Z_3 results.

These results are printed in the auxiliary sheet A.2. The auxiliary sheet A.6 can be used to calculate the S-parameter matrix for the T-circuit, which in turn is printed in auxiliary sheet A.2.

2.3 Circuit Design of Two-Ports via Scattering Parameters

In RF technology, similar to optics, circuits are used, which are connected (serially) in series. Transmitters, receivers and various other sub-systems are created using the necessary two ports (components).

Many circuits that are used in these systems can be calculated from the tables in the auxiliary sheets A.2 and A.3, which can be found in the appendix. It is very advantageous that the scattering parameters can be specified for all the circuits shown. Note that all admittances and impedances of the columns for the scattering parameters are normalized to Z_0 or Z_L. This method only takes into account two-port circuits where both ports are related to the same impedance.

The usefulness of these help sheets will now be demonstrated using a so-called attenuator. An attenuator is used in RF technology when

1. the power of a signal is to be reduced.

2. the mismatch is to be improved by twice the attenuation value.

The second point is explained in more detail with the introduction of the signal flow diagrams in chapter 10.

An attenuator should be ideally matched ($S_{ii} = 0$). Furthermore, an attenuator should have a certain transmission loss that is as frequency-independent as possible. We choose 3 dB.

Realizing an attenuator from resistors[4], it is advantageous to use a PI or T circuit with three resistors each. Fig. 2.14 illustrates the attenuator with the component designations according to auxiliary sheet A.2.

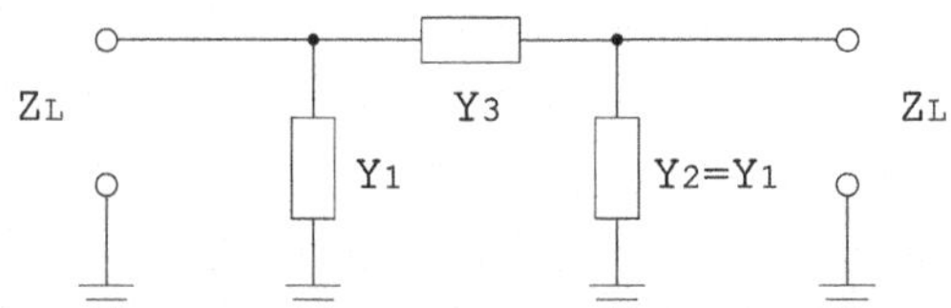

Fig. 2.14: Illustration of an attenuator based on a PI circuit with resistors

——— *Calculation PI attenuator* ———

Given: $S_{ii} = 0$; $IL = 3\,dB$; $i, j = 1, 2$; *PI structure:* $Y_2 = Y_1$; $Z_L = Z_0 = 50\,\Omega$

Find: $Z_1 = \frac{1}{Y_1}$; $Z_3 = \frac{1}{Y_3}$

1. *The numerator of S_{ii} of the PI circuit according to auxiliary sheet A.2 must be be zero: z_3 can be specified as a function of z_1.*

$$2 \cdot y_1 - z_3 \cdot (1 - y_1^2) = 0 \qquad \Rightarrow \qquad z_3 = \frac{2 \cdot z_1}{z_1^2 - 1} \tag{2.29}$$

2. *The transmission loss results according to Eq. (2.23):*

$$S_{ij} = \alpha = 1/10^{\frac{3}{20}} = 0.7079 \quad . \tag{2.30}$$

3. *The transmission value must be S_{ij} of the PI circuit according to auxiliary sheet A.2: z_1 can be specified as a function of the attenuation α.*

$$\alpha = \frac{2}{2 + 2 \cdot y_1 + z_3 \cdot (1 + 2 \cdot y_1 + y_1^2)} \tag{2.31}$$

If Eq. (2.29) is used, the following applies:

$$\frac{1}{\alpha} = 1 + y_1 + y_1 \cdot \frac{1 + y_1}{1 - y_1} \qquad \Rightarrow \qquad z_1 = \frac{1/\alpha + 1}{1/\alpha - 1} = 5.85 \quad . \tag{2.32}$$

From the Eq. (2.29) the value for the series resistance $z_3 = 0.352$ is obtained. Denormalization to $Z_0 = 50\,\Omega$ results in: $Z_1 = 292\,\Omega$ and $Z_3 = 17.6\,\Omega$.

——— *Calculation PI attenuator* ———

[4]In practice, so-called SMD components are used (SMD: surface mounted device) of the 0402 design (approx. 0.5 mm * 1.0 mm * 0.5 mm) up to around 4 GHz.

2.4 Scattering Matrices of Networks with Special Properties

To assess measurements and characteristics of components, modules, test sockets and much more, it is useful to have knowledge of the general properties of the scattering parameters. This knowledge also helps when you want to develop circuits and need to determine the number of unknown parameters.

Matching

The variable that is most frequently optimized in the RF range is the matching. This means either that the input reflection coefficient of a component should be zero ($S_{ii} = 0$) or that it should correspond to the conjugate complex value of the upstream component (e.g.: $Sb_{11} = Sa_{22}^{\star}$, Fig. 2.15). The first case ($S_{ii} = 0$) is also referred to as wave matching or system matching. The second case is a power matching.

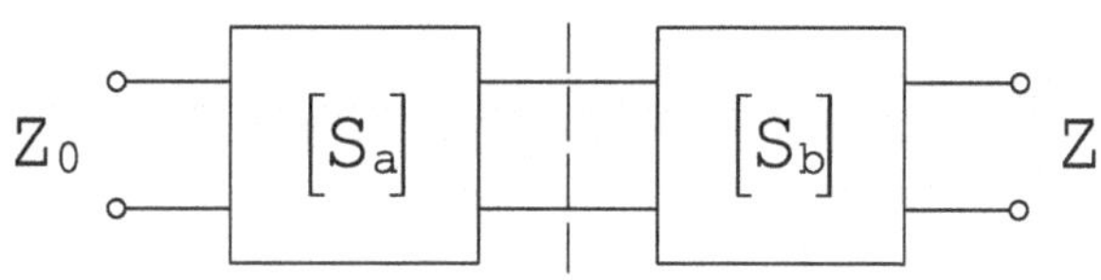

Fig. 2.15: Two matched two-ports in series connection: $Sa_{11} = 0$, $Sb_{22} = 0$ and $Sb_{11} = Sa_{22}^{\star}$

In the circuit in Fig. 2.15, the outer ports are matched to the system impedance and the two inner ports are matched to the conjugate complex impedance of the opposite port.

2.4.1 Properties of the Scattering Parameters of Passive Components

Reciprocity

If passive components do not contain materials with direction-dependent (anisotropic) properties, the components have a `reciprocal` behavior. This means that the forward transmission corresponds to the backward transmission. The following applies for a reciprocal three-port:

$$S_{21} = S_{12} \ , \qquad S_{31} = S_{13} \ , \qquad S_{32} = S_{23} \ . \tag{2.33}$$

An example of an anisotropic material is a ferrite with premagnetization.

The majority of passive components are reciprocal!

Examples of reciprocal components: SMD coils, capacitors, resistors, cables, connectors, line components, antennas, measuring tips etc..

The best known non-reciprocal active RF component is the amplifier.

Symmetry

Symmetry is when the input reflection coefficients of a circuit are identical to the output reflection coefficients:

$$S_{11} = S_{22} \quad . \tag{2.34}$$

If the two-port scattering matrix of a single reciprocal (concentrated) component is invested, which is connected between the two ports either in series or to ground, it is usually a symmetrical two-port. Fig. 2.16 shows symmetrical two-ports of this simplest category.

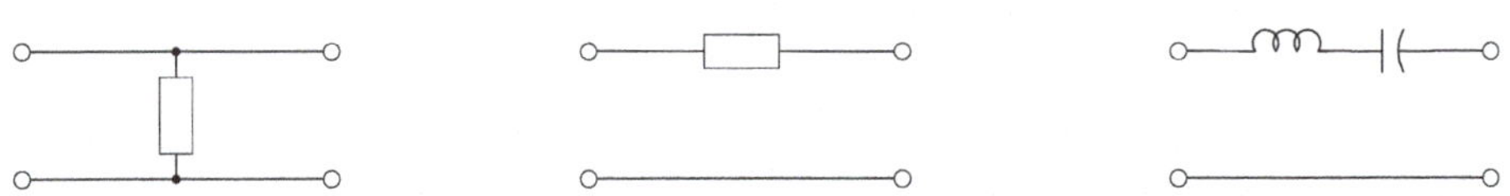

Fig. 2.16: Representation of three symmetrical two-ports, each with a respectively one or two passive components

A series resonant circuit is shown in the right-hand image of Fig. 2.16. This is also symmetrical, as these are concentrated components without spatial expansion and, for example, the coil can be replaced by two coils with half the value, which are located to the right and left of the capacitor.

Furthermore, all two- and multi-ports that have a line of symmetry and are reciprocal are symmetrical networks (example: Fig. 2.17).

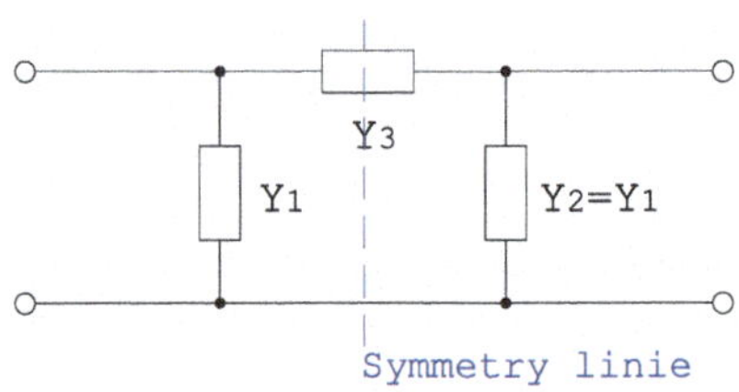

Fig. 2.17: Representation of a symmetrical two-port containing three passive SMD components

In this case too, the forward transmission corresponds to the backward transmission. Consequently, symmetrical networks are always reciprocal.

Symmetric networks can therefore always be characterized by only two scattering parameters (e.g. S_{11} and S_{21})!

Losses

In addition to the properties of matching, reciprocity and symmetry presented above, a generalized loss statement is of the greatest practical importance.

The following applies to a passive component:

$$|S_{ij}| \leq 1 \quad \text{resp.} \quad \leq 0\,\text{dB} \quad . \tag{2.35}$$

The procedures of linear amplifiers are used (chapter 8), the maximum available gain (MAG)) is a parameter that specifies the losses for passive components and circuits when both sides are matched.

In other words, the MAG specifies the maximum S_{21} value that can ideally be achieved with this passive component. This maximum transmission value can only be achieved through a so-called renormalization, which is presented in detail in the next section and is shown in Fig. 2.18.

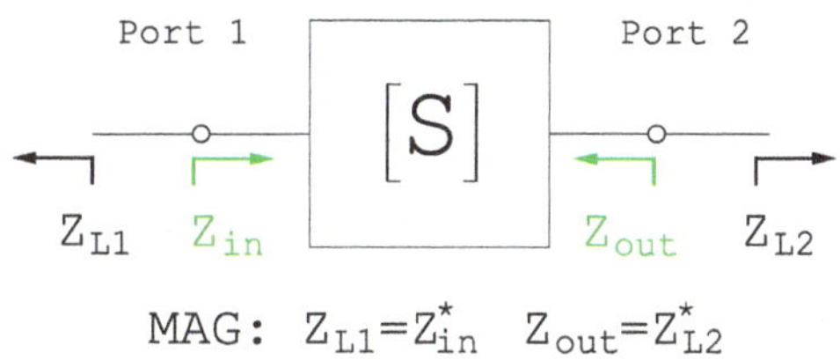

Fig. 2.18: Representation of the mathematically generated matching

Mathematically, the MAG can be calculated as follows:

$$\mathrm{MAG} = \left| \frac{S_{21}}{S_{12}} \left(k - \sqrt{k^2 - 1} \right) \right| \tag{2.36}$$

$$\text{with} \quad k = \frac{1 + |\det S|^2 - |S_{11}|^2 - |S_{22}|^2}{2\, |S_{12}\, S_{21}|} \tag{2.37}$$

$$\text{and} \quad \det S = S_{11} \cdot S_{22} - S_{12} \cdot S_{21} \quad . \tag{2.38}$$

Each RF circuit simulator can specify the MAG value (Ansoft serenade: GMAX) of a two-port directly based on the Eqs. (2.36) and (2.37).

The MAG values of two 3.3 nH coils are given as an example. Fig. 2.19 shows an expensive coil from Toko, which is realized using a wire, and a low-cost coil from TDK, which is manufactured using thin-film technology.

Ideally, this example should provide two loss curves, which decrease monotonically with frequency. The existing ripple, which even extends into the range $S_{21} > 0\,\mathrm{dB}$, comes from measurement errors. Nevertheless, it can be seen that the Toko coils have noticeably lower losses.

2.4.2 Properties of the Scattering Parameters of Passive Lossless Components

Many passive RF components have very low transmission losses. For a better assessment of these components, it is very helpful to know the characteristics of lossless passive components.

It can be shown that, in general, for a passive lossless multi-port the so-called *unitarity condition*

$$[\mathbf{S}]^T\ [\mathbf{S}]^* = [\mathbf{I}] \tag{2.39}$$

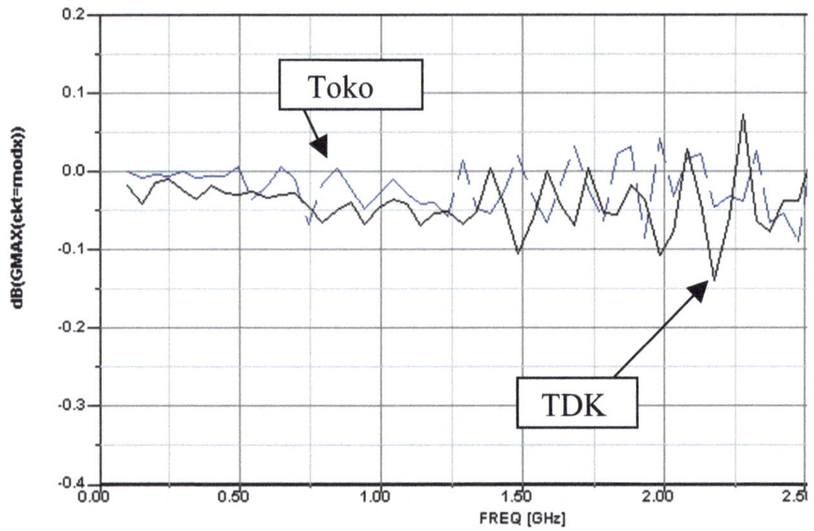

Fig. 2.19: MAG values of two 0402-SMD coils calculated from measured values

holds. In other words, multiplying the transposed scatter matrix by the conjugate complex scatter matrix gives the unit matrix ($[\mathbf{I}]$). From this general unitarity condition, the following can be derived for the important statements for two- and three-ports.

Lossless Two-Ports

For a lossless, reciprocal and/or symmetrical two-port, the following applies:

$$|S_{21}|^2 = 1 - |S_{11}|^2 \quad . \tag{2.40}$$

The results of this simple calculation are printed on the auxiliary sheet of the page 440. This table is very helpful when assessing electrical circuits.

The practitioner likes to divide a low-loss circuit into two parts:

a lossless circuit and a matched attenuator.

This is visualized graphically in Fig. 2.20.

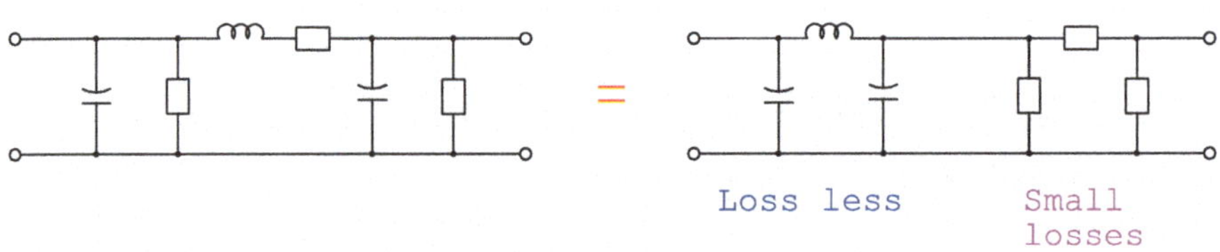

Fig. 2.20: Representation of the decomposition of a component (here low-pass filter) into a lossless circuit and a matched attenuator

The following approximate calculation is intended to illustrate this procedure, which is very often required in practice.

——— *Exercise: Loss Prediction* ———

Given: *During the first production run, a filter had IL=1.5 dB transmission loss and a reflection loss of RL=10 dB. It is assumed that the filter in the second run has a improved reflection loss (matching) of 15 dB.*

Find: *What transmission loss will the filter have after the second pass?*

1. *Attenuation due to the mis-matching after the 1st run from the table:* *0.46 dB*

2. *Attenuation with an improved matching of 15 dB from the table:* *0.14 dB*

3. *Calculation of the losses for the 2nd run:*

$$IL = 1.50\,dB - 0.46\,dB + 0.14\,dB = 1.18\,dB$$

4. *Explanation:*

Approximately, the total attenuation IL is made up of the ohmic attenuation IL_{ohm} and the attenuation due to the lack of matching $IL_{reflect}$. The IL_{ohm} component is estimated at 1.50 dB-0.46 dB. The influence of the lack of matching is reduced by 0.46 dB-0.14 dB.

Run	*RL*	*IL*	$IL_{reflect}$	IL_{Ohm}
1	*10 dB*	*1.5 dB*	*0.46 dB*	*1.04 dB*
2	*15 dB*	*1.18 dB*	*0.14 dB*	*1.04 dB*

——— *Exercise: Loss Prediction* ———

This calculation is only an approximation based on the use of values from a lossless two-port. However, this calculation is sufficiently precise in practice for low-loss components!

Lossless Three-Ports

Simplified, the unitarity condition provides the result, that the absolute value squares of a row or a column of an S-matrix must have the value 1. For example, the following applies to a three-port:

$$1 = |S_{11}|^2 + |S_{21}|^2 + |S_{31}|^2 \quad . \tag{2.41}$$

This can be shown for three-ports, that the three properties

- Lossless ($[\mathbf{S}]^T\,[\mathbf{S}]^* = [\mathbf{I}]$),
- Reciprocity ($S_{ij} = S_{ji}$), $\quad i \neq j$
- Matching ($S_{ii} = 0$)

can *not* realize at the same time! One of these system properties cannot be fulfilled with a three-ports. This statement can be proven using the unitarity relationship ([48]).

2.4.3 Common Conversions of Scattering Parameters

Renormalization

A scattering parameter matrix is usually related to a reference impedance (also port impedance). The parameters are usually related to 50 Ω. However, conversions from one reference impedance to another (possibly complex) impedance are required for various reasons. Derivations and solutions for such calculations have been published, for example, under [48, 87]. In the following, a possible renormalization without derivation for the application is presented.

Given the scatter matrix $[\mathbf{S_a}]$ whose n port impedances are known and may be different and/or complex. The values (Z_{Lai}) of the individual port impedances are in the main diagonal of the otherwise with zero filled matrix:

$$[\mathbf{Z_{La}}] = \begin{pmatrix} Z_{La1} & 0 & 0 & \cdots & 0 \\ 0 & Z_{La2} & 0 & & \\ 0 & 0 & Z_{La3} & & \\ \vdots & & & \ddots & \\ 0 & & & & Z_{Lan} \end{pmatrix} \tag{2.42}$$

The Z-matrix $[\mathbf{z_a}]$ normalized to Z_{Lai} can be calculated using the specifications of the auxiliary sheet A.5 from the scatter matrix $[\mathbf{S_a}]$.

The absolute Z-parameters can be calculated from these normalized Z-parameters.

$$[\mathbf{Z}] = [\mathbf{Z_{La}}]^{1/2}\ [\mathbf{z_a}]\ [\mathbf{Z_{La}}]^{1/2} \tag{2.43}$$

Now, during renormalization, the aim is to set the scattering parameters to other port impedances, which are now in the matrix

$$[\mathbf{Z_{Lb}}] = \begin{pmatrix} Z_{Lb1} & 0 & 0 & \cdots & 0 \\ 0 & Z_{Lb2} & 0 & & \\ 0 & 0 & Z_{Lb3} & & \\ \vdots & & & \ddots & \\ 0 & & & & Z_{Lbn} \end{pmatrix} \tag{2.44}$$

are filed, can be obtained. To do this, the Z-matrix normalized to the new values must first be calculated.

$$[\mathbf{z_b}] = [\mathbf{Z_{Lb}}]^{-1/2}\ [\mathbf{Z}]\ [\mathbf{Z_{Lb}}]^{-1/2} \tag{2.45}$$

Finally, with the help of the unit matrix $[\mathbf{I}]$ from

$$[\mathbf{S_b}] = [\mathbf{I}] - 2\,([\mathbf{z_b}] + [\mathbf{I}])^{-1} \tag{2.46}$$

calculate the required renormalized S-parameter matrix.

These conversions are implemented in common RF simulators and network analyzers and can be performed under renormalization.

——— *Exercise: Renormalization* ———

Given.: *From the Z-matrix of the circuit, which can be easily determined using the auxiliary sheet A.2 of the circuit according to Fig. 2.21, the scatter matrix is to be determined for the case that port 1 is closed with 50 Ω and port 2 with 100 Ω, be calculated using Matlab.*

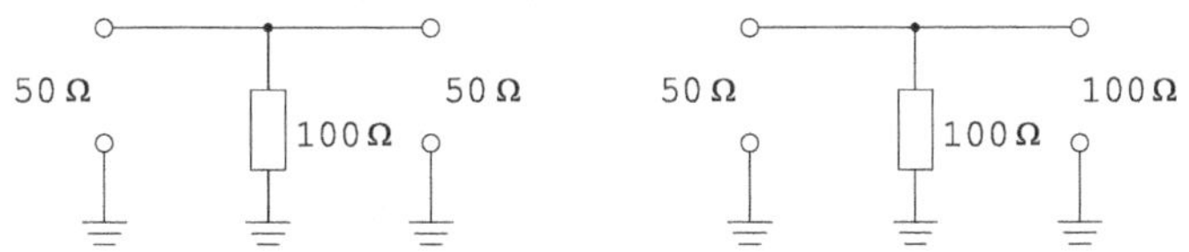

Fig. 2.21: Two-port to be renormalized

The solution can be found directly in the corresponding derivation:

```
Z=[100 100; 100 100];          Zlb=[50 0; 0 100];

inv_Zlb=Zlb^-0.5               zb=inv_Zlb*Z*inv_Zlb

I=[1 0;0 1];                   Sb=I-2*inv(zb+I)
```

The corresponding Matlab output looks like this:

```
inv_Zlb =   0.1414          0
            0               0.1000

zb =        2.0000    1.4142
            1.4142    1.0000

Sb =        0.0000    0.7071
            0.7071   -0.5000
```

The perfect matching of Sb_{11} can be checked directly.

——— *Exercise: Renormalization* ———

De-embedding

De-embedding is already self-explanatory. This means that the actual device under test (DUT) is "embedded" (Fig. 2.22) and must be calculated from the series connection of the desired two-port and at least one other two-port.

The scattering parameters of the surrounding networks are known for this procedure. These are often lines. It is particularly easy to mathematically "embed" these surrounding networks by using the so-called transmission or chain parameters. This mathematics is explained in detail in the next section.

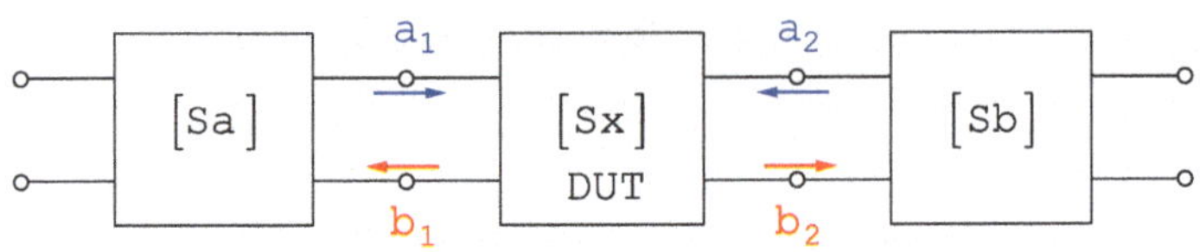

Fig. 2.22: The measurement object (DUT) is "embedded" between two networks.

2.5 Transmission, Chain and other Matrices

2.5.1 The Transmission Matrix Σ

Many RF circuits consist of the series connection of two ports. Examples of this are transmitter and receiver paths of transmission links. The scattering parameters have the disadvantage that for series-connected of two-ports the resulting scattering matrix cannot be specified in a simple manner.

The transmission matrix Σ provides a remedy. In the transmission matrix, the outgoing and returning waves are are assigned differently than in the scattering matrix (compere Eqs. (2.15-2.17) with Eqs. (2.47-2.49)). The transmission matrix is defined as:

$$\begin{aligned} a_1 &= \Sigma_{11}\, b_2 + \Sigma_{12}\, a_2 \\ b_1 &= \Sigma_{21}\, b_2 + \Sigma_{22}\, a_2 \end{aligned} \tag{2.47}$$

or as a matrix equation:

$$\begin{pmatrix} a_1 \\ b_1 \end{pmatrix} = \begin{pmatrix} \Sigma_{11} & \Sigma_{12} \\ \Sigma_{21} & \Sigma_{22} \end{pmatrix} \begin{pmatrix} b_2 \\ a_2 \end{pmatrix} \quad . \tag{2.48}$$

The following applies in abbreviated form

$$\begin{pmatrix} a_1 \\ b_1 \end{pmatrix} = [\boldsymbol{\Sigma}] \begin{pmatrix} b_2 \\ a_2 \end{pmatrix} \quad . \tag{2.49}$$

Due to this assignment, the output wave vector of a two-port corresponds to the input wave vector of a two-port connected in series. Therefore, the transmission matrix of several series-connected two-ports Σ_{ges} results from the product of the individual Σ matrices. As an example, the following results for Fig. 2.23.

$$[\boldsymbol{\Sigma_{ges}}] = [\boldsymbol{\Sigma_1}] \cdot [\boldsymbol{\Sigma_2}] \cdot [\boldsymbol{\Sigma_3}] \quad . \tag{2.50}$$

In addition to the simple case of wanting to calculate the entire scattering matrix as shown in Fig. 2.23, it is often the case that the scattering parameters of $n-1$ individual two-ports and the total matrix are known and the scattering matrix of a two-port in the chain is to be calculated (e.g. for "de-embedding").

In both cases, it is necessary to be able to convert the scattering matrix into transmission parameters and vice versa.

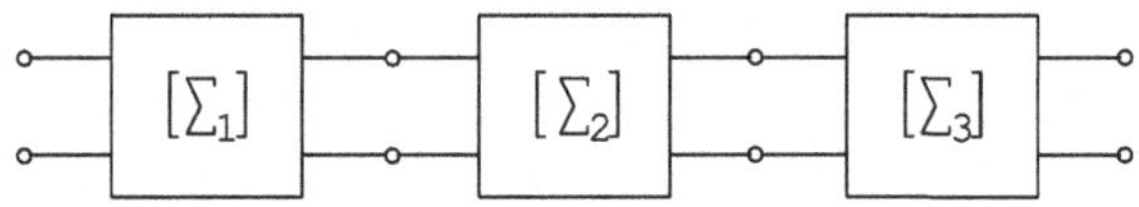

Fig. 2.23: Σ-Matrix of two-ports connected in series

The conversions are as follows:

$$\Sigma_{11} = \frac{1}{S_{21}} \qquad \Sigma_{12} = -\frac{S_{22}}{S_{21}} \qquad \Sigma_{21} = \frac{S_{11}}{S_{21}} \qquad \Sigma_{22} = -\frac{\det(S)}{S_{21}} \tag{2.51}$$
$$\text{with} \quad \det(S) = S_{11} \cdot S_{22} - S_{12} \cdot S_{21} \quad ,$$

or in matrix notation:

$$[\mathbf{\Sigma}] = \frac{1}{S_{21}} \begin{pmatrix} 1 & -S_{22} \\ S_{11} & -\det(S) \end{pmatrix} \tag{2.52}$$

and for the reverse case in matrix notation

$$[\mathbf{S}] = \frac{1}{\Sigma_{11}} \begin{pmatrix} \Sigma_{21} & \det(\Sigma) \\ 1 & -\Sigma_{12} \end{pmatrix} \quad \text{with} \quad \det(\Sigma) = \Sigma_{11} \cdot \Sigma_{22} - \Sigma_{12} \cdot \Sigma_{21} \quad . \tag{2.53}$$

The physical meaning of the Σ parameters will not be discussed further. The transmission matrices should only be used for the calculation of series circuits.

The "De-embedding" case according to Fig. 2.22 can be used as a typical application.

Step 1: The known scattering parameters of the matrices $[\mathbf{Sa}]$ and $[\mathbf{Sb}]$ as well as the overall circuit ($[\mathbf{Sges}]$) are converted into transmission parameters using the Eq. (2.52) into transmission parameters are converted.

Step 2: The Σ parameters of the DUT ($[\mathbf{Sx}]$) are calculated via the Eq. (2.54), which results from Eq. (2.50) and multiplication of left and right with the embedding networks:

$$[\mathbf{\Sigma_x}] = [\mathbf{\Sigma_a}]^{-1} \cdot [\mathbf{\Sigma_{ges}}] \cdot [\mathbf{\Sigma_b}]^{-1} \quad . \tag{2.54}$$

Step 3: Using Eq. (2.53), $[\mathbf{\Sigma_x}]$ is used to calculate the scatter matrix $[\mathbf{Sx}]$ of the searched DUT is calculated.

Matching

A very important special case in practice is that all two-ports can be regarded as perfectly matched in approximation. In this simple case, all transmission parameters can be multiplied or added together:

$$Sges_{21} = Sa_{21}\, Sx_{21}\, Sb_{21} \quad \text{resp.} \quad Sges_{21}^{dB} = Sa_{21}^{dB} + Sx_{21}^{dB} + Sb_{21}^{dB} \quad . \tag{2.55}$$

——— *Exercise: Improving matching* ———

Given: *The input resistance of an amplifier with an amplification of* S_{V21}^{dB}=*20 dB is* R_{in}=*150 Ω. To improve the input matching, an attenuator with* $S_{A11} = 0$ *is connected upstream of the amplifier. This measure is intended to achieve an adjustment of* S_{ges11}^{dB}=*-15 dB for the entire circuit.*

Find:

(a) *Sketch the chain circuit, entering the input reflection designations* S_{ges11}, S_{A11} *for the attenuator and/or* S_{V11} *for the amplifier.*

(b) *Calculate* S_{V11}^{dB} *from* R_{in}.

(c) *What attenuation* (S_{A21}^{dB}) *must the attenuator have?*

(d) *What is the gain of the new amplifier adapted on the input side? Please specify* S_{ges21}^{dB} *as a function of* S_{V21}^{dB} *and* S_{A21}^{dB}.

Calculation:

To (a)

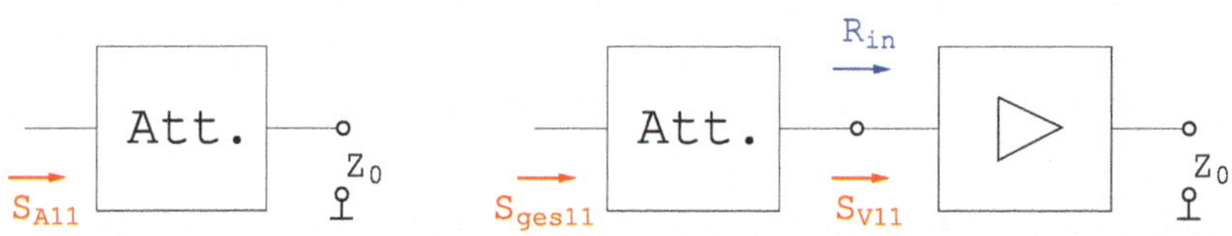

Fig. 2.24: Sketch of the chain circuit to improve input matching

To (b)

$$S_{V11} = \frac{R_{in} - Z_0}{R_{in} + Z_0} \quad \text{resp.} \quad S_{V11} = 0.5 \quad \text{or} \quad S_{V11}^{dB} = -6.0\,\text{dB} \quad , \tag{2.56}$$

To (c)

$$S_{ges11}^{dB} = 2\,S_{A21}^{dB} + S_{V11}^{dB} \quad \text{resp.} \tag{2.57}$$

$$S_{A21}^{dB} = \frac{1}{2}\left(S_{ges11}^{dB} - S_{V11}^{dB}\right) \quad \text{resp.} \quad S_{A21}^{dB} = -4.5\text{dB} \quad , \tag{2.58}$$

To (d)

$$S_{ges21}^{dB} = S_{A21}^{dB} + S_{V21}^{dB} \quad \text{resp.} \quad S_{ges21}^{dB} = 15.5\,\text{dB} \quad . \tag{2.59}$$

——— *Exercise: Improving matching* ———

Multi-Ports

Multi-ports are becoming increasingly common in RF technology and are almost standard in high-speed digital technology. Fig. 2.25 shows the case of the chain connection of three four-port, which is important in practice. For the conversion into transmission

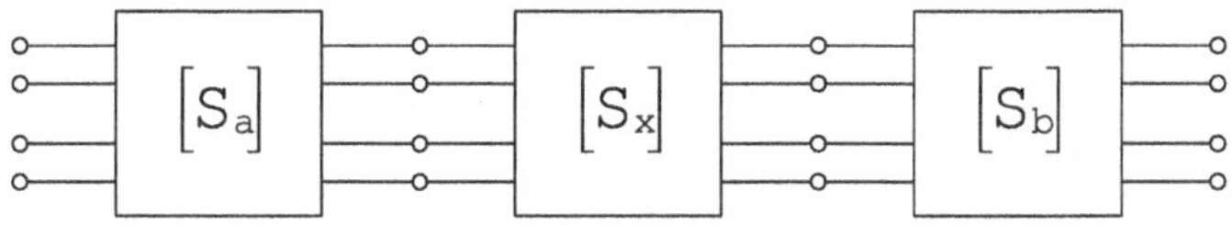

Fig. 2.25: Series connection of three four-ports

parameters and vice versa, the S and Σ− matrices must be broken down into four sub-matrices must be broken down. The following applies to the four-port case

$$\begin{pmatrix} S_{11} & S_{12} & S_{13} & S_{14} \\ S_{21} & S_{22} & S_{23} & S_{24} \\ S_{31} & S_{32} & S_{33} & S_{34} \\ S_{41} & S_{42} & S_{43} & S_{44} \end{pmatrix} = \begin{pmatrix} [\mathbf{S_{11}}] & [\mathbf{S_{12}}] \\ [\mathbf{S_{21}}] & [\mathbf{S_{22}}] \end{pmatrix} \tag{2.60}$$

and

$$\begin{pmatrix} \Sigma_{11} & \Sigma_{12} & \Sigma_{13} & \Sigma_{14} \\ \Sigma_{21} & \Sigma_{22} & \Sigma_{23} & \Sigma_{24} \\ \Sigma_{31} & \Sigma_{32} & \Sigma_{33} & \Sigma_{34} \\ \Sigma_{41} & \Sigma_{42} & \Sigma_{43} & \Sigma_{44} \end{pmatrix} = \begin{pmatrix} [\mathbf{\Sigma_{11}}] & [\mathbf{\Sigma_{12}}] \\ [\mathbf{\Sigma_{21}}] & [\mathbf{\Sigma_{22}}] \end{pmatrix} \quad . \tag{2.61}$$

These new sub-matrices can be used in the equations (2.51) to (2.53), as was done there with the pure parameters, and thus carry out the back and forth transformation. Furthermore, this procedure can also be applied to any other even-numbered multi-port.

2.5.2 The Chain Matrix [A] and the ABCD Matrix

In particular, the chain matrix [**A**] or the ABCD matrix is useful for more complex circuit designs as described in Chapter 2.3. Both matrices link the input currents and voltages with the output currents and voltages. The only difference is that the ABCD matrix operates with referenced values and the A-matrix operates with unreferenced values, see Fig. 2.26.

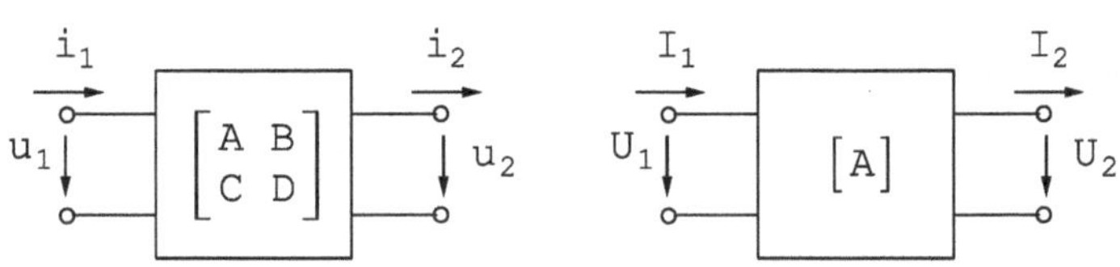

Fig. 2.26: Counting arrows for currents and voltages in the definition of the ABCD chain matrix (left) and the A chain matrix (right)

The matrices are defined as follows:

$$\begin{pmatrix} u_1 \\ i_1 \end{pmatrix} = \begin{pmatrix} A & B \\ C & D \end{pmatrix} \begin{pmatrix} u_2 \\ i_2 \end{pmatrix} \quad , \quad \begin{pmatrix} U_1 \\ I_1 \end{pmatrix} = \begin{pmatrix} A_{11} & A_{12} \\ A_{21} & A_{22} \end{pmatrix} \begin{pmatrix} U_2 \\ I_2 \end{pmatrix} \quad , \tag{2.62}$$

with the quantities related to the real reference impedance Z_0

$$u = \frac{U}{\sqrt{Z_0}} \quad , \quad i = I \cdot \sqrt{Z_0} \tag{2.63}$$

also applies:

$$u = a + b \qquad \text{and} \qquad i = a - b \quad . \tag{2.64}$$

Useful two-port specifications of components of the A-matrix can be found in the auxiliary sheets A.2, A.3 and A.4. The conversion to scattering parameters and other network parameters is given in the auxiliary sheets A.5 and A.6.

The ABCD matrix is frequently used, particularly in American literature. If the above-mentioned auxiliary sheets are used, then when converting A-chain parameters into scattering parameters, one also works de facto with the ABCD matrix due to the normalization to Z_0 required there.

The chain matrix and the ABCD matrix can be used in the same way to calculate series-connected connected two-ports as shown with the transmission matrix. For example, the three series-connected two-ports with the chain matrices $[\mathbf{A_1}]$, $[\mathbf{A_2}]$ and $[\mathbf{A_3}]$ the chain matrix of the entire circuit $[\mathbf{A_{ges}}]$ can be calculated via:

$$[\mathbf{A_{ges}}] = [\mathbf{A_1}] \cdot [\mathbf{A_2}] \cdot [\mathbf{A_3}] \ . \tag{2.65}$$

Only the A-parameters are specified in the auxiliary sheets of the appendix. However, the auxiliary sheet A.6 contains the normalized A-parameters, from which result in the ABCD parameters:

$$A = a_{11} \ , \quad B = a_{12} \ , \quad C = a_{21} \ , \quad D = a_{22} \ . \tag{2.66}$$

Chapter 3

Passive RF Components Made from Concentrated Components

In many products that contain RF circuits, the passive modules are realized from components that can be described very precisely by means of equivalent circuits of concentrated elements. Examples include all circuits with semiconductor components for use up to 100 GHz and so-called SMDs (short for Surface-Mounted Devices) for use up to approx. 10 GHz, Fig. 3.1. In these circuits, only connecting cables between individual components can be regarded as distributed elements. The distributed elements are introduced in the next chapter.

Fig. 3.1: Illustration of a 2.45 GHz receiver circuit with a large number of coils and capacitors in SMD design

This chapter first describes concentrated components and their equivalent circuit diagrams. Passive two-, three- and four-port circuits are then introduced, which can be produced from concentrated components are then presented. The scattering parameters

H. Heuermann, *Radio Frequency Technologies*,
https://doi.org/10.1007/978-3-658-50170-9_3

of these components are used to describe their functionality in radio frequency systems.

3.1 Concentrated Elements and Components

3.1.1 Ideal Concentrated Elements

Ideal concentrated elements are also referred to in the literature as clamping elements. An ideal element consists of an impedance (or admittance), which can be purely real and *has no geometric extension.* This means that the lengths specified in Fig. 3.2 must be zero. Such components do not even exist to a good approximation for the radio frequency

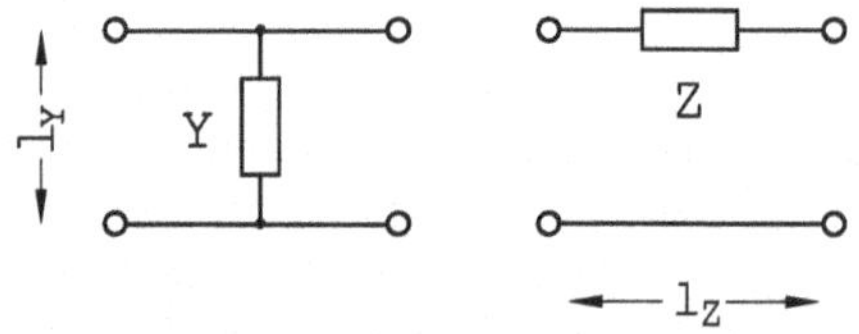

Fig. 3.2: Illustration of two clamping elements with length specification

range. What is available in practice, however, are components that can be precisely modeled using several ideal concentrated elements.

In circuit simulators, ideal concentrated elements such as

the capacitor or the capacitance

the coil or the inductance

the resistance or the reactance

are preferably used.

To derive more complex modules, the first step is also based on the fact, that it is assumed that ideal components are available. Only after this first modelling of the properties of the real components, a module can be designed.

3.1.2 Real Concentrated Components

Real concentrated components[1] can be achieved through the core element and a series of parasitic elements. The equivalent circuit diagrams of these components depend on the type of component and very much from the manufacturing process,

technology used and the

installation position.

[1] The better term is quasi-concentrated components.

A very impressive example of the installation position are the implementations of two 470 pF chip capacitors, which are connected via an inductor (consisting of the parallel connection of three individual bonding wires) to a planar coupled to a planar RF line, to see (Fig. 3.3).

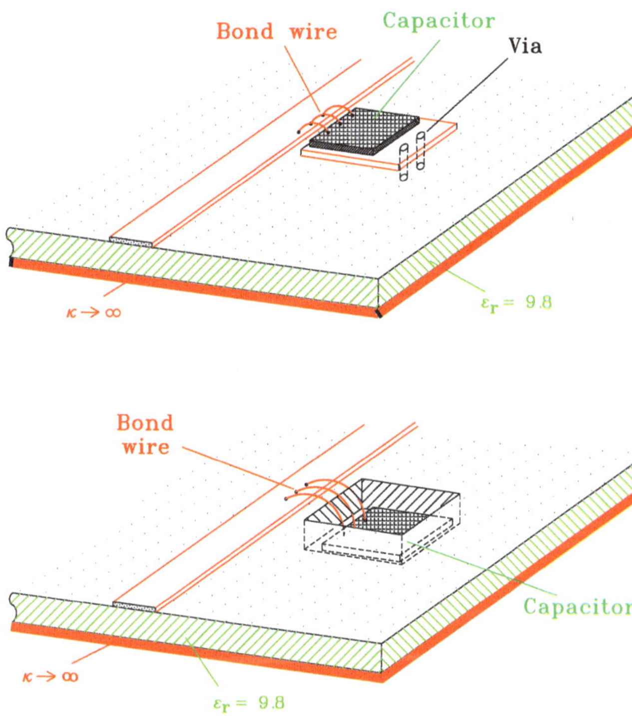

Fig. 3.3: Illustration of two chip capacitors connected to a to a microstrip line in different installation positions. Top: off-line; Bottom: in-line

These circuits are a so-called *bypass*. A low-frequency signal should be able to pass through undisturbed, while a RF signal is blocked.

The equivalent circuit diagrams of both realizations with the same capacitors are shown in Fig. 3.4. The equivalent circuit of the off-line realization has a series resonance with Lb1 and C and a parallel resonance with Cp and Ld. In contrast, the equivalent circuit of the in-line realization only has a series resonance. The two measurement results shown in

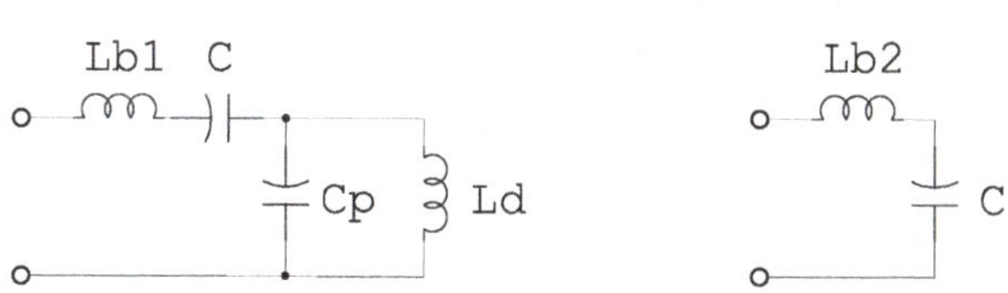

Fig. 3.4: Equivalent circuit diagrams for Fig. 3.3: Left off-line; Right in-line

Fig. 3.5 show the series resonance at around 0.5 GHz and the parallel resonance at 18 GHz for the off-line realization. In the in-line realization, it can be seen that from 0.5 GHz only the series inductance ($|j\,\omega\,Lb2|$) has an influence on the electrical behaviour.

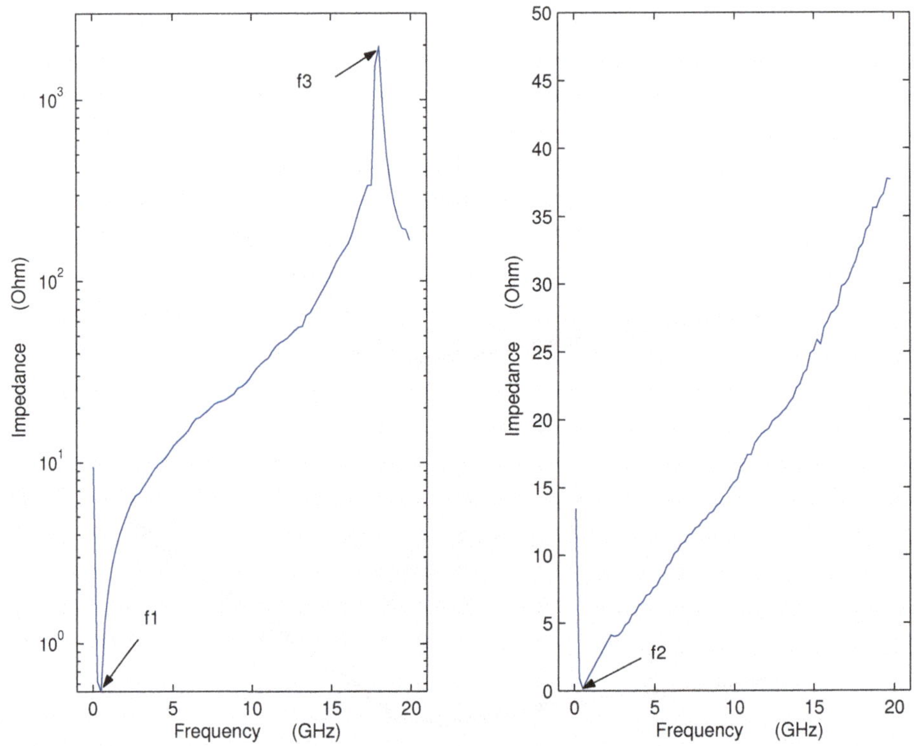

Fig. 3.5: Measured values of the amounts of the impedances of an off-line mounted capacitor (left) and an in-line mounted capacitor (right)

The fact that the installation position has a noticeable influence on the electrical behavior of a concentrated component means that the electrical specifications of the manufacturers of concentrated components generally have a systematic "error" for use in the circuit.

Through his own modeling, the RF engineer quickly develops a database that helps him to estimate the parasitic variables.

If a maximum magnitude value of 20 Ω is assumed for sufficient functionality of a bypass, the Fig. 3.5 provides the statement that the off-line setup can be used up to around 5 GHz and the in-line setup up to around 11 GHz.

——— *Exercise: Bypass* ———

Given: *The core element, capacitor C, has a value of 338 pF. The bond wire 2 is 1.5 times longer than the bond wire 1 and therefore the approximation applies:* $L_{b2} = 1.5\,L_{b1}$. *The impedance minima in the off-line and in-line set-up are* $f_1 = 0.3\,GHz$ *and* $f_2 = 0.5\,GHz$. *The maximum of the off-line setup is* $f_3 = 18\,GHz$.

The following approximations are valid: 1. $\frac{1}{\omega_3 C} \approx 0\,\Omega$ *at 18 GHz, 2.* $\frac{1}{\omega_1 C_p} \to \infty\,\Omega$ *at 0.3 GHz.*

Find: *Calculate the values of the elements* C_p, L_d, L_{b1} *and* L_{b2}.

$$L_{b2} = \frac{1}{\omega_2^2\, C} = 0.3\,nH \qquad \text{resp.} \qquad L_{b1} = L_{b2}\ /\ 1.5 = 0.2\,nH \tag{3.1}$$

$$\omega_1^2 = \frac{1}{C(L_{b1} + L_d)} \qquad \Rightarrow \qquad L_d = \frac{1}{\omega_1^2\, C} - L_{b1} = 0.63\,nH \tag{3.2}$$

$$C_p = \frac{1}{\omega_3^2\, L_d} = 0.12\,pF \tag{3.3}$$

——— *Exercise: Bypass* ———

3.1.2.1 Quality of Coils and Capacitors

Real components exhibit losses. In the case of concentrated components, it is particularly interesting to quantify the losses of coils and capacitors. A resistor is used to model the losses, which is connected in series for an inductor and in series or as a parallel element for a capacitor (Fig. 3.6).

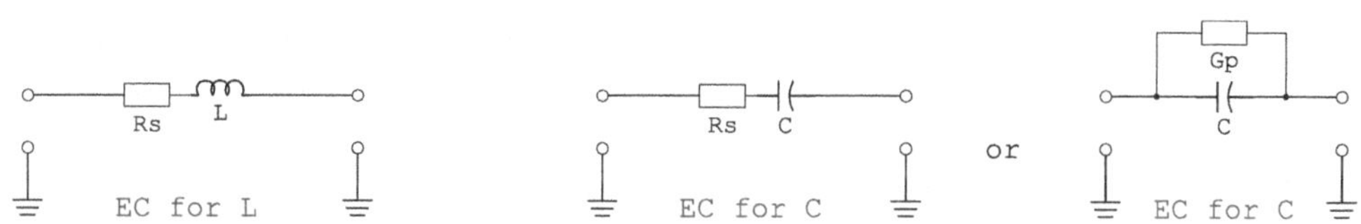

Fig. 3.6: Equivalent circuit (EC) real capacitors and coils for the consideration of losses

The Q-factor (quality factor) is used in RF technology to describe the quality of components and reflects the ratio of stored energy to losses. This definition is explained in detail in chapter 6. Mathematically, there are simple relationships for this quality definition:

$$\text{Coils:} \quad Q = \frac{\omega L}{R_s}\,, \qquad \text{Capacities:} \quad Q = \frac{1}{\omega C\, R_s} \quad \text{or} \quad Q = \frac{\omega C}{G_p}\,. \tag{3.4}$$

This quality definition is the standard in data sheets and simulators. On closer inspection, it can be seen that the value of the series resistance of a coil increases with increasing frequency for a fixed quality value Q ($R_s = Q \cdot \omega L$). This definition used to make sense for coils, as large components with thick wires or foils only had a current flow in the penetration depth range and therefore had a frequency-dependent resistance.

For air capacitors, according to this definition, the series resistance would have to decrease for a fixed quality value Q with $R_s = 1/(Q \cdot \omega C)$, which is physically nonsensical. Most of today's components are manufactured using SMD or semiconductor technology and very often have only thin metallization thicknesses. For this reason, the values of the series resistance are often constant over a wide frequency range.

This classic quality definition Q is therefore
only correct for the corresponding frequency!

An improved definition of quality is therefore increasingly being used by (male and female) specialists. This takes into account any frequency dependence of the component and loss value:

$$\text{Coils:} \quad Qlr(f) = \frac{1\,\text{GHz} \cdot L(f)}{R_s(f)} \,, \quad \text{Capacities:} \quad Qcr(f) = \frac{1}{1\,\text{GHz} \cdot C(f)\; R_s(f)} \,. \tag{3.5}$$

Fig. 3.7 shows the results of the two definitions for a semiconductor coil based on measured values.

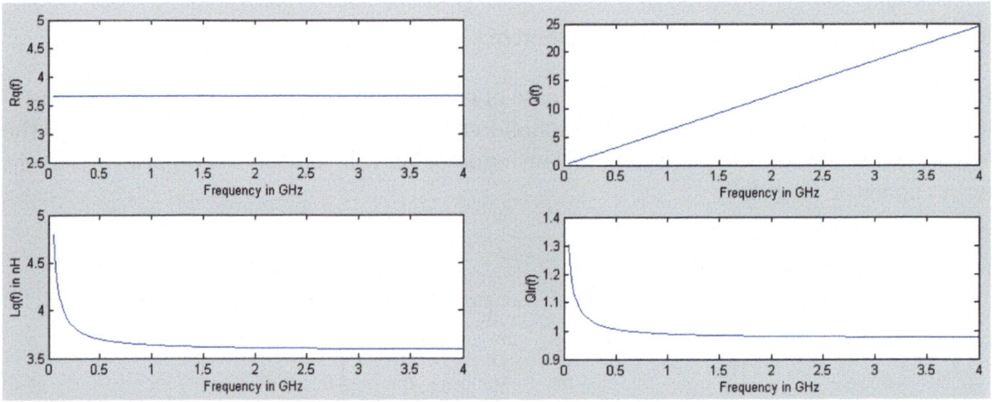

Fig. 3.7: Calculation results of a real semiconductor coil for the consideration of losses ($L = L_q$; $R_p = R_q$)

In the left half of the Fig. 3.7, there is hardly any frequency dependence for the series resistance and a noticeable frequency dependence of the inductance value of the semiconductor coil.

The classical quality factor Q suggests a very good value at 4 GHz. In contrast, Qlr indicates values that allow an assessment over the entire frequency range.

As an alternative to the definition according to the equation (3.5), the following definition of quality factors is practicable:

$$\text{Coils:} \quad Ql(f) = \frac{L(f)\,(\text{in nH})}{R_s(f)\,(\text{in }\Omega)} \,, \quad \text{Capacities:} \quad Qc(f) = \frac{1}{C(f)\,(\text{in pF})\; R_s(f)\,(\text{in }\Omega)} \,. \tag{3.6}$$

The disadvantage of both definitions is the need for a data list. Therefore, more complex equivalent circuits are a very interesting solution for modeling the losses. These are discussed in detail in chapter 9 using the example of an inductor.

3.1.2.2 SMDs

Surface mounted devices or SMD components are of great importance in the everyday life of electronics and RF circuit technology up to 10 GHz is of great importance. Over 90 % of the components mounted in so-called handsets are SMDs. The dimensions of SMD components are shown in Fig. 3.8.

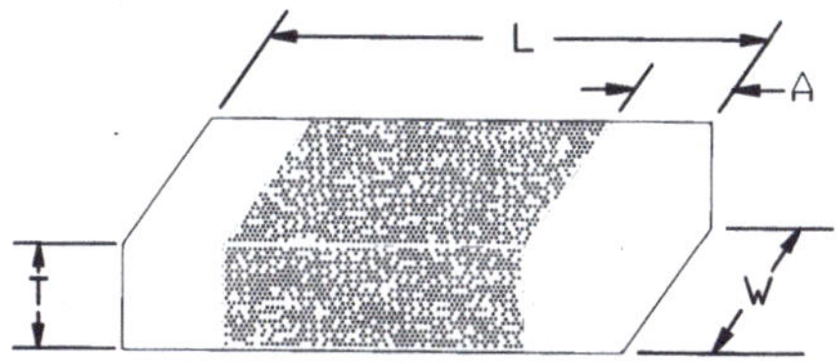

Fig. 3.8: Dimensions of SMD components

The designations of the different SMD components originate from the size in *inch*. The following table shows the common sizes with the typical dimensions:

Tab. 3.1: Geometric properties of SMD components

Bauform	L (inch)	W (inch)	L (mm)	W (mm)	T (mm)	A (mm)
0201	0.02	0.01	0.5	0.25	0.3	0.15
0402	0.04	0.02	1.0	0.5	0.55	0.25
0603	0.06	0.03	1.6	0.8	0.9	0.45
0805	0.08	0.05	2.0	1.25	1.35	0.55

A visualization of the designs and sizes is shown in the Fig. 3.9.

Fig. 3.9: Visual size comparison of SMD components

SMD components are mainly used on laminate boards (FR4 boards) in modern circuits for cost reasons. The layout of the metallization surfaces (often also metallization pads) is referred to as *footprint*. A typical footprint is shown in Fig. 3.10. Dimensioning is usually based on specifications for the soldering process. This footprint is to be seen as component of an SMD component. This means that in the case of RF, the manufacturer's data can only be used to a limited extent. In practice, the SMD components are much more reproducible than the manufacturers state (but do not guarantee).

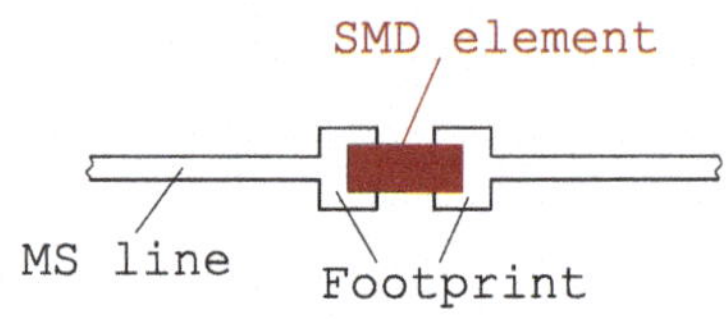

Fig. 3.10: Top view of an *footprint* including SMD component

FR4 boards with a layer height of just 0.2 mm are used in mass applications. In this case, the parasitic elements for the equivalent circuit diagram of a series-connected SMD component can be specified as shown in Fig. 3.11 depending on the design. The impedance

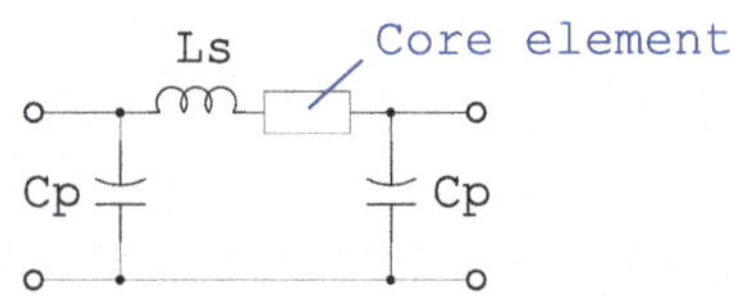

Fig. 3.11: Equivalent circuit diagram of serial SMD components (core element = impedance)

values specified by the manufacturers should relate to the core elements. However, as the measurement technology is not entirely trivial, the data often has a systematic error over the entire series. The developer should be aware of this error of often 20 %[2].

Tab. 3.2: Parasitic values of mounted SMD components for substrates with 0.2 mm layer height

Design	Ls in nH	Cp in fF
0201	0.35	90
0402	0.55	130
0603	0.80	200
0805	1.2	350

There are a large number of technologies in which SMD components are manufactured. The advantages of SMD components over older designs are so numerous that listing them would would go beyond the scope of this format. The disadvantages, on the other hand, should be known: Reflow soldering briefly exposes the SMD components to a high temperature of more than 200 degrees.

This can cause problems with aluminum electrolytic capacitors, supercapacitors or rechargeable batteries. In the case of semiconductor components with connections on the underside (such as BGAs), the solder joints can only be checked using X-rays.

The most common problem is bending stress on the circuit board. These lead to component cracks that are often not visually recognizable.

[2]In [118, 119] the necessary RF measurement technology is discussed.

SMD-Coils

Thin-film technology is often used for resistors in componenticular, but also for coils. With such coils, the parasitic values change depending on whether the thin-film coil is located between the carrier and the FR4 board or whether the coil is the top element in the multilayer structure. Coils that are cut from a thick copper foil using a laser cut (Fig. 3.12) do not have this disadvantage. Furthermore, the series resistance of these coils is lower and the Q-factor higher. SMD coils with even lower losses are made with a

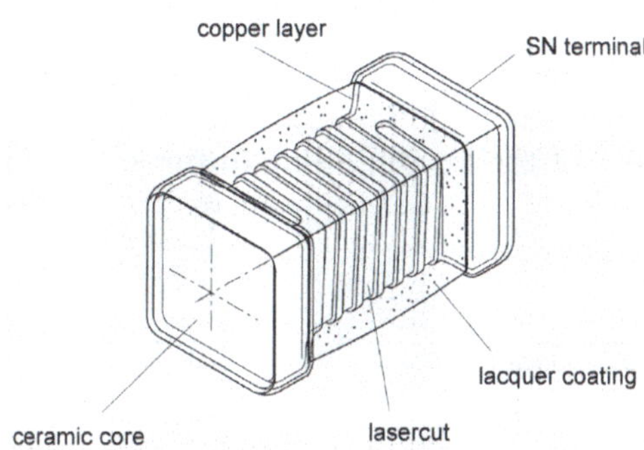

Fig. 3.12: SMD coils produced with laser cutting in the design 0402

wire that is wound around a round ceramic carrier or ferrite. The smallest wire coils are available in sizes up to 0402 and are only used for special applications due to their high cost.

The image provides an overview of SMD coils from various manufacturers 3.13.

	Stelco	TDK	Johanson	Würth	Coilcroft
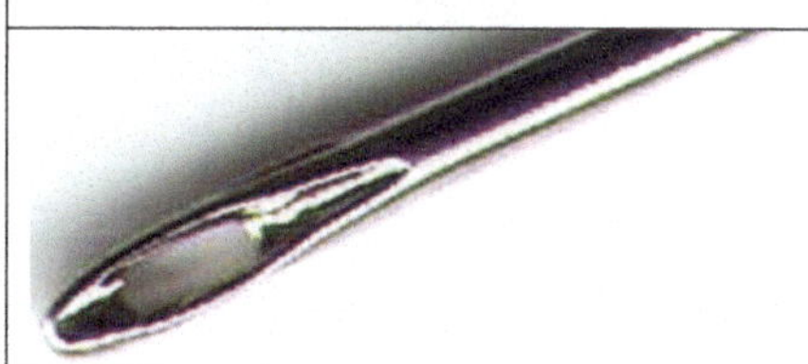			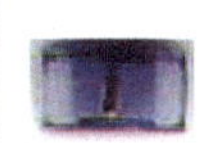		

Fig. 3.13: Various SMD coils in the design 0402

SMD capacitors

Capacitors for electronics and RF technology are preferably manufactured using multilayer ceramic technologies in the classic block capacitor design. For very small capacitance values and high quality (low losses = low series resistance), the capacitances between copper lines are used on planar ceramic carriers.

The image 3.14 provides a brief overview of SMD capacitors from some manufacturers.

The typical costs for high-volume use of individual SMD components are below 1 cent for resistors, 1.5 cent for capacitors and over 2 cent for coils.

Fig. 3.14: Various SMD capacitors in the design 0402

SMD-Semiconductors

SMD technology is the preferred form of use for semiconductors in electrical engineering. The pins are often led out of the housing. Due to the smaller parasitic sizes, packages in which the connections are below the semiconductor are increasingly being used in RF technology, Fig. 3.15.

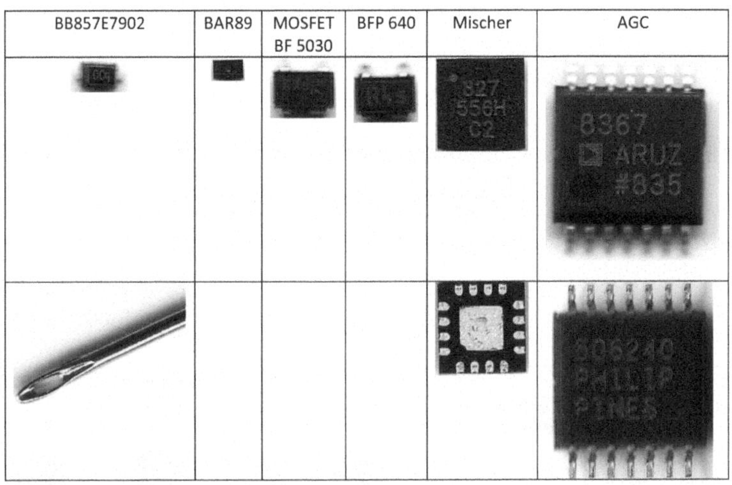

Fig. 3.15: Various SMD semiconductors in different designs

3.1.2.3 Semiconductor Components

Large circuit parts of modern high-frequency circuits are realized in semiconductors. For assemblies up to 6 GHz, silicon (Si) as a SiGe semiconductor or even as a cost-effective CMOS technology is replacing the "traditional" RF semiconductor gallium arsenide (GaAs).

Fig. 3.16 shows a comparison of the properties of passive components realized in GaAs and Si technologies.

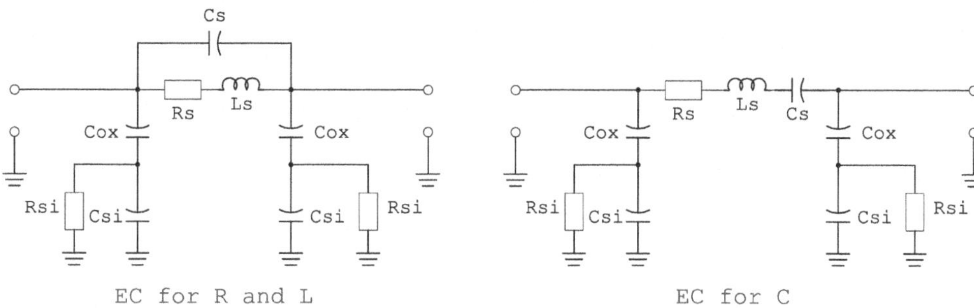

Fig. 3.16: Equivalent Circuits (ECs) for passive components on semiconductors

Tab. 3.3: Properties of passive concentrated components in gallium arsenide (GaAs) and silicon (Si)

	GaAs	**Si**
Resistors	$R_{si} \approx C_{si} \approx C_{ox} \approx C_s \approx L_s \approx 0$ almost ideal $1 \ldots 10^4\ \Omega$	$C_s \approx 0$ large capacitive parasitics $1 \ldots 10^4\ \Omega$
Inductors	$R_{si} \approx C_{si} \approx C_s \approx 0$ metallic losses $< 10\,\text{nH}$ Q(2 GHz) ≈ 20	$C_s \approx 0$, great otherwise. C's metallic and substrate losses $< 10\,\text{nH}$ Q(2 GHz) ≈ 8
Capacities	$R_{si} \approx C_{ox} \approx C_{si} \approx 0$ almost ideal $< 40\,\text{pF}$ Q(2 GHz) ≈ 100	 large capacitive parasitics $< 50\,\text{pF}$ Q(2 GHz) ≈ 40
Transmission lines	almost ideal metallic losses 0.1 dB/mm at 2 GHz	large capacitive parasitics metallic and substrate losses 1 dB/mm at 2 GHz

In the meantime, new Si technologies based on Cu metallization using high-resistance Si base material are being developed, which allow the realization of components with similar quality to GaAs. There is also a pronounced trend towards multilayer arrangements in semiconductors. Such processes can be used to produce improved inductors in particular.

Capacitors manufactured using semiconductor technologies are usually plate capacitors with a very low dielectric strength.

With coils, the developer sometimes has many degrees of freedom. The geometry (size, winding direction and type, conductor width, etc.) for a fixed layer structure can be optimally adjusted for the given requirements.

Electromagnetic field simulators (2.5D and 3D) are used in particular for coils to predict the component element values.

3.1.2.4 Multilayer Modules in LTCC and Laminates

Nowadays, you won't find a smartphone without so-called LTCC (low temperature cofired ceramic) modules. LTCC modules consist of a multilayer ceramic (up to 20 layers) as a carrier and have passive and/or active SMDs and/or pure semiconductor chips on the top layer and possibly in so-called cavities.

The multilayer ceramics are produced extremely flat. Therefore, all components manufactured in these layers have very large (parasitic) capacities against mass. The dielectric losses are negligible, whereas the conductive paths realized with thick-film technology exhibit noticeable metallic losses.

Since the dielectric values of the ceramics are very constant throughout production, field simulation can be used to realize the passive components if these values are well known.

Laminates are also used for the realization of modules and as carrier materials for consumer products in wireless communication. The dielectric losses are noticeably greater than with LTCC. However, this is partly compensated for by the lower metallization losses (Cu foils). The disadvantage of laminates is that it is not possible to produce such thin layers and fine structures as with LTCC. In terms of price, laminates are significantly cheaper for small quantities and on a par with LTCC for large quantities.

The following Fig. 3.17 shows a multiband transmitter module of a smartphone output stage and illustrates the high packing density that is possible with LTCC.

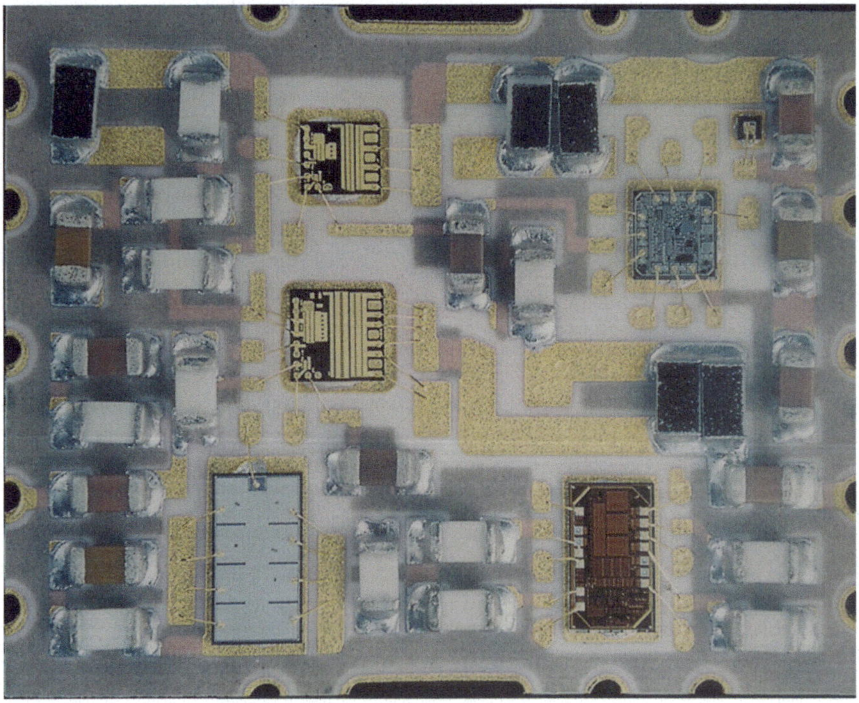

Fig. 3.17: Top view of an LTCC multilayer module with 0402 SMD and IC assembly (amplifier, drain switch, NVG, control electronics)

3.1.2.5 Coupled Inductors

If you have the design of the coils in your own hands, it is highly attractive to realize a circuit not only with discrete coils, but to use coupled coils.

Fig. 3.18 shows a bandpass filter that can be realized using two coupled coils. This filter is discussed in detail in chapter 5.

Fig. 3.18: Left: Bandpass filter realized from two parallel resonant circuits which are coupled via an inductive current coupling. Right: Equivalent circuit diagram

This example illustrates that hardly any more volume is required to manufacture the coils for these bandpass filters than a single coil takes up.

Especially as coils in semiconductor and LTCC technology take up considerably more space than capacitors, the percentage volume reduction of the circuit is almost 50 %. In addition, the cable lengths can be significantly reduced, which means that the losses are reduced, the parasitic variables are lower and the cut-off frequency increases. An example of a so-called DIFFERENTIAL COIL is printed in [122]. However, compared to the use of individual inductors, the amount of development work and therefore the development costs increase.

3.2 Two-Ports

Circuits will be presented in the following sub-chapters, which can be advantageously realized with the aid of passive concentrated components.

RF filter circuits are not included in these sub-chapters, as these are dealt with in detail in the following chapter 5.

3.2.1 Attenuators

The scattering matrix of an ideal attenuator with attenuation a (in dB) is

$$||[\mathbf{S_A}]|| = \begin{pmatrix} 0 & A \\ A & 0 \end{pmatrix} \qquad \text{with} \quad A = 10^{-\frac{a}{20\,dB}} \quad . \tag{3.7}$$

An attenuator based on a π resistor circuit was already presented in the section 2.3. The fact that attenuators can be calculated even more easily is shown using a T resistor circuit in section 5.4. The calculation equations for these two important structures are therefore given.

The choice of topology depends on the parasitic elements of the components: If you have a large capacitance to ground, the T-circuit is more advantageous, but the series inductances predominate, the π-circuit is more advantageous.

An excellent theory with the aid of resistance layers, which provides very good results in the production of attenuators is given in [109]. Fig. 3.19 shows the realization of two possible structures.

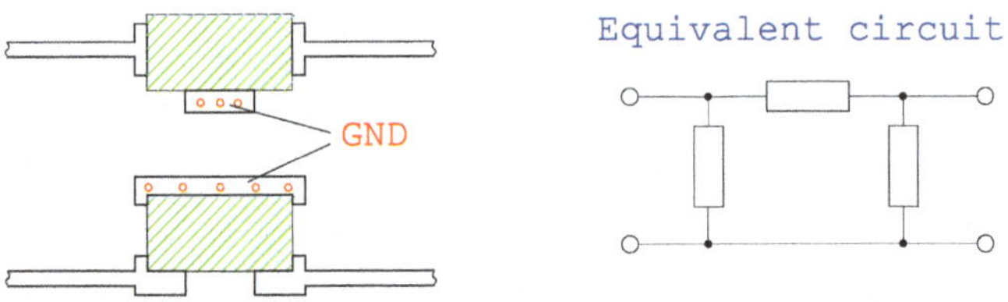

Fig. 3.19: Duality of attenuators with resistance layers (hatched areas) and associated equivalent circuit with individual resistors

This theory can also be used to design highly complex switchable attenuators quite easily.

3.2.2 Impedance Transformers

The task of an impedance transformer is to match two different impedances to each other without reflections by means of a two-port circuit. The most common special case is that the two impedances are real resistances. This section is limited to this case.

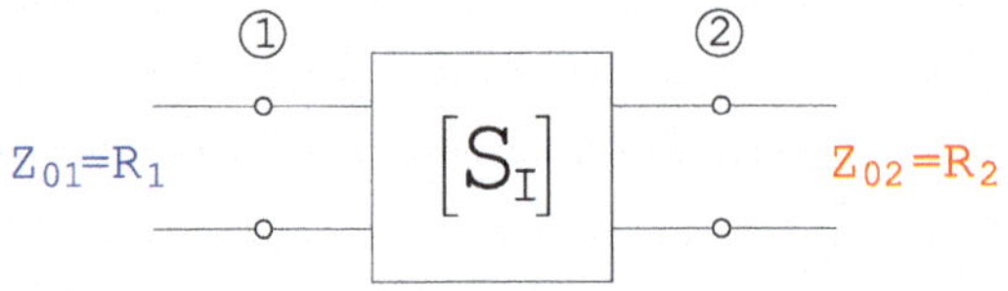

Fig. 3.20: Two-port circuit of an impedance transformer for the signal conversion from the real resistor value R_1 to R_2

The scatter matrix belonging to Fig. 3.20 for this case is as follows:

$$|[\mathbf{S_I}]| = \begin{pmatrix} 0 & 1 \\ 1 & 0 \end{pmatrix} \quad . \tag{3.8}$$

Consequently, the S-matrix of an ideal impedance transformer almost corresponds to that of a direct through-connection. Compared to the latter, a phase shift is permissible.

An impedance transformer transforms the input resistance R_1 with the transformation ratio t to the terminating resistor R_2:

$$t = \frac{R_1}{R_2} \quad . \tag{3.9}$$

An ideal impedance transformer is lossless and broadband matched.

Winding Transformer

In the electronics sector and occasionally in the RF sector, a winding transformer (transformer for short) is used for this purpose. The following applies to the transformation ratio (short: turns ratio) a of the currents and voltages for the transformer:

$$\mathrm{a} = \frac{I_2}{I_1} = \frac{U_1}{U_2} = \sqrt{\frac{R_1}{R_2}} = \sqrt{t} \quad . \tag{3.10}$$

In the RF range, transformers are rarely used due to the high costs, losses due to stray inductances and other parasitic elements as well as various problems with the use of ferromagnetic materials.
Winding transformers offer advantages with regard to the often large bandwidth. This is why they are also used for special components and in measurement technology.

Γ-Impedance Transformer

The impedance transformer most commonly used in the areas of fast electronics and RF circuits is the Γ-transformer (Fig. 3.21). It consists, for example, of a capacitor C, which is connected to ground (hence shunt capacitor), and a series coil L.

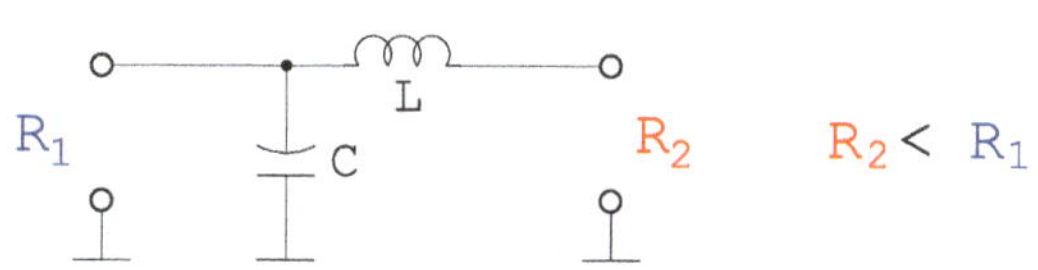

Fig. 3.21: Γ-*transformer: Impedance transformer consisting of a series inductor and a shunt capacitor*

In the specified configuration, the Γ-transformer transforms a low-impedance resistor R_2 into a higher-impedance input resistor R_1. This means that the following must apply for this arrangement: $t > 1$.

The conditions apply to this transformer:

$$\mathrm{Re}\{Y_1\} = G_1 = \frac{1}{R_1} = \frac{1}{R_2\, t} \quad , \qquad \mathrm{Im}\{Y_1\} = 0 \quad . \tag{3.11}$$

The analysis of the circuit according to Fig. 3.21 provides:

$$\begin{aligned} Y_1 &= j\omega C + 1 \,/\, [j\omega L + R_2] \quad , \\ \mathrm{Re}\{Y_1\} &= G_2 \,/\, [1 + (\omega L G_2)^2] \quad , \\ \mathrm{Im}\{Y_1\} &= \omega C - \omega L G_2^2 \,/\, [1 + (\omega L G_2)^2] \quad . \end{aligned} \tag{3.12}$$

Equating the two equations under (3.11) and (3.12) results in the two equations for determining the components C and L to be dimensioned as a function of the terminating resistor R_2 $(= 1/G_2)$ and transformation ratio t, which fulfill the matching and transformation properties exactly at the center frequency ω_m.

$$\omega_m C = \sqrt{(1 - 1/t)/t} \cdot G_2 \quad , \qquad \omega_m L = \sqrt{t - 1} \cdot R_2 \quad . \tag{3.13}$$

The advantages of the Γ-transformer are

Very low costs, small space requirement,

low proportion of components and few losses.

A disadvantage is the finite bandwidth, but this can be compensated for by a multi-stage design. Another way to improve the bandwidth is to use an additional shunt capacitor, which results in the so-called `Collins` filter ([48], book volume 1).

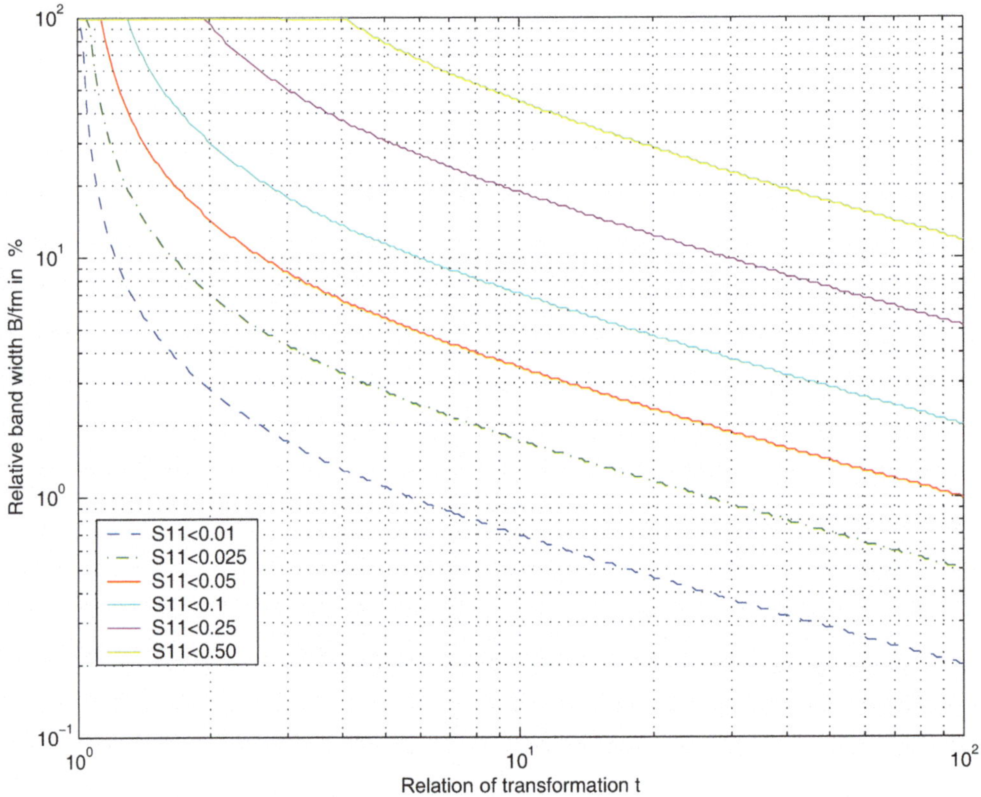

Fig. 3.22: Γ-transformer: Relative bandwidth B/f_m of the impedance transformer as a function of the transformation ratio t and the maximum input reflection factor S_{11} in %

The relative bandwidth B_r is calculated from the bandwidth $B = f_2 - f_1$ (with the lower frequency limit f_1 and the upper frequency limit f_2) divided by the center frequency $f_m = \frac{f_1+f_2}{2}$.

When designing a Γ-transformer, it is helpful to use the Fig. 3.22 for an estimate of the relationships between relative bandwidth, matching and transformation ratio.

As will be shown in chapter 4, a small inductance can be replaced by a short high-impedance cable. In this extremely cost-effective form, the Γ-transformer is used very frequently. In particular, the fact that a high-impedance cable can be used and that the optimum position of the capacitor can be determined by displacement makes this Γ transformer very popular with practitioners.

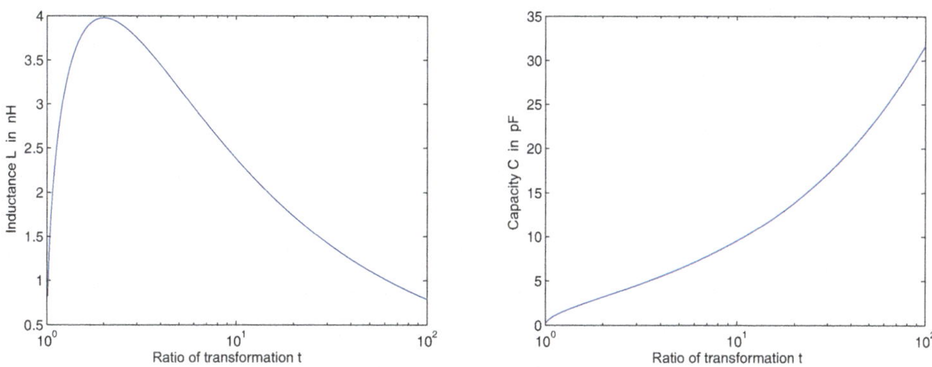

Fig. 3.23: Resulting values for a Γ-transformer at 1 GHz via the transformation ratio t

Fig. 3.23 shows how differently the values of inductance and capacitance behave over the transformation ratio. It is clear that for a small t the series inductance must be very small and the shunt capacitor must also have small values.

Multi-Stage Γ-Transformers

If you want to achieve a large relative bandwidth and a large impedance transformation, you have to connect several Γ-transformers in series.

The resulting transformation ratio is calculated by multiplying the individual stages ($t = t_1 \cdot t_2 \cdot ...$). It is best to use the same transformation t_{st} for each stage ($t = t_{st}^x$ with x: number of stages).

The following example shows the application of the design of a Γ-transformer.

——— *Exercise: Output Matching GSM Amplifier* ———

Given: *The output impedance of a GSM amplifier is 2 Ω. The matching (here reflection losses) should be 20 dB, so that this amplifier is operated with a high degree of efficiency. The input resistance of the antenna is 50 Ω.*

Find: *What must the impedance transformation look like, so that this amplifier can be used for the European GSM band (TX: 880-915 MHz) and the American IS-54 band (824-849 MHz)?*

1. *A matching of 20 dB corresponds to the maximum reflection factor r_m of*

$$S_{11} = 10^{-\frac{20\,dB}{20\,dB}} = 0.1 \tag{3.14}$$

and the required transformation ratio is $t = 25$ and the required relative bandwidth is

$$B/fm = (915 - 824) \; / \; \frac{824 + 915}{2} = 0.105 = 10.5\,\% \quad . \tag{3.15}$$

2. *The Fig. 3.22 shows that for a relative bandwidth of around 10 % and an adjustment of* $|S_{11}| = 0.1$ *a transformation ratio of maximum 6 can be achieved. Consequently, the transformation network can only be achieved using a two-stage Γ-transformer with ratios of* $t = 5$ *in each case.*

First stage $2\,\Omega$ *to* $10\,\Omega$*:*

$$C_1 = \frac{\sqrt{(1-0.2)\cdot 0.2}\ /\ 2\,\Omega}{2\cdot\pi\cdot 869.5\,MHz} = 36.6\,pF\ , \quad L_1 = \frac{\sqrt{(5-1)}\cdot 2\,\Omega}{2\cdot\pi\cdot 869.5\,MHz} = 0.73\,nH\ . \tag{3.16}$$

Second stage $10\,\Omega$ *to* $50\,\Omega$*:*

$$C_2 = \frac{\sqrt{(1-0.2)\cdot 0.2}\ /\ 10\,\Omega}{2\cdot\pi\cdot 869.5\,MHz} = 7.3\,pF\ , \quad L_2 = \frac{\sqrt{(5-1)}\cdot 10\,\Omega}{2\cdot\pi\cdot 869.5\,MHz} = 3.66\,nH\ . \tag{3.17}$$

——— *Exercise: Output Matching GSM Amplifier* ———

Alternative Γ-Transformer

The second possible structure using a capacitor and an inductor to realize a Γ-transformer is shown in Fig. 3.24.

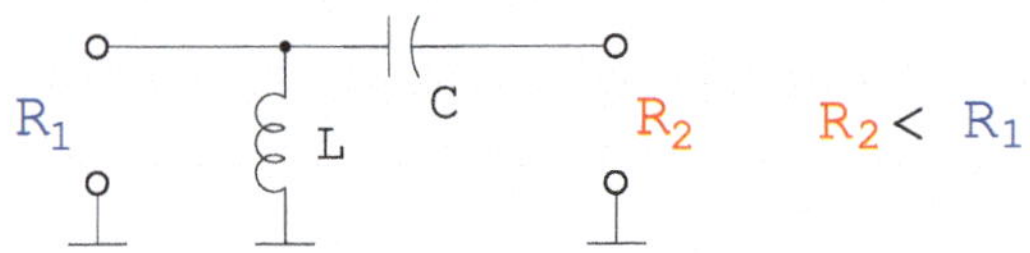

Fig. 3.24: Γ-impedance transformer with series capacitor

This structure with a series capacitor and a shunt inductor has the advantage that the capacitor realizes a direct current separation (DC bias). This structure is very frequently used in the circuitry of resonators and filters.

The components are calculated for the center frequency ω_m via:

$$C_t = \frac{1}{\omega_m}\sqrt{\frac{1\ /\ R_2}{R_1\ -\ R_2}}\quad , \qquad L_t = R_1\ R_2\ C_t \quad . \tag{3.18}$$

This transformer has similar characteristics in terms of bandwidth and can be estimated approximately using the information in Fig. 3.22.

3.2.3 ±90°-Phase Shifter

An ideal phase shifter is perfectly adapted and realizes a constant frequency-independent phase shift by the value $\angle(S_{ij}) = -\varphi$:

$$[\mathbf{S}_{-\varphi}] = \begin{pmatrix} 0 & e^{-j\varphi'} \\ e^{-j\varphi'} & 0 \end{pmatrix} \qquad \text{whereby} \tag{3.19}$$

$$(0^\circ \leq \varphi \leq 360^\circ) \qquad \text{and} \qquad \varphi' = \frac{\pi}{180^\circ}\varphi \quad . \tag{3.20}$$

The name of the phase shifter is derived from the phase shift of the transmission value ($\angle(S_{ij})$). The following therefore applies to a $\pm 90^\circ$ phase shifter:

$$[\mathbf{S}_{\pm 90^\circ}] = \begin{pmatrix} 0 & \pm j \\ \pm j & 0 \end{pmatrix} \quad . \tag{3.21}$$

In practice, it is usually sufficient if phase shifters can be implemented in a narrow band. Phase shifters are rarely required as individual moduls in the system. However, the $\pm 90^\circ$ phase shifters are very important as sub-module in modules such as signal dividers, balancing elements, filters, couplers and much more.

As shown in chapter 4, $\pm 90^\circ$ phase shifters at center frequency correspond to so-called $n \cdot \lambda/4$ lines ($n = 1$ for -90°, $n = 3$ for 90°). Many passive RF frequency circuits can be synthesized in a simple way with $\lambda/4$ lines. As a concentrated replacement solution, modern RF frequency technologies offer highly integrated alternatives for $\pm 90^\circ$ phase shifters.

The simplest $\pm 90^\circ$ phase shifters are realized by a π or T-structure with reactive components.

To dimension the π structure according to Fig. 3.25, use the auxiliary sheet A.2 with the two conditions $S_{11} = 0$ and $S_{21} = \pm j$. Alternatively, a much simpler form of calculation is presented in section 5.4.

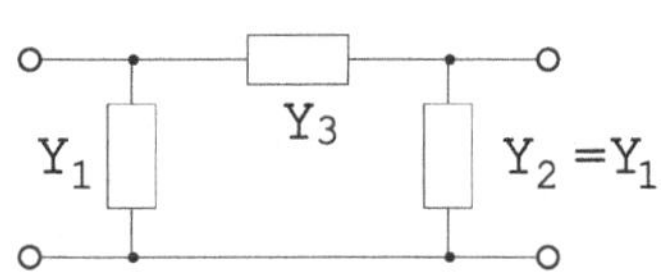

Fig. 3.25: *Realization of a phase shifter with three reactive elements*

The two resulting phase shifters have high-pass and low-pass characteristics respectively (Fig. 3.26).

Fig. 3.26: Circuits for $\pm 90^\circ$ phase shifting

The solution obtained with $Y_1 = jB_1$:

$$B_1 = \pm 1 \quad \text{and thus for} \tag{3.22}$$

$$\angle(S_{ij}) = -90^\circ: \quad \omega_m C_{LP} = \frac{1}{Z_0}; \qquad \angle(S_{ij}) = 90^\circ: \quad \omega_m L_{HP} = Z_0 \quad . \tag{3.23}$$

Furthermore, the normalized series element with $Z_3 = \frac{1}{Y_3} = jX_3$ corresponds to the normalized conductance B_1:

$$X_3 = \pm 1 \quad \text{and thus for} \tag{3.24}$$

$$\angle(S_{ij}) = -90^\circ : \quad \omega_m L_{LP} = Z_0 \; ; \qquad \angle(S_{ij}) = 90^\circ : \quad \omega_m C_{HP} = \frac{1}{Z_0} \; . \tag{3.25}$$

The complementary values result for the T-structure.

Example: A -90°-π phase shifter ($\angle(S_{ij}) = -90^\circ$) delivers the values $C_1 = 3.18\,pF$ and $L_2 = 7.96\,nH$ in the $50\,\Omega$ system at 1 GHz. The magnitude in dB and the phase in degrees of the scattering parameter S_{21} over the frequency are shown in Fig. 3.27.

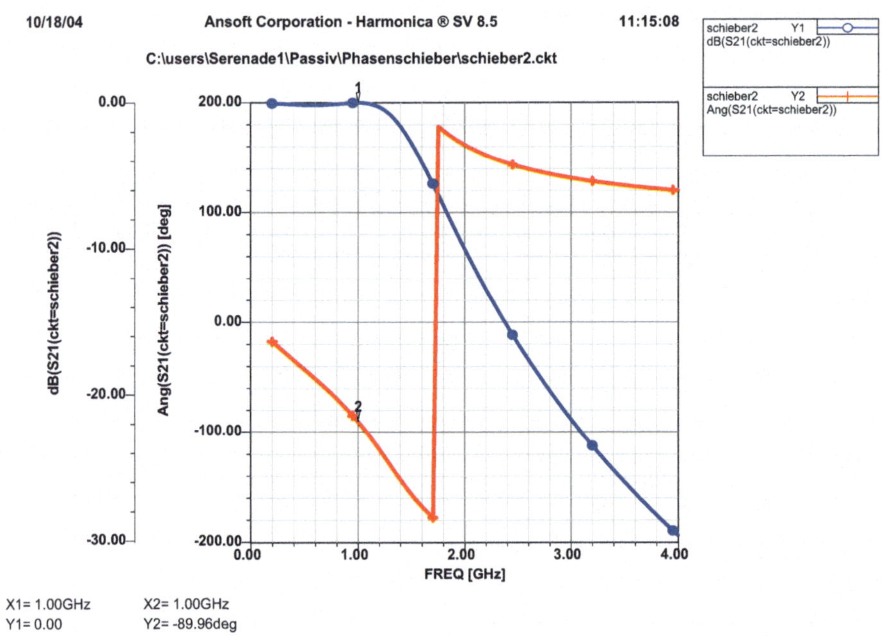

Fig. 3.27: Realization of a -90° phase shifter with two shunt capacitances and one series inductance

This -90° phase shifter has a low-pass character. The 90° phase shifter, on the other hand, has a high-pass character. With this circuit topology, the ideal S-parameters of the phase shifter are only maintained for one frequency point.

If you want to achieve a higher bandwidth, you must increase the number of components. Examples of this are given in [103]. A simplified analysis form of this symmetrical circuit is given in section 5.4.

3.3 Three and Four Ports

The electrotechnical literature contains a very large number of implementations of passive modules in the RF range, such as the so-called signal dividers, baluns and couplers, which can be produced with $n \cdot \lambda/4$ lines. As already mentioned in the previous section, these components can also be manufactured with concentrated elements using phase shifters.

This chapter presents three- and four-port modules that can be constructed using only concentrated components.

3.3.1 DC and Control Signal Supply

The first three-port presented only contains two classic RF frequency ports. In this direct current feed circuit (short: Bias-T and more rarely DC feed), a direct current or a low-frequency signal is to be fed to port 1 alone via port 3. This circuit is also used to feed in control signals.

Fig. 3.28 shows a realization that can be used in the upper MHz and lower GHz range and is easy and inexpensive to manufacture using SMD components.

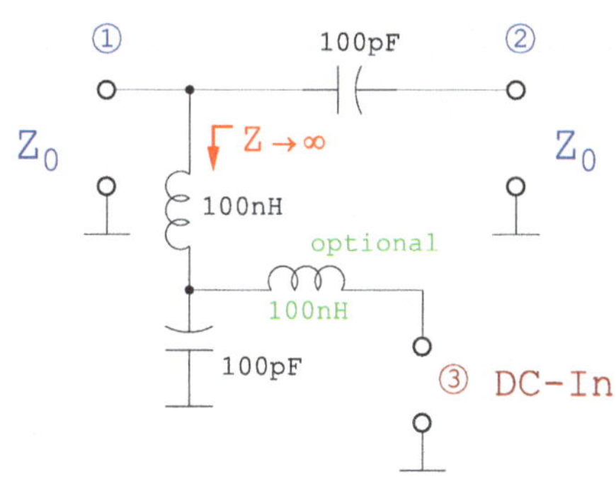

Fig. 3.28: Bias-T or DC or control signal feed for the MHz to GHz range

The series capacitor isolates port 2 from the direct current. The combination of series coils and shunt capacitors realizes a low-pass filter. If the series coil is selected as the first element, the input resistance is very high and the RF signal is hardly affected. The corresponding S-matrix is for values in the MHz and GHz range:

$$||[\mathbf{S_{BT}}]|| = \begin{pmatrix} 0 & 1 & 0 \\ 1 & 0 & 0 \\ 0 & 0 & 1 \end{pmatrix} \quad . \tag{3.26}$$

The capacitance and inductance values shown in Fig. 3.28 can be increased, which is particularly beneficial for functionality at lower frequencies. However, the materials used by the manufacturers then often exhibit greater losses, so that the desired freedom from losses is no longer given. In addition, larger components have a low maximum frequency due to parasitic effects.

The structure of this DC supply can be found in a very large number of circuits. However, as the functionality can also be maintained with smaller component values, the components of the DC supply are often used for other electrical circuit purposes. This is shown in the design of linear amplifiers (chapter 8).

3.3.2 Circuits Around Semiconductors

The wiring between semiconductors and homogeneous lines, as described in the next chapter, is characterized or realized by

semiconductor housings or test fixtures or probes.

These short connection distances can be modeled by electrical networks. These networks in turn can be modeled using concentrated components (usually only capacitors and coils). An example is shown in Fig. 3.29, [95].

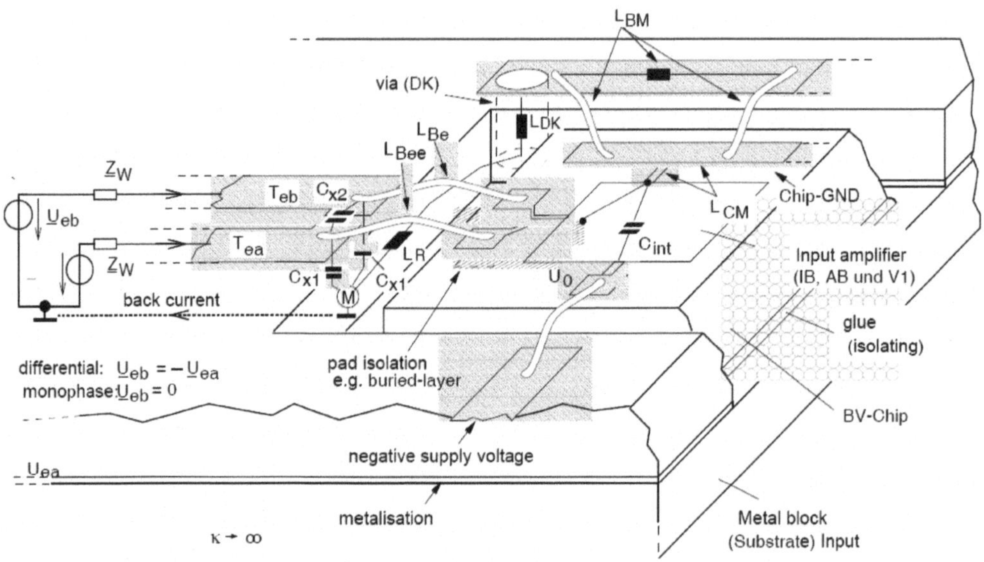

Fig. 3.29: Representation of the input circuit of the test socket of an amplifier IC (BV chip) with identification of the parasitic components, [95]

Based on the content of the basics in chapter 4, the RF developer should be able to determine a network for such circuits. Idealized networks such as short circuits, open circuits and through-connections are then used for the IC. This is done either in field simulators or in real setups. The S-parameters for the outer feeds with different inner terminations are obtained from both. In the final step, the S-parameters can be fitted using the selected networks. How this is done can be found in chapter 9.

3.3.3 Resistive Signal Divider

An ideal signal splitter should be well matched on all sides and split a signal broadband into two parts. As a rule, the signal splitter is limited to a symmetrical division of the energy (attenuation is $3\,\text{dB} \mathrel{\hat{=}} 0.707$). However, it is often helpful in practice, if the two continuing paths are decoupled from each other (see Eq. (3.27); $S_{23} = S_{32} = 0$).

$$[\mathbf{S_{ST}}] = \begin{pmatrix} 0 & 0.7 & 0.7 \\ 0.7 & 0 & 0 \\ 0.7 & 0 & 0 \end{pmatrix} \quad . \tag{3.27}$$

Based on the unitarity relationship (Eq. (2.39)), it can be shown that such a universally adapted and lossless three-port *not* can be realized.

If a broadband signal divider is required, the following two solutions can be used.

Divider Matched on the Input Side

The simplest signal divider with input-side matching consists of two resistors, each corresponding to the reference impedance (Fig. 3.30).

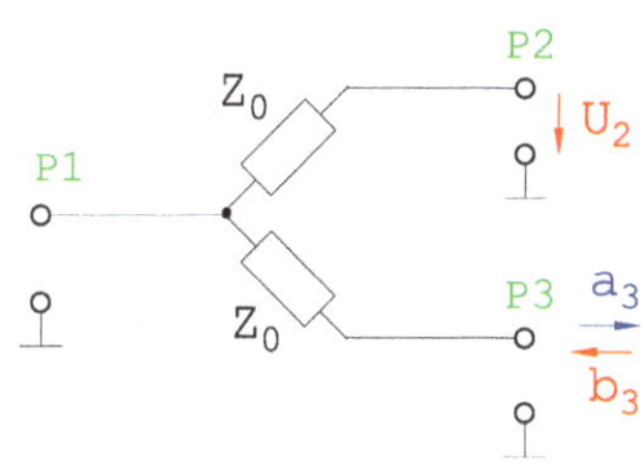

Fig. 3.30: Input-side matched lossy signal divider in single-pole representation

The scatter matrix of this signal splitter is as follows:

$$[\mathbf{S_{ST2}}] = \begin{pmatrix} 0 & 0.5 & 0.5 \\ 0.5 & 0.25 & 0.25 \\ 0.5 & 0.25 & 0.25 \end{pmatrix} \quad . \tag{3.28}$$

It can be seen that the divider 6 dB ($\hat{=}\,0.5$) has transmission losses per path (and therefore has 3 dB losses) and is only matched with 12 dB ($\hat{=}\,0.25$) at ports 2 and 3. It can also be seen that the isolation between the output ports is 12 dB and is therefore sufficient for many applications.

The design is based on the required entrance adjustment at port 1:

$$Z_{in} \stackrel{!}{=} Z_0 = 2\,Z_0\,//\,2\,Z_0 \quad . \tag{3.29}$$

Fig. 3.31 illustrates how the transmission scattering parameters (e.g. here S_{21}) of this three-port are calculated by considering the third port by means of a resistor using the known auxiliary means (auxiliary sheet 2).

This signal divider has the interesting property for measurement technology that the normalized voltage at one output port (e.g.: U_2/Z_0) corresponds to the outgoing wave of the other output port (e.g.: a_3). The derivations and applications can be found in the section 10.4.1.

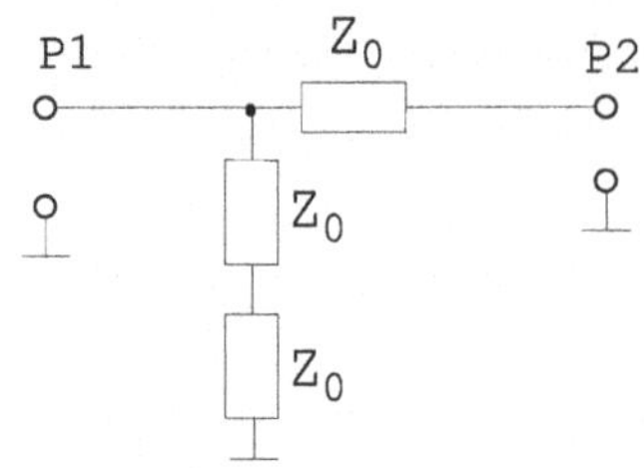

Fig. 3.31: Representation of the three-port two-port reduction for analysis of the input-side matched signal divider

Divider Matched on All Sides

The signal divider with universal matching consists of three resistors (Fig. 3.32). The following must apply to the resistors Z: $Z = Z_0/3$.

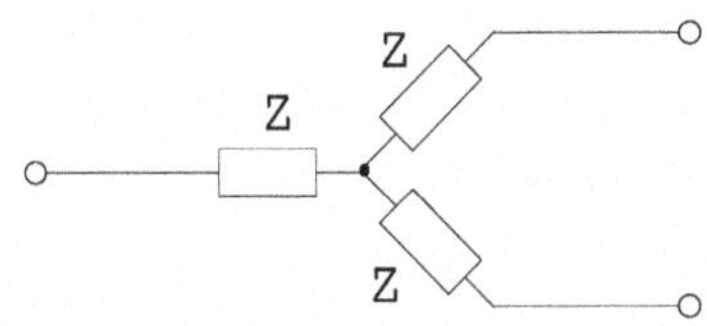

Fig. 3.32: *Lossy signal divider matched on all sides in single-pole representation*

The scatter matrix of this signal splitter is as follows:

$$[\mathbf{S_{ST3}}] = \begin{pmatrix} 0 & 0.5 & 0.5 \\ 0.5 & 0 & 0.5 \\ 0.5 & 0.5 & 0 \end{pmatrix} \quad . \tag{3.30}$$

It can be seen that the divider also has 6 dB transmission loss per path and is perfectly matched to all ports. It can also be seen that the isolation between the output ports is only 6 dB.

This simple signal divider is calculated via the matching:

$$Z_{in} \stackrel{!}{=} Z_0 = Z + (Z + Z_0) \,//\, (Z + Z_0) \quad . \tag{3.31}$$

3.3.4 Reactive Signal Divider

Reactive signal dividers only have the reactive components coils and capacitors.

——— *Exercise: Lossless 3 dB signal divider* ———

Given: *A lossless 3 dB signal divider (for the forward direction) with a relative bandwidth of 70 % can be constructed according to the Fig. 3.33.*

Find:

1. *Divide the circuit into two two-port circuits with impedance transformation.*

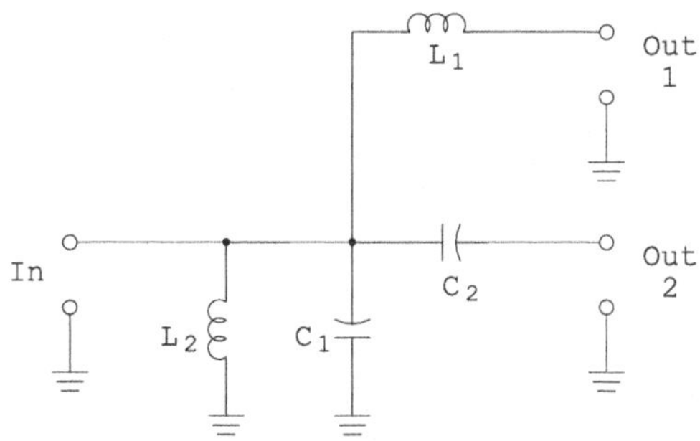

Fig. 3.33: Illustration of the lossless 3 dB signal divider based on Γ-transformers

2. *Calculate the 4 elements C_1, L_1, C_2 and L_2 for $f = 1.0\,GHz$.*

To 1. *Two Γ-transformers connected in parallel (as shown in Fig. 3.34) must be used.*

Fig. 3.34: Decomposition of the lossless 3 dB signal divider into two Γ-transformers

To 2. *With the transformation ratio $t = 2$ applies:*

$$C_1 = \frac{\sqrt{(1-0.5)\cdot 0.5}\ /\ 50\,\Omega}{2\cdot\pi\cdot 1.0\,GHz} = 1.59\,pF\ , \quad L_1 = \frac{\sqrt{(2-1)}\cdot 50\,\Omega}{2\cdot\pi\cdot 1.0\,GHz} = 7.96\,nH \tag{3.32}$$

and

$$C_2 = \frac{1}{2\,\pi\,1.0\,GHz}\sqrt{\frac{1/50}{100-50}} = 3.18\,pF\ , \quad L_2 = 50\,\Omega\cdot 100\,\Omega\cdot 3.18\,pF = 15.9\,nH\ . \tag{3.33}$$

——— *Exercise: Lossless 3 dB signal divider* ———

This lossless 3 dB signal divider rotates the phase of S_{21} by -45° and the phase of S_{31} by +45° at center frequency. The output signals therefore have a difference phase of 90°. This differential phase of 90° (which is given over a very large relative bandwidth) is desirable in some circuits. However, very often you want to have 0° as the differential phase.

This often desired lossless 3 dB signal divider with 0° difference phase at the output can be realized by using a single Γ-transformer of the previous exercise. This signal divider therefore has either a low-pass or a high-pass characteristic and very few components, Fig. 3.35.

For the magnitude values, the signal dividers based on the Γ-transformer have the following S-matrix at center frequency:

$$|[\mathbf{S_{ST4}}]| = \begin{pmatrix} 0 & 0.7 & 0.7 \\ 0.7 & 0.5 & 0.5 \\ 0.7 & 0.5 & 0.5 \end{pmatrix} \quad . \tag{3.34}$$

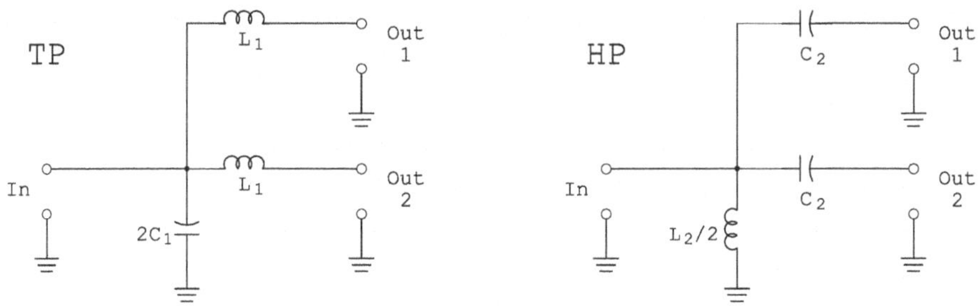

Fig. 3.35: Representation of the lossless 3 dB signal divider with 0° phase difference at the output

Another reactive signal divider that requires 2 coils and 2 capacitors is presented in chapter 6 as a ±90°balun. With this signal divider, the differential phase at the output is 180°.

Other Signal Dividers

Based on the simple work with the Γ-transformers, you can easily derive many other signal dividers:

- For a 1-to-4 signal divider, each input of the four Γ-transformers must be 200 Ω and each output must have 50 Ω.

- Unbalanced signal dividers can be derived from different input resistances.

——— *Exercise: Lossless 1 to 4 signal divider* ———

Given: *A broadband lossless 1 to 4 signal splitter (for the forward direction) can be constructed according to the Fig. 3.36.*

Find:

1. *Enter all S-parameters for this 5-port.*

Solution: $S_{11} = 0$ *applies, since a wave entering port 1 is split but not reflected.*

It applies $S_{22} = S_{33} = S_{44} = S_{55} = 0.0625$. *This value is calculated from the double attenuation due to the divider matched on all sides (0.5 each) and the reflection of the divider matched on the input side at the output port (0.25).*

This also results in $S_{21} = S_{31} = S_{41} = S_{51} = 0.25$. *Both divider types have a transmission loss of 0.5.*

Isolating the dividers matched on all sides with 0.5 results directly in the values: $S_{54} = S_{32} = 0.5$.

The isolation of the other outputs is calculated in the same way as the rear reflection values: $S_{53} = S_{43} = S_{52} = S_{42} = 0.0625$.

——— *Exercise: Lossless 1 to 4 signal divider* ———

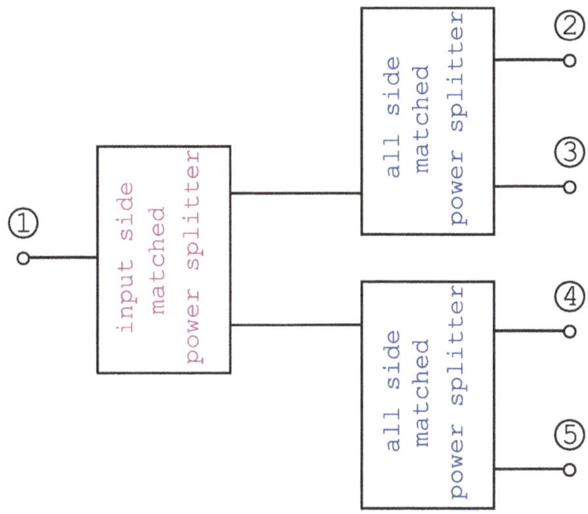

Fig. 3.36: Representation of the lossless 1 to 4 signal divider

3.3.5 Various Couplers and Baluns

A large number of so-called couplers and balun circuits can be realized very advantageously using concentrated components. These components can also be used to split the signals with low loss. As a rule, these are not broadband components.

However, the so-called multimode scattering parameters, which are first explained in chapter 6, are required to describe them. For this reason, these components are also only dealt with in detail in chapter 6.

Couplers are synthesized by first calculating couplers using lines. In the second step, these lines are synthesized using phase shifters, which are constructed using concentrated components.

The example in Fig. 3.37 shows how easy it is to implement a four-port coupler consisting of four so-called $\lambda/4$ lines. As shown in the following chapter, these $\lambda/4$ lines can alternatively be realized using -90°phase shifters.

The magnitude values of the scatter matrix of this coupler are as follows:

$$|[\mathbf{S_{Hyb}}]| = \begin{pmatrix} 0 & 0 & 0.707 & 0.707 \\ 0 & 0 & 0.707 & 0.707 \\ 0.707 & 0.707 & 0 & 0 \\ 0.707 & 0.707 & 0 & 0 \end{pmatrix} \quad . \tag{3.35}$$

This hybrid coupler fulfills the requirements of an ideal signal splitter! The disadvantage is that the components required are relatively large, if the -90° lines are made from concentrated components, as shown in Fig. 3.38. If coupled inductors are also used here, this also results in a highly integrated passive circuit.

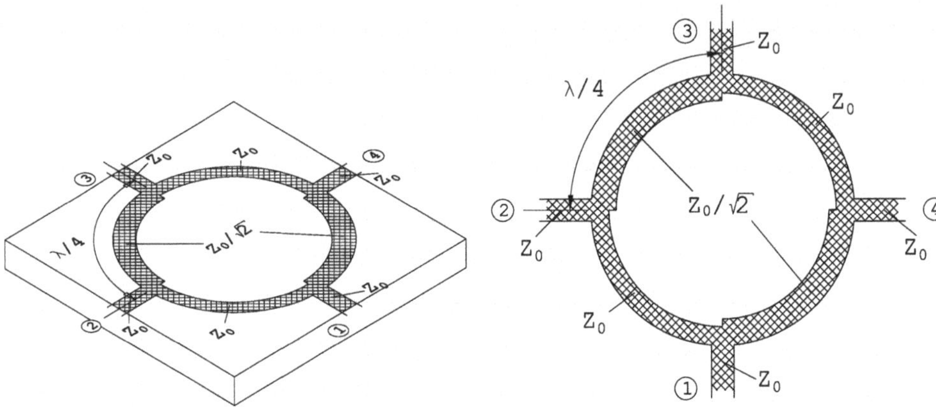

Fig. 3.37: Hybrid coupler for splitting the signals into two branches realized in so-called microstrip line technology (chapter4)

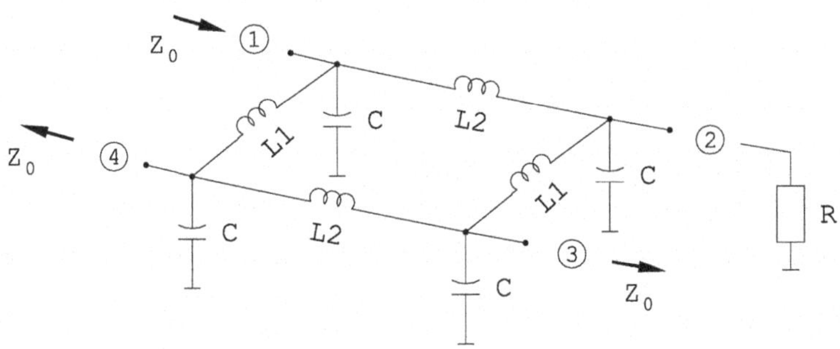

Fig. 3.38: Hybrid coupler for splitting the signals into two branches realized from concentrated elements and terminated port 2 ($R = Z_0$)

3.4 Multifunctional Modules

A long-standing trend in research in the field of passive RF circuits is to increase functionality in such a way that the circuits:

- can be used over large bandwidths and /or
- can be used in several frequency ranges and/or
- have several functionalities at the same time.

The following circuit is a simple example of a multifunctional component in practice.

<u>90° Phase Shift, DC Feed and Impedance Transformation</u>

In the ±90° phase shifters presented here, each element rotates the phase by ±30° at the operating frequency. If a phase shifter is realized with only two elements each element must rotate the phase by ±45°. This is feasible but always requires an impedance transformation. These phase shifters have the same structure as the Γ-transformer. These phase shifters can be calculated using Eq. (2.27).

If you now want to combine the three functionalities, 90° phase shifting, DC injection and impedance transformation, you can do this excellently with just four components, as shown in Fig. 3.39.

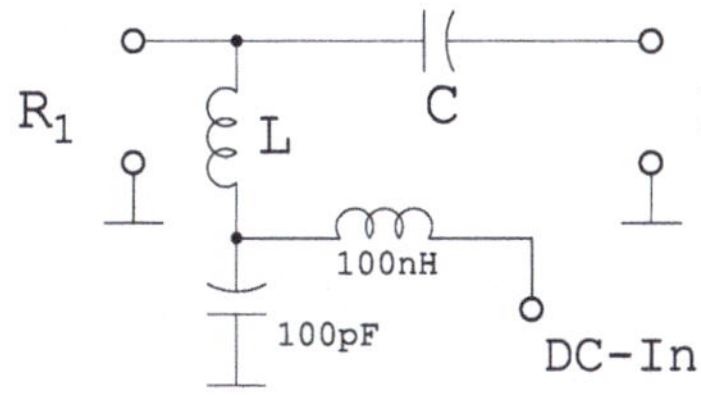

Fig. 3.39: Realization of a three-port circuit with 90° phase shift, DC injection and impedance transformation

The auxiliary sheets from page 446 support the design of circuits that have an impedance transformation. The procedure is the same as when working with the auxiliary sheet A.2.

Dual-Band Functions

A simple procedure for realizing special functionalities for two frequency bands in one circuit is the use of series and/or parallel resonant circuits. These resonant circuits can be used to implement frequency-dependent coils and capacitors on the one hand and to turn a capacitor into a coil and vice versa on the other. One possible application is presented on the page 184.

The electrical behavior of a series resonant circuit can be illustrated graphically over the frequency as shown in Fig. 3.40 with the abbreviations OC for open circuit and SC for short circuit.

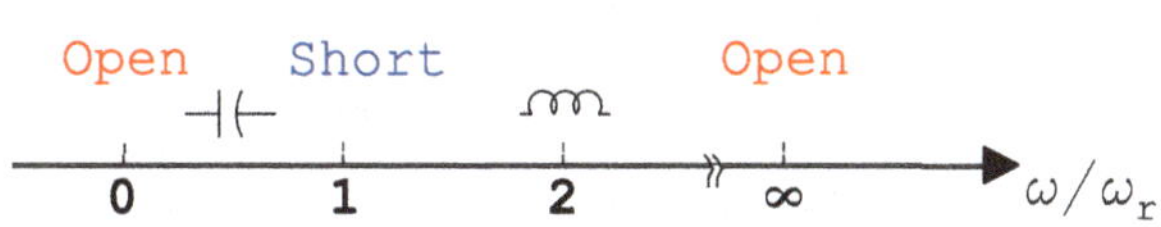

Fig. 3.40: Representation of the series impedance of a series resonant circuit over the frequency

To calculate these replacement elements at the frequencies f_i, it makes sense to use the detuning ν_i and the resonance frequency ω_r. The resulting inductance L_i and capacitance C_i at f_i can be calculated using

$$L_i = \frac{\nu_i\,\omega_r}{\omega_i}\,L_s \quad \text{resp.} \quad C_i = -\frac{\omega_r}{\nu_i\,\omega_i}\,C_s \quad \text{for the series circuit} \tag{3.36}$$

$$L_i = -\frac{\omega_r}{\nu_i\,\omega_i}\,L_p \quad \text{resp.} \quad C_i = \frac{\nu_i\,\omega_r}{\omega_i}\,Cp \quad \text{for the parallel circuit} \tag{3.37}$$

$$\text{with} \quad \nu_i = \frac{\omega_i}{\omega_r} - \frac{\omega_r}{\omega_i} \quad \text{and} \quad \omega_r^2 = \frac{1}{L_s\,C_s} = \frac{1}{L_p\,C_p} \quad . \tag{3.38}$$

The following table 3.4 illustrates which module can be realized at which frequency.

Tab. 3.4: Properties of resonant circuits as frequency-dependent components

	Series resonant circuit	Parallel resonant circuit
$f_1 < f_2 < f_r$	C_1 and C_2 $C_1 < C_2$	L_1 and L_2 $L_1 < L_2$
$f_1 < f_r < f_2$	C_1 and L_2	L_1 and C_2
$f_r < f_1 < f_2$	L_1 and L_2 $L_1 < L_2$	C_1 and C_2 $C_1 < C_2$

Unfortunately, it is not possible to use capacitances and inductances with the properties, that the associated values decrease with increasing frequency.

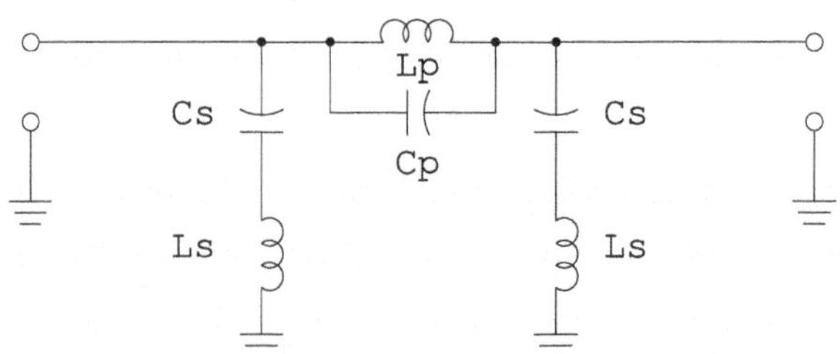

Fig. 3.41: Realization of a ±90°phase shifter for two frequency bands

Examples of the application of this theory can be found in the RF technique, which can be downloaded at [70]. Exam K5, for example, contains the calculation of a dual-band phase shifter with ±90° shift. The corresponding circuit diagram can be seen in the Fig. 3.41.

Chapter 4

RF Transmission Lines: Theory, Line Types and Applications

In the RF sector, transmission lines are not only used for data and signal transmission and for energy transport, but also as so-called "quasi-concentrated" and "distributed" components used. For electronics engineers, it is always surprising how many phenomena transmission lines, which are only regarded as wires, exhibit at higher frequencies.

Nevertheless, the study of RF transmission lines is extremely important for all electronics engineers who want to realize a data transmission line in the MHz or even GHz range. Especially for the transmission of digital signals with clock rates in the Mbit and Gbit range, the transmission lines cause very high bit error rates, as in these cases they are broadband transmission lines that can have different propagation times for lower-frequency and higher-frequency signal components.

Transmission lines in the RF range can be divided into two classes:

- Waveguide
  (Examples of waveguides are rectangular waveguides and fiber optic cables),
- Multi-wire conductor
  (The important ones are the two- and three-wire transmission lines).

The results of transmission line theory are applicable to both types. Waveguides and waveguide components are rarely used in RF electronics. Due to their complexity and main areas of application such as RF large signal transmitters, they are not discussed further and reference is made to the numerous literature, e.g. [21, 48, 131]. The so-called three-wire transmission lines are introduced in chapter 6.

Examples of two-wire transmission lines are shown in Fig. 4.1. These two-wire transmission lines are also referred to as Lecher-type transmission lines in older literature.

In contrast to waveguides, the transverse dimensions of two-wire transmission lines are very small and are only in the range of a few degrees of the wavelength (compare equation

H. Heuermann, *Radio Frequency Technologies*,
https://doi.org/10.1007/978-3-658-50170-9_4

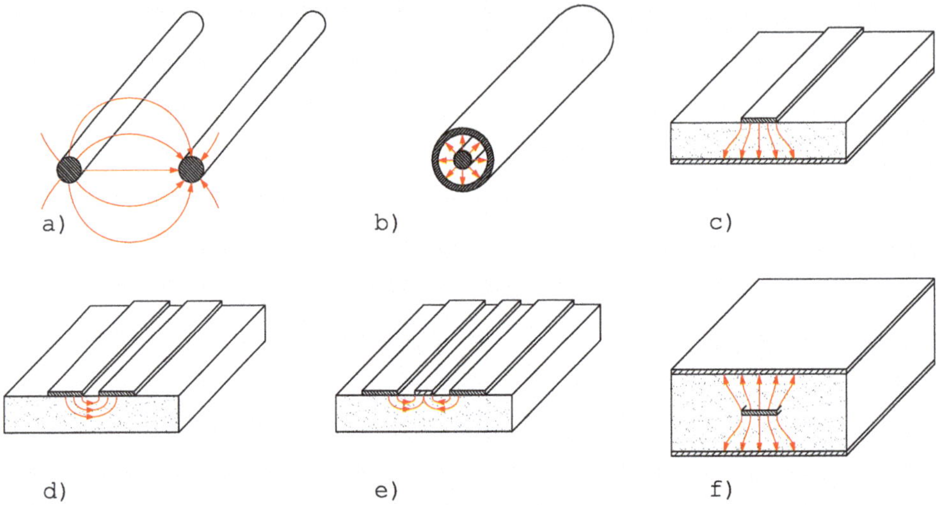

Fig. 4.1: Typical two-wire transmission lines with the corresponding occurring electric fields: a) Parallel wire, two-wire or twisted pair transmission line; b) Coaxial cable or transmission line; c) Microstrip line; d) Coplanar two-band line; e) Coplanar line; f) Stripline (or triplate line)

(2.7)). Therefore, for two-wire transmission lines it is permitted and possibly advantageous to operate with the terms current and voltage.

Two-wire transmission lines consist of either a balanced or an unbalanced transmission line pair. In a symmetrical transmission line, one transmission line is the forward transmission line and the other is the return transmission line of the current.

In an unbalanced arrangement, the larger transmission line usually has the function of a ground and thus represents the return transmission line for the current. A ground is a very wide transmission line (sometimes closed, as with the coaxial cable), which has a potential of 0 V. The latter may be achieved by connecting it to an operating ground or earth.

Restriction to TEM and Quasi-TEM Modes

This chapter describes RF transmission lines that have (almost) only E- and H-field components transverse to the direction of propagation. If longitunal components occur, they should be significantly smaller than the transverse components. The latter wave types are referred to as quasi-TEM modes[1] and are treated in a similar way to TEM modes.

TEM and quasi-TEM transmission lines are by far the most important in the practice of RF technology.

[1] TEM wave: transverse electromagnetic wave.

Overview

First, the general transmission line theory and important general variables are of transmission lines are introduced.

TEM Wave guides are presented below. With the introduction of the microstrip transmission line, which is frequently used in practice, the difference between TEM and quasi-TEM wave guides is illustrated. The following overview of the technology for manufacturing transmission lines for planar circuits is important.

Using the example of a coaxial cable, it is shown how precisely the characteristic impedance can be calculated using mechanical errors and, in particular, the finite penetration of the electromagnetic wave.

Subsequently, the input resistances of transmission lines with various terminations, transmission line transformations and the transmission line diagram (Smith chart) are considered.

Finally, the realization of components with short transmission lines (so-called quasi-concentrated transmission line components) is discussed.

4.1 The General Transmission Line Theory

The current and voltage waves propagating on a two-wire line build up electromagnetic fields around the transmission lines. The energy stored in these fields can be attributed to a capacitance C' and an inductance L' per unit length. The equivalent circuit diagram (ECD) for a short piece of transmission line in the widely used differential representation is shown in Fig. 4.2.

The losses of the short transmission line element (per unit length) can be described with the series resistance R' and the shunt conductance G'. These differential elements are also referred to as components per unit length (e.g. capacity per unit length).

To simplify the analysis, we will use the equivalent circuit diagram (ECD) for a short section of transmission line in the difference representation, as shown in Fig. 4.3.

As will be shown at the end of this chapter, a short transmission line can be used to manufacture inductors and capacitors. The equivalent circuit diagrams already show that a transmission line that is designed in such a way that the capacitance layer is very small, acts as a series inductor. If, on the other hand, the geometries of the transmission line are selected in such a way that the inductance per unit length can be neglected, the ECD according to Fig. 4.3 reflects a capacitance against ground.

Derivation of the Telegraph Equation

Just as with the derivation of the scattering parameters, we are only interested in sinusoidal processes in the steady state, i.e. the complex calculation can be applied again, [11]. Kirchhoff's mesh rule applied to the ECD yields

$$U(x) - U(x + \Delta x) - \Delta U(x) = 0 \quad \text{or} \tag{4.1}$$

$$U(x) - U(x + \Delta x) - I(x) \cdot (R' + j\omega L') \cdot \Delta x = 0 \ . \tag{4.2}$$

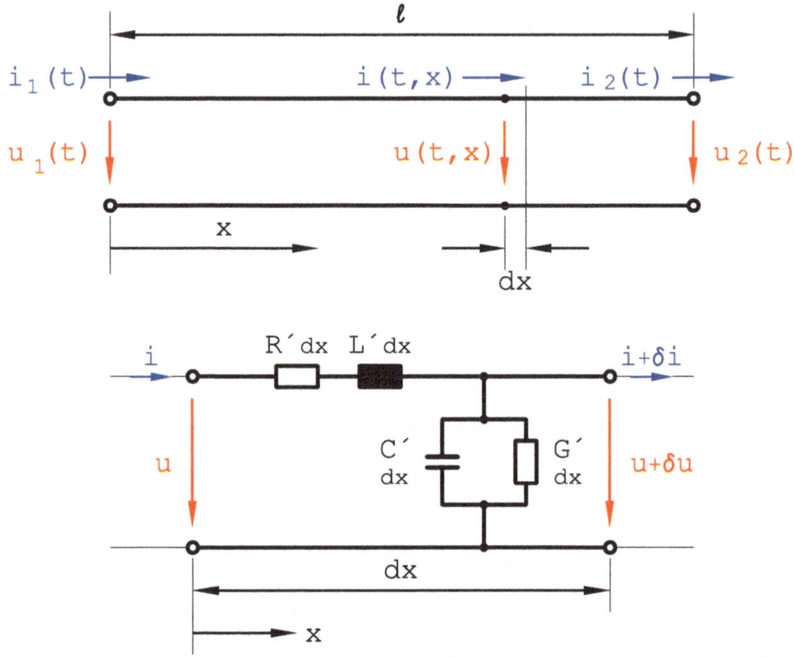

Fig. 4.2: Transmission line and ECD of a transmission line segment: Differential representation

For the borderline case $\Delta x \rightarrow 0$, i.e. the difference quotient becomes the differential quotient:

$$\lim_{\Delta x \rightarrow 0} \frac{U(x) - U(x+\Delta x)}{\Delta x} - I(x) \cdot (R' + j\omega L') = 0 \quad , \tag{4.3}$$

$$\text{bzw.} \quad -\frac{dU(x)}{dx} = I(x) \cdot (R' + j\omega L') \quad . \tag{4.4}$$

For the further derivation, the derivative of the Eq. (4.4) to dx is required in rearranged form:

$$-\frac{dI(x)}{dx} = \frac{d^2U(x)}{dx^2} / (R' + j\omega L') \quad . \tag{4.5}$$

If Kirchhoff's node rule is applied to the ECD according to Fig. 4.3, the result is

$$I(x) - I(x+\Delta x) - \Delta I(x) = 0 \quad \text{resp.} \tag{4.6}$$

$$I(x) - I(x+\Delta x) - U(x+\Delta x) \cdot (G' + j\omega C') \cdot \Delta x = 0 \quad . \tag{4.7}$$

For the limiting case $\Delta x \rightarrow 0$ applies:

$$\lim_{\Delta x \rightarrow 0} \frac{I(x) - I(x+\Delta x)}{\Delta x} - \lim_{\Delta x \rightarrow 0} U(x+\Delta x) \cdot (G' + j\omega C') = 0 \quad , \tag{4.8}$$

$$\text{bzw.} \quad -\frac{dI(x)}{dx} = U(x) \cdot (G' + j\omega C') \quad . \tag{4.9}$$

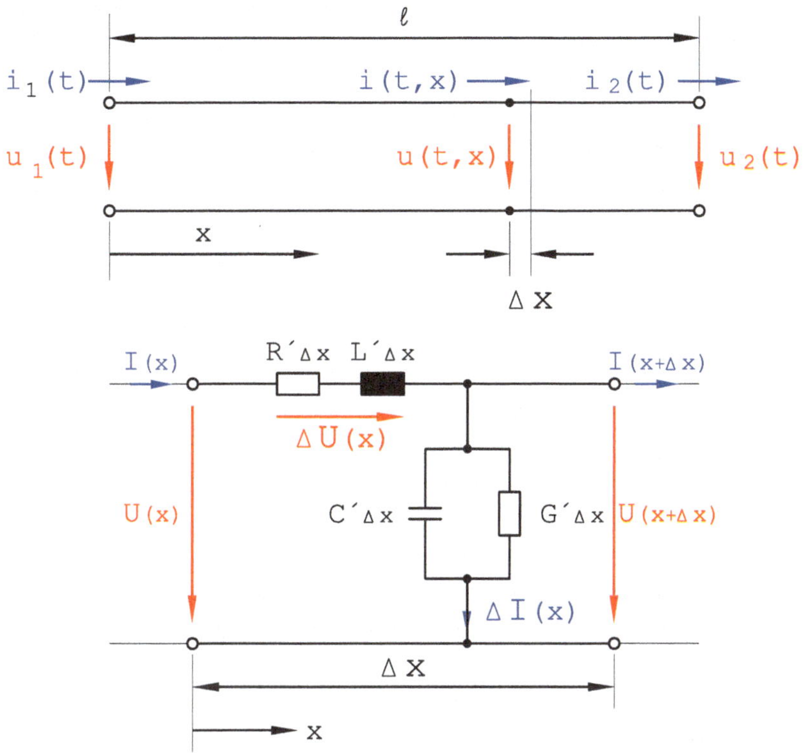

Fig. 4.3: Transmission line and ECD of a transmission line segment: Difference-representation

In the next step, the differential current in the Eqs. (4.5) and (4.9) is to be eliminated:

$$\frac{d^2U(x)}{dx^2} - (G' + j\omega C') \cdot (R' + j\omega L') \cdot U(x) = 0 \quad . \tag{4.10}$$

This second-order differential Eq. (4.10) thus derived to describe a short transmission line segment is also called

`telegraph equation`

(or telegrapher's equations).

Solution of the Telegraph Equation

To solve the telegraph equation, it is advantageous to consider the important variables in transmission line theory:

- γ: Propagation constant
- α: Attenuation constant
- β: Phase constant

Their meaning is explained below. The physical units are $1/m$. The following abbreviation should be used with these quantities:

$$\gamma = \alpha + j\beta = \sqrt{(G' + j\omega C') \cdot (R' + j\omega L')} \quad . \tag{4.11}$$

If you insert the abbreviations into the Eq. (4.10), the following applies:

$$\frac{d^2U}{dx^2} - \gamma^2 \cdot U = 0 \quad . \tag{4.12}$$

The differential Eq. (4.12) has the general solution approach,

$$U(x) = U_p e^{-\gamma x} + U_r e^{\gamma x} \quad , \tag{4.13}$$

which is also referred to as `first transmission line equation`.

The voltages U_p and U_r are mathematically complex integration constants that can be determined from the boundary conditions. Physically, they are the voltages of the propagating (U_p) and returning (U_r) voltage waves[2].

The `second transmission line equation` can be derived from the rearranged Eq. (4.4).

$$I(x) = -\frac{1}{R' + j\omega L'} \cdot \frac{dU}{dx} \tag{4.14}$$

If Eq. (4.13) is used for the differential quotient, the following applies:

$$I(x) = -\frac{1}{R' + j\omega L'} \cdot (-\gamma) \cdot (U_p\, e^{-\gamma x} - U_r\, e^{\gamma x}) \tag{4.15}$$

and with Eq. (4.11)

$$I(x) = \sqrt{\frac{G' + j\omega C'}{R' + j\omega L'}} \cdot (U_p\, e^{-\gamma x} - U_r\, e^{\gamma x}) \quad . \tag{4.16}$$

The inverse root expression is the most important variable for the RF transmission line. It is the `characteristic impedance`

$$Z_L = \sqrt{\frac{R' + j\omega L'}{G' + j\omega C'}} \quad . \tag{4.17}$$

This results in the two transmission line equations from Eq. (4.13) and Eq. (4.16) with Eq. (4.17):

$$U(x) = U_p\, e^{-\gamma x} + U_r\, e^{\gamma x} \quad , \tag{4.18}$$

$$I(x) Z_L = U_p\, e^{-\gamma x} - U_r\, e^{\gamma x} \quad . \tag{4.19}$$

The two transmission line equations provide the result that

$$U(x) = I(x) Z_L$$

is valid for the case when U_r is zero.

In this case, only the outgoing wave U_p propagates and the so-called `matching` is present.

The existence of these two voltage waves was used in chapter 2 to introduce the two voltage waves a and b normalized to $\sqrt{Z_L}$.

[2]The indices r and p stand for the English designations *reflection* and *propagation*.

Determination of the Integration Constants U_p and U_r

The integration constants U_p and U_r are determined by boundary conditions (ratios at the start and end of the transmission line). It is shown below that the currents ($I_{(l)}$) and voltages ($U_{(l)}$) along the transmission line can be calculated using the currents (I_0) and voltages (U_0) at the terminating resistor. To simplify matters, we therefore place the zero point of the coordinate system at the end of the transmission line and define $l = -x$ so that we do not have to calculate with negative coordinate values (Fig. 4.4).

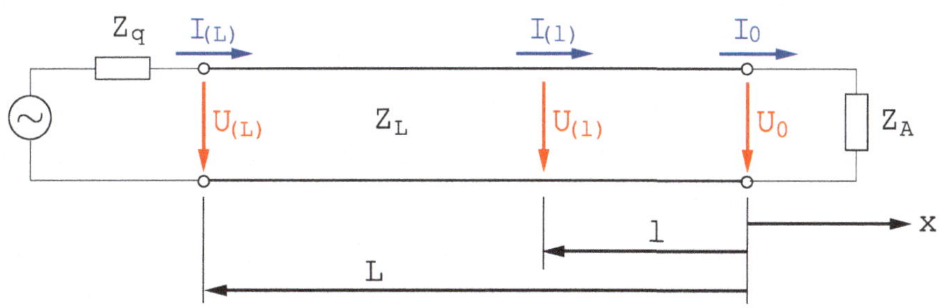

Fig. 4.4: Coordinates and designations of currents and voltages along the transmission line

For the coordinate zero point, the Eqs. (4.18) and (4.19):

$$U_{(l=0)} = U_0 = U_p + U_r \tag{4.20}$$

$$\text{and} \quad I_{(l=0)} \cdot Z_L = I_0 \cdot Z_L = U_p - U_r \quad . \tag{4.21}$$

From this follows for the boundary conditions sought:

$$U_p = \frac{1}{2} \cdot (U_0 + I_0 \cdot Z_L) \tag{4.22}$$

$$\text{and} \quad U_r = \frac{1}{2} \cdot (U_0 - I_0 \cdot Z_L) \quad . \tag{4.23}$$

Next, the Eqs. (4.22) and (4.23) are substituted into the Eqs. (4.18) and (4.19) for the point $-x = l$.

$$U(l) = \frac{1}{2} \cdot (U_0 + I_0 \cdot Z_L)\, e^{\gamma l} + \frac{1}{2} \cdot (U_0 - I_0 \cdot Z_L)\, e^{-\gamma l} \tag{4.24}$$

$$I(l) \cdot Z_L = \frac{1}{2} \cdot (U_0 + I_0 \cdot Z_L)\, e^{\gamma l} - \frac{1}{2} \cdot (U_0 - I_0 \cdot Z_L) e^{-\gamma l} \tag{4.25}$$

With the known transformations

$$\sinh(\gamma l) = \frac{1}{2}\left(e^{\gamma l} - e^{-\gamma l}\right) \quad \text{and} \quad \cosh(\gamma l) = \frac{1}{2}\left(e^{\gamma l} + e^{-\gamma l}\right) \tag{4.26}$$

of the hyperbolic functions results:

$$U(l) = U_0 \cdot \cosh(\gamma l) + I_0 \cdot Z_L \cdot \sinh(\gamma l) \quad , \tag{4.27}$$

$$I(l) = I_0 \cdot \cosh(\gamma l) + U_0 / Z_L \cdot \sinh(\gamma l) \quad . \tag{4.28}$$

The assignment of the current and voltage ratios is preferably specified in the chain matrix representation (A-parameter):

$$\begin{pmatrix} U(l) \\ I(l) \end{pmatrix} = \begin{pmatrix} \cosh(\gamma l) & Z_L \cdot \sinh(\gamma l) \\ 1/Z_L \cdot \sinh(\gamma l) & \cosh(\gamma l) \end{pmatrix} \begin{pmatrix} U_0 \\ I_0 \end{pmatrix} \quad . \tag{4.29}$$

If $U_{(l=L)}$ and $I_{(l=L)}$ are used at the input of the transmission line according to Fig 4.4 and if $Z_q = Z_A = Z_L$ applies, the scattering parameters for this adapted lossy transmission line can be calculated using auxiliary sheet 6 (page 445) to calculate the scattering parameters:

$$[\mathbf{S_{Line}}] = \begin{pmatrix} 0 & e^{-\gamma l} \\ e^{-\gamma l} & 0 \end{pmatrix} \quad . \tag{4.30}$$

Consequently, the logarithmized transmission value of a transmission line can be calculated from

$$S_{21}^{dB} = 20 \lg\left(\left|e^{-\gamma l}\right|\right) \quad . \tag{4.31}$$

4.1.1 Low-Loss Transmission Lines

Approximations are now to be introduced which have no serious impact on the results of technically meaningful transmission lines, but which simplify the analytical treatment immensely.

The approximations are based on the fact that the losses per unit length (R' and G') are significantly smaller than the stored energies per unit length (in ωL' and/or ωC').

Consequently, the propagation constant according to Eq. (4.11):

$$\gamma = \alpha + j\beta = \sqrt{(G' + j\omega C')(R' + j\omega L')} \quad , \tag{4.32}$$

$$\gamma = \sqrt{j\omega C' \left(1 + \frac{G'}{j\omega C'}\right) j\omega L' \left(1 + \frac{R'}{j\omega L'}\right)} \quad , \tag{4.33}$$

$$\gamma \approx j\omega \sqrt{L'C'} \sqrt{1 + G'/(j\omega C') + R'/(j\omega L')} \tag{4.34}$$

$$\text{mit} \quad \sqrt{1+x} \approx 1 + x/2 \quad ,$$

$$\gamma \approx j\omega \sqrt{L'C'} \left(1 + G'/(j2\omega C') + R'/(j2\omega L')\right) \quad , \tag{4.35}$$

$$\gamma \approx \underbrace{j\omega \sqrt{L'C'}}_{=j\beta} + \underbrace{\frac{R'}{2}\sqrt{C'/L'} + \frac{G'}{2}\sqrt{L'/C'}}_{=\alpha} \quad . \tag{4.36}$$

The Eq. (4.17) for calculating the characteristic impedance can be simplified in the same way:

$$Z_L = \sqrt{\frac{R' + j\omega L'}{G' + j\omega C'}} = \sqrt{\frac{j\omega L'}{j\omega C'}} \cdot \sqrt{\left(1 + \frac{R'}{j\omega L'}\right) / \left(1 + \frac{G'}{j\omega C'}\right)} \approx \sqrt{\frac{L'}{C'}} \quad . \tag{4.37}$$

The most important result of this approximation is that the characteristic impedance Z_L can be assumed to be purely real to a good approximation. This means that weak losses hardly influence the characteristic impedance. It has thus been shown that the losses with respect to the voltage drop on the transmission line (and thus with respect to the complex characteristic impedance) are so small, that the longitude components of the field can be neglected for the propagation phenomena.

If the losses for the attenuation cannot be neglected, this is referred to as a quasi-TEM wave guide. Quasi-TEM guides have small longitude and large transverse field components.

Substituting the result of Eq. (4.37) into the solution for the attenuation measure α (Eq. (4.36)), we obtain

$$\alpha \approx \frac{R'}{2\,Z_L} + \frac{G'\,Z_L}{2} \quad . \tag{4.38}$$

It can be shown via the so-called power loss method [48] that generalized applies:

$$\alpha = \frac{1}{2\;\mathrm{Re}\{Z_L\}}\left(R' + G'\;|Z_L|^2\right) \quad . \tag{4.39}$$

As will be shown later, the approximate solutions of the phase constant and the characteristic impedance

$$\beta \approx \omega \cdot \sqrt{L'C'} \qquad \text{and} \qquad Z_L \approx \sqrt{\frac{L'}{C'}} \tag{4.40}$$

correspond to the exact solutions for the lossless case (= TEM wave guides).

4.1.2 Attenuation of a Transmission Line

In practice, a transmission line attenuation is used, which is specified in dB. This transmission line attenuation is calculated from the logarithmic attenuation constant α^{dB} and the corresponding transmission line length l. From the calculation equation (2.23) of the insertion loss applies:

$$\frac{\alpha^{dB}\,l}{dB} = 20\;\lg\frac{1}{e^{-\alpha l}} \quad . \tag{4.41}$$

The magnitude values of the transmission parameters of the scattering parameters and the attenuation differ only in the sign and therefore also apply to the logarithmized values of a transmission line piece:

$$\alpha^{dB}\,l = -\,S_{21}^{dB} \quad . \tag{4.42}$$

Consequently, the logarithmized transmission value of a transmission line of length l can also be calculated from

$$S_{21}^{dB} \;=\; 20\;\lg\left(e^{-\alpha l}\right) \quad . \tag{4.43}$$

With the property of the logarithmn $\lg a^n = n\,\lg a$ and the equation reduction of length l applies:

$$\frac{\alpha^{dB}}{dB/m} = 8.686\,\frac{\alpha}{Np/m} \quad , \tag{4.44}$$

where the introduced abbreviation Np stands for the *dimensionless* unit Neper. In contrast to a logarithmic specification, no conversion takes place. It is therefore equivalent whether the unit for the damping constant α is specified in Neper or nothing is specified.

4.1.3 Transmission Line Theory of Lossless Transmission Lines

In the loss-less case, the loss coatings R' and G' are zero. Fig. 4.5 illustrates the simplified ECD of a transmission line segment.

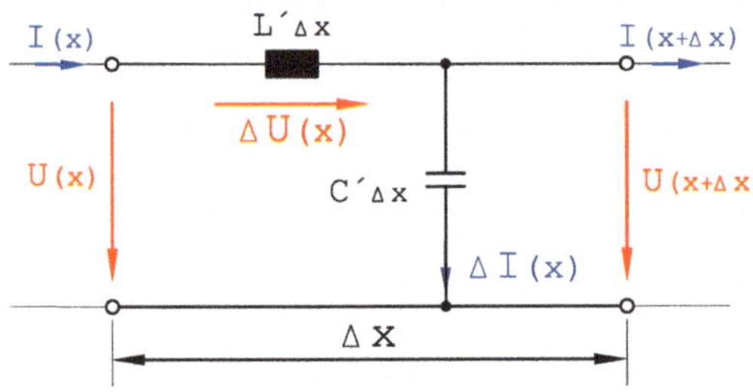

Fig. 4.5: ECD of a transmission line segment without losses

This simplifies the Eqs. (4.11) and (4.17):

$$\beta = \omega \cdot \sqrt{L'C'} \qquad \text{and} \qquad Z_L = \sqrt{\frac{L'}{C'}} \quad . \tag{4.45}$$

The attenuation constant α is zero!

Since the propagation constant γ is now purely imaginary, the hyperbolic functions can be simplified into trigonometric functions via Eq. 4.46.

$$\cosh(jx) = \cos(x) \qquad \text{and} \qquad \sinh(jx) = j\sin(x) \tag{4.46}$$

This results in the following simplifications for the lossless case for Eqs. (4.27) to (4.29):

$$U(l) = U_0 \cdot \cos(\beta l) + j\, I_0 \cdot Z_L \cdot \sin(\beta l) \quad , \tag{4.47}$$

$$I(l) = I_0 \cdot \cos(\beta l) + j\, U_0/Z_L \cdot \sin(\beta l) \quad . \tag{4.48}$$

$$\begin{pmatrix} U(l) \\ I(l) \end{pmatrix} = \begin{pmatrix} \cos\beta l & j\, Z_L \cdot \sin(\beta l) \\ j\, 1/Z_L \cdot \sin(\beta l) & \cos(\beta l) \end{pmatrix} \begin{pmatrix} U_0 \\ I_0 \end{pmatrix} \quad . \tag{4.49}$$

If $U_{(l=L)}$ and $I_{(l=L)}$ are used in the Eq. (4.49) of the A-chain matrix according to Fig. 4.4 and $Z_q = Z_A = Z_L$, then for this adapted lossless transmission line, the scattering parameters can be calculated using auxiliary sheet 6 (page 445):

$$[\mathbf{S_{Line}}] = \begin{pmatrix} 0 & e^{-j\beta l} \\ e^{-j\beta l} & 0 \end{pmatrix} \quad . \tag{4.50}$$

Before these equations from (4.46) and (4.49) are further used to derive the input resistance, important variables of the transmission line theory should be introduced.

——— *Exercise: Overland Power Line* ———

Given: *A homogeneous and loss-free 500 km long transmission line for energy transmission is to be investigated. The transmission line was dimensioned for a characteristic impedance of $Z_L = 10\,k\Omega$. The operating frequency is 50 Hz. C=167 nF was measured as the input capacitance. The approximate speed of light is $c_0 = 2.99 \cdot 10^8 \frac{m}{s}$ and for air $\epsilon_r = 1$.*

Find:
1. Calculate the capacitance value per unit length C' and the transmission line value per unit length L'.
2. Specify an equivalent circuit diagram (ECD) for the transmission line. For this purpose, the very good approximation used in practice is that the ECD of a transmission line with an electrical length $p = \omega\,\ell\,180°/(c_0\,\pi)$ is valid up to around 10° and the values per unit length can be replaced by components.
3. What amounts of the S-parameters result from a simulation with port impedances of 10 kΩ? Calculate the values per unit length of the S-parameters of the equivalent circuit up to 1 kHz using a circuit simulator (ADS, Serenade, ...).

To 1. *Due to the homogeneity, the capacitance per unit length C' can be calculated directly from the measured value of the total capacitance C:*

$$C' = C/\ell = \frac{167\,nF}{500\,km} = 0.33\,\frac{pF}{m} \quad . \tag{4.51}$$

The transmission line value per unit length can be calculated directly from Eq. (4.40):

$$L' = Z_L^2\,C' = 10000^2\,\Omega^2\,0.33\,\frac{pF}{m} = 33\,\frac{\mu H}{m} \quad . \tag{4.52}$$

To 2. *The 500 km long transmission line causes a phase rotation at 50 Hz according to:*

$$p = \omega/c_0\;\ell\;180°/\pi = \frac{2\,\pi\,50\,1/s}{2.99\,10^8\,m/s}\,500\cdot 10^3\,m\,\frac{180°}{\pi} = 30.1° \quad . \tag{4.53}$$

If we calculate with the approximation that around 10of electrical length can be replaced by an ECD consisting of a series inductance Lx and a capacitance against ground Cx, this results in the simple equivalent circuit shown in Fig. 4.6

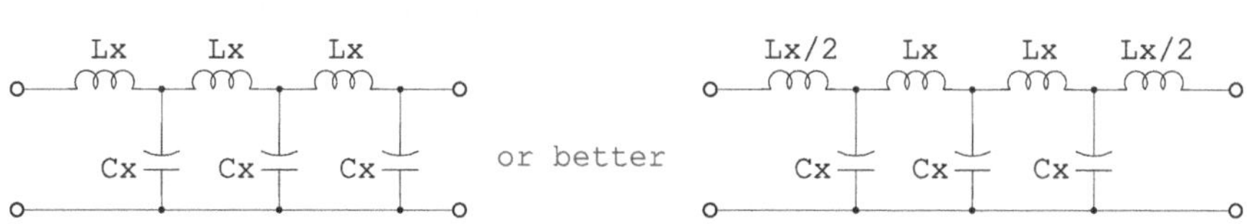

Fig. 4.6: ECD of the 500 km long overland transmission line

with the values:

$$Cx = \frac{C'\,500\cdot 10^3\,m}{3} = 55\,nF \quad \text{and} \quad Lx = \frac{L'\,500\cdot 10^3\,m}{3} = 5.5\,H \quad . \tag{4.54}$$

Zu 3. *The transmission line with the characteristic impedance of 10 kΩ is very well matched at -27 dB. Therefore, the following must apply:* $S_{11} = 0$. *Since the transmission line was assumed to be lossless, the following also applies:* $S_{21} = 1$. *The results of the simulation are shown in Fig. 4.7.*

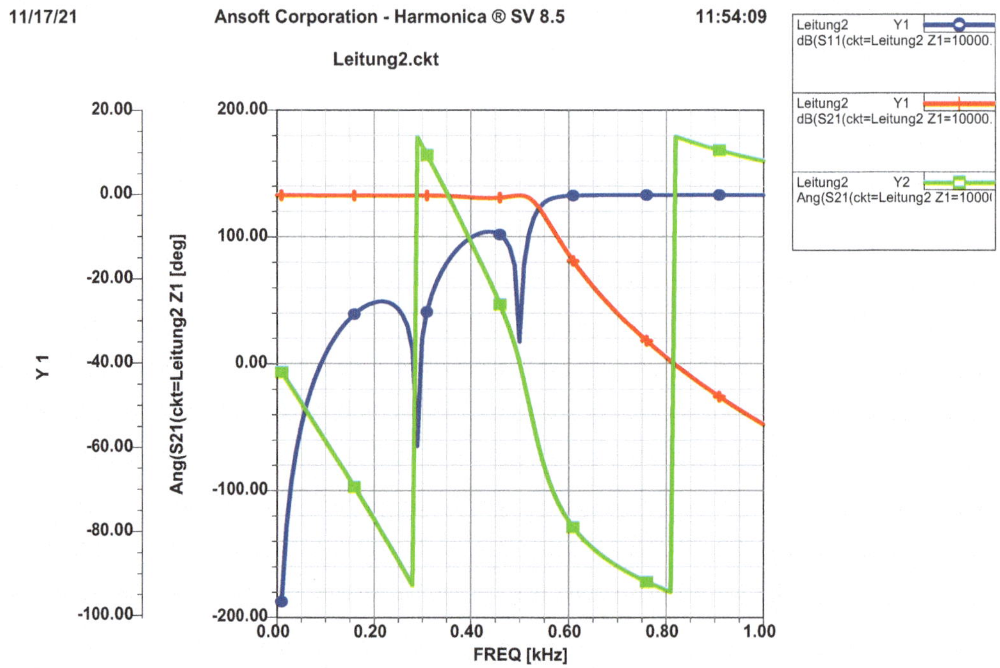

Fig. 4.7: Amounts of the reflection and transmission values in the usual logarithmic representation and phase representation of the transmission values (between -180° and 180°) of the ECD with LX/2 of an overland transmission line

You can clearly see that this ECD can only be used up to 300 Hz with fitting values that are better than -20 dB.

——— *Exercise: Overland Power Line* ———

In section 6.2 (page 195) it is shown that low-pass and high-pass filters can be developed from this transmission line equivalent circuit diagram.

4.1.4 Wavelength and Phase Velocity

Fig. 4.8 shows a transmission line extending infinitely to the right, on which a wave propagating in the $+x$ direction is excited by the generator at $x = 0$. At $x = 0$, the polarity (e.g. of the voltage) changes sinusoidally over time.

These polarity changes propagate from the generator at the propagation speed c. If you "sit" on the wave at a point of constant phase, so to speak, then you move along with the wave at the phase velocity v_p.

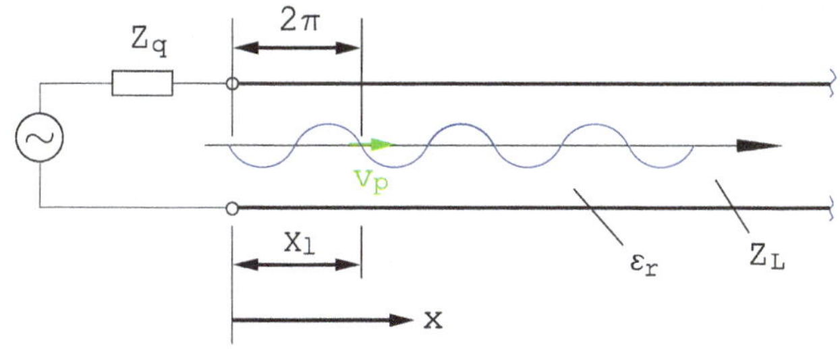

Fig. 4.8: Progressive wave on a homogeneous transmission line

The phase of $x = 0$ appears delayed by the time $t_l = \frac{x_l}{v_p}$ at x_l.

A wavelength λ corresponds to the path of the length x_l traveled within the time $t_l = \frac{1}{f}$. Furthermore, with the TEM wave, the phase moves at the same speed as the propagation speed c. The following therefore applies:

$$v_p = c = \frac{x_l}{t_l} = \frac{\lambda}{1/f} = \lambda f \quad . \tag{4.55}$$

In local terms, the delay is 2π per wavelength λ. The latter is also referred to as phase delay, rotation or shift.

The phase constant β indicates the phase rotation per unit length (meter). Therefore also applies:

$$\beta = \frac{2\pi}{\lambda} \quad . \tag{4.56}$$

If the Eq. (4.56) is expanded in the numerator and denominator with the frequency f, the following follows:

$$\beta = \frac{2\pi f}{\lambda f} = \frac{\omega}{v_p} \quad . \tag{4.57}$$

Rearranged with Eq. (4.45)

$$v_p = \frac{\omega}{\beta} = \frac{\omega}{\omega\sqrt{L'C'}} = \frac{1}{\sqrt{L'C'}} \quad . \tag{4.58}$$

In technical transmission lines, a dielectric material with the dielectric constant (permittivity) ϵ_r is often inserted between the conductors. The following applies in this case:

$$v_p = \frac{c_0}{\sqrt{\epsilon_r}} \tag{4.59}$$

and

$$\lambda = \frac{\lambda_0}{\sqrt{\epsilon_r}} \qquad \text{with} \qquad \lambda_0 = \frac{c_0}{f} \quad . \tag{4.60}$$

c_0 and λ_0 are the free space constants.

The group velocity v_g indicates the speed at which the energy transport takes place on a transmission line. This corresponds to the speed of an envelope curve of a wave packet.

The general definition for the group velocity is

$$v_g = \frac{d\omega}{d\beta} \quad . \tag{4.61}$$

For a TEM wave, the phase velocity v_p and the group velocity v_g are equal.

4.1.5 Group Delay

The group delay τ_g is the delay required by a group of frequency-adjacent harmonic time functions to traverse a network (e.g. two-ports). If all existing signals above the frequency have the same propagation time in a network, there is no distortion of the signals.

Mathematically, the group delay τ_g results from the negative derivative of the transmission phase (e.g. $\angle S_{21}$) of the system according to the angular frequency ω. The following therefore applies to a two-port:

$$\tau_g = -\frac{d\angle S_{21}}{d\omega} \qquad \left(\text{general:} \quad \tau_g = -\frac{d\phi_{(\omega)}}{d\omega}\right) \quad . \tag{4.62}$$

If τ_g is constant over the frequency range used, the system is distortion-free. This assumes a linear phase response of the two-port. Such an ideal transmission line has a phase response as shown in Fig. 4.9.

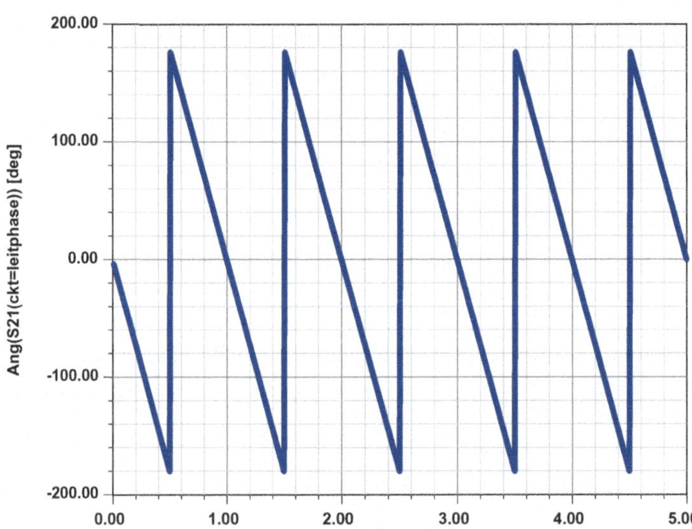

Fig. 4.9: Phase progression of the transmission phase of S_{21} of an ideal matched transmission line ($\epsilon_r = 1\,;\ \ell = 30\,\text{cm}$)

For a transmission line, the group delay corresponds to the signal delay through the transmission line, as long as no dispersion (frequency dependence of the transmission line's transmission behavior) occurs.

This is the case for TEM transmission lines as long as no waveguide mode can propagate. The only TEM transmission line that does not have a waveguide mode is the so-called M-line, Fig. 4.10, [70].

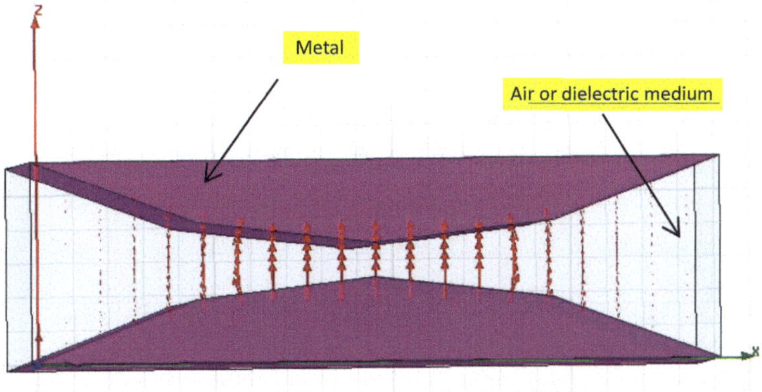

Fig. 4.10: Cross-section with E-field of a true mono-mode TEM transmission line (M-Line)

This M-Line, which is not widely used, is not one of the transmission lines that can be purchased or is technically easy to manufacture. In practice, these purchasable or technically easily producible transmission lines, only the coaxial cable (up to the first waveguide mode) can be regarded as a genuine TEM transmission line.

All other technical transmission lines belong to the group of quasi-TEM wave guides. These have a non-negligible dispersion.

The group routing behavior of a simple twisted pair transmission line (standard telephone cable with a length of 21 m) is shown in Fig. 4.11 is shown.

This transmission line can only be used up to a frequency of around 150 MHz. This means that at least the third harmonic must still be transmitted correctly when transmitting a square wave signal. For this reason, digital signals cannot have a data rate higher than 50 MHz.

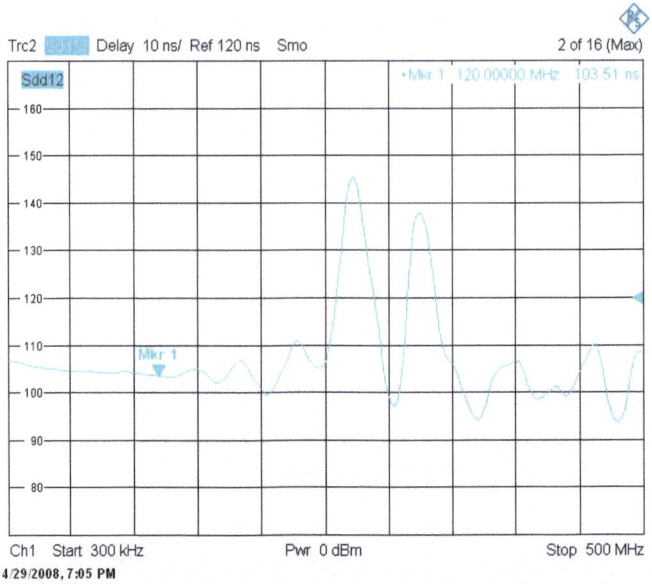

Fig. 4.11: Measured group delay of a telephone transmission line

4.1.6 Eye Diagram and BER Test

While RF technicians use the network analyzer to characterize components and measure the S-parameters, digital technicians prefer the oscilloscope. They use the eye diagrams to measure the bit error rate (BER).

The BER test resulting from the eye diagram is carried out by 1. placing a random bit sequence on the transmission path and 2. "counting" how often the "mask" is violated.

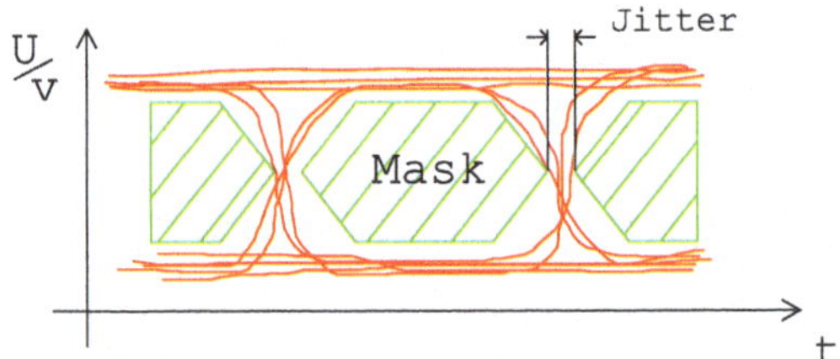

Fig. 4.12: Illustration of the BER test

The construction of the eye diagram is based on the superimposed measurement of the signal sequences in the Fig. 4.13.

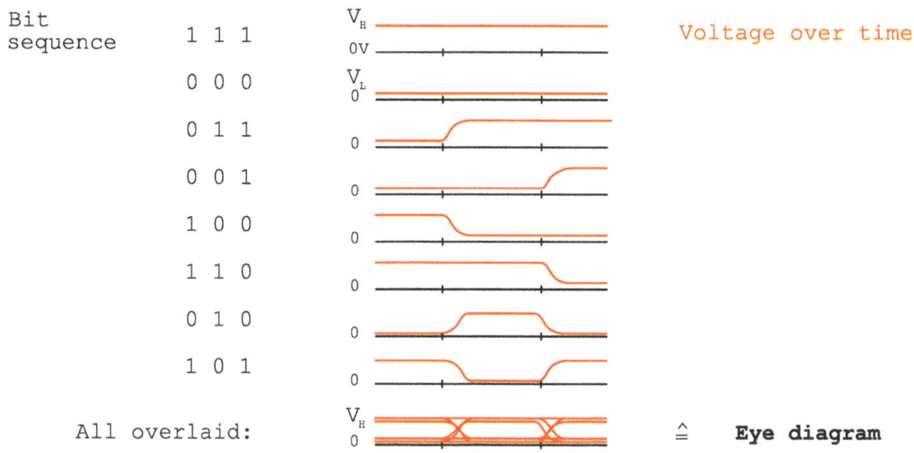

Fig. 4.13: Explanations of the eye diagram

4.2 TEM and Quasi-TEM Wave Guide

A TEM wave consists of an electric field ($\vec{E}$) and a magnetic field ($\vec{H}$), which are each directed transversely to the direction of propagation.

Strictly speaking, there is no physically realizable TEM transmission line, as every transmission line has losses and therefore there is a (very) small voltage drop of the propagating waves in the direction of propagation. Consequently, there is always a (very) small forward component of the electromagnetic field.

However, this effect is so small that we will refer to metallic two-wire conductors as TEM wave guides if both conductors are embedded between a homogeneous dielectric.

With the introduction of the microstrip transmission line, the problem of inhomogeneous dielectric fillings and the resulting quasi-TEM wave guide is addressed.

TEM Wave Guide and Capacity per Unit Length

It was calculated using the Eq. (4.37) for lossy wave guide in good approximation and for lossless transmission line exactly that applies:

$$Z_L = \sqrt{\frac{L'}{C'}} \quad . \tag{4.63}$$

Furthermore, by equating the Eqs. (4.58) and (4.59) after a conversion:

$$\sqrt{L'} = \frac{\sqrt{\epsilon_r}}{c_0\,\sqrt{C'}} \quad . \tag{4.64}$$

If Eq. (4.64) is inserted into Eq. (4.63), a very simple relationship is obtained: the characteristic impedance of a (TEM) transmission line depends only on the capacitance per unit length C', the relative dielectric constant ϵ_r and the speed of light c_0:

$$\boxed{Z_L = \frac{\sqrt{\epsilon_r}}{c_0\ C'}} \quad . \tag{4.65}$$

With $c_0 = 1/\sqrt{\mu_0\epsilon_0}$ and the introduction of the *field characteristic impedance of free space*

$$Z_f = \sqrt{\mu_0/\epsilon_0} = 120\ \pi\ \Omega \tag{4.66}$$

Eq. (4.65) can be converted to

$$Z_L = \frac{\sqrt{\epsilon_r}}{C'}\ \sqrt{\mu_0\ \epsilon_0} = \frac{\sqrt{\epsilon_r}}{C'}\ Z_f\ \epsilon_0 \ = \ \frac{\sqrt{\epsilon_r}}{C'}\ \epsilon_0\ 120\ \pi\ \Omega \quad . \tag{4.67}$$

The quintessence is the determination of the characteristic impedance of a homogeneous TEM transmission line via a capacitance measurement or calculation.

Even a fairly rough segmentation into individual capacitances enables very precise numerical calculations of the characteristic impedance. Based on the segmentation shown in Fig. 4.14, the following results for a coaxial cable characteristic impedance of 49.99 Ω (instead of 50.00 Ω). The rectangular segments shown were approximated by plate

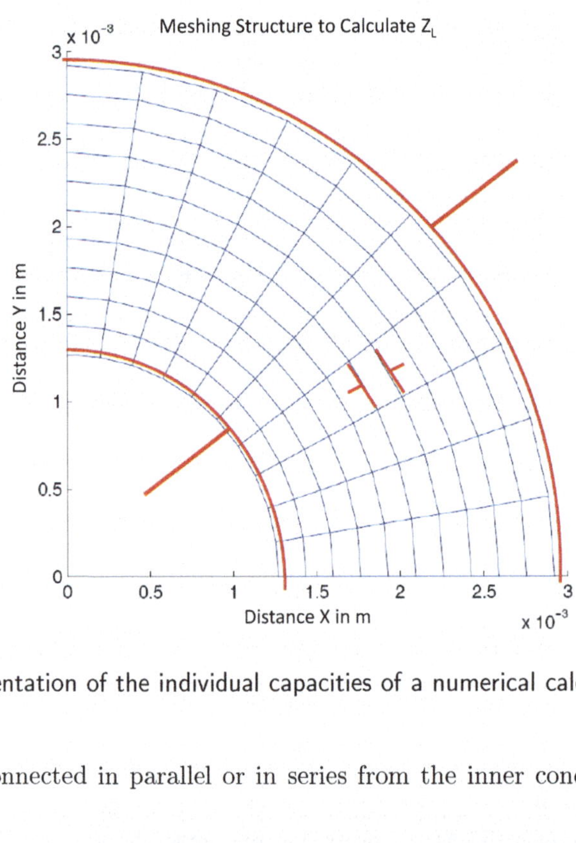

Fig. 4.14: Representation of the individual capacities of a numerical calculation program for Z_L

capacitors and connected in parallel or in series from the inner conductor to the outer conductor.

Losses from TEM Wave Guides

The losses α can be separated into metallic losses α_R and dielectric losses α_G. According to Eq. (4.38) applies:

$$\alpha \approx \frac{R'}{2\,Z_L} + \frac{G'\,Z_L}{2} = \alpha_R + \alpha_G \quad . \tag{4.68}$$

Metallic Losses

The metallic losses, which are taken into account in the resistor per unit length R', can be calculated from the circumferences (! we calculate with losses per unit length) or the widths of the current-carrying forward and return conductors (u_1, u_2) and the skin depth δ.

The skin depth δ can be calculated from the permeability constant μ and the specific resistance ρ as a function of the frequency f:

$$\delta = \frac{\sqrt{\rho}}{\sqrt{\pi\;\mu\;f}} \quad . \tag{4.69}$$

It is advantageous to introduce the sheet resistance R_{sq}:

$$R_{sq} = \frac{\rho}{\delta} = \sqrt{\pi\;\rho\;\mu\;f} \quad . \tag{4.70}$$

Using the sheet resistance, the resistance value per unit length can be calculated via

$$R' = R_{sq}\,(1/u_1 + 1/u_2) \quad . \tag{4.71}$$

When used in Eq. (4.68), the following results for the metallic losses

$$\alpha_R = \frac{R_{sq}\,(1/u_1 + 1/u_2)}{2\;Z_L} \tag{4.72}$$

with the conclusion that the metallic losses can be reduced if

- the metallic surface resistance is reduced.
- the surface is maximized.
- the characteristic impedance is maximized.
- the frequency is set as low as possible.

In practice, the roughness of the metal surface often has a very large influence on the sheet resistance. As a rule, the influence is a factor of 2 or more! Even semiconductors sometimes exhibit roughness that extends into the μm range. Fig. 4.15 shows a fairly large metal structure in a semiconductor with a relatively high roughness.

Fig. 4.15: Sectional view through the metallization layer of a semiconductor

Dielectric Losses

The cause of the losses of a conductor, caused by the so-called derivative per unit length G', are only dielectric losses[3], which are calculated via the loss factor

$$\tan\delta = \frac{\epsilon_r''}{\epsilon_r'} \qquad \text{with} \qquad \epsilon_r = \epsilon_r' - j \cdot \epsilon_r'' \quad . \tag{4.73}$$

The dissipation factor $\tan\delta$ of a dielectric is either specified by the manufacturer or can be measured. Typical values can be found in Tab. 4.2. The dissipation angle δ is often used when referring to the dissipation factor.

The leakage losses per unit length can be calculated using the capacitance value per unit length and the loss factor via

$$G' = \omega\ C'\ \tan\delta \tag{4.74}$$

([48], in the German version, part 1, page 58).

For the dielectric losses α_G applies to **each** TEM wave guide according to Eq. (4.68) with Eq. (4.65) and Eq. (4.74):

$$\underline{\underline{\alpha_G}} = \frac{Z_L}{2}\, G' = \frac{\omega\ C'\ \tan\delta}{2\ c_0\ C'}\sqrt{\epsilon_r} = \pi\ f\ \frac{\sqrt{\epsilon_r}}{c_0}\ \tan\delta = \underline{\underline{\frac{\pi}{\lambda}\ \tan\delta}} \quad . \tag{4.75}$$

Eq. (4.75) shows that the dielectric losses of any TEM wave guide depend neither on the geometry nor on its characteristic impedance! The losses depend only on the frequency (wavelength) and the loss factor of the material.

Note: Except for $\tan\delta$ calculations (loss investigations), the good approximation is used $\epsilon_r \approx \epsilon_r'$.

[3]This applies to frequencies greater than 100 Hz.

General: The following calculations of the characteristic values of the various transmission lines include the approximation according to Eq. (4.68). In practice, however, the solutions are sufficiently accurate.

4.2.1 The Coaxial Cable

The `coaxial cable` is very important, as it represents the RF standard transmission line between devices and housed components, as well as in measurement technology. The cross-section of a coaxial cable is shown in Fig. 4.16.

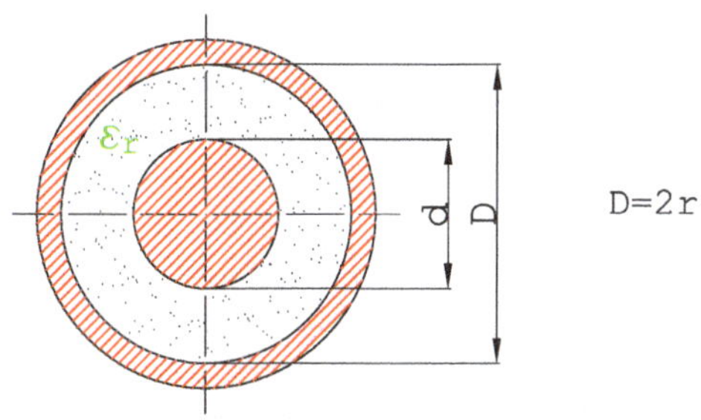

Fig. 4.16: Cross-section of a coaxial cable filled with dielectric

If the voltage U is applied to a coaxial cable with the capacitance value per unit length C', the charge value per unit length $Q' = C' \cdot U$ is set on the inner and outer conductor. From the requirement that the dielectric flux that penetrates a potential surface must be equal to the charge enclosed by it ($\oint D \, ds_\phi = Q'$), the following results for the coaxial cable with $ds_\phi = r \cdot d\phi$ via

$$\epsilon \oint E_r \, ds_\phi = Q' \quad \Longrightarrow \quad \epsilon_0 \, \epsilon_r \, E_r \, 2 \, \pi \, r = Q' \tag{4.76}$$

and thus for the electric field:

$$E_r = \frac{Q'}{2 \, \pi \, \epsilon_0 \, \epsilon_r} \, \frac{1}{r} \quad . \tag{4.77}$$

The voltage can be determined via the electric field using the following equation:

$$U = \int_{d/2}^{D/2} E_r \, dr = \frac{Q'}{2 \, \pi \, \epsilon_0 \, \epsilon_r} \int_{d/2}^{D/2} \frac{dr}{r} = \frac{Q'}{2 \, \pi \, \epsilon_0 \, \epsilon_r} \ln \frac{D}{d} \quad . \tag{4.78}$$

This gives the capacitance value per unit length of the coaxial cable via $C' = Q'/U$ with

$$C' = \frac{2 \, \pi \, \epsilon_0 \, \epsilon_r}{\ln \frac{D}{d}} \quad . \tag{4.79}$$

Eq. (4.79) inserted into Eq. (4.67) gives the required characteristic impedance of the coaxial cable:

$$Z_L = \frac{60 \, \Omega}{\sqrt{\epsilon_r}} \ln \frac{D}{d} \quad . \tag{4.80}$$

This solution is mathematically and in practice precise. Only for reference, calibration and calibration transmission lines are the effects due to the penetration depth calculated, which slightly influences the characteristic impedance. These effects are described in detail for the coaxial cable at the end of this section (from page 94).

For the metallic losses α_R the following applies according to Eq. (4.72):

$$\alpha_R = \frac{R_{sq}}{2\,Z_L}\,\frac{1}{\pi}\left(\frac{1}{d}+\frac{1}{D}\right) \quad . \tag{4.81}$$

The most common coaxial connectors can be divided into two classes:

1. Precision connectors (numbers often correspond to the outer diameter)

1 mm oder W	-110 GHz	PC-1.85	-65 GHz
PC-2.40 oder V	-50 GHz	PC-2.92 oder K	-40 GHz
PC-SL 40	-40 GHz	RPC-3.50	-26.5 GHz
PC-SL 26.5	-26.5 GHz	PC-SP oder BMA	-22 GHz
PC-N	-18 GHz	RPC-TNC	-18 GHz
PC-7	-18 GHz	PC-N75 Ohm	-4 GHz

2. Standard connectors (here only the most common types)

Micro-RF	SMP	MMCX	FME
IEC Antenna	MCX	SSMA	SMA
SSMB	SSMC	SMC	MCX 75 Ohm
SMG	BNC 50 Ohm	C 50 Ohm	N 50 Ohm
SnapN	UHF	D-Sub	TNC 50 Ohm
SMB (Snap on)	7-16	BNC 75 Ohm	C 75 Ohm
N 75 Ohm	F	TNC 75 Ohm	Twinax

Further details on these coaxial plug or screw connectors and the associated coaxial cables can be found in [117, 146].

4.2.2 Paired Strips and Parallel Wires Transmission Line

The `paired strips transmission line` (Fig. 4.17) had almost ceased to have any significance in the meantime. Now this transmission line type is increasingly being used as a low-loss transmission line in semiconductor circuits.

Under the condition $b \gg a$, the calculation of the capacitance load of the band transmission line is particularly simple. The capacitance layer C' results from the formula of the plate capacitor:

$$C' = \frac{b\,\epsilon}{a} \quad . \tag{4.82}$$

If C' is inserted into the Eq. (4.67), the characteristic impedance of a paired strips transmission line is obtained:

$$Z_L = \frac{Z_f}{\sqrt{\epsilon_r}}\,\frac{a}{b} \quad . \tag{4.83}$$

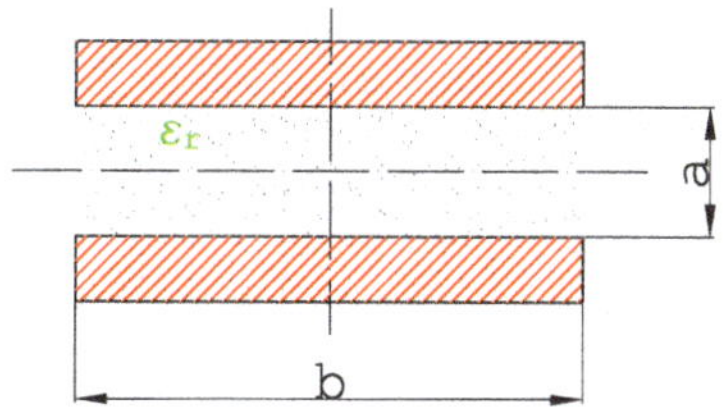

Fig. 4.17: Cross section of a transmission line

The metallic losses of the paired strips transmission line result with $u_1 = u_2 = b$ according to Eq. (4.72):

$$\alpha_R = \frac{R_{sq}}{b\ Z_L} = \frac{R_{sq}}{a\ Z_f}\sqrt{\epsilon_r} \quad . \tag{4.84}$$

The `parallel wires transmission line` (Fig. 4.18) is used in fast data transmission up to the GHz range for the transmission of symmetrical signals. As a so-called twisted pair transmission line, this parallel wires transmission line is in a little bit more complex manner increasingly being used in mass applications, see page 68 of [146].

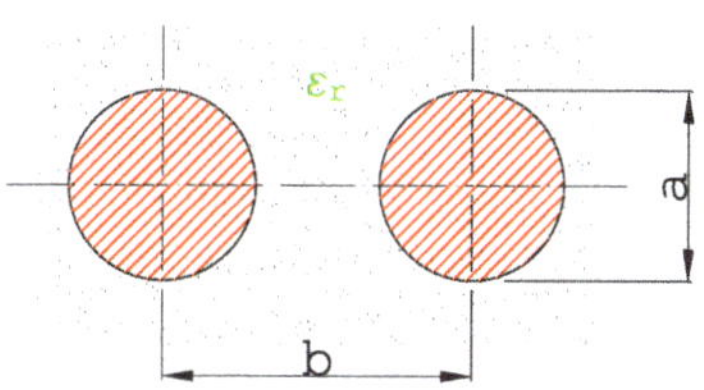

Fig. 4.18: Cross-section of a parallel wire transmission line

We will limit ourselves here to stating the result for the characteristic impedance. In the event that $b \gg a$, the result is the characteristic impedance:

$$Z_L = \frac{120\ \Omega}{\sqrt{\epsilon_r}} \ln\frac{2\ b}{a} \quad . \tag{4.85}$$

The metallic losses of the parallel wire transmission line result in an approximation[4] However, this was used in the approximation with $u_1 = u_2 = \pi a$ according to Eq. (4.72):

$$\alpha_R \approx \frac{R_{sq}}{Z_L}\frac{1}{\pi\ a} \quad . \tag{4.86}$$

4.2.3 The Stripline

The double-sided shielded `stripline` (also: sandwich line or Tri-plate[5]; Fig. 4.19) is used in the inner layers of all modern multilayer RF circuits.

[4]The current no longer flows homogeneously through the outer areas of the wires.
[5]Tri-plate is a registered trademark of Sanders Associates.

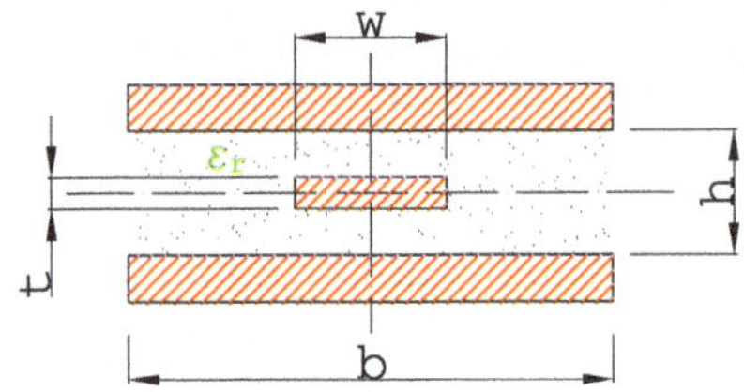

Fig. 4.19: Cross section of a shielded stripline

In the case that $b \gg h$, the characteristic impedance (also for conductor thicknesses $t \neq 0$) can be calculated exactly by conformal mappings via elliptic integrals ([48]):

$$Z_L = \frac{Z_f}{4\sqrt{\epsilon_r}} \frac{K_k}{K_{k'}} \quad \text{with} \quad k' = \sqrt{1-k^2} \quad \text{and} \quad k = \tanh\left(\frac{\pi}{2}\frac{w}{h}\right) \quad . \tag{4.87}$$

The complete elliptic integrals K can be solved using Matlab and the command "*ellipke*", for example.

When calculating the metallic losses, the specification of the current-carrying conductor area (the ground areas) poses a problem. A large number of approximate solutions can be found in the literature. Meanwhile, the most accurate approximate solutions are based on numerical field simulation solutions of Maxwell's equations. A large number of simulations have been carried out for a wide range of geometrical conditions, conductivities and dielectric constants and adapted to this complex parameter space.

The very complex equations that "fit" these multidimensional problems, are contained in circuit simulator tools (e.g. the transmission line tool of the free Serenade SV version from Ansoft). A calculation example is printed in Fig. 4.20.

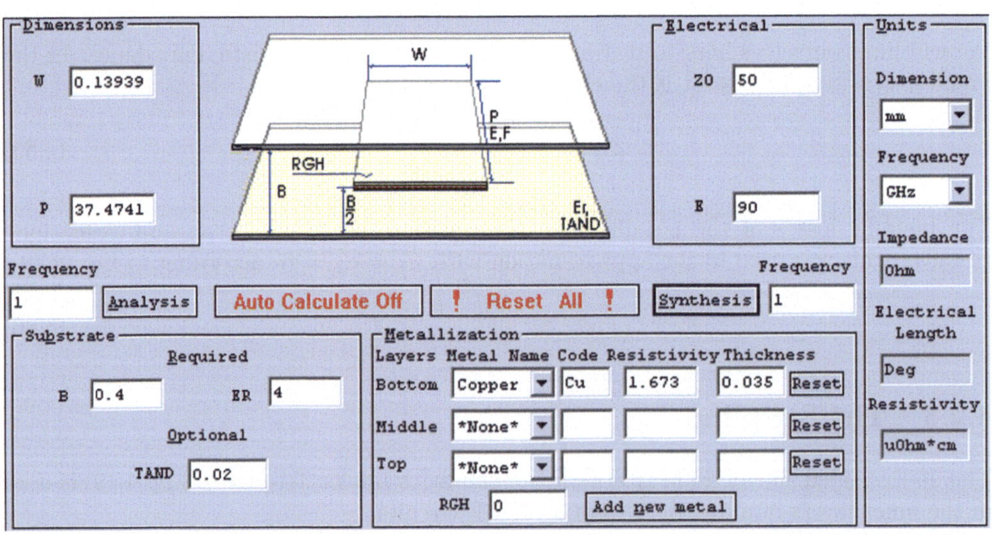

Fig. 4.20: Calculation of a 50 Ω-Stripline with Serenade

This calculation of the transmission line parameters using existing programs is also proposed below for the quasi-TEM transmission lines.

Calculation of the Dimensioning of the RF Cables Using AI

You can use also the AI for the calculation. To do this, it is necessary to make a correct request including the boundary conditions. This can then look as follows:

Analysis the length and the width of the inner transmission line of a stripline. The characteritic impedance is 50 Ohm, the electrical length is 90 degree, f = 1 GHz, DK-value = 4, distance between the ground plates is 0.4 mm.

The result only provides an insufficient value with a strip width (w) of 1.44 mm and a length of 37.5 mm. More accurate models are used in the circuit simulation software solutions.

General: Further exact solutions of a large number of (often exotic) TEM transmission lines are given in [146].

4.2.4 Quasi-TEM Waveguide: Microstrip and Coplanar Transmission Lines

For TEM transmission lines, we have only permitted a homogeneous dielectric between the two conductors. However, there are various cases in which this condition can no longer be met. The best known case is the so-called microstrip line (Fig. 4.21).

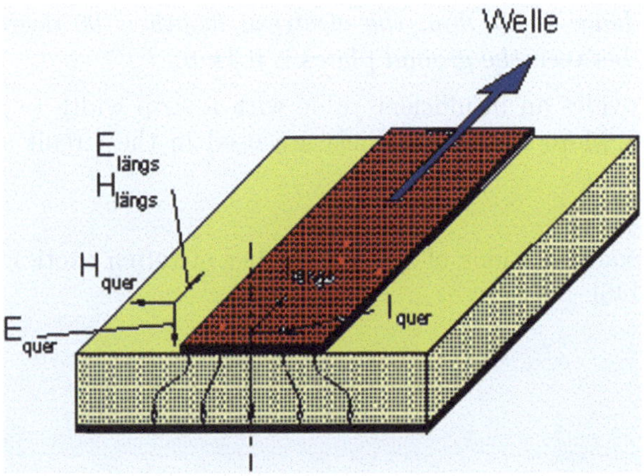

Fig. 4.21: The microstrip line with the field components

The layered dielectric is the cause that a `quasi-TEM wave` now propagates. Quasi-TEM waves are waves in which electrical and magnetic longitudinal components (in the direction of propagation) are also present in the lossless case. However, these longitudinal components are so small compared to the transverse components that they are primarily approximation can be neglected.

The most serious change compared to TEM waveguides is in the permittivity number. The constant value ϵ_r must be replaced by a $\epsilon_{r\,eff}$ that changes over the frequency (= dispersive). The effective permittivity, just like ϵ_r for the TEM wave, represents the link between the free-space wavelength and the conductor wavelength:

$$\epsilon_{r\,eff} = \left(\frac{\lambda_0}{\lambda}\right)^2 \quad . \tag{4.88}$$

If we now replace ϵ_r with $\epsilon_{r\,eff}$ in the contexts for the TEM waveguides, these can still be used for the quasi-TEM waveguides. For example, the following now applies to the phase velocity

$$v_p = \frac{c_0}{\sqrt{\epsilon_{r\,eff}}} \quad . \tag{4.89}$$

In summary, Tab. 4.1 shows the differences between TEM and quasi-TEM waveguides.

Tab. 4.1: Representation for the description of quasi-TEM transmission lines

TEM	$Z_L = f(\text{Geometry, Material})$	$\epsilon_r = f(\text{Material})$
Quasi-TEM	$Z_L = f(\omega, \text{Geometry, Material})$	$\epsilon_{r\,eff} = f(\omega, \epsilon_r, \text{Geometry})$

The microstrip line (Fig. 4.22) is de facto the standard line in modern RF electronics. Whether circuits are made up of just one layer or several layers, the top layer is often realized as a microstrip line.

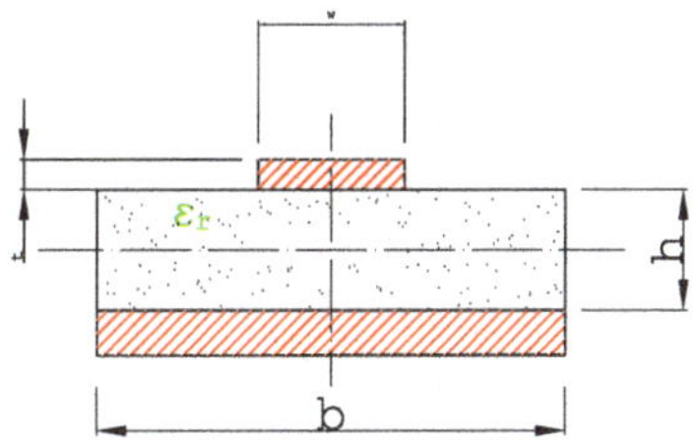

Fig. 4.22: Cross-section of a microstrip line

The advantages of the microstrip line are the high packing densities, the easy combinability with SMD components, low tolerance sensitivity and the very favorable possibility of realizing distributed and quasi-concentrated components from it.

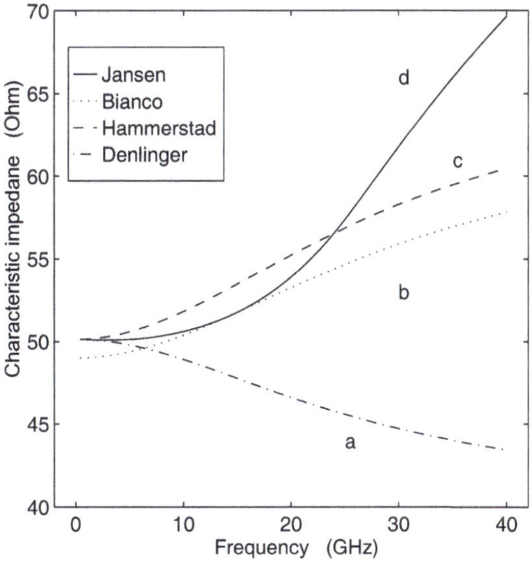

Fig. 4.23: Comparison of the dispersion of the characteristic impedance of a nominally 50 Ω microstrip line on an aluminum oxide substrate up to 40 GHz a) Calculation according to Denlinger ([20]); b) Model according to Bianco et al ([13]); c) according to Hammerstad/Jensen ([47]); d) according to Jansen/Kirschning ([77], now recommended)

To determine the characteristic impedance, losses and effective dielectric constant of a microstrip line, the same applies as for the losses of the stripline. As these values can only be calculated using complex formulas based on the evaluation of numerical methods, an available program should be used. The investigations, which are based on modern numerical programs, have been published by JANSEN and KIRSCHNING [77, 85] have been published. The "classic" publications, which are based on various approximations or definitions, can be found under [13, 20, 40, 41, 47, 129, 148, 149] and in the book [45], among others.

The comparison of the classic closed solutions for the effective dielectric constant shows a much more uniform picture, as can be seen in Fig. 4.24.

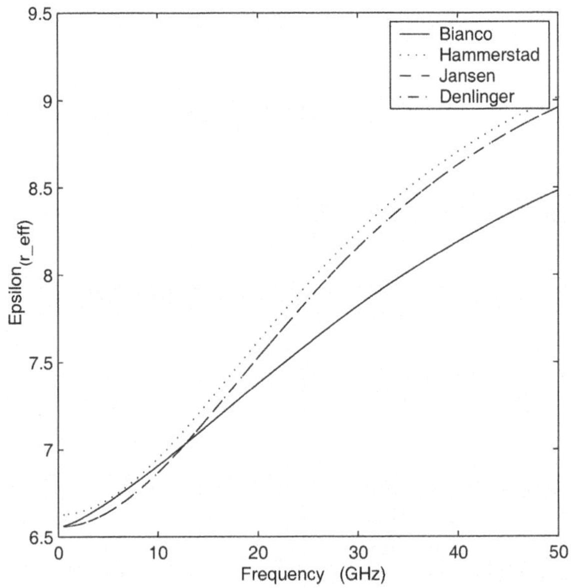

Fig. 4.24: Comparison of the dispersion of the effective dielectric constant of a nominally 50 Ω microstrip line on an aluminum oxide substrate (AlO2, ϵ_r=10) up to 40 GHz

It is important for the developer to recognize how much the dielectric constant changes with frequency. Quasi-TEM lines should therefore be avoided when implementing broadband systems.

The field images of the microstrip line (Fig. 4.25) for low and high frequencies show that the field is almost completely concentrated in the substrate at high frequencies and therefore the dispersive $\epsilon_{r\,eff}$ at high frequencies almost corresponds to the ϵ_r of the substrate.

At low frequencies, on the other hand, the field largely spreads out in the air, which means that the dispersive $\epsilon_{r\,eff}$ is significantly smaller there than at high frequencies.

If one compares these statements on the dispersions of the characteristic impedance with the effective permittivity, one comes to the conclusion that this no longer corresponds to the statements on the relationship between capacitance and characteristic impedance. If

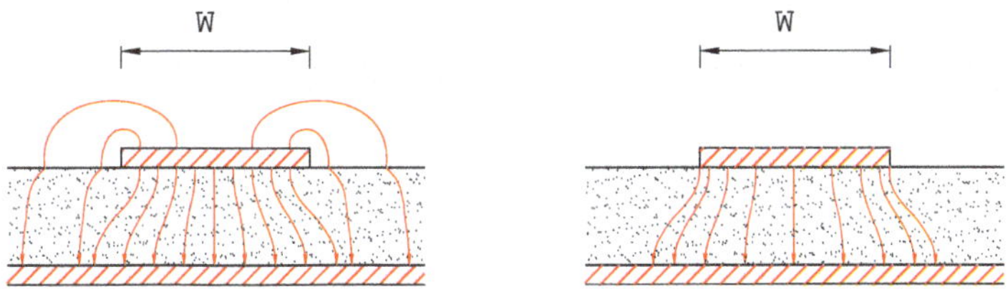

Fig. 4.25: The electric field of the microstrip line at low (left) and high frequencies (right)

the permittivity increases, then the characteristic impedance should decrease.

The reason why the value of the characteristic impedance increases at higher frequencies is that the current and thus also the electric field is "pushed" to the edge of the line at higher frequencies. No current flows in the inner area of the conductor, which is equivalent to a significant narrowing of the conductor width.

The coplanar line (Fig. 4.26) is used instead of the microstrip line when a very short connection between the signal conductor and the ground is required. This is the case, for example, when semiconductors without a housing are to be placed directly on the substrate using the so-called *flip-chip technique* are to be placed directly on the substrate. This technology is preferably used at very high frequencies (e.g. 77 GHz for automotive radar).

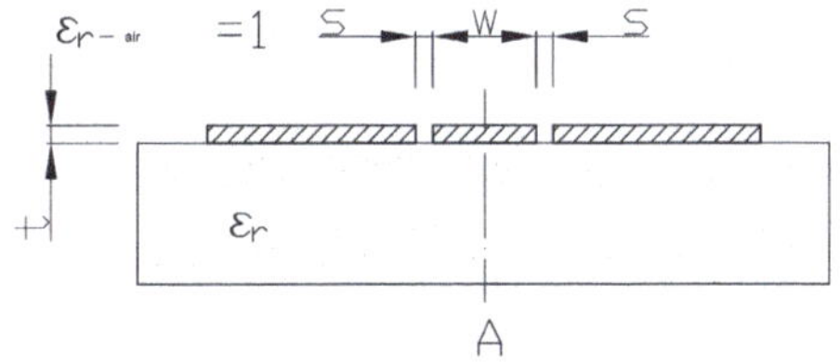

Fig. 4.26: Cross-section of a coplanar line

The characteristic variables can be calculated using complete elliptical integrals, [146, 48].

With coplanar lines, modes other than the desired slotted conductor mode often propagate at higher frequencies (above 20 GHz), provided a ground plane is in the immediate vicinity, which is very often the case in practice. Fig. 4.27 provides an overview of the possible modes.

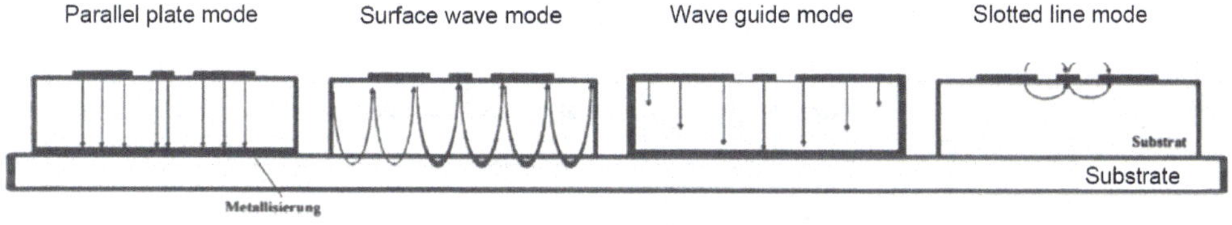

Fig. 4.27: Layouts of various parasitic modes

In particular, the parallel plate mode (which corresponds to the microstrip mode) is very often dominant. A very simple remedy was first published in [128] and is illustrated in Fig. 4.28. The chaotic ground trimming changes the characteristic impedance for the parallel plate mode. This therefore no longer propagates.

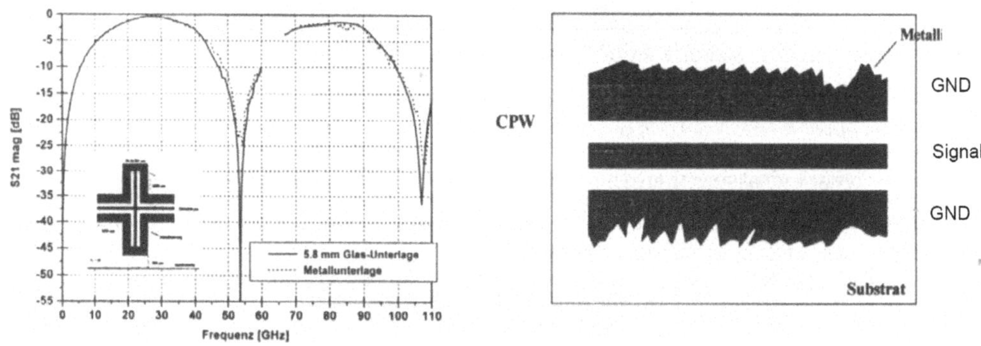

Fig. 4.28: Left: Measurement results of a coplanar cross-branch up to 110 GHz; Right: Measure to suppress moding problems

The measurement results up to 110 GHz on a metal chuck of a wafer prober illustrate the effectiveness of this very simple measure.

In the frequency ranges above 5 GHz and even below 40 GHz, a hybrid of a coplanar and a microstrip line is also very popular. The line has less dispersion and also less radiation effects than a microstrip line. However, the upper ground planes do not allow a very high packing density. A field simulator such as HFSS or CST[6] is required to calculate the line parameters. The electrical field diagram of this mixed form is shown in Fig. 4.29.

In the meantime, this problem has also been recognized by a large number of researchers. At frequencies approaching and exceeding 100 GHz, microstrip transmission lines encounter significant challenges due to mode conversion phenomena. As the frequency increases, quasi-TEM microstrip modes can couple into parallel-plate or stripline-like modes, especially in tightly layered structures or when the substrate is thick relative to the wavelength.

This mode conversion leads to unintended dispersion, increased radiation losses, and degraded measurement accuracy in high-frequency circuits, [69]. Moreover, discontinuities, vias, and transitions in the layout can exacerbate this conversion by breaking the field symmetry required for pure microstrip propagation [100].

Experimental observations and electromagnetic simulations confirm that such multimode behavior becomes prominent beyond 70-100 GHz [107]. Careful design, including mode suppressing structures or transitioning to coplanar waveguide (CPW) or substrate-integrated waveguide (SIW) alternatives, is often necessary to ensure accurate measurements in this regime [3].

[6] Both are the most popular numeric field simulators.

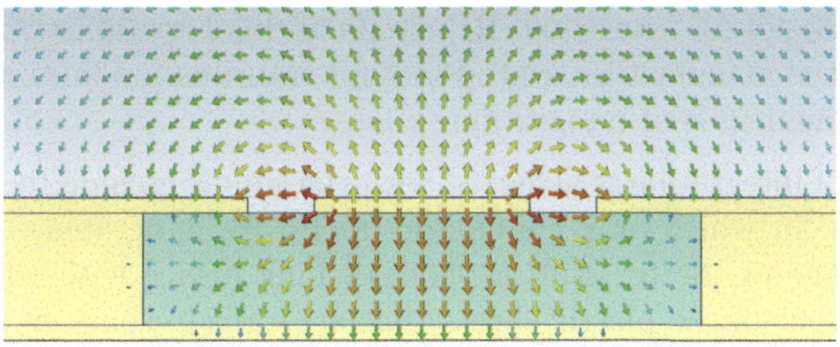

Fig. 4.29: Representation of the electric field of a line calculated using CST with a mode that lies between the microstrip and coplanar mode

4.2.5 Manufacturing Technology for Planar Circuits

RF integrated circuits generally consist of a dielectric material as a substrate material, which is provided with a single, double-sided or multi-layer metallization. As shown, the metallizations are structured in such a way that the arrangement can conduct electromagnetic waves. Various technologies are commonly used to manufacture integrated RF and microwave circuits, as shown in Table 4.2.

Tab. 4.2: Overview of important RF integrated circuit technologies with the abbreviations SL: Stripline and SC: Semiconductor

	Common etching technique	Thin-film hybrid technology	Thick-film hybrid technology	LTCC: Low temperature co-fired ceramic	Semiconductor on sapphire	monolithic MIC
Typical substrate material	FR4, (glass fiber reinforced) teflon ($\epsilon_r = 2.2$)	Al_2O_3-ceramics ($\epsilon_r = 9.8$)	Al_2O_3-ceramics ($\epsilon_r = 9.8$)	Mixed ceramics ($4 < \epsilon_r <$ 10 to 40)	Sapphire ($\epsilon_r = 11.6$ resp. 9.4)	Silicon ($\epsilon_r = 11.9$) GaAs ($\epsilon_r = 13$)
Integrated elements	Lt	Lt, R, (C)	Lt, R, C	Lt, R, C, semiconductors	Lt, R, C, semiconductors	Lt, R, C, semiconductors
Hybrid elements	R, C, L, semiconductors, connecting elements	Semiconductors, (C), connecting elements	Semiconductors, connecting element	R, C, L, semiconductors, connecting element	None	None
Typical substrate thicknesses	10mil = 0.254mm	25mil =0.635mm	25mil = 0.635mm	$n \cdot 0.1$mm	25mil = 0.635mm	0.15mm
Typical circuit area	50x100mm^2	25x25mm^2	25x25mm^2	60x60mm^2	25x25mm^2	2x2mm^2
Typical conductor material	Copper	Copper, gold	Silver palladium, gold palladium	Silver/copper	Gold	Gold
Smallest line and gap width	ca. 0.05mm	20-50μm	ca. 0.1mm	0.05mm	20-50μm (SL), ca. 1μm SC)	5-10μm (SL), ca. 0.5μm (SC)
Typical conductor thickness	17μm, 35μm	5-10μm	25μm	5-10μm	5-10μm (SL), 0.2μm (SC)	1-2μm (SL), 0.1μm (SC)
Hybridization process	Soldering, sticking	Soldering, sticking, alloying, bonding	Soldering, sticking, bonding	Soldering, sticking, alloying, bonding	None	None
Degree of integration	Low	Medium	Medium	High	High	Very high
Device investment	Very low	High	Moderate	Very high	Very high	Very high
Frequency range	$\leq$ 10GHz	$\leq$ 50GHz	$\leq$ 20GHz	$\leq$ 20GHz	$\leq$ 50GHz	$\leq$ 300GHz

The various technologies for integrated RF circuit technology differ primarily in the substrate material, in the manufacturing processes for transmission lines, resistor ranges and capacitors and in the degree of integration.

The lowest degree of integration is achieved with the simple etching technology of copper-clad circuit boards, in which only conductor tracks are integrated. All that is needed is a milling technique or, better still, a photo-etching device with which the conductor structure is etched out of the plastic sheets that are already commercially available with copper cladding on both sides.

Low-loss plastic substrates with typical thicknesses of 5 - 25 mil (0.127 mm - 0.635 mm) are used as the starting material for circuit production. A frequently used substrate for inexpensive consumer products is now FR4 (glass fiber reinforced epoxy resin, also known as laminate) with a relative dielectric constant ϵ_r in the range of 3.4 to 4.

Pure and glass fiber reinforced PTFE (Teflon) with the dielectric constant $\epsilon_r = 2.2$ for pure PTFE. FR4 has significantly greater dielectric losses than PTFE carriers. Commercially available metallization thicknesses are 17.5 μm. The metallization thickness is often specified in ounces per area (1 $\frac{\text{oz}}{\text{foot}^2}$ corresponds to a copper metallization thickness of 35 μm).

4.2.6 Coaxial Cable as Reference Standard

The coaxial transmission line without dielectric (as a so-called air line) is used as a reference standard in RF measurement technology for calibrating a network analyzer. This section shows how precisely such a coaxial air line must be designed so that it maintains the characteristic impedance with the desired accuracy.

It is known in electrical engineering that the electric field penetrates less into metallic conductors with increasing frequency. This physical effect causes dispersions (frequency dependencies) that affect the losses and the characteristic impedance of a metallic conductor.

In the following, the very precisely calculable and measurable coaxial air line will be used to show how dispersion affects the characteristic impedance of the coaxial line in comparison to other faults.

<u>Dispersions using the Example of a Coaxial Transmission Line</u>

Coaxial precision air lines are manufactured with the utmost mechanical precision and are available as reference standards for measurement technology. Inner and outer conductors are measured individually and thus paired in such a way that manufacturing tolerances are compensated.

Individual TRL/LRL calibration kits [118] for precision measurements using vector network analyzers (chapter 10) can be put together with air lines. They are also suitable for precision measurements in the time domain and for other special applications.

<u>Influence of Mechanical Faults on the Characteristic Impedance of a Coaxial Air Line</u>

The measurement accuracy that can be achieved with air lines depends to a large extent on compliance with the characteristic impedance Z_0. This in turn depends very much on the precision of the outer and inner conductor diameter.

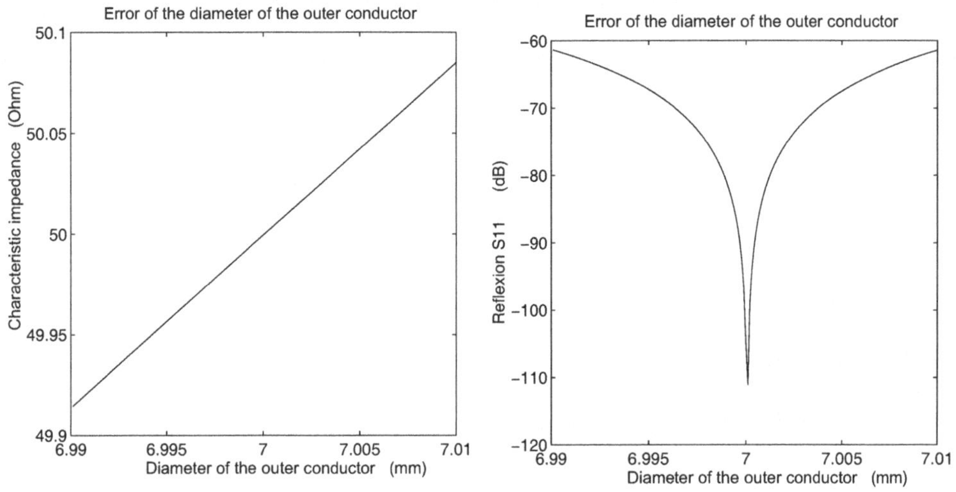

Fig. 4.30: Influence of deviations in the outer conductor diameter of the coaxial cable on the characteristic impedance

The characteristic impedance Z_0 of a coaxial air transmission line depends on the following relationship to the geometric dimensions, [48]:

$$Z_0 = \frac{1}{2\,\pi} \sqrt{\frac{\mu_0}{\epsilon_0\;\epsilon_r}} \cdot \ln\left(\frac{D}{d}\right) \quad . \tag{4.90}$$

The diameter of the outer conductor is described by D and the inner conductor diameter by d. The constants are:

- Dielectric constant: $\epsilon_0 = 8.8542\,pF/m$
- Permeability constant: $\mu_0 = 0.4\,\pi\,\mu H/m$
- Material constant, air: $\epsilon_r = 1.0006$

The influence of deviations in the outer conductor on the characteristic impedance of a 7 mm air line is shown in Fig. 4.30.

From Fig. 4.31 it can be seen that the requirements for the inside diameter exceed those for the outside diameter.

Precision manufacturers guarantee a mechanical precision including surface roughness of only $\pm\;5\,\mu m$ compared to the ideal value. This ensures a reflection loss that is better than 60 dB to 18 GHz. As mentioned, a further improvement is achieved by measuring the inner and outer conductors individually and pairing them in such a way that the reflection loss is minimized.

The specification of a reflection loss of more than 60 dB is met from a frequency of 200 MHz. It is shown below that the other effects on the characteristic impedance can be

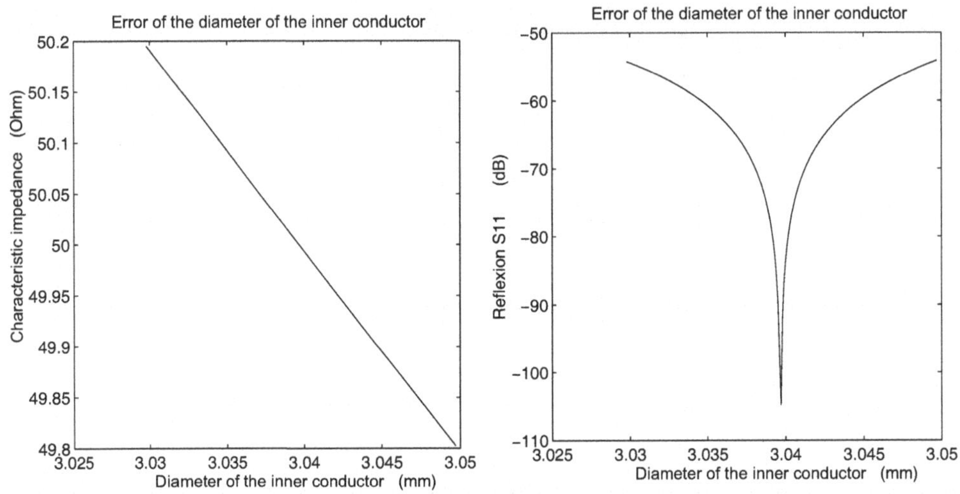

Fig. 4.31: Influence of deviations in the inner conductor diameter of the coaxial cable on the characteristic impedance

corrected, whereby the application range can be extended. Furthermore, more detailed investigations and measurements show that the other error effects no longer have any influence on the given specifications.

Influence of the Penetration Depth on the Characteristic Impedance of Coaxial Air Lines

If you want to measure at lower frequencies, the effects due to the skin depth δ must also be taken into account, [153]. According to [79], the most exact solution regarding the dependence of the characteristic impedance on the skin depth is given by [19].

This results in a corrected characteristic impedance Z_0' via

$$Z_0' = Z_0 \left(1 + (1 - j)\, \delta \, \frac{1/D + 1/d}{4 \, \ln (D/d)} \right) \tag{4.91}$$

with the penetration depth δ, which is calculated from the permeability constant μ_0 and the conductivity[7] σ of $16 \cdot 10^6 \, S/m$ calculated as a function of the frequency f:

$$\delta = \frac{1}{\sqrt{\pi \, \mu_0 \, \sigma \, f}} \quad . \tag{4.92}$$

The result is illustrated in the Fig. 4.32.

The right part of the Fig. 4.32 shows that the characteristic impedance values below 200 MHz are outside the specification range of -60 dB due to the effects of the penetration depth.

If this effect is taken into account, an air duct can also be used below 200 MHz.

[7] The gold-plated transmission air line is made of brass (Ms 58).

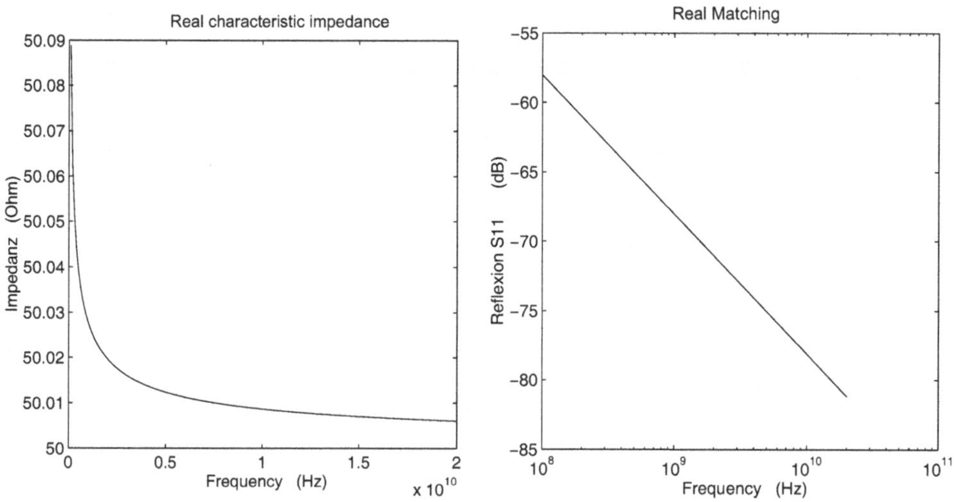

Fig. 4.32: Characteristic impedance of a 7 mm air line taking into account the skin depth δ

Further Influences on the Characteristic Impedance of a Coaxial Air Line

Another possible fault is the deflection of the inner conductor. The influence of a displacement x (unit as D) of the inner conductor on the characteristic impedance of an air line can be calculated using the following Eq. (e.g. [146]):

$$Z_0' = 60\,\Omega\ \cosh\left(\frac{D}{2\,d}(1-(x\cdot 2/D)^2)+\frac{d}{2\,D}\right) \quad . \tag{4.93}$$

With a 7 mm air duct, only displacements of the inner conductor from the center of more than 100 μm have a sufficiently large influence to push the values of the air line out of the specification range (Fig. 4.33). The maximum deflection for inner conductors is $5-30\,\mu m$ for cable lengths of $50-300\,mm$.

In addition to the mechanical precision of the inner and outer conductor, the reproducibility of the adapted connectors has the greatest influence on accuracy. As expected, the reproducibility of RPC-7 precision connectors[8] is very good. This can be seen for reflection measurements ($dS_{11} = Sa_{11} - Sb_{11}$) and transmission measurements ($dS_{21} = Sa_{21} - Sb_{21}$) in the Fig. 4.34. The air line was rotated by 30° for each reproducibility measurement. The measured values of each measurement were subtracted from the measured values of the 0° position.

The reproducibility of air lines with PC-N connectors is only approx. 5 dB worse than that of lines with RPC-7 precision connectors.

All other coaxial connectors (including the precision connectors) have significantly poorer reproducibility values than PC-N and RPC7. Therefore, only these two connector systems should be used for precision measurements.

[8] Connectors with the designation RPC are products from Rosenberger, [117].

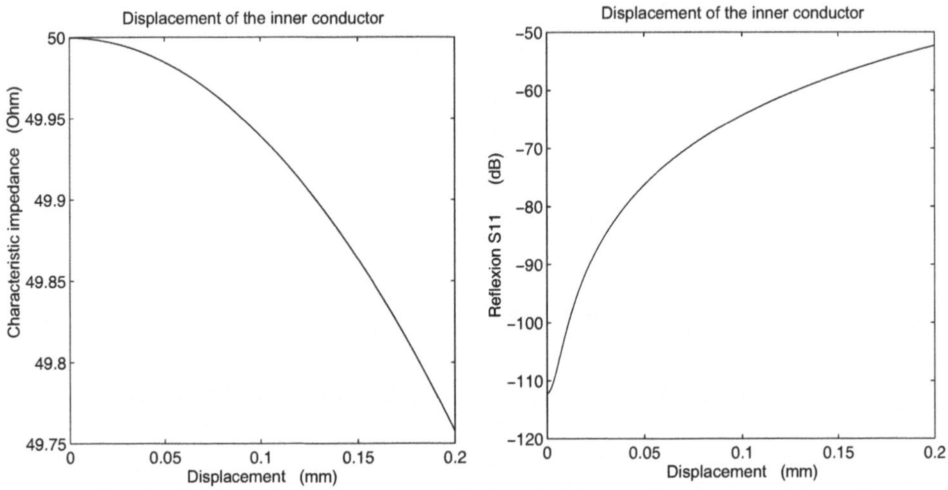

Fig. 4.33: Wave impedance change due to the displacement of the inner conductor from the center

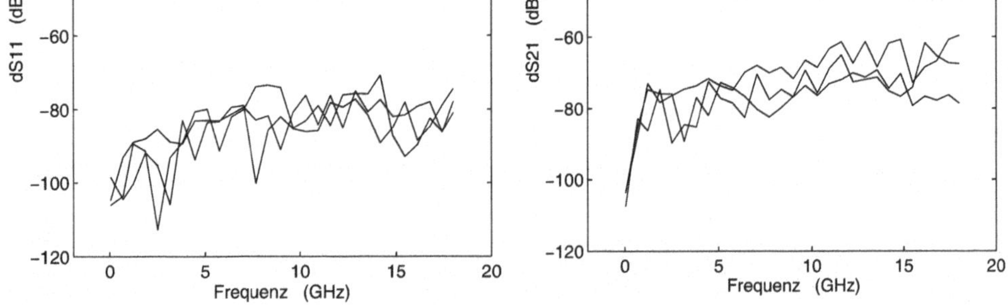

Fig. 4.34: Reproducibility of the reflection and transmission behavior of an air line with RPC-7 connectors after re-contacting and rotation

4.3 Impedance Transformation and Smith Chart

4.3.1 Input Resistance of a Transmission Line

Transmission line theory states that the voltage $U(l)$ and the current $I(l)$ at a distance l from the load can be specified as a function of the voltage U_0 and the current I_0 at the load, the characteristic impedance Z_L, the electrical length βl and, if applicable, the losses α (chapter 4.1.3).

By means of Eq. (4.29) we have a correlation for the lossy case and with Eq. (4.49), which is printed again as Eq. (4.94) for the lossless case.

$$\begin{pmatrix} U(l) \\ I(l) \end{pmatrix} = \begin{pmatrix} \cos(\beta l) & j\,Z_L \cdot \sin(\beta l) \\ j/Z_L \cdot \sin(\beta l) & \cos(\beta l) \end{pmatrix} \begin{pmatrix} U_0 \\ I_0 \end{pmatrix} \quad , \tag{4.94}$$

The next step is to investigate the input resistances for a terminated lossless line according to Fig. 4.35.

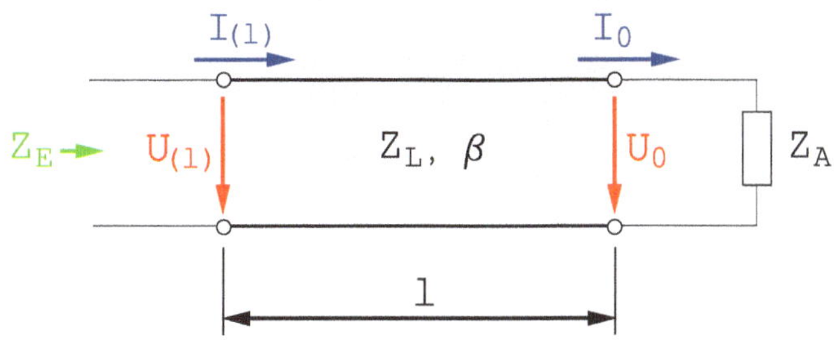

Fig. 4.35: Cable with impedance termination Z_A

With $Z_A = U_0/I_0$ the two linear equations from Eq. (4.94) can be transformed into Eq. (4.95).

$$\frac{U(l)}{U_0} = \cos(\beta l) + j\,\frac{Z_L}{Z_A} \cdot \sin(\beta l) \quad , \tag{4.95}$$

$$\frac{I(l)}{I_0} = \cos(\beta l) + j\,\frac{Z_A}{Z_L} \cdot \sin(\beta l) \quad . \tag{4.96}$$

With the definition for the input resistance

$$Z_E = \frac{U(l)}{I(l)} \tag{4.97}$$

and the division of the Eq. (4.95) by the Eq. (4.96) you get the input resistance of this "single port" with

$$Z_E = Z_A \frac{1 + j\frac{Z_L}{Z_A} \cdot \tan(\beta l)}{1 + j\frac{Z_A}{Z_L} \cdot \tan(\beta l)} \quad . \tag{4.98}$$

In the following, the special cases with terminations at $Z_A = Z_L$, $Z_A = 0\,\Omega$, $Z_A = \infty\,\Omega$ and $Z_A = jX_A$ will be discussed.

Matched Transmission Line and Field Image Jump

A line is considered matched if the line is extended to infinity or if the terminating resistor Z_A corresponds to the characteristic impedance of the line Z_L. The model of an infinitely long line is often used as a boundary condition in numerical simulators (field simulators). If the terminating resistor Z_A is set equal to the characteristic impedance of the line Z_L, the adjustment is based on a mathematical model, which is always well fulfilled in the circuit simulator (e.g. ADS, Serenade) and in practice also at low frequencies.

At RF (estimate: $\lambda_0 = 30\,\%$ of the conductor diameter), however, there is a so-called `field image jump`. The field pattern of the cable does not match the field pattern of the connecting elements (e.g. SMDs, connected to ground). Although both elements (e.g. cable and SMD resistor) have the same impedance, there is not a very good match! Fig. 4.36 shows two coaxial conductors that are not mounted centered.

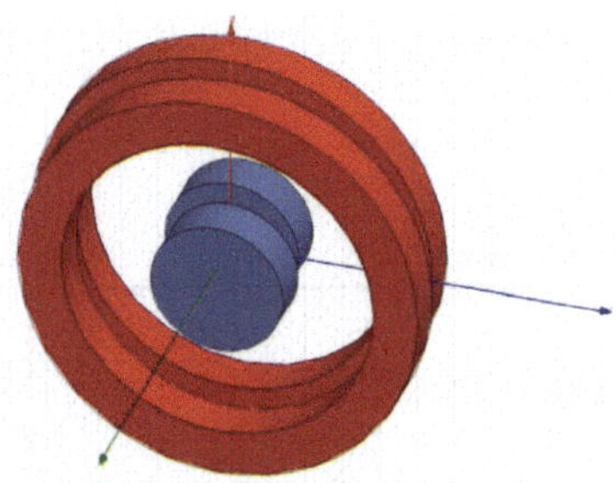

Fig. 4.36: Illustration of the `field image jump` due to a transition between two identical coaxial conductors

Although both coaxial conductors are perfectly designed for 50 Ω, there is a reflection at this transition point. This reflection is due to the fact that the field image of line 1 does not match the field image of line 2; there is a field image jump.

If the condition $Z_A = Z_L$ is entered into Eq. (4.98), this results in the following for the input resistance

$$Z_E = Z_L \frac{1 + j \cdot \tan(\beta l)}{1 + j \cdot \tan(\beta l)} = Z_L \quad . \tag{4.99}$$

Thus, regardless of the length of the cable, the input resistance corresponds to the characteristic impedance.

Short-Circuited Cable

If the condition of the termination $Z_A = 0\,\Omega$ is inserted in Eq. (4.98), this results in for the input resistance

$$Z_E = \frac{Z_A + j\ Z_L \cdot \tan(\beta l)}{1 + j\frac{Z_A}{Z_L} \cdot \tan(\beta l)} = j\ Z_L\ \tan(\beta l) = j\ X_E \quad . \tag{4.100}$$

The input resistance of the line therefore corresponds to a pure reactance, which is dependent on the length and the characteristic impedance of the line. Due to the lower expansion, only the range up to $\pi/2$ is used as reactance. For this range, this results in an inductance with the value

$$L = \frac{1}{\omega} Z_L \tan(\beta l) \quad . \tag{4.101}$$

Fig. 4.37 shows how a microstrip line with length l and characteristic impedance Z_L can be used to replace a shunt coil.

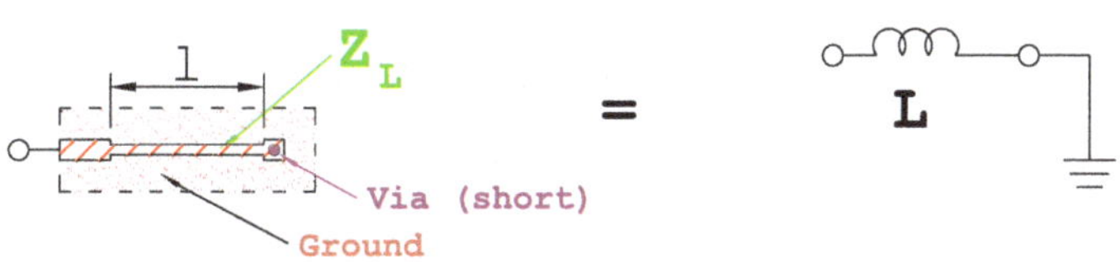

Fig. 4.37: Short microstrip line to realize an inductance connected to ground

As the inductance increases with the characteristic impedance, a transmission line with the highest possible impedance should be selected. The minimum conductor width (often 0.1 mm) and the conductor height determine the maximum characteristic impedance.

If you want to realize a series coil in planar technology, you have to remove the ground and use a symmetrical two-wire line (Fig. 4.38 with the coupling to a microstrip line).

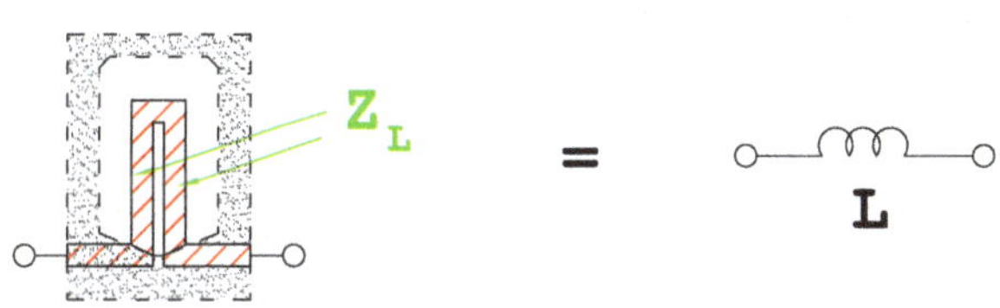

Fig. 4.38: Short symmetrical strip line short-circuited at the end to realize a series-connected inductance

As shown in Fig. 4.39, the reactance can assume values between $-\infty$ and $+\infty$ depending on the cable length.

A so-called `standing` wave forms on the line. In practice, a voltage probe can be used to measure the voltage curve as a function of location (=principle of the measuring line, [125]). With $U_0 = 0$ follows from Eq. (4.95):

$$U(l) = j\,Z_L \cdot \sin(\beta l)\,I_0 \quad \text{resp.} \quad I(l) = \cos(\beta l)\,I_0 \quad . \tag{4.102}$$

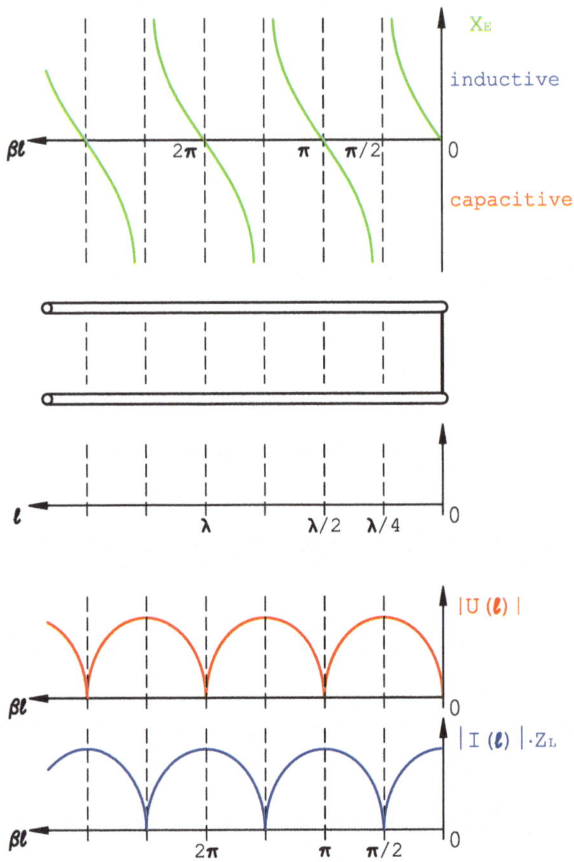

Fig. 4.39: Reactive resistances and current and voltage distributions of lines with a short circuit

Open-Circuited Cable

If you use the condition $Z_A = \infty\,\Omega$ in the Eq. (4.98), this results in for the input resistance

$$Z_E \;=\; \frac{1 + j\,\frac{Z_L}{Z_A} \cdot \tan(\beta l)}{\frac{1}{Z_A} + j\frac{1}{Z_L} \cdot \tan(\beta l)} \;=\; -\,j\;Z_L\;\cot(\beta l) \;=\; j\;X_E \quad . \tag{4.103}$$

Consequently, the input resistance of the line also corresponds to a pure reactance here, which is dependent on the length and the characteristic impedance of the line. Here too, only the range up to $\pi/2$ is used as a component with reactance due to the lower expansion. This results in a capacitance with the value

$$C \;=\; \frac{1}{\omega\;Z_L\;\cot(\beta l)} \quad . \tag{4.104}$$

Fig. 4.40 shows how a microstrip line with length l and characteristic impedance Z_L can be used to replace a shunt capacitor.

Fig. 4.40: Short microstrip line for realizing a to ground connected capacitance

In practice, this shunt capacitor is used very frequently. However, the practitioner applies the open-circuit cable twice at half its length. He also adds additional metal surfaces (as can be seen in Fig. 4.41) in order to increase the capacitor value via a short solder bridge. The value can be reduced using a small hand cutter.

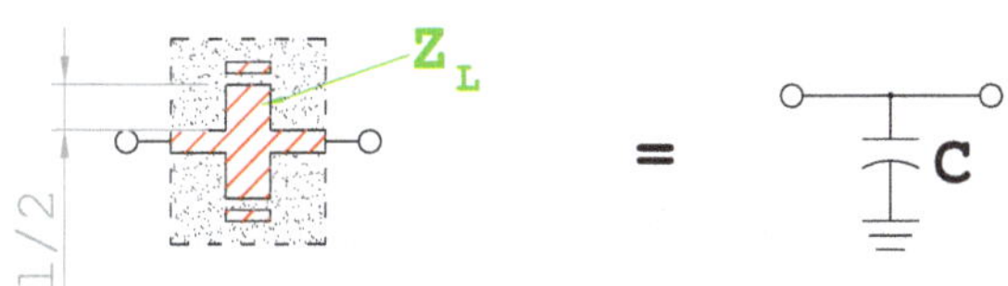

Fig. 4.41: The practical solution of short microstrip lines to realize a shunt capacity

If you want to realize a series capacitor in planar technology, you also have to remove the ground here and realize a pure two-wire conductor (Fig. 4.42 with the coupling to a microstrip line).

Fig. 4.42: Short coupled stripline for the realization of a series-connected capacitance

As shown in Fig. 4.43, the reactance can assume values between $-\infty\,\Omega$ and $+\infty\,\Omega$ depending on the cable length. With $I_0 = 0\,\mathrm{A}$ results from Eq. (4.95):

$$U(l) = \cos(\beta l)\, U_0 \quad \text{bzw.} \quad I(l) = j\,\frac{1}{Z_L}\sin(\beta l)\, U_0 \quad . \tag{4.105}$$

NOTE:

> Transmission lines in the range of $0 < \ell < \lambda/8$ are used as quasi-concentrated components. An inductance can be realized by means of a short-circuited line and a capacitance by means of an open-circuited line.

Quasi-concentrated components are discussed separately at the end of the chapter.

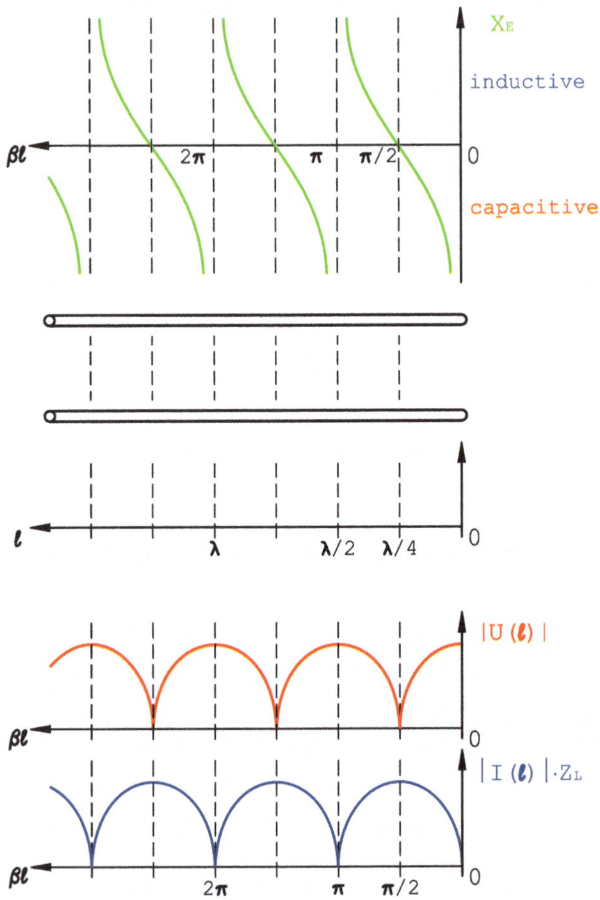

Fig. 4.43: Reactive resistances and current and voltage distributions of lines with an open circuit

Cable Termination with Reactance

If the condition $Z_A = jX_A$ is inserted into Eq. (4.98), the following results for the input resistance

$$Z_E = j\,X_A \frac{1 + \frac{Z_L}{X_A} \cdot \tan(\beta l)}{1 - \frac{X_A}{Z_L} \cdot \tan(\beta l)} = j\,X_E \quad . \tag{4.106}$$

Consequently, here too, the input resistance of the line corresponds to a pure reactance, which is dependent on the length and the characteristic impedance of the line and the reactance X_A.

Similar to the Figs. 4.39 and 4.43 showing the reactance for short circuit and open circuit, the input reactance can assume values between $-\infty\,\Omega$ and $+\infty\,\Omega$ when terminated with X_A depending on the line length. This means that short-circuits and open circuits are "degenerate" reactances, as no active power is converted there either.

4.3.2 Cables as Impedance Transformers

Impedance transformers realized with concentrated components have already been introduced in chapter 3.2.2. The impedance transformers presented here have the advantage that they can handle much more power and the circuit board does not have to be mounted with SMD components. A disadvantage in the single-digit GHz range is the relatively large PCB area required.

Impedance transformers consisting only of a lossless series line will now be described.

In practice, the assumption of no loss is sufficient for dimensioning (length and characteristic impedance) is sufficient. However, to assess the losses of a component, the losses of the cable must be taken into account again.

$\lambda/2$-Transformer

The condition $l = n\lambda/2$ with $n = 1, 2, 3, ...$ yields for the electric length

$$\beta l = \frac{2\,\pi}{\lambda}\,n\,\frac{\lambda}{2} = n\,\pi \tag{4.107}$$

and therefore $\tan(\beta l) = 0$.

Therefore, with Eq. (4.98) the following Eq. is valid: $Z_E = Z_A$.

The input resistance of a $\lambda/2$ long cable therefore corresponds to the output resistance, regardless of the characteristic impedance of the cable. Consequently, the following applies to the reflection and transmission parameters

$$S_{11} = 0 \quad \text{and} \quad |S_{21}| = 1 \quad . \tag{4.108}$$

This special behavior is often combined with the behavior of a $\lambda/4$ line[9].

It should be noted in the application that the bandwidth of the $\lambda/2$ transformation becomes smaller and smaller with increasing deviation of the port impedances. Fig. 4.44 shows two different $\lambda/2$ transformers.

Fig. 4.44: Representation of $\lambda/2$ transformers with low and high impedance transformation line

The different bandwidths are illustrated in Fig. 4.45. The ports have 50 Ω and the transformation lines have 20 Ω, 40 Ω, 60 Ω and 80 Ω.

[9]This can be the same line at half the frequency.

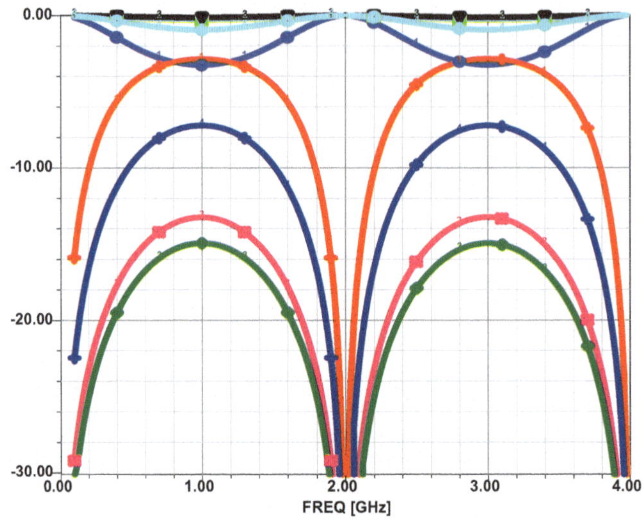

Fig. 4.45: Representation of the reflection and transmission behavior of $\lambda/2$ transformers for 2 GHz with low and high impedance characteristic impedances

It can be seen that such lines deviating from 50 Ω can be used as transmission lines over a relatively large bandwidth.

λ/4-Transformer

The condition $l = k\lambda/4$ with $k = 1, 3, 5, ...$ results in the following for the electrical length

$$\beta l = \frac{2\,\pi}{\lambda}\,k\,\frac{\lambda}{4} = k\,\frac{\pi}{2} \tag{4.109}$$

and consequently $\tan(\beta l) = \infty$.

Used in Eq. (4.98) applies: $Z_E = Z_L^2/Z_A$.

The output resistance Z_A of a $\lambda/4$ long line can therefore be transformed to the input resistance Z_E. The result (Z_E) depends on the characteristic impedance of the line.

The bandwidth of a $\lambda/4$-transformer depends on the difference between the two resistances to be transformed. If the differences are small, the bandwidth is very large.

$\lambda/4$-transformers are very frequently used in RF circuits. They are very often used to filter signals, in particular to suppress the first harmonic. For this purpose, a $\lambda/4$-line short-circuited at the end of the line is connected as a stub, Fig. 4.46.

4.3.3 The Reflection Factor r

Reflection Factor at the End of a Cable

With the introduction of the scattering parameters (section 2.2.2), the reflection coefficients of two-ports (S_{11} and S_{22}) were discussed in detail. When considering a one-port,

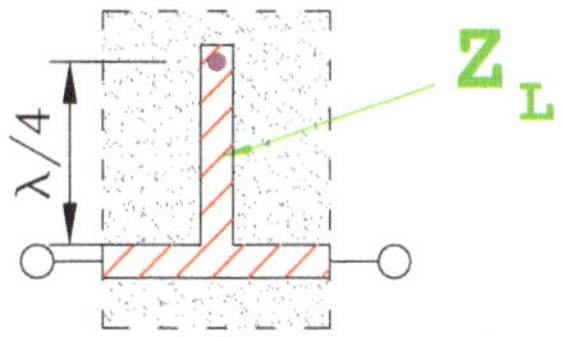

Fig. 4.46: Illustration of the use of a $\lambda/4$- transformer for harmonic filtering

it is common to denote the reflection coefficient of this one-port by r. Other common designations are ρ and Γ.

The reflection factor r is defined the same as S_{ii} via

$$r = \frac{b}{a} = \frac{\text{reflected wave}}{\text{incident wave}} \tag{4.110}$$

Furthermore, the both wave sizes a and b were introduced as normalized voltage waves (Eq. (2.10)). Thus also applies:

$$r = \frac{U_r}{U_p} = \frac{\text{reflected voltage wave}}{\text{incident voltage wave}} \quad . \tag{4.111}$$

We obtained the solutions (Eqs. (4.22) and (4.23)) in the section 4.1 under the point *determination of the integration constants* U_r *and* U_p:

$$U_p = \frac{1}{2} \cdot (U_0 + I_0 \cdot Z_L) \tag{4.112}$$

and

$$U_r = \frac{1}{2} \cdot (U_0 - I_0 \cdot Z_L) \tag{4.113}$$

With $Z_A = U_0/I_0$, the reflection factor at the end of a line is thus obtained:

$$r_A = \frac{U_0 - I_0 \cdot Z_L}{U_0 + I_0 \cdot Z_L} = \frac{Z_A - Z_L}{Z_A + Z_L} \quad . \tag{4.114}$$

For passive lossy terminations, $\mathrm{Re}\{Z_A\} > 0$ applies. For this (frequent) case, the value of the reflection coefficient is always less than 1.

<u>Reflection Factor at the Input of a Line</u>

If we consider the voltage at the input of a lossless line (Fig. 4.47) with Z_L, the following applies according to Eq. (4.18):

$$U(x) = U_p\, e^{-j\beta x} + U_r\, e^{j\beta x} \quad . \tag{4.115}$$

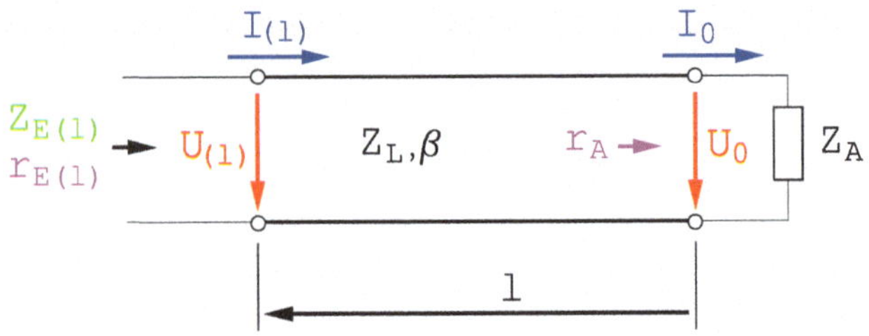

Fig. 4.47: Reflection factors $r_{E(\ell)}$ and r_A at the input and termination of a transmission line

Note that ℓ and x have a different direction: $\ell = -x$. The voltage waves are shifted by the electrical length $\beta\ell$ compared to the waves at the termination. The following therefore applies to the reflection factor at the input of the line:

$$r_{E(\ell)} = \frac{U_r \, e^{-j\beta\ell}}{U_p \, e^{j\beta\ell}} = \frac{U_0 - I_0 \cdot Z_L}{U_0 + I_0 \cdot Z_L} \, e^{-2j\beta\ell} = \frac{Z_A - Z_L}{Z_A + Z_L} \, e^{-2j\beta\ell} = r_A \, e^{-2j\beta\ell} \quad . \quad (4.116)$$

Reflection Factor at the Input of a Cable for Different Terminations

A real reflection factor results for a real termination ($Z_A = R_A$). A distinction is made between two cases:

Case 1: $R_A \geq Z_L$ applies here: $0 \leq r_A \leq 1$,

Case 2: $R_A \leq Z_L$ applies here: $-1 \leq r_A \leq 0$.

These results are shown for $r_A = r_{E(0)}$ in Fig. 4.48.

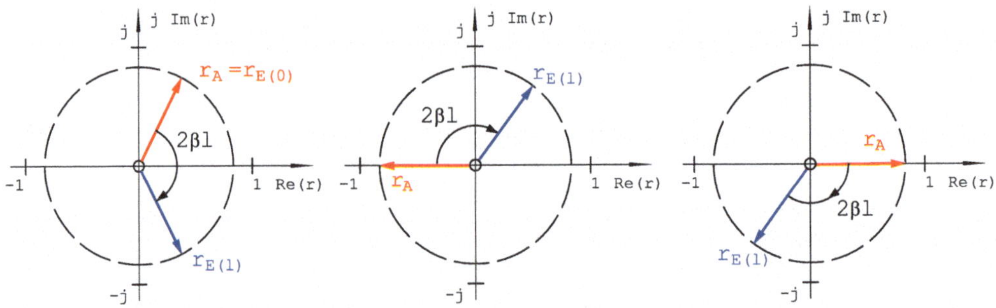

Fig. 4.48: Representation of $r_A = r_{E(0)}$ and of $r_{E(\ell)}$ in the complex plane for the three cases: Z_A = complex; Z_A = real = $R_A < Z_L$; Z_A = real = $R_A > Z_L$

Fig. 4.48 shows the ranges for any complex and real reflection coefficients, as well as the behavior of the input reflection coefficient $r_{E(\ell)}$ as a function of the termination r_A and the electrical line length $\beta\ell$.

If you want to determine the input reflection coefficient $r_{E(\ell)}$, the pointer of the final reflection coefficient r_A is rotated clockwise (mathematically negative sense) by the angle $2\,\beta\ell$ (!).

For the three special cases open circuit, short circuit and matching, the reflection coefficients on the real axis again result according to Eq. (4.114):

- Open: $Z_A \to \infty\,\Omega \quad r_{LL} = 1 \quad ,$
- Short: $Z_A = 0\,\Omega \quad r_{KS} = -1 \quad ,$
- Matching: $Z_A = Z_L \quad r_{AP} = 0 \quad .$

The results are shown in Fig. 4.49.

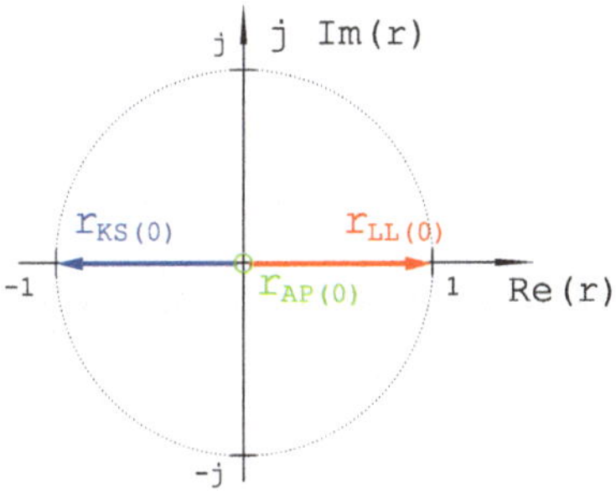

Fig. 4.49: The three special cases (short circuit, open circuit and matching) for r_A

4.3.4 Voltage Standing Wave Ratio and Matching Factor

It has already been shown in the section 4.3.1 that a "standing wave" propagates at a termination of a line that does not correspond to the characteristic impedance of the line. Fig. 4.50 illustrates the current and voltage levels along a line in the event that there is a termination with impedance Z in the (left-hand) plane at ℓ=0 m whose reflection is between 0 and 1.

In the past (until around 1980), the maxima and minima were determined using voltage probes in slotted cables and the complex reflection factor of the reflection object was calculated from their ratio and the distance from the reflection object.

This arrangement is called a "measuring line" and is currently still used for learning purposes, [125].

However, working with the voltage ratios in the form of the matching factor m and in particular in the form of the standing wave ratio VSWR (also ripple factor s) has continued to this day.

The following applies:

$$\text{VSWR} = s = \frac{1}{m} = \frac{|U_{max}|}{|U_{min}|} = \frac{1+|r|}{1-|r|} \quad . \tag{4.117}$$

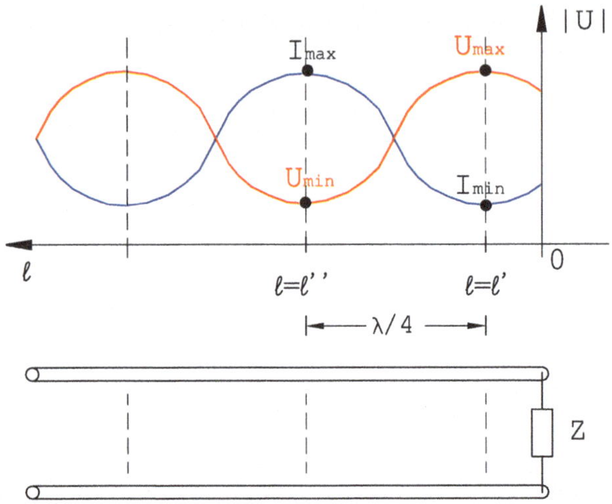

Fig. 4.50: Progression of the voltage and current magnitudes along a loss-free line for a reflection termination between 0 and 1

For some components, it is common to specify the VSWR (Voltage Standing Wave Ratio) instead of the reflection coefficient.

4.3.5 The Smith Chart

In the past, the Smith chart was the most important graphical aid for the RF technician. The Smith chart was a powerful graphical tool used extensively in RF and microwave engineering for the analysis and design of transmission lines and matching networks.

It allows engineers to represent complex impedances and reflection coefficients on a normalized, two-dimensional chart.

By plotting impedance or admittance values, the Smith chart simplifies the process of impedance matching, which is crucial for maximizing power transfer and minimizing signal reflection.

It is particularly useful for visualizing how impedance changes along a transmission line with varying electrical length.

The chart also supported the design of single- and multi-stage matching networks using lumped or distributed elements.

In addition to impedance matching, it was used to interpret S-parameters, VSWR values, and other key RF performance metrics.

Overall, the Smith chart remained an essential tool in both educational and industrial applications, bridging theoretical analysis with practical RF circuit design.

Today, it is still of outstanding importance in the representation of reflection and, where applicable, transmission values.

Furthermore, the Smith chart is used for circuit synthesis in software tools, such as Serenade or ADS to realize transformation networks.

This approach has completely replaced the "classic" graphic design with compass and pencil. For qualified work with these tools, the construction of the Smith chart is presented in detail below.

The "classic" graphical work with the Smith chart is only used very rarely in practice. For this reason, reference is made here to the numerous literature.

This graphical working method is explained in detail in almost every basic book on RF technology.

Construction of the Smith Chart

For the Eq. (4.114) we can also generally use

$$r = \frac{Z - Z_L}{Z + Z_L} = \frac{z-1}{z+1} \quad \Longleftrightarrow \quad z = \frac{1+r}{1-r} \quad \text{resp.} \quad Z = Z_L \, \frac{1+r}{1-r} \tag{4.118}$$

with the normalized complex impedance $z = Z/Z_L$.

The reflection factor r and the normalized impedance z can be divided into real and imaginary parts:

$$r = \mathrm{Re}\{r\} + j\ \mathrm{Im}\{r\} = u + j\,v \quad , \quad z = \mathrm{Re}\{z\} + j\ \mathrm{Im}\{z\} = w + j\,x \tag{4.119}$$

The Smith chart is the graphical representation of the relationship between the complex variables r and z according to Eq. (4.118). This "mapping" of the z values to the r values is also referred to as a conformal mapping. Fig. 4.51 shows how individual impedances

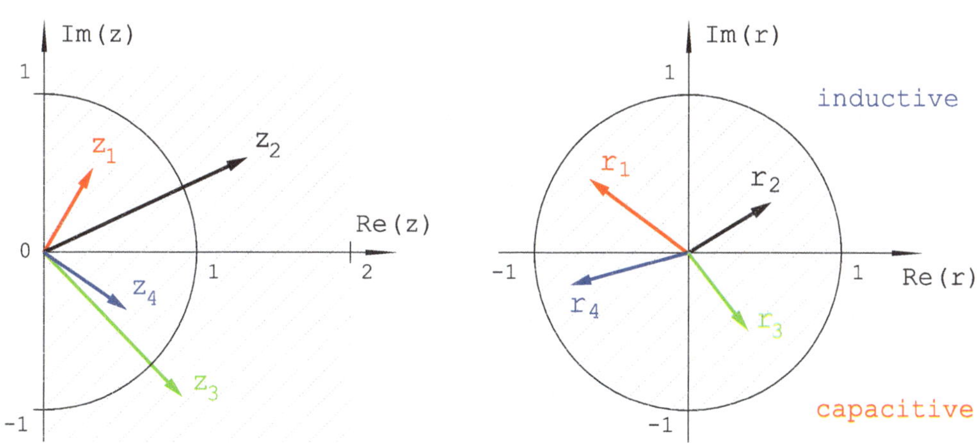

Fig. 4.51: Mapping of normalized impedances to the reflection values

of the right z half plane (= passive components) are mapped to the inside of the unit circle in the r plane.

To construct the Smith chart, map the lines $w = const.$ and $x = const.$ in the r-plane ($= u, v$-plane) (based on the real and imaginary parts of the equation (4.118)). The lines are described by the following circular equations:

$$\left(u - \frac{w}{w+1}\right)^2 + v^2 = \frac{1}{(w+1)^2} \quad , \qquad (u-1)^2 + \left(v - \frac{1}{x}\right)^2 = \frac{1}{x^2} \quad . \tag{4.120}$$

For certain values of w ($= w_0 = const.$), the locus curves are shown in Fig. 4.52.

The center coordinates u_M and v_M and the radius R of the resulting circles in the r-plane are determined for the lines $w_0 = const.$ in the z-plane using the equations

$$u_M = \frac{w_0}{w_0 + 1} \quad , \qquad v_M = 0 \quad , \qquad R = \frac{1}{w_0 + 1} \quad . \tag{4.121}$$

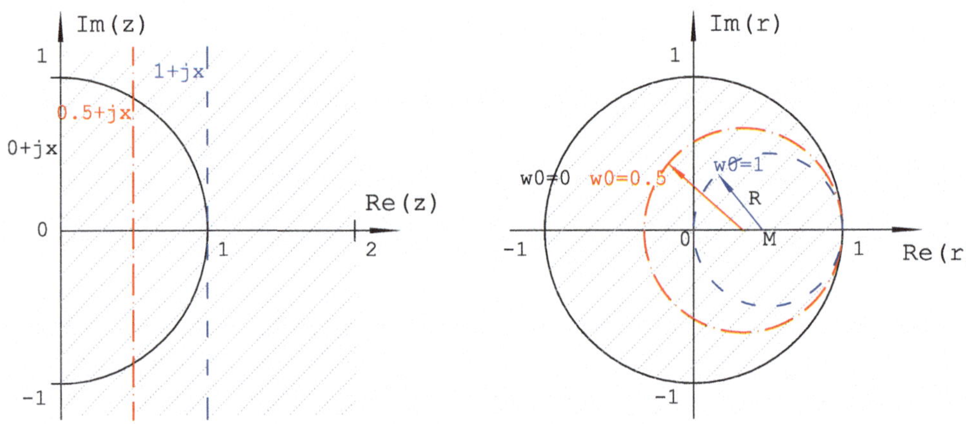

Fig. 4.52: Mapping of impedances to reflection values for constant real parts of z

For constant imaginary parts of z, the circles result with

$$u_M = 1 \quad , \qquad v_M = \frac{1}{x_0} \quad , \qquad R = \frac{1}{|x_0|} \quad . \tag{4.122}$$

These circles are also entered in the complete Smith chart in Fig. 4.53 and the constant values for x_0 are shown.

A Smith chart, as used for graphical solutions, is shown in the appendix as Fig. A.12.

It was shown how to construct auxiliary lines for impedances in the reflection plane. In the same way, auxiliary lines can be constructed for the admittances in the reflection plane. This results in an image that is mirrored on the Y-axis. The circle that passes through the zero point for the admittances is referred to below as the y-unit circle.

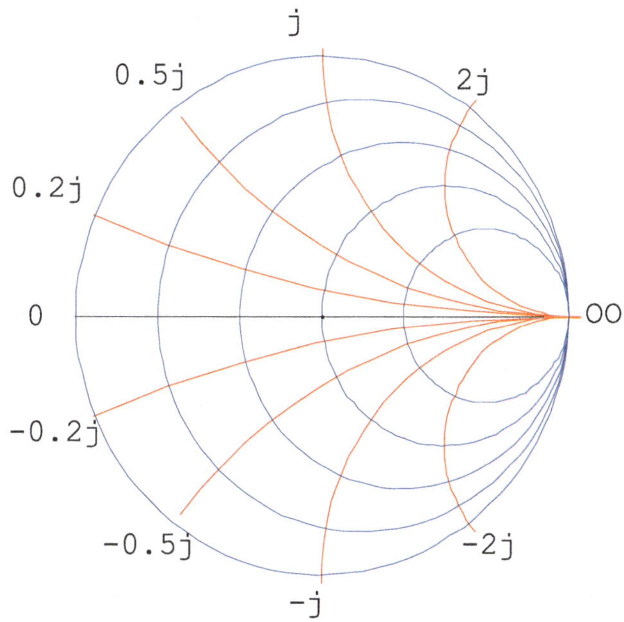

Fig. 4.53: Smith chart for constant imaginary parts of z (marked by values)

Circuit Synthesis with the Smith Chart

The Smith chart is a powerful tool for circuit synthesis, particularly in RF and microwave engineering. It allows for intuitive visualization of complex impedance transformations, making impedance matching and network design more efficient.

By using the chart, engineers can easily determine matching network components such as inductors and capacitors without extensive calculations. The normalized impedance and admittance representation simplifies the design of transmission line networks and matching circuits.

Additionally, the Smith chart aids in stability analysis and optimization, helping engineers create more efficient and reliable high-frequency circuits.

We will now explain in more detail why the Smith chart is a very important tool for synthesizing small transformation circuits. It also helps to understand the functionality of components.

In a transformation, a reflection value is to be converted into another value (often $r = 0$). If this is desired, you need to know which components change the reflection value and how.

These properties of the components are shown in Fig. 4.54 and in the corresponding auxiliary sheet A.11. The assignments can be found in the following list.

Apart from the series cable, all components rotate on circuits that end in the open circuit or short circuit point. The series line with $Z_L = Z_0$ rotates the reflection value on a circle whose center corresponds to the zero point of the Smith chart.

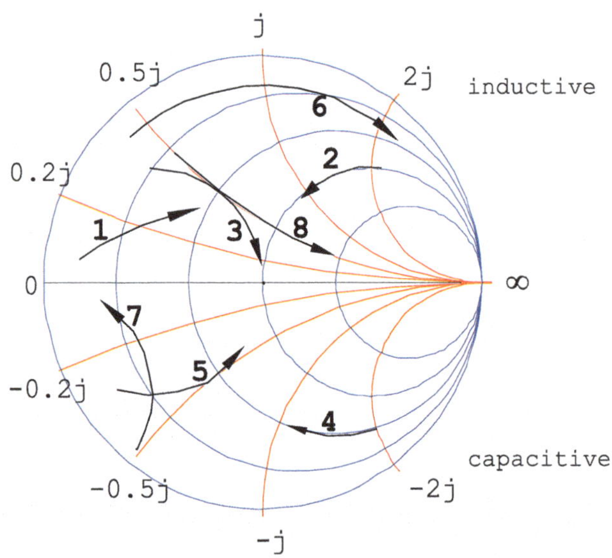

Fig. 4.54: Transformation behavior of various components

1. Transformator
2. Series-C
3. Shunt-C or open ended cable
4. Series-L
5. Shunt-L or short-circuited cable
6. Series line
7. Shunt-R
8. Series-R

The result in Fig. 4.55 should serve as an example of the usefulness of the Smith chart for circuit synthesis. The task is to transform an impedance with the normalized value

$$z_A = 0.4 - j\,0.6$$

to the matching point ($z = 1,\ r = 0$) for 600 MHz. The first element to be connected is a short-circuited stub line.

This is optimized so that the 50 Ω circuit[10]. In other words, it is normalized: $\mathrm{Re}\{z\} = 1 = const.$.

The result is a 50 Ω transmission line with an electrical length of 53.8 degrees. A series coil is then added and designed so that the matching point is reached.

For this coil, 7.27 nH are determined. The transformation path is shown in Fig. 4.55.

The resulting network is shown in Fig. 4.56.

[10]This is referred to below as the z-unit circuit.

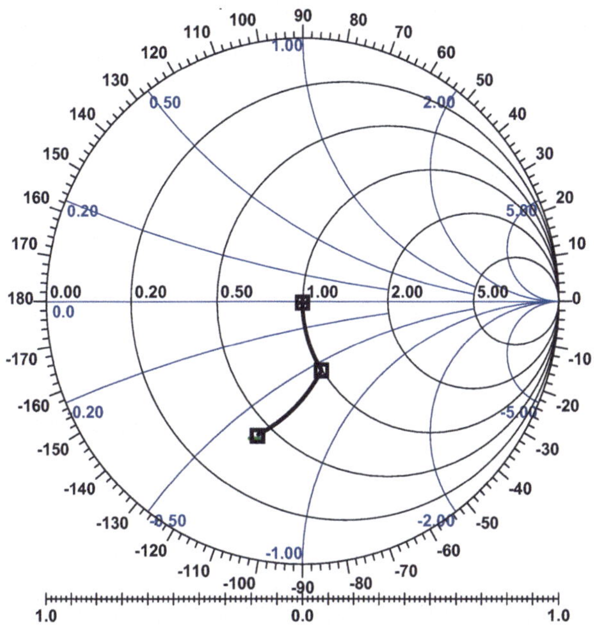

Fig. 4.55: Using the Smith tool from Serenade

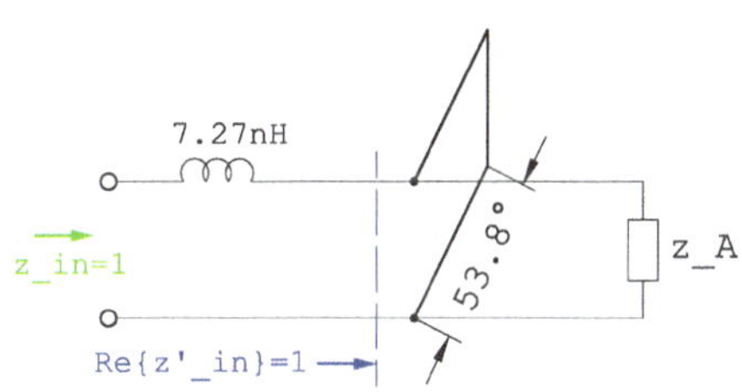

Fig. 4.56: Resulting network of the application example for the Smith chart

The 50 Ω-circle should be referred to as a z-unit circle and is usually used as an important intermediate step in the transformation, as in the example shown.

Analytical Approach I:

The analytical method is of particular interest for broadband adjustments. The analytical solution can be used to perform an optimization over the frequency using *Matlab*, for example.

The solution shown in Fig. 4.55 is also used for this solution: The first component is used to transform the arbitrary complex terminating impedance to the z- or y-unit circuit. This means that it applies to the input of the termination including the upstream component:

$$\mathrm{Re}\{z'_{in}\} = 1 \quad \text{or} \quad \mathrm{Re}\{y'_{in}\} = 1 \quad .$$

The remaining imaginary part is compensated for by means of a second component (components 2, 3, 4 and 5 can be used).

All you have to do is sketch a Smith chart that contains the outer unit circle (for r) and the two inner unit circles (for y and z) (see left-hand illustration in Fig. 4.57). The networks are unique as long as you are in one of the 8 segments of the right-hand illustration of Fig. 4.57).

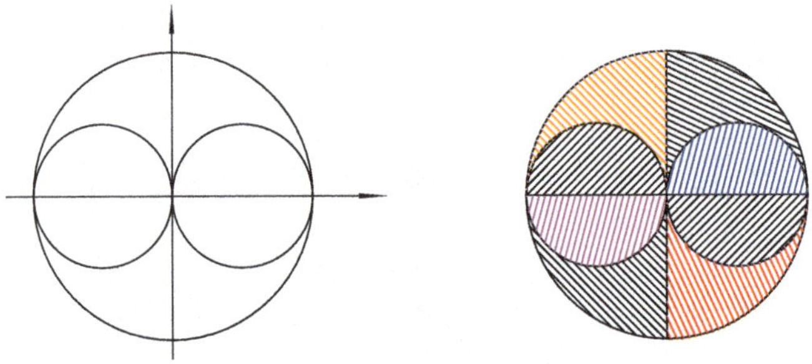

Fig. 4.57: Simple auxiliary images with y and z unit circle for working with the Smith chart

It is helpful to bear in mind that working on the y-unit circuit corresponds to setting up a parallel resonant circuit and working on the z-unit circuit corresponds to setting up a series resonant circuit, Fig. 4.58.

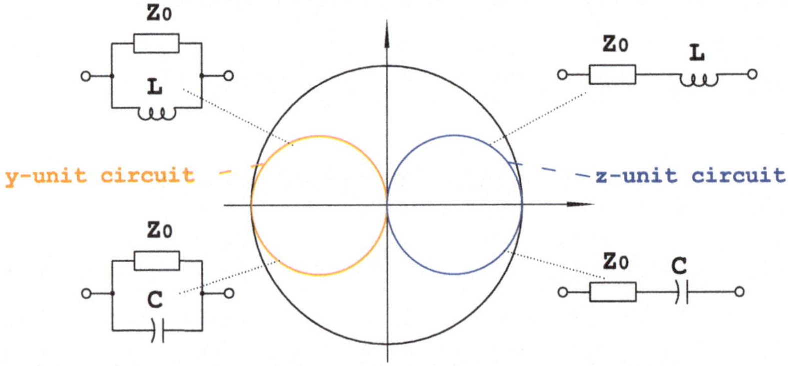

Fig. 4.58: Notes on the y- and z-unit circle in the Smith chart

Exercise: Matching

Given: *The so-called s2p data and thus the scattering parameters of a transistor for 1 GHz are given:*

```
#  GHz  S  MA  R  50
! f          S11          S21          S12          S22
! GHz     Mag    Ang    Mag    Ang    Mag    Ang    Mag    Ang
 1.00    0.56   -90    13.2   180    0.01   0.0    0.62   -90
```

Using lossless coils, the input and output of the transistor are to be transformed to the system impedance $Z_0 = 50\,\Omega$ at 1 GHz and thus an RF amplifier is to be realized.

Find:

(a) *Specify the two possible transformation networks and paths for the input and output.*

(b) *Calculate the inductance values for the input and output in the case that there is a shunt inductance at the RF amplifier ports (approximation: $S_{12} = 0$ and therefore the transistor is feedback-free).*

To a)

As S_{11} and S_{22} hardly differ, the same networks can be used. Fig. 4.59 shows the two solutions. Continue with the solution 2.

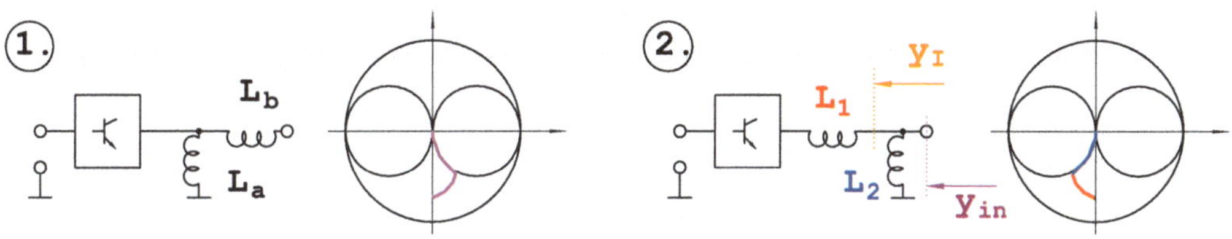

Fig. 4.59: Two possible transformation paths and networks with inductances

To b)

First, the input impedances are calculated:

$$z_A = \frac{Z_a}{Z_0} = \frac{1 + S_{ii}}{1 - S_{ii}} \qquad \text{delivers on} \tag{4.123}$$

Input: $z_A' = 0.523 - j0.853$ *and at the output:* $z_A'' = 0.445 - j0.896$.

Now the calculation of y_I.

$$z_I = j\omega\ell_1 + z_A \qquad \text{with} \qquad z_A = x_A + jb_A \qquad \text{and} \qquad \omega\ell_1 = \frac{\omega L_1}{Z_0} \tag{4.124}$$

$$z_I = j\underbrace{(\omega\ell_1 + b_A)}_{=k} + x_A \tag{4.125}$$

$$y_I = \frac{1}{x_A + jk} = \frac{x_A - jk}{x_A^2 + k^2} \equiv 1 + j\ \mathrm{Im}\{y_I\} \tag{4.126}$$

The real part of this Eq. (4.126) provides the values for L_1.

$$x_A = x_A^2 + k^2 \qquad \text{resp.} \qquad 0 = \omega^2\ell_1^2 + 2b_A\omega\ell_1 + x_A^2 - x_A + b_A^2 \tag{4.127}$$

$$\omega\ell_1 = -b_A \pm \sqrt{-x_A^2 + x_A} \quad \text{with a negative sign, since } b_A < 0 \tag{4.128}$$

and the shortest transformation path is sought.

Input: $L_1' = 2.81\,\mathrm{nH}$ *and at the output:* $L_1'' = 3.18\,\mathrm{nH}$.

The imaginary part of the Eq. (4.126) provides via

$$y_{in} = 1 \qquad \text{and} \qquad 1 = \frac{1}{j\omega\ell_2} + y_I \tag{4.129}$$

and

$$\frac{j\,k}{x_A^2 + k^2} = \frac{1}{j\omega\ell_2} \quad \Rightarrow \quad L_2 = -\frac{x_A^2 + k^2}{\omega\,k}\,Z_0 \tag{4.130}$$

the values for L_2.

Input: $L_2' = 8.33\,nH$ *and at the output:* $L_2'' = 7.12\,nH$.

——— *Exercise: Matching* ———

This way of calculating the matching is the classic way. The following second method is based on the Γ impedance transformer (Γ-trafo for short) introduced in Chapter 3.

Analytical Approach II:

The advantages and disadvantages of the second analytical way compared to first way:

Pro: More broadband, less calculation/computing effort;

Con: Only for fitting values within the unit circles more broadband and one component more.

This path has the simple condition as a first step:

$$\text{Im}\{z_{in}'\} = 0 \quad \text{or alternatively} \quad \text{Im}\{y_{in}'\} = 0 \quad .$$

The first component (z_1 or y_1) is calculated via the condition. The input impedance (z_{in}') is then purely real. To transform to $Z_{in} = Z_0$, a Γ-trafo is used in the second step.

The following Fig. 4.60 illustrates in the left half the case with a first component z_1, which is connected in series to the termination.

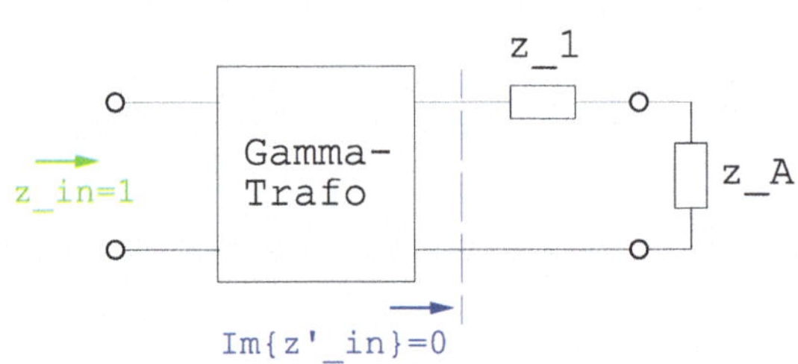

Fig. 4.60: General network for analytical path II with the Γ-transformer

Fig. 4.61 specifies two cases for the transformation of the first component under a. The left-hand case moves away from the fitting point (center point) and should therefore not be selected. Under b, the transformation (red) with a first component and the two possible Γ transformation paths are specified.

4.4 Quasi-Concentrated Transmission Line Components

In the frequency range from around 4 GHz to around 20 GHz, short microwave lines can be used to produce components that have a comparable electrical behavior to concentrated components. As these components have a finite and non-negligible length and width, these line components are referred to as quasi-concentrated elements. So far, only one

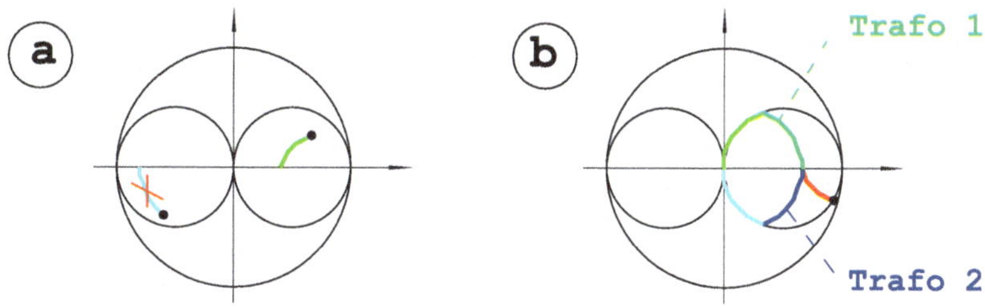

Fig. 4.61: Illustrations of analytical path II - on the left for the first component and on the right for all three components for the both possible Γ-transformers

shunt-L and one shunt-C have been presented in the derivations from Eq. (4.105) for the microstrip line while maintaining the mass. Now a series L and an alternative shunt C are to be derived.

The circuit design corresponds to the design with concentrated elements, but the component design is carried out with distributed lines. In the circuit simulation, the circuit can then only be optimized with distributed lines.

The following must apply for the derivation $\lambda \gg \ell$.

The approximate calculation for a high-impedance short line, which can be regarded as a quasi-concentrated coil, is shown below.

According to Eq. (4.98), the following applies to the input resistance of a line

$$Z_E = Z_A \frac{1 + j\frac{Z_L}{Z_A} \cdot \tan(\beta l)}{1 + j\frac{Z_A}{Z_L} \cdot \tan(\beta l)} \quad . \tag{4.131}$$

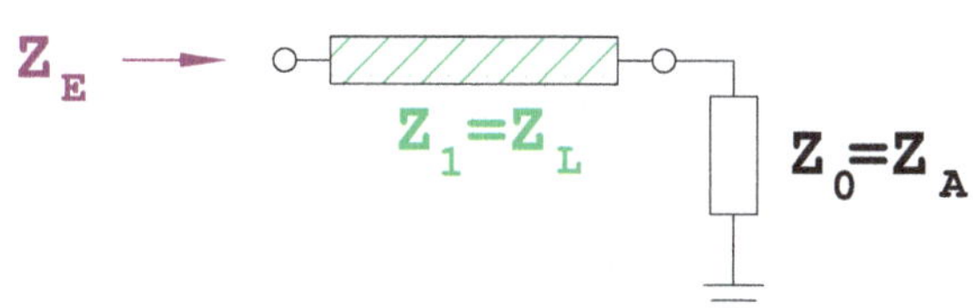

Fig. 4.62: Representation for calculating Z_E according to Eq. (4.131)

We now consider a short (high-impedance) line with the characteristic impedance $Z_L = Z_1$ and the length $l_1 = \ell_1$. For the termination, we use the system impedance Z_0 (usually: $Z_0 = 50\,\Omega$). Thus, the Eq. (4.131) can be converted into

$$Z_E = \frac{Z_0 \cdot \cos(\beta l_1) + jZ_1 \cdot \sin(\beta l_1)}{\cos(\beta l_1) + j\frac{Z_0}{Z_1} \cdot \sin(\beta l_1)} \quad . \tag{4.132}$$

Since $\beta l_1 = 2\pi l_1/\lambda \ll 1$ should apply, the approximations

$$\cos(\beta l_1) \approx 1 \quad , \qquad \sin(\beta l_1) \approx \beta l_1 \tag{4.133}$$

can be used. With these approximations, Eq. (4.132):

$$Z_E \approx \frac{Z_0 + jZ_1 \cdot \beta l_1}{1 + j\frac{Z_0}{Z_1} \cdot \beta l_1} \quad . \tag{4.134}$$

With the further approximation $\frac{1}{1+x} \approx 1 - x$, which is very well fulfilled for $x \ll 1$ with $Z_1 \gg Z_0$, it follows:

$$Z_E \approx (Z_0 + jZ_1 \cdot \beta l_1) \cdot (1 - j\frac{Z_0}{Z_1} \cdot \beta l_1) \quad . \tag{4.135}$$

Multiplied out results in:

$$Z_E \approx \underbrace{Z_0}_{=Z_A} + \underbrace{Z_0\,(\beta l_1)^2}_{\approx 0} + \underbrace{j\,Z_1 \cdot \beta l_1 \cdot (1 - \left(\frac{Z_0}{Z_1}\right)^2)}_{=\omega L_1} \quad . \tag{4.136}$$

This gives the input resistance Z_E of a series circuit consisting of an inductance L_1 and a resistor Z_0, which corresponds to the connection Z_A. The inductance can be calculated from this as follows:

$$L_1 = \frac{1}{\omega}\,Z_1 \cdot \frac{2\pi}{\lambda}\,l_1 \cdot \left[1 - \left(\frac{Z_0}{Z_1}\right)^2\right] = Z_1 \cdot \frac{\sqrt{\epsilon_r}\,l_1}{c_0} \cdot \left[1 - \left(\frac{Z_0}{Z_1}\right)^2\right] \tag{4.137}$$

with $\lambda = c_0/f/\sqrt{\epsilon_r}$. For practical use, you need Eq. (4.137) solved to Z_1:

$$Z_1 = \frac{L_1\,c_0}{2\,\sqrt{\epsilon_r}\,l_1} \pm \sqrt{\left(\frac{L_1\,c_0}{2\,\sqrt{\epsilon_r}\,l_1}\right)^2 - Z_0^2} \quad . \tag{4.138}$$

The sign decision is made using the criterion $Z_1 > Z_0$.

If the same calculation is carried out for a low-resistance cable Z_2 with the length l_2 to realize a capacitor C_2 which is connected to ground (shunt-C, shunt capacitance), the following result is obtained, [78]:

$$C_2 = \frac{1}{Z_2} \cdot \frac{\sqrt{\epsilon_r}\,l_2}{c_0} \cdot \left[1 - \left(\frac{Z_2}{Z_0}\right)^2\right] \quad . \tag{4.139}$$

For practical use, you need Eq. (4.139) solved to Z_2:

$$Z_2 = \frac{C_2\,Z_0^2\,c_0}{2\,\sqrt{\epsilon_r}\,l_2} \pm Z_0\,\sqrt{\left(\frac{C_2\,Z_0\,c_0}{2\,\sqrt{\epsilon_r}\,l_2}\right)^2 - 1} \quad . \tag{4.140}$$

The sign decision is made using the criterion $Z_2 < Z_0$.

The following example illustrates the extent to which the approximations are valid:

Given: $Z_0 = 50\,\Omega,\ p_l = 36°,\ f = 1\,\mathrm{GHz},\ Z_1 = 100\,\Omega,\ \epsilon_r = 1$

For these values, a microstrip line with a characteristic impedance of $100\,\Omega$ and a substrate height of 1 mm has the the following geometrical dimensions:

$$w = 1.6\,mm\ , \qquad l = 30\,mm \quad .$$

Following Eq. (4.137), you obtain for the inductance

$$\mathrm{L} = 7.5\,\mathrm{nH} \quad .$$

For verification and further discussion, the simulation of this approximated coil and that of an ideal coil is shown in the Fig. 4.63 and 4.64.

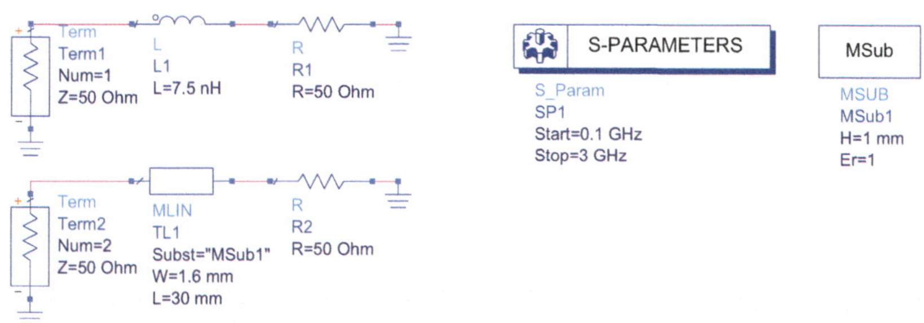

Fig. 4.63: Circuit diagram for the comparative simulation of an ideal and an approximated coil

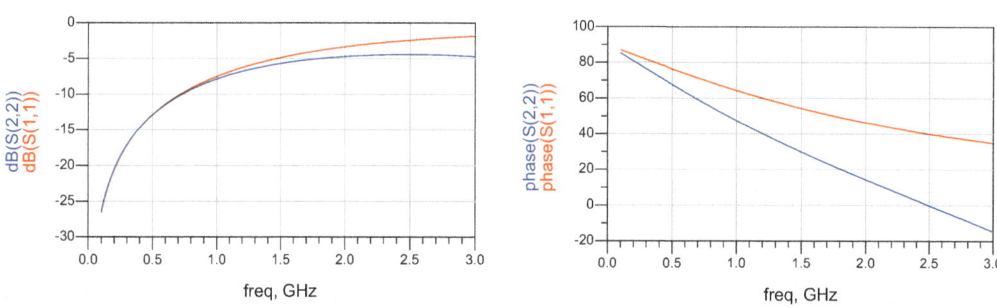

Fig. 4.64: Results of the comparative simulation of an ideal and an approximated coil

It can be seen that the magnitude values agree well up to frequencies of 1 GHz. However, there are already large differences in the phase values at this frequency.

The latter is the reason why the line elements must always be taken into account in optimizing circuit simulations!

An example of these quasi-concentrated components in a low-pass filter, which are very popular in practice, is shown in Fig. 4.65.

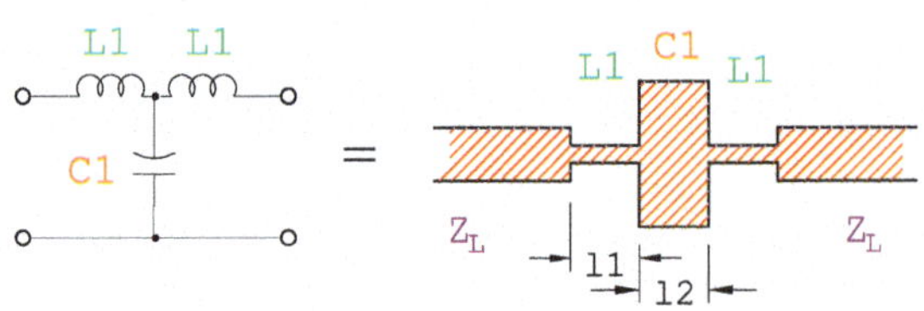

Fig. 4.65: Low-pass filter in microstrip line technology realized with quasi-concentrated components

Chapter 5

Circuit Theory and Synthesis with Common and Differential Mode Quantities

In the circuits and components discussed so far, only one electrical mode has been permitted. However, in many modern RF components and high-speed digital circuits in TEM systems in particular, two or even three modes occur. The different modes are distinguished by the electrical field patterns and possibly different propagation characteristics.

With the rise of mobile communications, the use of differential (push-pull) signals became more common due to their advantages in high-frequency signal transmission. Differential signaling reduces electromagnetic interference (EMI) and crosstalk, which is crucial in compact and densely packed mobile devices. It also improves signal integrity by minimizing the impact of noise and external interference.

However, differential circuits require balanced components and precise PCB layout, increasing design complexity. Additionally, differential signaling can lead to higher power consumption due to the need for additional circuitry like differential amplifiers and transformers. Despite these challenges, its benefits in noise immunity and signal robustness make it indispensable in modern RF and communication systems.

This chapter first explains in more detail which modes are involved. New "scattering parameters" are then introduced for the most important two- and three-mode systems. Based on these mixed-mode parameters (M-parameters for short), the so-called symmetry transformers as well as mode blockers and mode converters can be introduced in a simple manner.

The application of these M-parameters is then discussed. It is shown how differential components such as antennas and high-speed ICs are adapted. High-speed transceivers for digital signals are introduced. The use of so-called ECL ports in RF circuits is discussed. This is followed by the presentation of a new circuit synthesis technique for purely symmetrical networks. Based on these results, radio frequency coupler circuits, which are used in practice as pure four-port networks, can be precisely derived.

H. Heuermann, *Radio Frequency Technologies*,
https://doi.org/10.1007/978-3-658-50170-9_5

5.1 Introduction of the Mixed-Mode TEM-System

Radio frequency circuits have been built in unbalanced conductor systems in the last century, with a few exceptions. The most important lines are microstrip, coplanar, coaxial and stripline. These unbalanced cables all have a signal conductor and a ground segment.

On the other hand, symmetrical lines (e.g. parallel wire) were rarely used, as phenomena occur with these lines in practice that require detailed analysis. This problem is dealt with in detail in this chapter.

A major advantage of balanced lines and all balanced circuits over unbalanced systems is that electromagnetic radiation theoretically has no influence on signal transmission. Fig. 5.1 shows two planar lines and their electric fields under the influence of interference radiation.

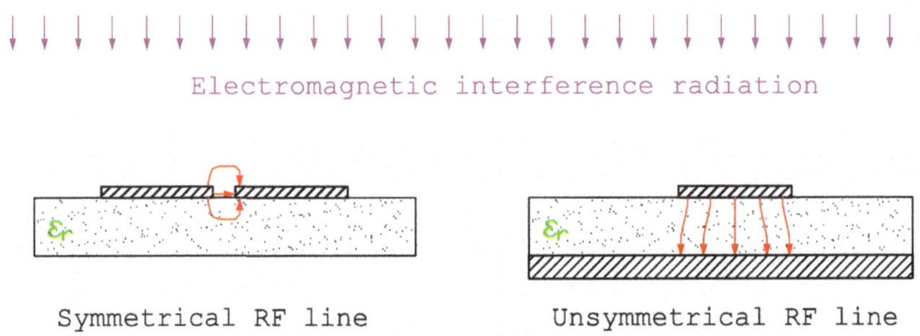

Fig. 5.1: Planar symmetrical and asymmetrical line with electric fields of the line modes and interference radiation

If the interference radiation is so great that it can induce a voltage of 100 mV in both conductor systems, then this voltage is added to the signal conductors. In the symmetrical system, the voltage of the forward and return conductors is increased by this 100 mV. However, this increase in potential has no effect on the signal characteristics!

In the unbalanced system, however, the transmission signal is raised by 100 mV to ground in this case, and the interferer is coupled into the signal in full magnitude.

Another advantage of symmetrical systems[1] is that they require voltage amplitudes (relative to ground) that are smaller by a factor of 2 to generate the required power. Although twice the number of transistors is required, this does not mean any real additional effort for integrated circuits[2].

The advantage is that with a maximum voltage amplitude U_{max} given by the semiconductor technology, which is proportional to the breakdown voltage, four times the output power P_{out} is achieved.

$$P_{out}^{unsym} = \frac{U_{max}^2}{Z_L} \quad , \qquad P_{out}^{sym} = \frac{(2 \cdot U_{max})^2}{Z_L} = 4 \cdot P_{out}^{unsym} \tag{5.1}$$

[1]In digital technology, the term differential systems is preferred.

[2]The important thing is that the space required to provide the power is almost the same.

These two advantages - the higher power yield and the higher immunity to interference - have ensured that balanced systems have been used almost exclusively in mass-produced RF electronics in recent years.

In the meantime, all semiconductor components for the front electronics in smartphones and for optical transmission systems are now only manufactured using symmetrical technology. The number of 4-port network analyzers sold, which can be used to measure symmetrical technology, already exceeds the number of two-port devices sold.

Narrow-band small signal components that operate in the higher GHz range, such as so-called Low Noise Converters (LNC) for receiving television satellite signals at 11 GHz, have a high interference suppression due to bandpass filtering and do not require power amplification. Such systems are still built using unbalanced technology.

5.1.1 Unbalanced Mode and Balance Mode in Two-Wire TEM Systems

In unbalanced two-wire TEM systems such as the microstrip line or the coaxial line, the only TEM mode that occurs is the `unbalanced mode`, which is closely related to the so-called `common mode`. Characteristic of this unbalanced mode is that a measurement of the signal (potential to ground or earth) only shows a result for the upper (signal) conductor of the microstrip line or only for the inner conductor of the coaxial line. The second conductor has the function of an earth and therefore has a potential of 0 V. Circuits for these unbalanced conductor systems are designed asymmetrically, as shown in Fig. 5.2 using the example of a transmission path with low-pass filtering.

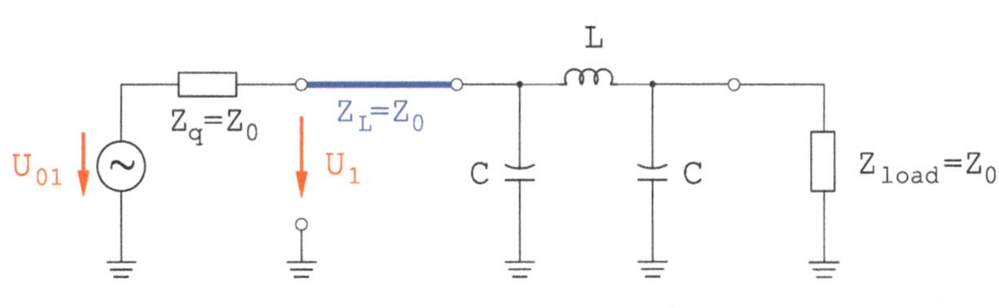

Fig. 5.2: Unbalanced two-wire system

A voltage meter would display the maximum value $U_{mess} = U_1 = U_{01}/2$ for Fig. 5.2 for a matched circuit at the input of the line.

In symmetrical two-wire systems, such as the parallel wire line, the only TEM mode that occurs is the so-called push-pull mode. This mode is characterized by the fact that a measurement of the signal (to ground or earth) gives a result on the upper conductor and on the lower conductor of the parallel wire line. Circuits for these balanced conductor systems are designed symmetrically, as illustrated in Fig. 5.3 using the example of a transmission path with low-pass filtering.

To prevent the potentials of the circuit from "drifting away", one point in the circuit is connected to earth in practice. This was indicated in Fig. 5.3 with a center tap of

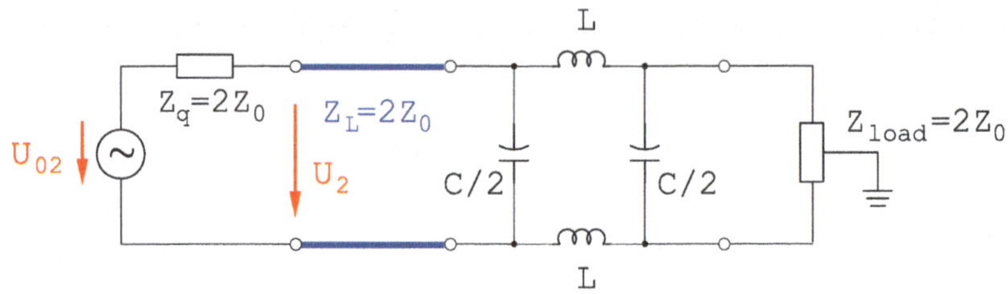

Fig. 5.3: Balanced two-wire system with push-pull mode via mirroring of the unbalanced circuit at the ground plane (practice: $Z_0 = 50\,\Omega$)

the load impedance. A voltage meter would display $U_{mess} = U_{02}/4$ as the maximum value for Fig. 5.3 for a matched circuit at the lower and upper input of the line, whereby $U_2 = U_{02}/2$ still applies.

It should be expressly emphasized that the two circuits according to Fig. 5.2 and Fig. 5.3 have the same transmission behavior and thus the same scattering parameters despite different voltages and input resistances. In practice, the scattering parameters of the unbalanced circuit are normalized to $50\,\Omega$ and those of the balanced circuit to $100\,\Omega$.

Any unbalanced circuit of the $50\,\Omega$ system can be mirrored at the ground plane and then immediately results in a balanced circuit (exactly a ground-balanced circuit) for a $100\,\Omega$ system. The ground connections can also be omitted, as shown in Fig. 5.3, which has the advantage that the shunt components can be combined to form a cross component (as was done with the capacitors).

A second way to create a balanced circuit directly from the design of an unbalanced circuit is shown in Fig. 5.4.

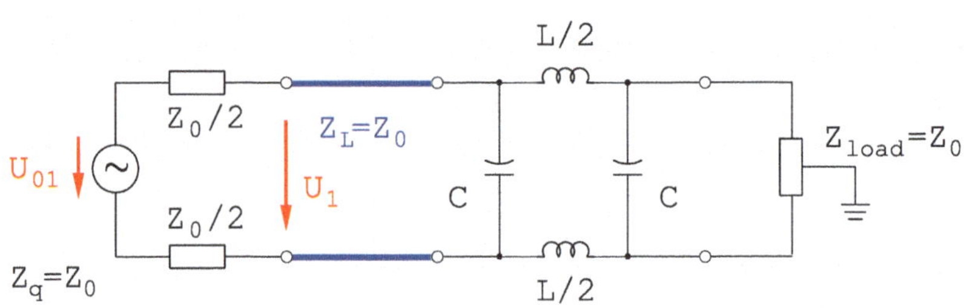

Fig. 5.4: Symmetrical two-wire system with push-pull mode (practice: $Z_0 = 100\,\Omega$)

The longitudinal or series components must be halved in the impedance value and implemented in the lower branch. This circuit behaves not only with regard to the transfer function, but also with regard to the current and voltage behavior, like the unbalanced circuit shown in Fig. 5.2.

Synthesis of Symmetrical Circuits

It is obvious that symmetrical circuits have more components with the same functionality. Balanced circuits are designed using the upper unbalanced half of a reference circuit with half the input impedance and half the line impedance. After designing the half circuit, the components are combined in the cross branches. This means that an unbalanced circuit according to Fig. 5.2 with fewer components and the same transmission behavior is used to calculate a balanced circuit according to Fig. 5.3. Alternatively, the unbalanced reference circuit can be designed for the same impedance and only the values of the series components need to be halved, as shown in Fig. 5.4.

5.1.2 Common and Differential Modes in Three-Wire Systems

The disadvantage of the symmetrical line is that in practice the conductor pair cannot be considered without an earth in the room.
If two wires are used for signal transmission in a device, there is usually a housing wall so that the parallel wire line system becomes a THREE-WIRE SYSTEM according to Fig. 5.5.

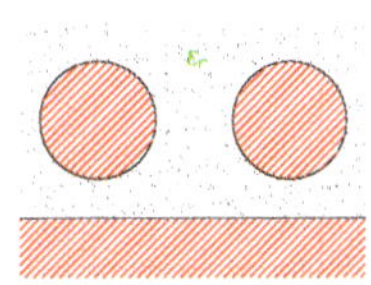

Fig. 5.5: Three-wire conductor: Two-wire conductor on an enclosure wall

In modern multilayer PCBs, ground layers are required to isolate different signals. In this case, symmetrical lines in inner layers are realized as two strip lines. Two microstrip lines are used on the top layer.

In practice, the use of two planar cables that have little influence on each other, a characteristic impedance of 50 Ω and the same electrical length has become established.

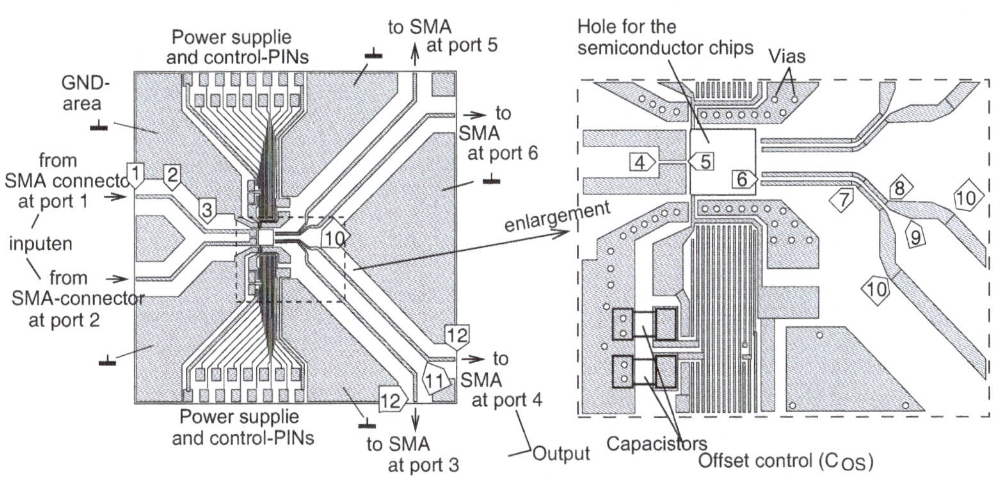

Fig. 5.6: Application board of a broadband amplifier in symmetrical microstrip technology

This is illustrated in Fig. 5.6, which shows the top view of an application board of a broadband amplifier for optical data transmission (13 Gbit/s) with one input and two outputs is shown, [95].

Two different TEM waves can propagate in a three-wire system, both of which must generally be taken into account:
En the other hand, the symmetrical conductor system fashion is spreading. This is the "'desired"' mode. This mode is also referred to as odd-mode.

On the other hand, a common mode is spreading. Energy that is fed into this conductor system as electromagnetic interference occurs as an even-mode. Both modes are shown

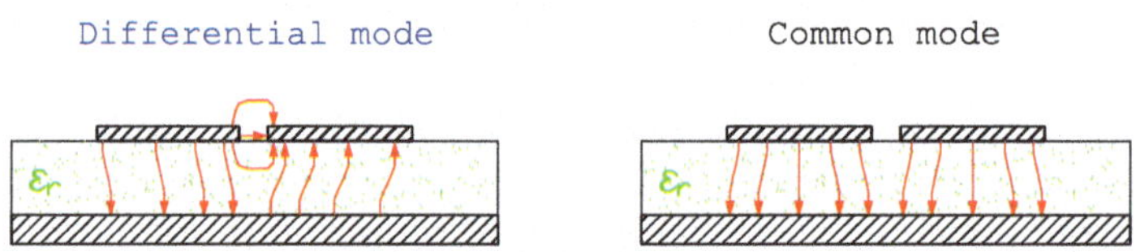

Fig. 5.7: Modes and electric fields in a planar symmetrical strip line scenario

for two closely adjacent microstrip lines in Fig. 5.7.

Both modes are quasi-TEM waves. However, as both modes are distributed differently in the air and in the dielectric, they generally have different effective permittivities ($\epsilon_{e_{eff}}$ and $\epsilon_{o_{eff}}$) and therefore different propagation constants.
For pure three-wire TEM waveguides (e.g. Fig. 5.5), the propagation constant of both modes is the same.

However, the characteristic impedances Z_{even} and Z_{odd} of both modes are almost always different. They can be calculated using the same methods as those presented in chapter 4 for TEM and quasi-TEM conductor systems.

Especially when using field simulators for their calculation, but also when working with components with symmetrical properties, it is very advantageous to use electric or magnetic walls.

Fig. 5.8 shows that "halving" the conductor system (using electric and magnetic walls) halves the push-pull impedance and doubles it of the common mode impedance.

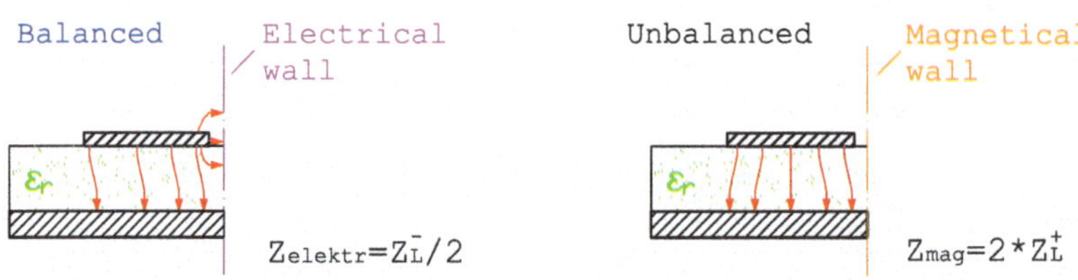

Fig. 5.8: Use of electric and magnetic walls when working with common-mode and differential-mode

By introducing the electric or magnetic wall, the "bisected" three-wire conductor, only the desired mode can propagate.

Furthermore, it is shown how a precise circuit design can be made despite the two modes at the input and/or output of RF circuits.
For circuit design and to perform accurate measurements, it is necessary to have a scattering parameter description of these two-mode systems. This scattering parameter description is presented in the following section.

5.2 Components with Three-Wire Systems at the Input and Output

Components with a balanced (or differential) input and a balanced output (such as the amplifier in Fig. 5.6) should only be operated with the push-pull waves a_i^- and b_i^-.

However, since the common mode at the inputs and outputs influences the electrical behavior, the wave sizes a_i^+ and b_i^+ of this mode must also be taken into account in the circuit design.

For push-pull signals, the voltage differences between two input and two output ports are considered. Here, the port is understood to be a two-wire line (in our example Fig. 5.6 a microstrip line). Two two-wire lines with a common earth are joined together to form a three-wire line. The two ports of the two-wire lines that are linked in this way are referred to as a `port pair` or `mixed-mode port` (MM-port for short).

To simplify the illustration, only one component with a balanced input port pair and a balanced output port pair will be considered below. The wave sizes at the input and output of the "two-port pair" can be linked again via scattering parameters. As the scattering parameters also describe properties between two different modes, these parameters are generally referred to as mode conversion parameters, mixed mode parameters or M-parameters for short. Fig. 5.9 illustrates such a two-port pair, which is also alternatively referred to as `MM-two-port` (short for mixed-mode two-port), [95].

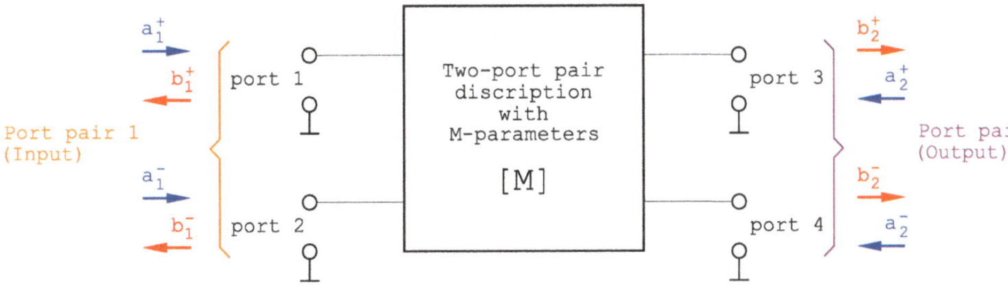

Fig. 5.9: Representation of a two-port pair (MM-two-port) with mode conversion parameters as well as common and differential mode waves

The M-parameters of the two-port pair are shown below as an M-matrix. The M-matrix is defined as follows:

$$\begin{pmatrix} b_1^- \\ b_2^- \\ -- \\ b_1^+ \\ b_2^+ \end{pmatrix} = \left[\begin{array}{cc|cc} M_{11}^- & M_{12}^- & M_{11}^{-+} & M_{12}^{-+} \\ M_{21}^- & M_{22}^- & M_{21}^{-+} & M_{22}^{-+} \\ \hline M_{11}^{+-} & M_{12}^{+-} & M_{11}^+ & M_{12}^+ \\ M_{21}^{+-} & M_{22}^{+-} & M_{21}^+ & M_{22}^+ \end{array}\right] \begin{pmatrix} a_1^- \\ a_2^- \\ -- \\ a_1^+ \\ a_2^+ \end{pmatrix} , \tag{5.2}$$

mit

$$M_{11}^- = \frac{b_1^-}{a_1^-}\Big|_{a_2^-=a_1^+=a_2^+=0} , \quad M_{21}^- = \frac{b_2^-}{a_1^-}\Big|_{a_2^-=a_1^+=a_2^+=0} , \quad M_{12}^{+-} = \frac{b_1^+}{a_2^-}\Big|_{a_1^-=a_1^+=a_2^+=0} \quad \text{usw..} \tag{5.3}$$

The M-parameters, which only have one value (+ or −) in the exponent, are so-called *intrinsic parameters.* Intrinsic parameters are the same as the scattering parameters described in chapter 2. These are therefore the reflection and transmission parameters for the common mode (+) or differential mode (−).

The M parameters that have two values (+− or −+) in the exponent are called *conversion parameters.* These parameters describe the reflection or transmission properties with simultaneous mode conversion. I.e. if the parameter M_{11}^{-+} has the value 0.7, then 50 % of the energy of a common mode wave incident at port pair 1 is reflected and converted into a push-pull wave.

Unfortunately, the spelling of the later introduced American nomenclature has become established, especially in measurement technology. In this notation, the + is replaced by a C or cm and the - by a D or dm. This results in the Eq. (5.2):

$$\begin{pmatrix} b_{dm1} \\ b_{dm2} \\ -- \\ b_{cm1} \\ b_{cm2} \end{pmatrix} = \left[\begin{array}{cc|cc} S_{DD11} & S_{DD12} & S_{DC11} & S_{DC12} \\ S_{DD21} & S_{DD22} & S_{DC21} & S_{DC22} \\ \hline S_{CD11} & S_{CD12} & S_{CC11} & S_{CC12} \\ S_{CD21} & S_{CD22} & S_{CC21} & S_{CC22} \end{array}\right] \begin{pmatrix} a_{dm1} \\ a_{dm2} \\ -- \\ a_{cm1} \\ a_{cm2} \end{pmatrix} . \tag{5.4}$$

This nomenclature is also used in the now published reference book on this topic [23]. In the following, however, the nomenclature of the M-parameters is used.

The use of signal flow graphs, which are presented in chapter 10 for scattering parameters, provides a clear illustration of the relationships between waves, modes and port pairs. In order to be able to introduce this form of representation here, the physically *not* existing DC and push-pull ports are used, as shown in Fig. 5.10.

The only difference to Fig 5.9 is the interpretation of the ports. Physically, the DC and push-pull waves propagate in the three-wire systems. In the following, these ports of the three-wire systems are to be identified as mixed-mode ports with the abbreviation MM-port and by labeling them with a number in the rectangular frame (see Fig. 5.10).

In abstract terms, one can imagine that the two wave types propagate in two different systems, which are coupled via the conversion parameters. For this general case, the signal flow graph[3] is shown in Fig. 5.11. This shows the common and differential mode system

[3]These are introduced in detail in the section 10.2.

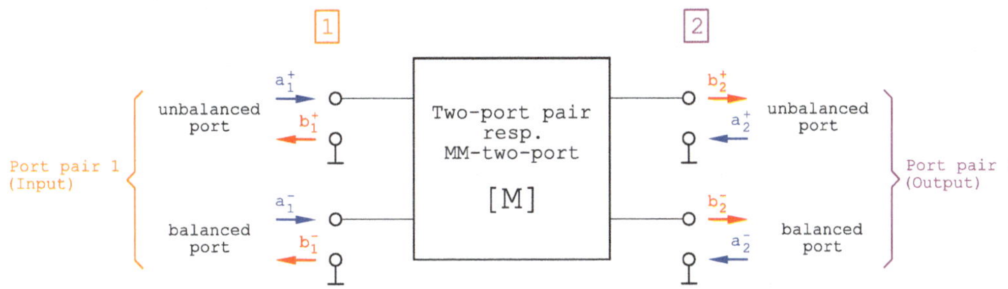

Fig. 5.10: Two-port pair or MM-two-port with abstract common and push-pull ports (mixed-mode-ports, MM-ports for short)

(green background), as well as the conversion system (diagonals, light background) in between.

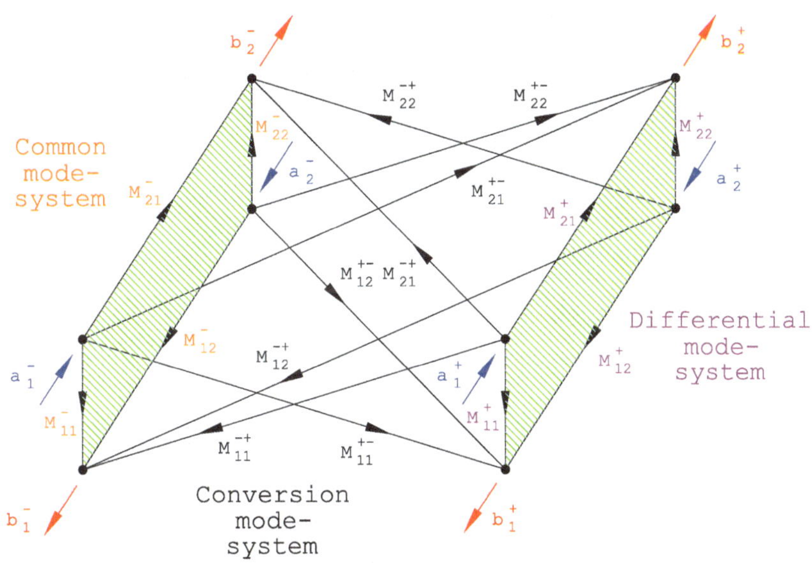

Fig. 5.11: Signal flow graph for a two-port pair with M-parameters with intrinsic mode parameters highlighted in green

If you have the freedom to keep the conversion parameters small in the circuit design, you will do so. Ideally, the two systems are separate, which is the case for a symmetrical two-port pair. The resulting simplified signal flow graph is shown in Fig. 5.12.

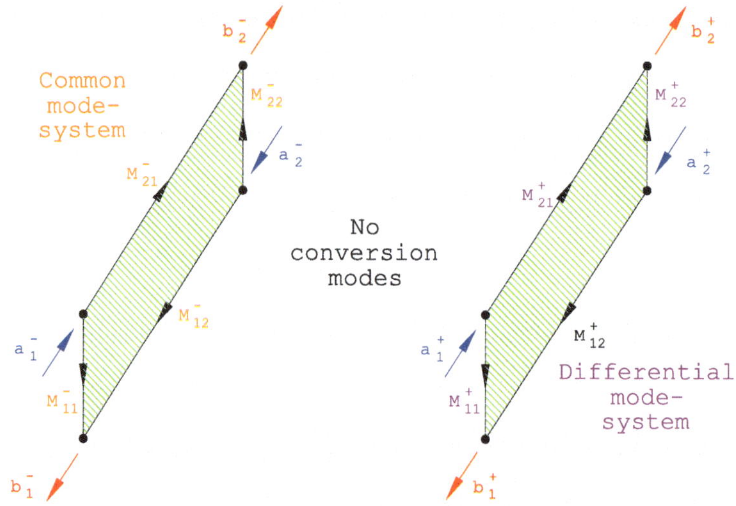

Fig. 5.12: Signal flow graph for a two-port pair without conversion parameters

Special Case: Three-Wire System Consisting of Two *Pure* Ground-Coupled Two-Wire Systems

It has already been pointed out at the beginning of this chapter that two microstrip lines, which are only coupled via the ground to form a three-wire system, have become established in practice (see Fig. 5.6).

We will limit ourselves to this special case, which is illustrated in Fig. 5.13, in this section.

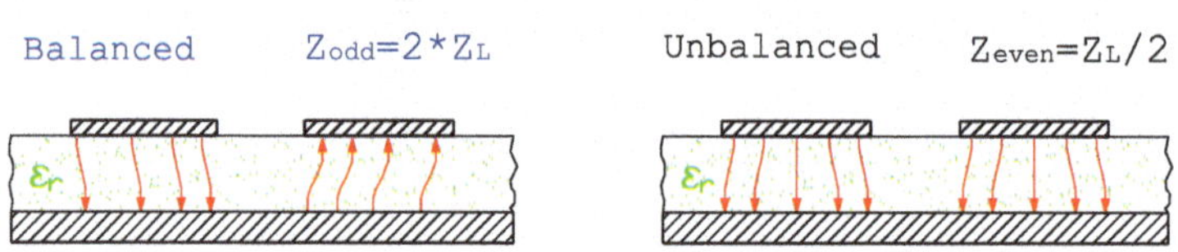

Fig. 5.13: Common mode and push-pull characteristic impedance for purely ground-coupled two-wire systems

This special case allows the network to be viewed mathematically and metrologically in a simple manner as a four-port with the scattering parameter matrix $[\mathbf{S}]$. Although the four-port scattering parameters do not physically provide any information about the transmission behavior, this approach helps to analyze and synthesize dual-mode systems using existing software and hardware tools based on scattering parameters.

Bild 5.14 shows the special two-port pair as a four-port. The dashed line is a reminder of the symmetry that exists in the ideal case.

In this special case, the electromagnetic waves of the four-port can be calculated from the waves of the common and differential mode wave and vice versa:

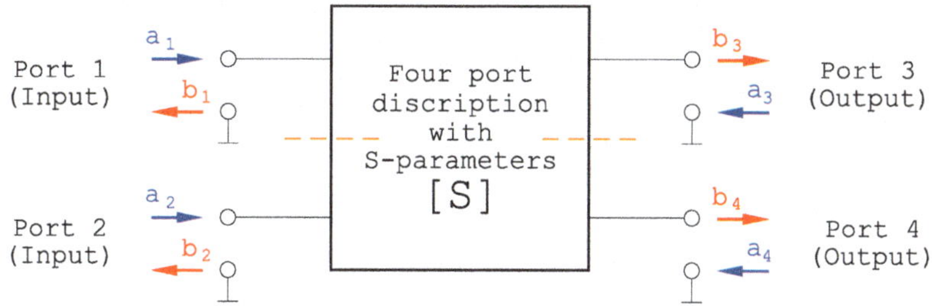

Fig. 5.14: Representation of a two-port pair with purely ground-coupled two-wire systems at the input and output as a four-port in scattering parameter nomenclature

$$a_1^+ = \frac{a_1+a_2}{\sqrt{2}}, \quad a_1^- = \frac{a_1-a_2}{\sqrt{2}}, \quad b_1^+ = \frac{b_1+b_2}{\sqrt{2}}, \quad b_1^- = \frac{b_1-b_2}{\sqrt{2}}, \tag{5.5}$$

$$a_2^+ = \frac{a_3+a_4}{\sqrt{2}}, \quad a_2^- = \frac{a_3-a_4}{\sqrt{2}}, \quad b_2^+ = \frac{b_3+b_4}{\sqrt{2}}, \quad b_2^- = \frac{b_3-b_4}{\sqrt{2}}, \tag{5.6}$$

$$a_1 = \frac{a_1^+ + a_1^-}{\sqrt{2}}, \quad a_2 = \frac{a_1^+ - a_1^-}{\sqrt{2}}, \quad b_1 = \frac{b_1^+ + b_1^-}{\sqrt{2}}, \quad b_2 = \frac{b_1^+ - b_1^-}{\sqrt{2}}, \tag{5.7}$$

$$a_3 = \frac{a_2^+ + a_2^-}{\sqrt{2}}, \quad a_4 = \frac{a_2^+ - a_2^-}{\sqrt{2}}, \quad b_3 = \frac{b_2^+ + b_2^-}{\sqrt{2}}, \quad b_4 = \frac{b_2^+ - b_2^-}{\sqrt{2}}. \tag{5.8}$$

Based on the linking of the electromagnetic waves of the asymmetric scattering parameter system with the equal and differential mode system, a conversion of the scattering parameter matrix $[\mathbf{S}]$ into the M- parameter matrix $[\mathbf{M}]$ can also be carried out[4].

[4]The derivation is similar to the shorter derivation between S- and M-parameters with three modes, which is printed in the following section.

Intrinsic parameter for the common mode:

$$M_{11}^{+} = \tfrac{1}{2}\left(S_{11} + S_{22} + S_{12} + S_{21}\right)\,, \quad M_{12}^{+} = \frac{1}{2}\left(S_{13} + S_{24} + S_{14} + S_{23}\right)\,, \tag{5.9}$$

$$M_{21}^{+} = \tfrac{1}{2}\left(S_{31} + S_{42} + S_{41} + S_{32}\right)\,, \quad M_{22}^{+} = \frac{1}{2}\left(S_{33} + S_{44} + S_{34} + S_{43}\right)\,. \tag{5.10}$$

Intrinsic parameter for differential mode:

$$M_{11}^{-} = \tfrac{1}{2}\left(S_{11} + S_{22} - S_{12} - S_{21}\right)\,, \quad M_{12}^{-} = \frac{1}{2}\left(S_{13} + S_{24} - S_{14} - S_{23}\right)\,, \tag{5.11}$$

$$M_{21}^{-} = \tfrac{1}{2}\left(S_{31} + S_{42} - S_{41} - S_{32}\right)\,, \quad M_{22}^{-} = \frac{1}{2}\left(S_{33} + S_{44} - S_{34} - S_{43}\right)\,. \tag{5.12}$$

Conversion parameters for differential mode to common mode:

$$M_{11}^{+-} = \tfrac{1}{2}\left(S_{11} - S_{22} - S_{12} + S_{21}\right)\,, \quad M_{12}^{+-} = \frac{1}{2}\left(S_{13} - S_{24} - S_{14} + S_{23}\right)\,, \tag{5.13}$$

$$M_{21}^{+-} = \tfrac{1}{2}\left(S_{31} - S_{42} + S_{41} - S_{32}\right)\,, \quad M_{22}^{+-} = \frac{1}{2}\left(S_{33} - S_{44} - S_{34} + S_{43}\right)\,. \tag{5.14}$$

Conversion parameters for common mode to differential mode:

$$M_{11}^{-+} = \tfrac{1}{2}\left(S_{11} - S_{22} + S_{12} - S_{21}\right)\,, \quad M_{12}^{-+} = \frac{1}{2}\left(S_{13} - S_{24} + S_{14} - S_{23}\right)\,, \tag{5.15}$$

$$M_{21}^{-+} = \tfrac{1}{2}\left(S_{31} - S_{42} - S_{41} + S_{32}\right)\,, \quad M_{22}^{-+} = \frac{1}{2}\left(S_{33} - S_{44} + S_{34} - S_{43}\right)\,. \tag{5.16}$$

The same statements apply to the M-parameters with regard to fitting, losslessness reciprocity and symmetry as we have already made in the Chap. 2 (introduction of the scattering parameters). However, there are additional points to consider.

Matching

With regard to matching: $\quad M_{ii}^{*} \to 0 \quad$ with $\quad * = -, +, -+, +-\ .$

With regard to matching, the most common optimization variable is: $\quad M_{11}^{-} \to 0\ .$

The Eq. (5.11) provides four possible scenarios, including the reciprocity that often exists in practice:

1. $S_{11} = S_{21} = S_{12} = S_{22} = 0 \quad ,$
2. $S_{11} = S_{21} = S_{12} = S_{22} \quad ,$
3. $S_{11} = -S_{22}$ and $S_{12} = S_{21} = 0 \quad ,$
4. $S_{11} + S_{22} = S_{12} + S_{21} \quad .$

An important special case that can be derived from this conversion for reciprocal components states that a matching for the common and push-pull mode of the two-ports i and j can only be realized if the following applies: $S_{ii} = S_{ij} = S_{ji} = S_{jj} = 0$. This relationship is used in the following to calculate couplers (four-port circuits).

Reciprocity

The following applies with regard to reciprocity: $M^*_{21} = M^*_{12}$
with $* = -, +, -+, +-$.

Reciprocity can also be easily checked using the conversion with the scattering parameters. For example, the M_{ij} parameters (for $j \neq i$) are calculated exclusively from transmission scattering parameters, all of which are reciprocal for a passive and isotropic network.

Symmetry

The following applies with regard to symmetry: $M^*_{11} = M^*_{22}$
with $* = -, +, -+, +-$.

In general, symmetry significantly reduces the number of unknowns in the M-matrix. Further details can be found in chapter 2.4.1. The symmetry of normal two-ports allows a very simplified circuit synthesis, see chapter 5.4. The same can also be done with the M-parameters, which can lead to new results in many cases.

Losses

Just as with the S-parameters, you can check the loss-free nature of the M-parameters from the squared values of a row or a column of the M-matrix, which must add up to 1. The following applies to the losslessness, for example

$$\left|M^{-}_{21}\right|^2 = 1 - \left|M^{-}_{11}\right|^2 - \left|M^{+-}_{11}\right|^2 - \left|M^{+-}_{21}\right|^2 \quad . \tag{5.17}$$

Mode Conversion and Semi-Symmetry

The semi-symmetry was already indicated in Fig. 5.14, which is emphasized again in Fig. 5.15.

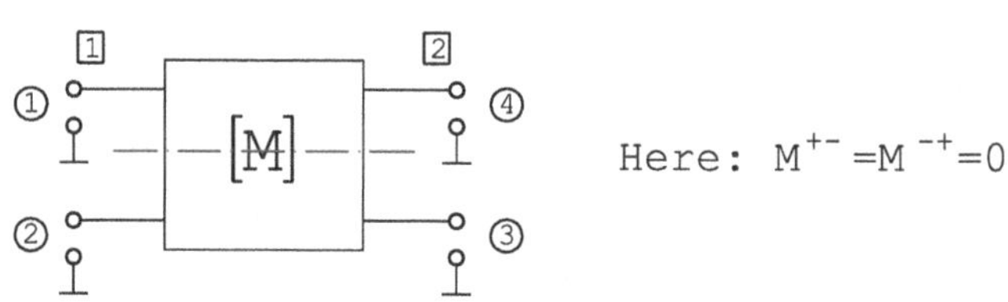

Fig. 5.15: MM-circuit with symmetry in the upper and lower branches

For such circuits, where the upper and lower branches are identical, the mode conversion parameters (all $M^{+-}-$ and $M^{-+}-$ values) are zero. This case is always aimed for in practice. The conversion parameters are therefore only observed in the measurement technology to ensure that they are below -20 dB and therefore no noticeable losses occur due to mode conversion.

Double Symmetry

Many networks exhibit the semi-symmetry described above and also the double symmetry described above. These networks are described in more detail in chapter 5.4.2.

Components in RF Circuit Technology

If you want to design a circuit in a balanced line system and the ground plane is in the immediate vicinity, the M-parameters must be used. If these parameters are not used at least for the calculation of the symmetrical properties, then any circuit simulation will generally fail. Ideally, all circuit components should be simulated in this three-wire system. In practice, this is suppressed by assuming a non-existent mode conversion.

The M-parameters are very important in measurement technology for verifying the developed circuits and components. The M-parameters provide information on the extent to which a component converts energy into the common mode.

This representation in M-parameters thus runs through all the components of RF technology that are already known for unbalanced circuits (mono-mode systems). In addition, there are other mixed-mode components that are not present in mono-mode systems. These are presented in the 5.2.3 chapter.

——— *Exercise: Dual-Mode Cable* ———

Given: *Given two lossless 50 Ω microstrip lines with an electrical length of 180° at 1 GHz (Fig. 5.16).*

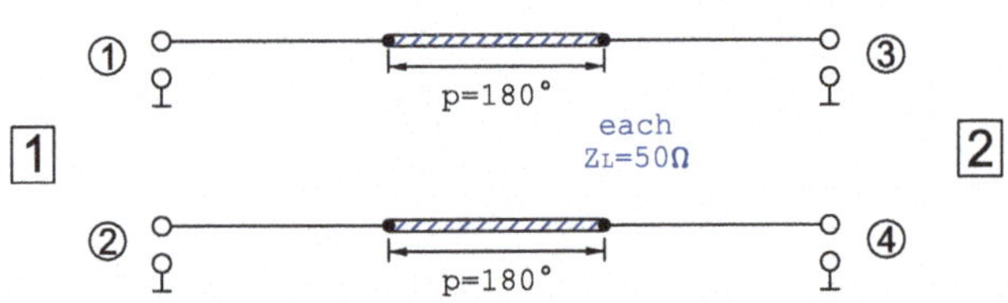

Fig. 5.16: Construction of a dual-mode cable from two microstrip cables

Find: *Specify the M-matrix of this duale-mode line for 1 GHz.*

Calculation:

The S-parameters of these two decoupled lines (line pair) are as follows:

$$[\mathbf{S_{Linepair}}] = \begin{bmatrix} 0 & 0 & -1 & 0 \\ 0 & 0 & 0 & -1 \\ -1 & 0 & 0 & 0 \\ 0 & -1 & 0 & 0 \end{bmatrix} \quad . \tag{5.18}$$

For the S-parameters, it can be seen that the reflection variables are all 0, the transmission variables $S_{31}, S_{13}, S_{42}, S_{24}$ are -1 for 180° phase rotation and the remaining transmission variables are 0 due to the isolation.

The M-parameters can be calculated from the S-parameters using the Eqs. (5.9-5.15).

$$[\mathbf{M_{DM-line}}] = \left[\begin{array}{cc|cc} 0 & -1 & 0 & 0 \\ -1 & 0 & 0 & 0 \\ \hline 0 & 0 & 0 & -1 \\ 0 & 0 & -1 & 0 \end{array}\right] \quad . \tag{5.19}$$

The result is that both the common mode and the differential mode are adapted. Both signals are rotated by 180° in the transmission phase. Mode conversion does not take place. This means that all mode conversion parameters (all $M^{+-}-$ and $M^{-+}-$ values) are also 0.

——— *Exercise: Dual-Mode Cable* ———

5.2.1 Transceiver for Digital Data Transmission

LVDS (Low Voltage Differential Signaling) has been the standard for digital transmission technology for many years.
LVDS is characterized by the following features:

- differential signal transmission,
- up to 3.25 GBit/s,
- small voltage signals ($\pm 350\,mV$),
- matched, at the end.

LVDS is an IEEE standard and is defined in IEEE Standard 1596.3-1996. We recommend [102] as basic documentation. The two new technologies LVPECL and CML, which allow transmission rates of up to 10 GBit/s, are also presented here.

Increasingly, the various new LV standards are also being differentiated based on their circuit technologies such as LVNECL, LVCMOS and LVTTL. It should be noted that ICs of these different technologies can generally be combined, but different offset values must be used, particularly with regard to DC voltage settings. These are hardware transceivers that can be used to build bus systems.

The standards for digital bus and data transmission systems for high-speed signals including their protocols are

> PCI-Express (full: Peripheral Component Interconnect Express) and HyperTransport.

However, both are now exclusively matched to the system impedance on both sides.

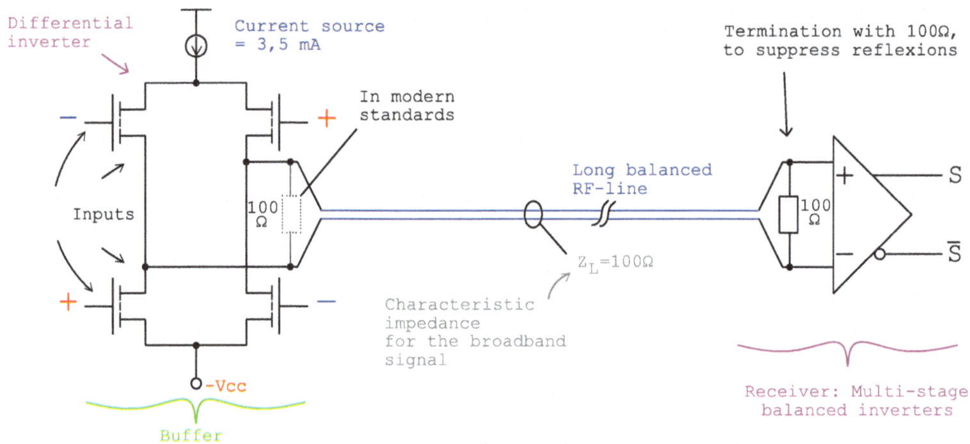

Fig. 5.17: Setup of a unidirectional transmission path

Transmission Line Architectures

A unidirectional transmit-receive set-up shows the Fig. 5.17.

Ideally, it is assumed that the CMOS transistors are ideal switches and that they are terminated with resistive 100 Ω resistors.

The bidirectional, so-called half-duplex architecture is illustrated in Fig. 5.18.

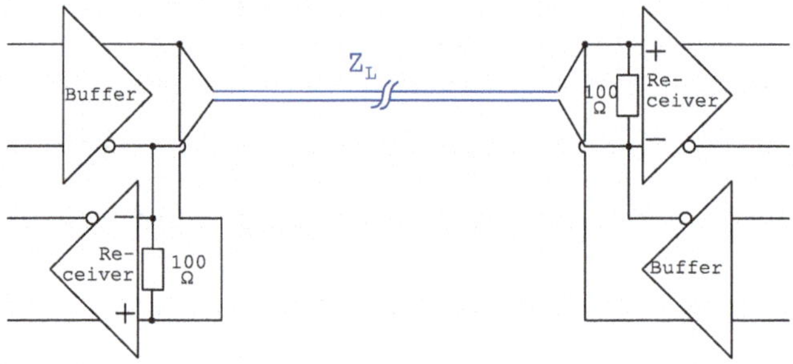

Fig. 5.18: Bidirectional transmission of digital signals in the RF range

With this half-duplex architecture, transmission is only possible in the forward or reverse direction.

In digital technology, there are still receivers for tapping lines with high impedance. The associated multidrop architecture is shown in Fig. 5.19.

Here, all connection lengths on the cables must be extremely short and the input impedance as high as possible (input capacitance therefore as low as possible).

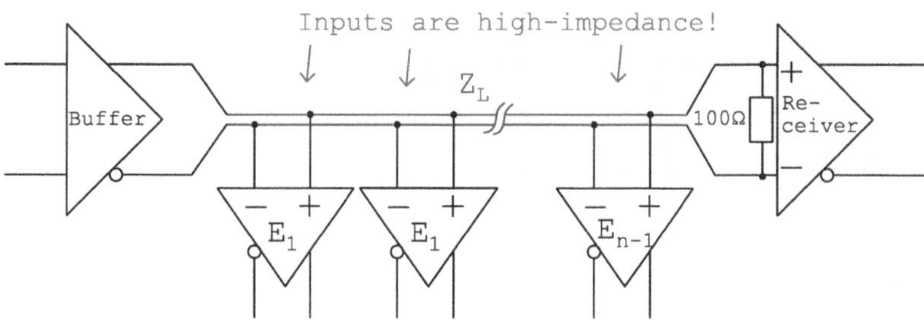

Fig. 5.19: Multidrop transmission with one-sided termination

5.2.2 Differential Transmission Line Technology for Digital Data Transmission

In the field of electronic components for high-speed data transmission, the state of the art is so advanced that bus systems for $n{\cdot}25\,\text{GBit/s}$[5] are available. These are offered by the leading processor manufacturers for so-called "chip to chip" (short C2C) connections.

Modern PC motherboards incorporate a variety of bus systems whose data rates vary significantly depending on the interface and technological standard. For instance, PCI Express Gen 5 offers up to 32 gigatransfers per second per lane, which, in a typical x16 configuration, translates into a theoretical bandwidth of roughly 64 GB/s in each direction. USB interfaces, such as USB 3.2 Gen 2x2, can achieve data rates of up to 20 Gbit/s, while Thunderbolt 3 is capable of transmitting up to 40 Gbit/s.

In contrast, SATA III-commonly used for storage devices provides a maximum throughput of 6 Gbit/s, although actual speeds are typically somewhat lower due to protocol overheads. Overall, these advanced bus systems enable data transfers ranging from a few gigabits per second to several dozen gigabytes per second, thereby meeting the demands of high-performance computing and modern digital applications.

The transmitting and receiving electronics allow even higher data rates. The data rates are limited by the fact that only the two "classic" conductor systems microstrip line and stripline are used.

The current state of research is two coupled microstrip lines on the top layer and two coupled strip lines (Fig. 5.20) in inner layers.

The frequently used microstrip line in particular has a highly dispersive behavior, see section 4.2.4, page 88. The phenomena of a single microstrip line have been researched over many decades [45, 47, 85, 129, 148, 149]. This dispersive behavior means that signal components with different spectral distributions also have different propagation times and therefore cause bit errors. In the GHz-range, microstrip lines are generally only used with a small relative bandwidth and therefore provide satisfactory properties.

Characteristic impedances of 100 Ω for the differential useful signal (push-pull signal) have established themselves as the state of the art.

[5] n: number of differential data line pairs.

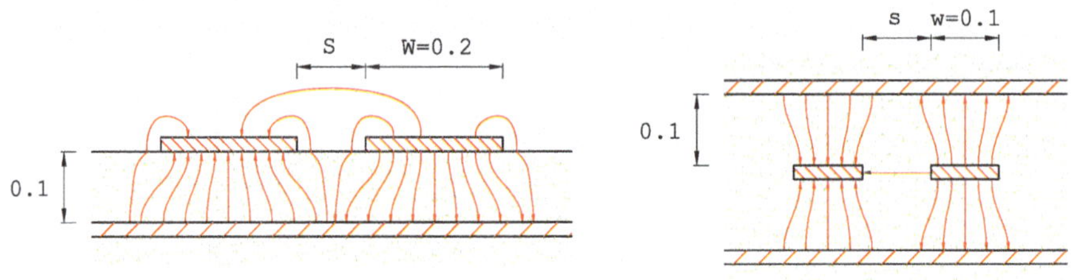

Fig. 5.20: Planar data line for differential signal transmission consisting of two coupled microstrip lines (left) and two coupled strip lines (right) with the electrical field diagram for the push-pull wave

The dimensions in the Fig. 5.20 are typical values for the push-pull characteristic impedance of 100 Ω and the standard material (FR4). These currently used cables are far from optimal for use in broadband differential data transmission. The points of criticism are that the transmission lines

1. cause too many bit errors due to the dispersive behavior and
2. therefore do not allow high data rates and
3. require too much space and
4. have too low a characteristic impedance for the common-mode waves.

Other planar line systems such as the coplanar line, the slotted line [146, 48] and many more are completely unsuitable for the realization of differential data lines, as they require a lot of space and are often difficult to connect.

New conductor systems must therefore also be included. For example, as an improved double microstrip line, one can imagine a symmetrical line with medium mass as shown in Fig. 5.21.

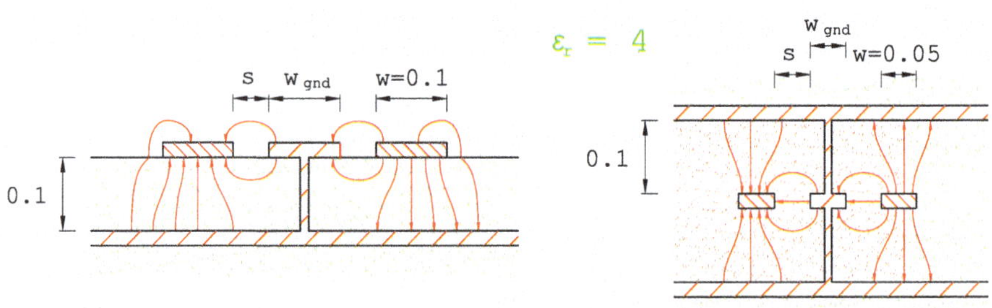

Fig. 5.21: Modified microstrip line (left) and stripline (right) with the electric field of the differential wave and typical dimensions

A double microstrip line modified in this way would potentially have the following advantages:

- less coupling between several differential data lines.
- smaller geometric dimensions with the same push-pull characteristic resistances.
- lower sensitivity to common mode interference.
- less electromagnetic radiation.
- lower dispersive behavior.
- shorter electrical length.

A slight improvement to the double stripline is also possible. Here too, an internal ground, as illustrated in Fig. 5.21, has positive effects in the form of lower coupling between several differential data lines and smaller dimensions with the same push-pull wave resistances.

Paired strips (section 4.2.2) are also conceivable as internal cables. These have the additional advantage of interference-free radial routing due to the principle, but also the disadvantage that these cables require an additional layer.

Numerical electromagnetic simulations are used in practice to investigate individual interference scenarios. The list of possible interference with data lines in computers is as long as you like. The Fig. 5.22 is intended to visualize some of the largest sources of interference for data lines.

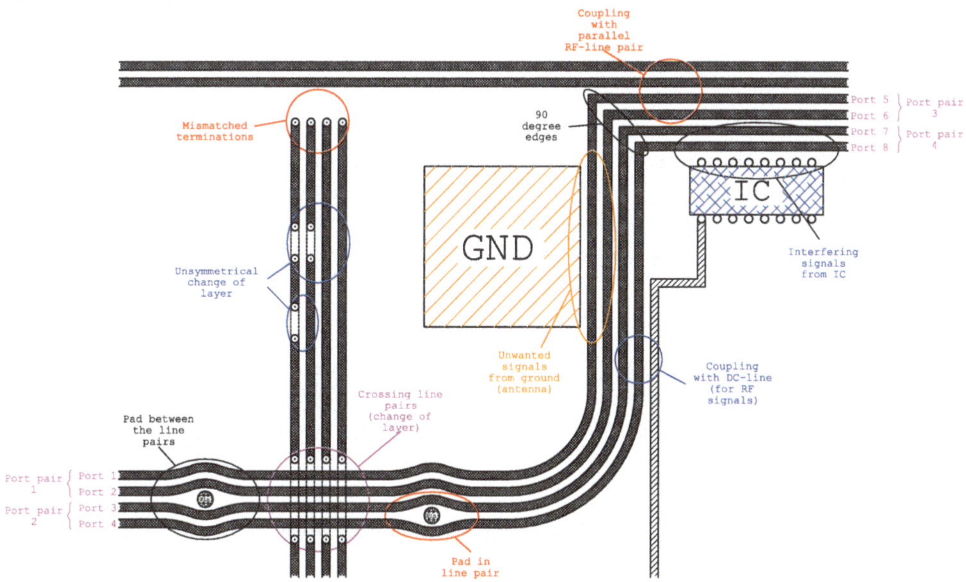

Fig. 5.22: Various interferences on the differential data lines in the transmission of fast broadband signals

These problems, which can be optimized using numerical tools, also include the transitions between different FR4 layers.

The most important measure for layout optimization for corners is optimization from the same electrical length, Fig. 5.23.

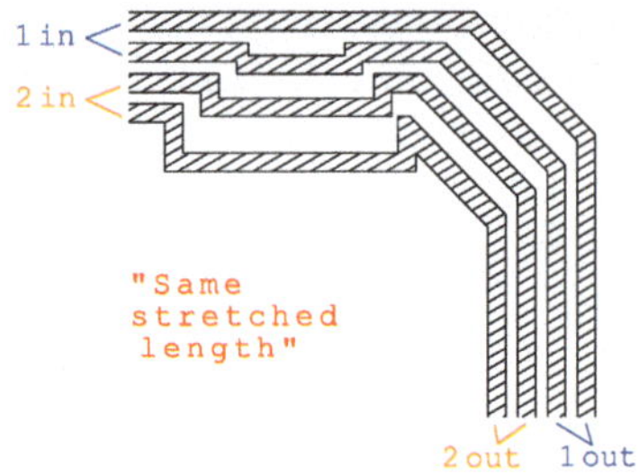

Fig. 5.23: Routing differential strip line around corners

5.2.3 Mode Blocker

Particularly in modern, highly integrated symmetrical circuits that operate at low operating voltages, great importance is attached to suppressing a common mode.

In the following, structures are presented that suppress unwanted modes by mismatching as components, or even just as elements of a component. At least at center frequency, none of the presented mode blockers exhibit mode conversion. For this reason, they will not be discussed below.

Among other things, this chapter teaches you how to deal with the common and differential modes. In general, the push-pull mode can be suppressed by inserting a short circuit between the conductors at the port. The common mode can be suppressed by removing the ground plane. Both are shown in the Fig. 5.24.

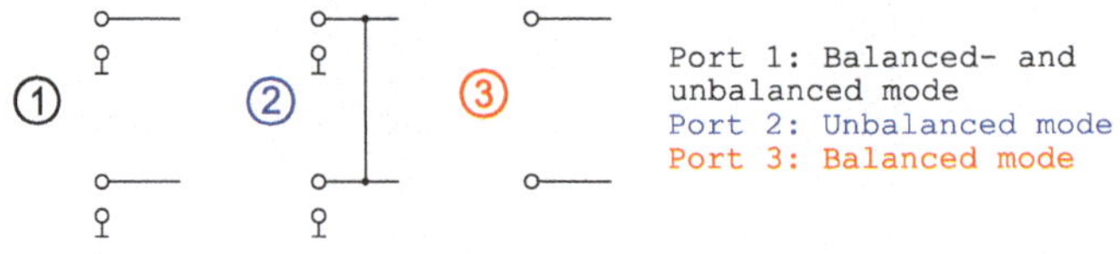

Fig. 5.24: Illustration of the mixed mode port and the measures for mode suppression

Broadband Mode Blockers based on Transformers

Transformers are excellently suited as broadband mode blockers in circuit simulation. In contrast, real transformers in the GHz range often have inadequate electrical properties due to their very large parasitic dirt effects. In addition, they are expensive and extremely large.

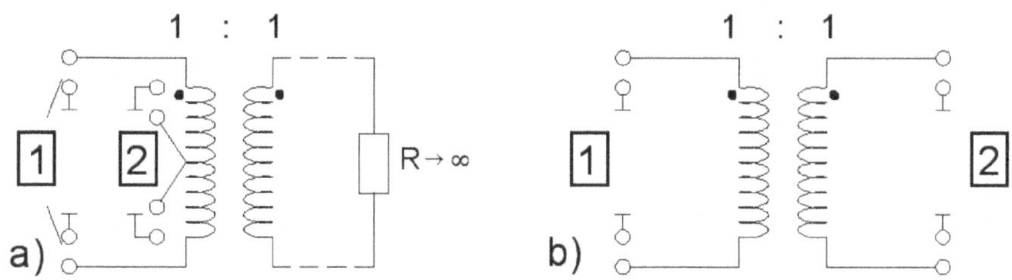

Fig. 5.25: Mode blocker based on ideal transformers: a) Differential mode blocker, b) Common-mode mode blocker

The Fig. 5.25 illustrates a common mode and a differential mode blocker.

In the differential blocker, the output with the MM-port 2 is located at the virtual midpoint of the lines. In general, an impedance connected at this point has no influence on the differential signal. It is therefore also possible that an MM-port is located at this midpoint. The push-pull signal is not present at this point and therefore cannot be transmitted. The push-pull signal only sees the impedance connected to the second winding. If, as shown in Fig. 5.25 a), a resistor is used whose resistance value approaches infinity, the second currentless winding could also be omitted and in this case the reflection factor $M_{ii}^{-} = 1$ is obtained. A finite resistance has no effect whatsoever on the matching and transmission of the common mode signal, $M_{ii}^{+} = 0\,,\ M_{ij}^{+} = 1.$[6]

The common mode blocker in part b) of the Fig. 5.25 in the form of a classic transformer is well known. The common mode signal is reflected with the reflection factor $M_{ii}^{+} = 1$.

Mode Blocker based on Strip Lines

Transmission lines in the GHz range can be realized with very good quality on PCBs or can also be simulated as phase shifters using SMDs or in integrated circuits.

The Fig. 5.26 illustrates one common and one differential mode blocker.

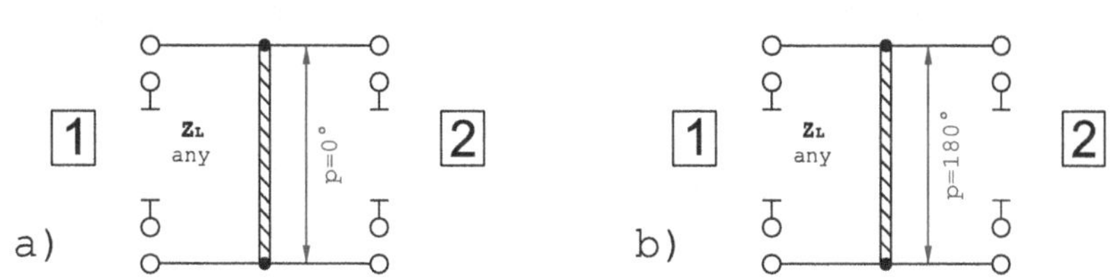

Fig. 5.26: Mode blocker based on lines: a) common mode blocker, b) differential mode blocker

In the differential-mode blocker, the line length is 0°, which is why it is also referred to as a 0°-blocker. This means that all signal conductors are directly short-circuited.

[6]Unless otherwise defined, the following should always apply: i and j can take the values 1 or 2 and $i \neq j$.

This blocker is very easy to implement in practice and works over an extremely wide bandwidth. A short circuit occurs in the junction for the push-pull signal ($M_{ii}^{-} = -1$). The common mode signal is not disturbed in any way. The excellent properties of this push-pull blocker can be used in push-pull systems by converting the useful signal locally into a common mode signal, filtering it with this blocker and converting it back again, [57]. The mode converters required for this are described in the next section.

Furthermore, this simple mode blocker is an important subcomponent in novel dual-mode resonator circuits, which are described in detail in chapter 5.

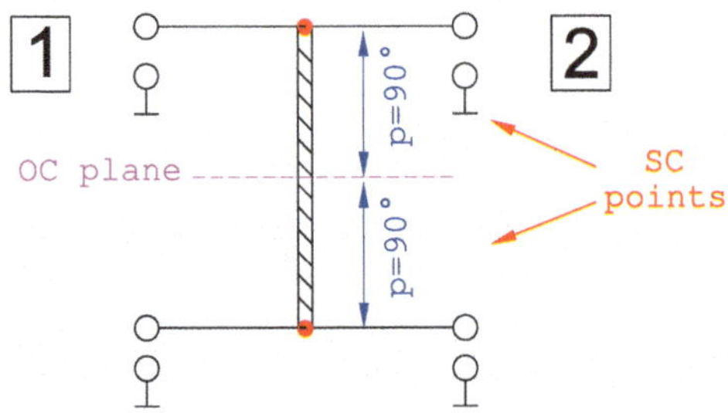

Fig. 5.27: Common-mode mode blocker based on lines with OC level: Open-Circuit level and Short Circuit points

The common-mode blocker according to Fig. 5.26 b) is shown in more detail in Fig. 5.27 and can only be realized in a narrow band. It works on the same principle as the push-pull blocker, but with an additional phase rotation of 90° from the center plane, which in this case represents an open plane. These open circuits are transformed via the $\lambda/4$ line into the two red short-circuit points shown. Thus, the common-mode signal is also reflected with a short circuit ($M_{ii}^{+} = -1$) and the push-pull signal remains unaffected, $M_{ij}^{-} = 1$.

Mode Blocker based on Resonant Circuits

Blockers with resonant circuits are very interesting for applications at low frequencies and especially for integrated circuits. A common and a differential mode blocker are shown in generalized form in Fig. 5.28.

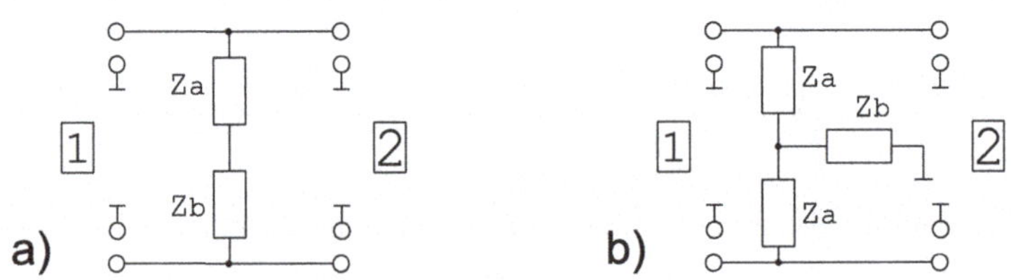

Fig. 5.28: Mode blocker based on resonators: a) Differential mode blocker, b) Common mode blocker

The push-pull blocker can be set up once using an LC series resonant circuit. It is therefore a frequency-selective 0°-blocker.

The common mode blocker with resonant circuits is much more interesting in practice. This is the most frequently used blocker in circuits. Here too, Z_a and Z_b can each be replaced by an inductor and a capacitor. The DC component is again blocked via a series resonance by a low-impedance termination, ($M_{ii}^+ = -1$).

This blocker is designed for the center frequency ω_{block} with, for example, two inductors with $Z_b = j\omega L$ and a shunt capacitor with $Z_a = \frac{1}{j\omega C}$ via $L = \frac{2}{\omega_{block}^2 C}$, Fig. 5.29.

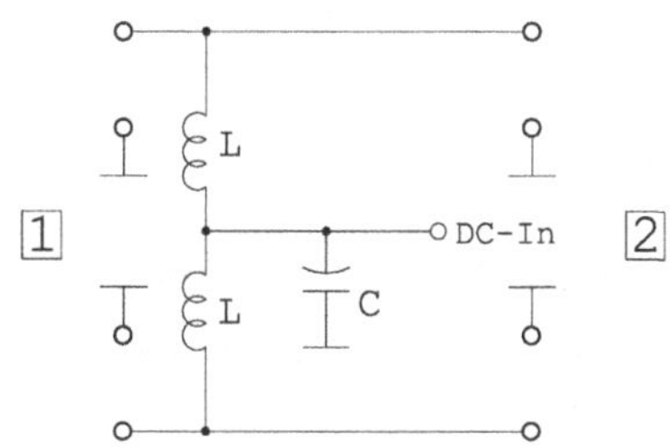

Fig. 5.29: DC-Feed with common mode rejection

This common mode blocker now has no ideal properties for the push-pull signal. The impedance $2 \cdot Z_a$ is contained in the push-pull branch. This blocker is therefore used in circuits in which these two components (Z_a) are included due to a different functionality. In this case, only the one additional component Z_b is necessary.

This means that the blocker function can be integrated into filters, amplifiers and many other components with the use of just one component.

The following blocker is intended to show that it is even possible without increased component to achieve the blocker functionality.

The Cross Blocker

If you evaluate the mixed-mode help sheets (appendix from page 453), you will come to the conclusion that crossed circuits can have mode blocker properties. These common-mode blockers and a push-pull blocker are shown in Fig. 5.30.

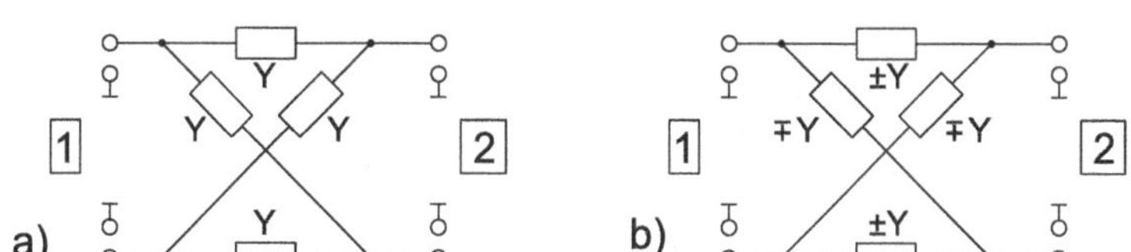

Fig. 5.30: Mode blocker based on crossed circuits: a) push-pull mode blocker, b) common mode blocker

In simple terms, this blocker can be explained as follows: If there is a positive and a negative voltage with the same amplitude on one side, the voltage on the other side of the blocker is zero via the voltage division at each terminal. No power is transmitted for the push-pull signal.

The push-pull blocker is extremely relevant for use in balanced amplifier circuits. The Miller capacitance can be compensated with this cross blocker, [134, 143]. It also helps to develop improved switches (Chap. 7).

This blocker can be completely described by the M-parameters (Eq. (5.20)).

$$[\mathbf{M_{cross-}}] = \begin{bmatrix} \frac{1-2y}{1+2y} & 0 & 0 & 0 \\ 0 & \frac{1-2y}{1+2y} & 0 & 0 \\ 0 & 0 & \frac{1}{1+4y} & \frac{4y}{1+4y} \\ 0 & 0 & \frac{4y}{1+4y} & \frac{1}{1+4y} \end{bmatrix} \tag{5.20}$$

The push-pull transmission is zero regardless of the choice of a finite conductance Y. This blocker cannot always be matched and is therefore often part of a component.

Alternatively, general two-port devices can also be used for the terminal element. The broadband mode blocker functionality is retained as long as the four two-ports are the same. For example, $\lambda/4$ lines provide a mode blocker that is adapted for the common mode signal and reflects the push-pull signal with high impedance.

The common mode blocker is also of great practical relevance. On the one hand, this cross blocker can be matched and, on the other, it suppresses the common mode signal.

$$[\mathbf{M_{cross+}}] = \begin{bmatrix} \frac{1+4y^2}{1-4y^2} & \frac{\pm 4y}{1-4y^2} & 0 & 0 \\ \frac{\pm 4y}{1-4y^2} & \frac{1+4y^2}{1-4y^2} & 0 & 0 \\ 0 & 0 & 1 & 0 \\ 0 & 0 & 0 & 1 \end{bmatrix} \tag{5.21}$$

Eq. (5.21) shows that, on the one hand, regardless of the choice of admittance Y in the common-mode signal, it is always blocked by a low-impedance full reflection. On the other hand, a number of circuit properties can also be synthesized for the common-mode functionality. For matching, Y must be purely reactive on the one hand and have a value of $0.5/Z_0$ on the other.

The following example shows a possible synthesis for a mode blocker that covers 90° phase shift and an impedance transformation.

5.2.4 ±90°-Phase Shifter with Mode Blocker and Impedance Transformer Functionality

In many technologies, the simplest realization form of the cross blocker for suppressing the common mode signal is a ±90°-phase shifter, as shown in Fig. 5.31.

In the following, this phase shifter circuit with impedance transformation and mode blocker functionality will be derived. Eq. (5.21) describes the circuit based on the associated mixed-mode auxiliary sheet only for the same impedance at the input and output. Consequently, this equation is not easily applicable here.

If you want to synthesize this circuit, the auxiliary sheets for impedance-transforming two-ports (page 446) are a great help. The cross structure can be found there. The two-port auxiliary sheets can be used because the mode conversion has now become zero.

Fig. 5.31: Phase shifter with mode blocker and impedance transformer functionality: left: +90° -phase shifter, right: -90° -phase shifter

If we set $Z_b = 1/Y$ and $Z_a = -1/Y$, we immediately obtain

$$M_{21}^{-} = \frac{4\,Y\,\sqrt{Z_{L1}\,Z_{L2}}}{4\,Y^2\,Z_{L1}\,Z_{L2} - 1} \quad . \tag{5.22}$$

To achieve the matching and 90° phase shift, the admittance must be controlled by an inductor and a capacitor with the conditions

$$C = \frac{1}{2\sqrt{Z_{L1}\,Z_{L2}}\,\omega_m} \quad , \qquad L = \frac{2\sqrt{Z_{L1}\,Z_{L2}}}{\omega_m} \tag{5.23}$$

for the center frequency ω_m can be realized.

An interpretation for $\omega_m = 1\,\text{GHz}$ and $Z_{L1} = 20\,\Omega$ as well as $Z_{L2} = 50\,\Omega$ provides the two values: $C = 2.5\,\text{pF}$ and $L = 10.1\,\text{nH}$. The corresponding simulation results are printed in Fig. 5.32.

Similar results with associated measurements can be found in [26]. This multi functional component is again very interesting for use as a sub-component. An example of its use in a hybrid coupler can also be found in [26].

5.2.5 Mode Converters and Mode Duplexers

The task of a `mode converter` is to convert a common mode wave into a differential mode wave and vice versa. Ideally, a mode converter is adapted for both modes ($M_{ii}^{+} = M_{ii}^{-} = 0$). In practice, this can only be achieved for an MM-port if no complex mode converters are used. For a lossless network, the mode conversion parameters must have a value of 1 ($|M_{ij}^{+-}| = |M_{ji}^{-+}| = 1$ or $|M_{ij}^{-+}| = |M_{ji}^{+-}| = 1$). The remaining transmission parameters are zero.

Mode converters can be divided into two classes:

1.: Transmission converter, 2.: Reflection converter.

In the section 5.3.3 (baluns), *transmission converter* for the conversion of an asymmetrical wave into a push-pull wave is presented in detail. An unbalanced transmission line can be imagined as a short-circuited two-wire pair. Consequently, the results can also be applied to a transmission converter for a two-port pair. Many types of transmission converters can be derived from the large number of balancing elements.

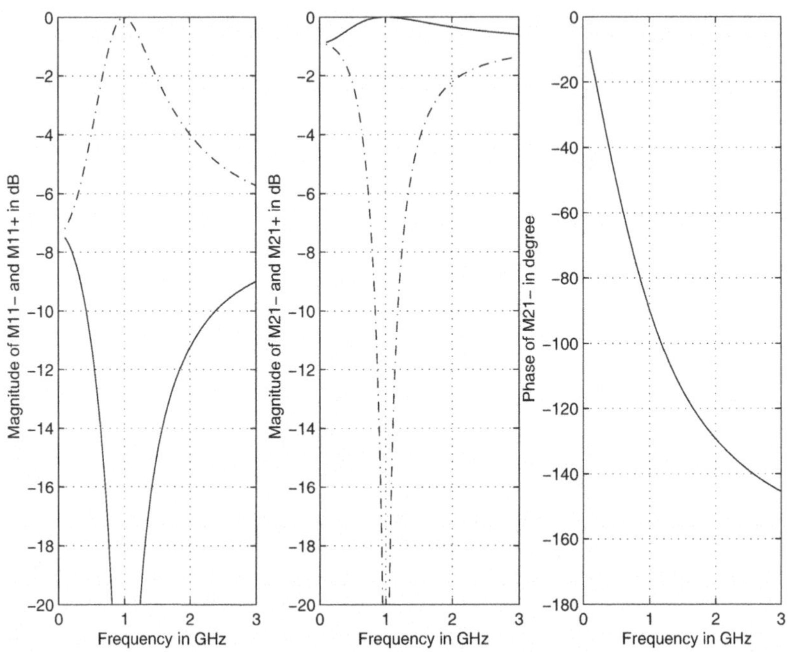

Fig. 5.32: M-parameter results for the -90°-phase shifter with impedance transformation and mode suppression for ω_m =1 GHz (common mode: dashed)

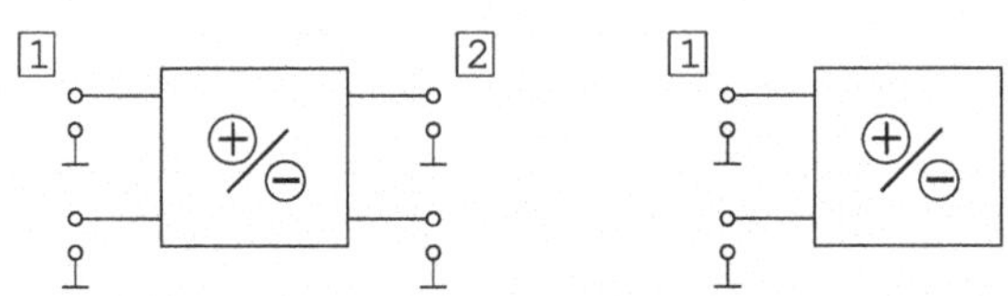

Fig. 5.33: Block diagrams of the transmission converter (left) and the reflection converter (right)

Reflection mode converters, on the other hand, are unknown in electrical engineering. Reflection converters are a single-port pair. For an ideal network, the four M parameters remain

$$M_{11}^{-} = M_{11}^{+} = 0 \quad \text{and} \quad \left|M_{11}^{+-}\right| = \left|M_{11}^{-+}\right| = 1 \quad .$$

However, the fact that reflection converters can be used very sensibly can be read in the following chapter on dual-mode resonators.

The simplest reflection converter consists of an open circuit and a short circuit, as shown in the Fig. 5.34.

If we consider the case of an incident common mode wave, it can be seen that some of the waves are reflected with a reflection factor of 1 and the other part of the waves with a reflection factor of -1 and thus as a push-pull wave. This simple converter is extremely broadband in practice.

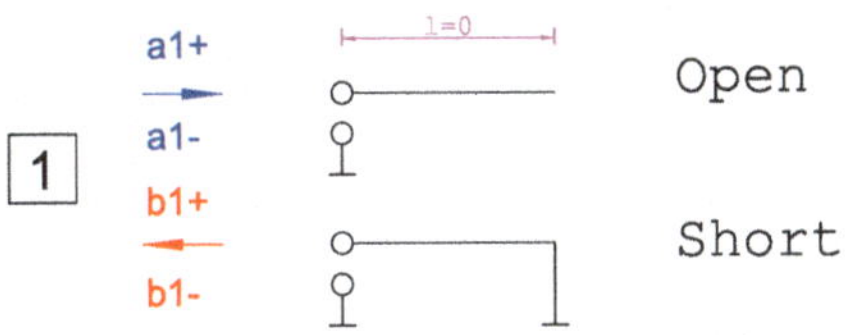

Fig. 5.34: Realization of a broadband reflection mode converter

The section 6.5 shows that narrowband reflection mode converters are also of great technical interest. If the parallel resonant circuit of the narrow-band mode converter from Fig. 5.35 is in resonance, this converter has the same properties as the broadband converter.

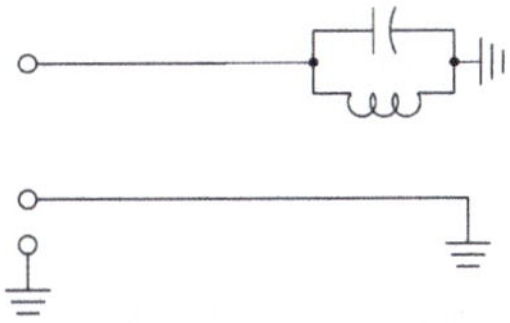

Fig. 5.35: Design of a narrowband reflection mode converter

Outside the resonant frequency, this mode converter only represents a better or worse short circuit for the common and differential mode.

If the short circuit of the broadband mode converter is replaced by a series resonant circuit, the result is another narrowband reflection mode converter that reflects the common and differential mode with high impedance outside the resonant frequency.

An **mode duplexer** must have three mixed-mode ports. Both modes are present at one port and only one mode can transmit to each of the other two ports.

A transformer can be used to implement a mode duplexer according to Fig. 5.36. This is very interesting for circuit simulation, among other things.

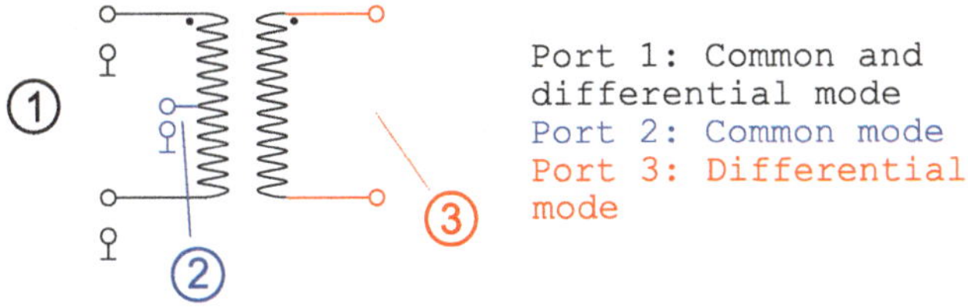

Fig. 5.36: Construction of a mode duplexer using a transformer

Fig. 5.37 shows a basic construction and the important M-parameters for an ideal and lossless mode duplexer.

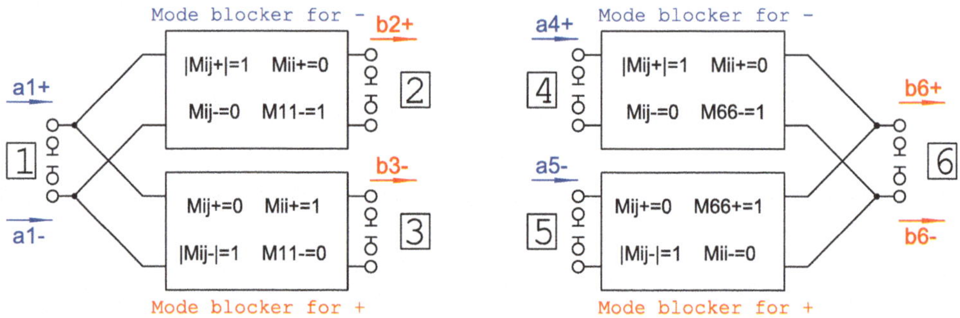

Fig. 5.37: Basic construction of a mode duplexer and ideal M-parameters

A mode duplexer therefore consists of two mode blockers that block the block mode as an open circuit.

As many mode blockers for the block mode reflect a short circuit, they may need to be extended with a $\lambda/4$ line. An example is shown in the Fig. 5.38.

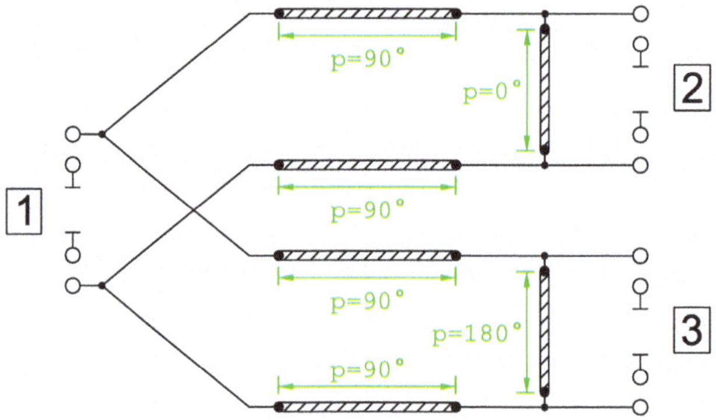

Fig. 5.38: Circuit diagram of a mode duplexer with cables

The corresponding simulation results with the most important M-parameters are printed in Fig. 5.39 for a center frequency of 1 GHz.

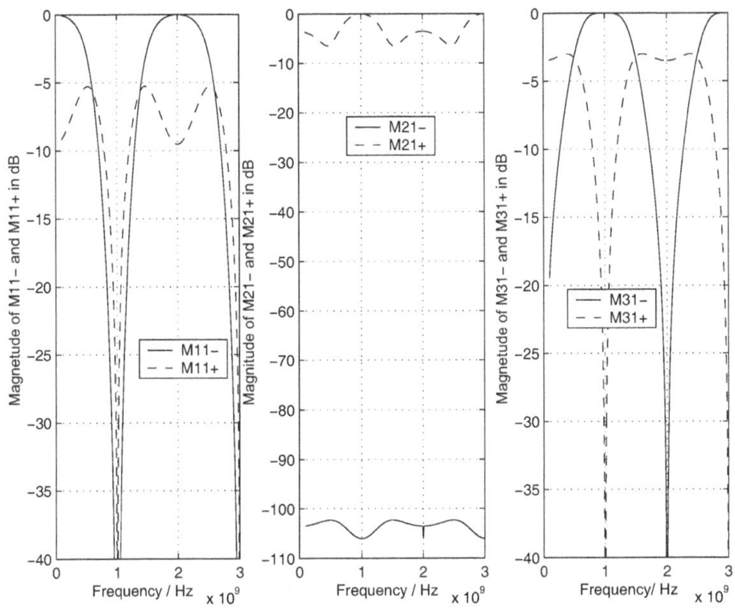

Fig. 5.39: M-parameters of a mode duplexer for 1 GHz, which is created via lines

It can be seen that the matching of the differential-mode wave at 1 GHz is much broader than the wave matching of the common-mode wave. This then also has a significant effect on the bandwidths of the associated transmission signals.

5.2.6 Propagation Matrix for Mixed-Mode Systems

The transmission matrix $[\Sigma]$ was introduced from the unbalanced circuit synthesis of two-ports. In contrast to the scattering parameters, transmission parameters do not link the incoming and outgoing wave quantities, but rather the quantities of a port.

This means that the scattering parameters of several networks connected in series can be calculated with little effort. An equivalent can be derived for dual-mode systems, which bears the name P-matrix (for propagation matrix). To calculate the overall M-

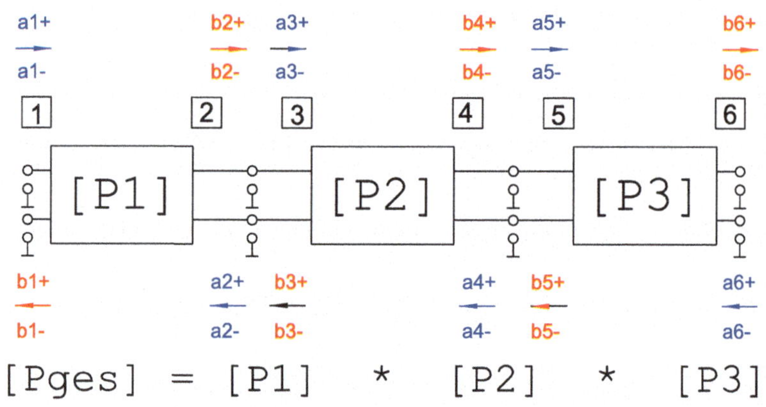

Fig. 5.40: Series connection of three MM-two-port pairs using P-matrix

matrix $[M_{overall}]$ from the M-matrices $[M_i]$ of the subcomponents, the parameters for all P-matrices $[P_i]$ are first determined. By using the identities of the wave sizes (see Fig. 5.40), the entire P-matrix $[P_{total}]$ can then be calculated using a matrix multiplication. Finally, $[M_{total}]$ is formed from $[P_{total}]$.

De-embedding of MM-two-port pairs is also possible, i.e. the extraction of a component surrounded by other networks. This requires the scattering parameters of the neighboring networks.

The matrix $[P_{tot}]$ formed from the matrix $[M_{tot}]$ is multiplied by the inverse of the P-matrices of the surrounding MM-two-port pairs.

The M-matrix associated with the de-embedding MM-two-port pair can then be obtained from the resulting P-matrix. The P-matrix is defined as follows:

$$\begin{pmatrix} a_1^+ \\ a_1^- \\ b_1^+ \\ b_1^- \end{pmatrix} = \begin{bmatrix} P_{11} & P_{12} & P_{13} & P_{14} \\ P_{21} & P_{22} & P_{23} & P_{24} \\ P_{31} & P_{32} & P_{33} & P_{34} \\ P_{41} & P_{42} & P_{43} & P_{44} \end{bmatrix} \begin{pmatrix} b_2^+ \\ b_2^- \\ a_2^+ \\ a_2^- \end{pmatrix} . \tag{5.24}$$

To convert the P-parameters from the M-parameters, the M-matrix from Eq. (5.2) must first be re-sorted:

$$\begin{pmatrix} b_1^+ \\ b_1^- \\ b_2^+ \\ b_2^- \end{pmatrix} = \begin{bmatrix} M_{11}^{+} & M_{11}^{+-} & M_{12}^{+} & M_{12}^{+-} \\ M_{11}^{-+} & M_{11}^{-} & M_{12}^{-+} & M_{12}^{-} \\ M_{21}^{+} & M_{21}^{+-} & M_{22}^{+} & M_{22}^{+-} \\ M_{21}^{-+} & M_{21}^{-} & M_{22}^{-+} & M_{22}^{-} \end{bmatrix} \begin{pmatrix} a_1^+ \\ a_1^- \\ a_2^+ \\ a_2^- \end{pmatrix} . \tag{5.25}$$

Vectors and matrices are divided into subvectors and submatrices for easy calculation.

$$\begin{pmatrix} \vec{b_1} \\ \vec{b_2} \end{pmatrix} = \begin{bmatrix} [M_{11}] & [M_{12}] \\ [M_{21}] & [M_{22}] \end{bmatrix} \begin{pmatrix} \vec{a_1} \\ \vec{a_2} \end{pmatrix} \tag{5.26}$$

with

$$\vec{b_1} = \begin{pmatrix} b_1^+ \\ b_1^- \end{pmatrix}, \quad \vec{b_2} = \begin{pmatrix} b_2^+ \\ b_2^- \end{pmatrix}, \quad \vec{a_1} = \begin{pmatrix} a_1^+ \\ a_1^- \end{pmatrix}, \quad \vec{a_2} = \begin{pmatrix} a_2^+ \\ a_2^- \end{pmatrix},$$

$$[M_{11}] = \begin{bmatrix} M_{11}^{+} & M_{11}^{+-} \\ M_{11}^{-+} & M_{11}^{-} \end{bmatrix}, \quad [M_{12}] = \begin{bmatrix} M_{12}^{+} & M_{12}^{+-} \\ M_{12}^{-+} & M_{12}^{-} \end{bmatrix},$$

$$[M_{21}] = \begin{bmatrix} M_{21}^{+} & M_{21}^{+-} \\ M_{21}^{-+} & M_{21}^{-} \end{bmatrix}, \quad [M_{22}] = \begin{bmatrix} M_{22}^{+} & M_{22}^{+-} \\ M_{22}^{-+} & M_{22}^{-} \end{bmatrix}.$$

Matrix (5.26) contains the equations

$$\vec{b_1} = [M_{11}]\,\vec{a_1} + [M_{12}]\,\vec{a_2} \quad , \tag{5.27}$$

$$\vec{b_2} = [M_{21}]\,\vec{a_1} + [M_{22}]\,\vec{a_2} \quad . \tag{5.28}$$

If Eq. (5.28) is converted to $\vec{a_1}$, the first equation for the P-matrix is obtained:

$$\vec{a_1} = [M_{21}]^{-1}\,\vec{b_2} - [M_{21}]^{-1}\,[M_{22}]\,\vec{a_2} \quad . \tag{5.29}$$

The second equation is obtained by inserting (5.29) into (5.27):

$$\vec{b_1} = [M_{11}]\,[M_{21}]^{-1}\,\vec{b_2} + \Big([M_{12}] - [M_{11}]\,[M_{21}]^{-1}\,[M_{22}]\Big)\,\vec{a_2} \quad . \tag{5.30}$$

From the Eqs. (5.29) and (5.30) result in the P-matrix. The following conversion variables can be determined by comparing the coefficients:

$$P_{11} = -P_x M_{21}^{-} \ , \quad P_{12} = P_x M_{21}^{+-} \ , \quad P_{21} = P_x M_{21}^{-+} \ , \quad P_{22} = -P_x M_{21}^{+} \ , \tag{5.31}$$
$$P_{13} = P_x(M_{21}^{-} M_{22}^{+} - M_{21}^{+-} M_{22}^{-+}) \ , \quad P_{14} = P_x(M_{21}^{-} M_{22}^{+-} - M_{21}^{+-} M_{22}^{-}) \ , \tag{5.32}$$
$$P_{23} = P_x(M_{21}^{+} M_{22}^{-+} - M_{21}^{-+} M_{22}^{+}) \ , \quad P_{24} = P_x(M_{21}^{+} M_{22}^{-} - M_{21}^{-+} M_{22}^{+-}) \ , \tag{5.33}$$
$$P_{31} = P_x(M_{11}^{+-} M_{21}^{-+} - M_{11}^{+} M_{21}^{-}) \ , \quad P_{32} = P_x(M_{11}^{+} M_{21}^{+-} - M_{11}^{+-} M_{21}^{+}) \ , \tag{5.34}$$
$$P_{41} = P_x(M_{11}^{-} M_{21}^{-+} - M_{11}^{-+} M_{21}^{-}) \ , \quad P_{42} = P_x(M_{11}^{-+} M_{21}^{+-} - M_{11}^{-} M_{21}^{+}) \ , \tag{5.35}$$

$$\begin{aligned} P_{33} = P_x(&-M_{12}^{+} M_{21}^{+} M_{21}^{-} + M_{12}^{+} M_{21}^{+-} M_{21}^{-+} + M_{22}^{+} M_{11}^{+} M_{21}^{-} \\ &-M_{22}^{+} M_{11}^{+-} M_{21}^{-+} - M_{22}^{-+} M_{11}^{+} M_{21}^{+-} + M_{22}^{-+} M_{11}^{+-} M_{21}^{+}) \ , \end{aligned} \tag{5.36}$$
$$\begin{aligned} P_{34} = -P_x(&M_{12}^{+-} M_{21}^{+} M_{21}^{-} - M_{12}^{+-} M_{21}^{+-} M_{21}^{-+} - M_{22}^{+-} M_{11}^{+} M_{21}^{-} \\ &+M_{22}^{+-} M_{11}^{+-} M_{21}^{-+} + M_{22}^{-} M_{11}^{+} M_{21}^{+-} - M_{22}^{-} M_{11}^{+-} M_{21}^{+}) \ , \end{aligned} \tag{5.37}$$
$$\begin{aligned} P_{43} = P_x(&-M_{12}^{-+} M_{21}^{+} M_{21}^{-} + M_{12}^{-+} M_{21}^{+-} M_{21}^{-+} + M_{22}^{+} M_{11}^{-+} M_{21}^{-} \\ &-M_{22}^{+} M_{11}^{-} M_{21}^{-+} - M_{22}^{-+} M_{11}^{-+} M_{21}^{+-} + M_{22}^{-+} M_{11}^{-} M_{21}^{+}) \ , \end{aligned} \tag{5.38}$$
$$\begin{aligned} P_{44} = -P_x(&M_{12}^{-} M_{21}^{+} M_{21}^{-} - M_{12}^{-} M_{21}^{+-} M_{21}^{-+} - M_{22}^{+-} M_{11}^{-+} M_{21}^{-} \\ &+M_{22}^{+-} M_{11}^{-} M_{21}^{-+} + M_{22}^{-} M_{11}^{-+} M_{21}^{+-} - M_{22}^{-} M_{11}^{-} M_{21}^{+}) \end{aligned} \tag{5.39}$$

with

$$P_x = \frac{1}{M_{21}^{+-} M_{21}^{-+} - M_{21}^{+} M_{21}^{-}} \ . \tag{5.40}$$

The conversion from P-parameters to M-parameters is done using a similar calculation to the one presented in Eqs. (5.24)-(5.30). This results in the following values:

$$M_{11}^{+} = M_x(P_{31}P_{22} - P_{32}P_{21}) \ , \quad M_{11}^{+-} = M_x(P_{32}P_{11} - P_{31}P_{12}) \ , \tag{5.41}$$
$$M_{11}^{-+} = M_x(P_{41}P_{22} - P_{42}P_{21}) \ , \quad M_{11}^{-} = M_x(P_{42}P_{11} - P_{41}P_{12}) \ , \tag{5.42}$$

$$\begin{aligned} M_{12}^{+} &= M_x(P_{33}P_{11}P_{22} - P_{33}P_{12}P_{21} - P_{13}P_{31}P_{22} + P_{13}P_{32}P_{21} + P_{23}P_{31}P_{12} - P_{23}P_{32}P_{11}), \\ M_{12}^{+-} &= M_x(P_{34}P_{11}P_{22} - P_{34}P_{12}P_{21} - P_{14}P_{31}P_{22} + P_{14}P_{32}P_{21} + P_{24}P_{31}P_{12} - P_{24}P_{32}P_{11}), \\ M_{12}^{-+} &= M_x(P_{43}P_{11}P_{22} - P_{43}P_{12}P_{21} - P_{13}P_{41}P_{22} + P_{13}P_{42}P_{21} + P_{23}P_{41}P_{12} - P_{23}P_{42}P_{11}), \\ M_{12}^{-} &= M_x(P_{44}P_{11}P_{22} - P_{44}P_{12}P_{21} - P_{14}P_{41}P_{22} + P_{14}P_{42}P_{21} + P_{24}P_{41}P_{12} - P_{24}P_{42}P_{11}), \end{aligned}$$

$$M_{21}^{+} = M_x P_{22} \ , \quad M_{21}^{+-} = -M_x P_{12} \ , \tag{5.43}$$
$$M_{21}^{-+} = -M_x P_{21} \ , \quad M_{21}^{-} = M_x P_{11} \ , \tag{5.44}$$
$$M_{22}^{+} = M_x(P_{12}P_{23} - P_{22}P_{13}) \ , \quad M_{22}^{+-} = M_x(P_{12}P_{24} - P_{22}P_{14}) \ , \tag{5.45}$$
$$M_{22}^{-+} = M_x(P_{21}P_{13} - P_{11}P_{23}) \ , \quad M_{22}^{-} = M_x(P_{21}P_{14} - P_{11}P_{24}) \ . \tag{5.46}$$

with

$$M_x = \frac{1}{P_{11}P_{22} - P_{12}P_{21}} \ . \tag{5.47}$$

The associated verifications and tests can be found in [27].

5.3 Components with Two- and Three-Wire Systems at the Input and Output

At the beginning of this chapter it was stated that integrated RF circuits mostly only have symmetrical inputs and outputs. These highly integrated ICs are used to convert low-frequency signals into RF signals. They supply a small signal for the transmit path and can process a small signal in the receive path. This part of a transmitting/receiving device is called *transceiver*.

In the chain circuit of an RF transmitting/receiving device, there are the so-called frontends with the module power amplifier, channel switch, transmit/receive switch and various other modules, all of which are realized with discrete components.

Since balanced circuits require 1.5 to 2 times more components than unbalanced circuits, these front-end circuits are generally realized using unbalanced technology.

Components that perform a mode conversion from the asymmetrical (also called unbalanced) to the symmetrical (also called balanced) mode and vice versa are called BALUN (from balanced to unbalanced).

In order to be able to describe such modules, the general M-parameters are derived. The general M-parameters contain the links between the unsymmetrical mode and the synchronous and push-pull modes.

5.3.1 Relationship between S- and General M-Parameters

Three modes are present on a balun or on modules that contain the balun functionality: unbalanced, common mode and differential mode. The common mode is in turn an interference mode. A balun should only convert energy between the unbalanced mode and the differential mode.

In general, such a two-port pair can be represented by an asymmetrical port 1 (with the applied wave sizes a_1 and b_1) and the port pair 2, which is composed of the two abstracted ports 2 and 3, as shown in Fig. 5.41. The common-mode waves a_2^+ and b_2^+ should be

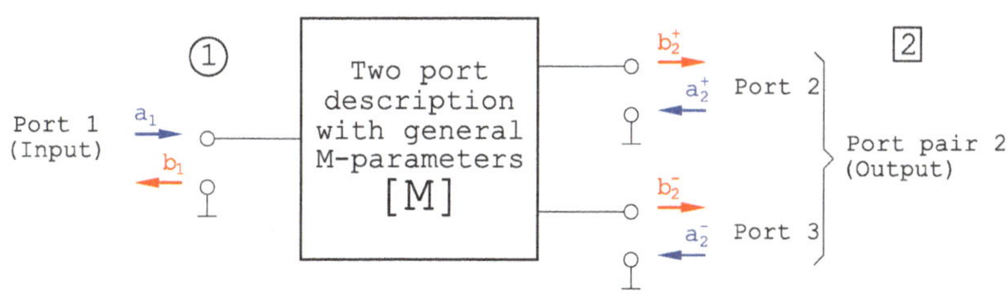

Fig. 5.41: Representation of a general MM two-port with an unbalanced input and a pair of common and differential ports at the output

present at port 2 and the push-pull waves a_2^- and b_2^- at port 3.

A second form of these balun circuits is a transmission mode converter, in which only one unsymmetrical line is attached to a port, Fig. 5.42. While a general transmission mode

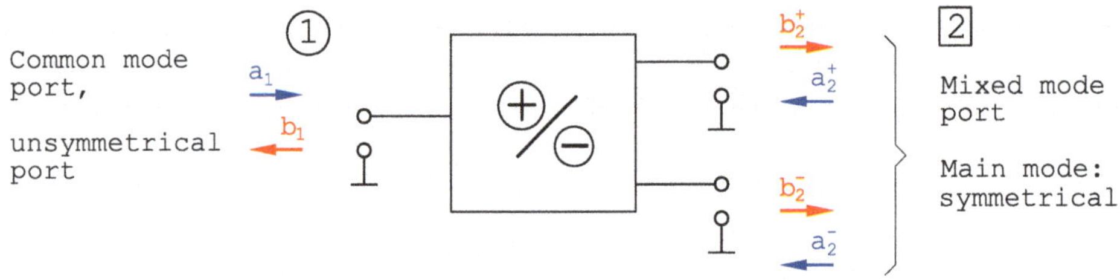

Fig. 5.42: Unidirectional transmission mode converter for the conversion of a common mode into a push-pull mode

converter can convert the modes bidirectionally, a balun can only do so unidirectionally. Only one special mode can be converted at the balun at an input port.

The relationship between the outgoing waves (b-waves) shown in Fig. 5.41 and the incoming waves (a-waves) is described by Eq. (5.48).

$$\begin{pmatrix} b_1 \\ b_2^- \\ b_2^+ \end{pmatrix} = \left[\begin{array}{cc|c} M_{11}^0 & M_{12}^{0-} & M_{12}^{0+} \\ M_{21}^{-0} & M_{22}^{-} & M_{22}^{-+} \\ \hline M_{21}^{+0} & M_{22}^{+-} & M_{22}^{+} \end{array}\right] \begin{pmatrix} a_1 \\ a_2^- \\ a_2^+ \end{pmatrix} \tag{5.48}$$

The separation between the desired four M-parameters and the remaining five M-parameters, which are linked to the common mode wave, is emphasized in Eq. (5.48).

In the further derivation, it will be shown that the M-parameters can be calculated from the S-parameters of the three-port, as was already stated for the M-parameters.

The following applies ([48]) for the relationship between the common-mode and differential-mode waves and the single-ended waves at ports 2 and 3:

$$a_2^+ = \frac{1}{\sqrt{2}}(a_2+a_3) \qquad a_2^- = \frac{1}{\sqrt{2}}(a_2-a_3) \qquad b_2^+ = \frac{1}{\sqrt{2}}(b_2+b_3) \qquad b_2^- = \frac{1}{\sqrt{2}}(b_2-b_3) \quad . \tag{5.49}$$

These links are used in the Eq. (5.50)

$$\begin{pmatrix} b_1 \\ \frac{1}{\sqrt{2}}(b_2-b_3) \\ \frac{1}{\sqrt{2}}(b_2+b_3) \end{pmatrix} = \begin{bmatrix} M_{11}^0 & M_{12}^{0-} & M_{12}^{0+} \\ M_{21}^{-0} & M_{22}^{-} & M_{22}^{-+} \\ M_{21}^{+0} & M_{22}^{+-} & M_{22}^{+} \end{bmatrix} \begin{pmatrix} a_1 \\ \frac{1}{\sqrt{2}}(a_2-a_3) \\ \frac{1}{\sqrt{2}}(a_2+a_3) \end{pmatrix} \quad . \tag{5.50}$$

In the following, the first line is shown in analytical representation as well as the sum and the difference of the 2nd and 3rd lines.

1. line:

$$b_1 = M_{11}^{0} \cdot a_1 + \frac{1}{\sqrt{2}}(M_{12}^{0+} + M_{12}^{0-}) \cdot a_2 + \frac{1}{\sqrt{2}}(M_{12}^{0+} - M_{12}^{0-}) \cdot a_3 \tag{5.51}$$

2. + 3. line:

$$\begin{aligned}\sqrt{2}\, b_2 = (M_{21}^{+0} + M_{21}^{-0}) \cdot a_1 &+ \frac{1}{\sqrt{2}}(M_{22}^{+} + M_{22}^{-+} + M_{22}^{+-} + M_{22}^{-}) \cdot a_2 \\ &+ \frac{1}{\sqrt{2}}(M_{22}^{+} + M_{22}^{-+} - M_{22}^{+-} - M_{22}^{-}) \cdot a_3\end{aligned} \tag{5.52}$$

3. - 2. line:

$$\begin{aligned}\sqrt{2}\, b_3 = (M_{21}^{+0} - M_{21}^{-0}) \cdot a_1 &+ \frac{1}{\sqrt{2}}(M_{22}^{+} - M_{22}^{-+} + M_{22}^{+-} - M_{22}^{-}) \cdot a_2 \\ &+ \frac{1}{\sqrt{2}}(M_{22}^{+} - M_{22}^{-+} - M_{22}^{+-} + M_{22}^{-}) \cdot a_3\end{aligned} \tag{5.53}$$

Comparing the three Eqs. (5.51-5.53) with the 3×3 scatter matrix:

$$\begin{pmatrix} b_1 \\ b_2 \\ b_3 \end{pmatrix} = \begin{bmatrix} S_{11} & S_{12} & S_{13} \\ S_{21} & S_{22} & S_{23} \\ S_{31} & S_{32} & S_{33} \end{bmatrix} \begin{pmatrix} a_1 \\ a_2 \\ a_3 \end{pmatrix} \quad , \tag{5.54}$$

the relationships between the 9 M- and the 9 S-parameters are as follows:

$$S_{11} = M_{11}^{0} \,, \tag{5.55}$$

$$S_{12} = \frac{1}{\sqrt{2}}(M_{12}^{0+} + M_{12}^{0-}) \,, \tag{5.56}$$

$$S_{13} = \frac{1}{\sqrt{2}}(M_{12}^{0+} - M_{12}^{0-}) \,, \tag{5.57}$$

$$S_{21} = \frac{1}{\sqrt{2}}(M_{21}^{+0} + M_{21}^{-0}) \,, \tag{5.58}$$

$$S_{22} = \frac{1}{2}(M_{22}^{+} + M_{22}^{-+} + M_{22}^{+-} + M_{22}^{-}) \,, \tag{5.59}$$

$$S_{23} = \frac{1}{2}(M_{22}^{+} + M_{22}^{-+} - M_{22}^{+-} - M_{22}^{-}) \,, \tag{5.60}$$

$$S_{31} = \frac{1}{\sqrt{2}}(M_{21}^{+0} - M_{21}^{-0}) \,, \tag{5.61}$$

$$S_{32} = \frac{1}{2}(M_{22}^{+} - M_{22}^{-+} + M_{22}^{+-} - M_{22}^{-}) \,, \tag{5.62}$$

$$S_{33} = \frac{1}{2}(M_{22}^{+} - M_{22}^{-+} - M_{22}^{+-} + M_{22}^{-}) \,. \tag{5.63}$$

The 9 equations can also be solved according to the M-parameters.

The following applies for an unbalanced input and a balanced output:

Intrinsic parameter for the unbalanced mode:

$$M_{11}^{0} = S_{11} \tag{5.64}$$

Intrinsic parameter for the common mode:

$$M_{22}^{+} = \frac{1}{2}(S_{22} + S_{23} + S_{32} + S_{33}) \tag{5.65}$$

Intrinsic parameter for the differential mode:

$$M_{22}^{-} = \frac{1}{2}(S_{22} - S_{23} - S_{32} + S_{33}) \tag{5.66}$$

Conversion parameter for the common mode to the unbalanced mode:

$$M_{12}^{0+} = \frac{1}{\sqrt{2}}(S_{12} + S_{13}) \tag{5.67}$$

Conversion parameter for the differential mode in the single-ended mode:

$$M_{12}^{0-} = \frac{1}{\sqrt{2}}(S_{12} - S_{13}) \tag{5.68}$$

Conversion parameter for unbalanced mode to common mode:

$$M_{21}^{+0} = \frac{1}{\sqrt{2}}(S_{21} + S_{31}) \tag{5.69}$$

Conversion parameter for the asymmetric mode in the differential mode:

$$M_{21}^{-0} = \frac{1}{\sqrt{2}}(S_{21} - S_{31}) \tag{5.70}$$

Conversion parameters for the common mode to differential mode:

$$M_{22}^{-+} = \frac{1}{2}(S_{22} + S_{23} - S_{32} - S_{33}) \tag{5.71}$$

Conversion parameters for the differential to the common mode:

$$M_{22}^{+-} = \frac{1}{2}(S_{22} - S_{23} + S_{32} - S_{33}) \tag{5.72}$$

The transmission M_{21}^{-0} is the parameter to be optimized for a balun. The S-parameter solution shows that M_{21}^{-0} is at a maximum when the absolute values of S_{21} and S_{31} result in $\frac{1}{\sqrt{2}}$ from the equal signal division and the phases of both transmission quantities are exactly at an angle of 180° to one another.

In this case, not only $\left|M_{21}^{-0}\right| = 1$, but also $M_{21}^{+0} = 0$. This means that no energy of the unbalanced mode is converted to the common mode.

5.3.2 Lossless and Wide-Band Balanced Signal Divider

Before we use the equations derived for balun circuits, it should be shown that these equations can be used in general. In addition, a very important signal divider will be introduced.

Fig. 5.42 indicates that a three-port of this kind is only used for the mode conversion from single-ended mode to push-pull mode. This is the main application, but not the only one. Without any component effort, a lossless and input-matched signal splitter can be

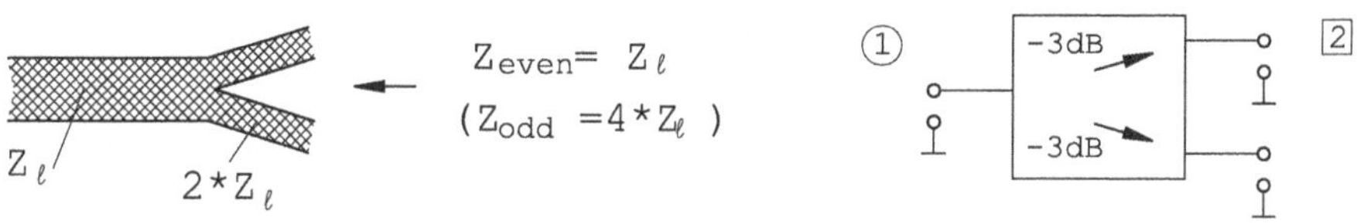

Fig. 5.43: Wide-band balanced signal divider with 100 Ω outputs in microstrip conductor technology (top view)

realized by means of the mode converter from the unsymmetrical mode into a common mode wave, Fig. 5.43.

The complete M-matrix of this circuit can also be found under the Eq. (5.74).

5.3.3 Balanced-to-Unbalanced Transformer

The most important circuits in RF electronics for converting an unbalanced line mode into a balanced line mode are presented below. There are many other possible solutions in RF-technology, particularly in antenna technology, but these relate to implementation forms with line components. Unfortunately, the majority of these circuits cannot be implemented using concentrated elements, as sheath currents from the outer conductors of the coaxial lines are used, among other things.

In an asymmetrical transmission line, positive and negative voltages arise on the conductor against the ground. Currents flow in both the signal line and the ground, which have the same directional dependency. The current distribution is asymmetrical in favor of the ground.

In a symmetrical line, the currents and voltages of both lines are equal and in phase opposition. In purely symmetrical operation, no ground current flows.

In general, a balanced-to-unbalanced transformer (also called a balun) is shown in Fig. 5.44.

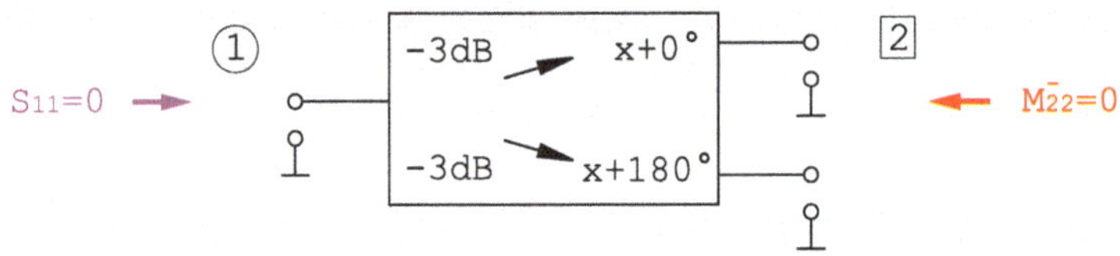

Fig. 5.44: Generalized representation of a balun

Balun with λ/2-Bypass

The most obvious way to realize a circuit that fulfills the condition under the Eq. (5.70) is to connect two unsymmetrical lines with the wave impedance $2 \cdot Z_\ell$ to the end of an unsymmetrical RF line with the impedance Z_ℓ two unsymmetrical lines with the impedance $2 \cdot Z_\ell$ are connected and one of the two lines is extended by 180° of electrical length (=$\lambda/2$).

Assuming negligible losses, the same voltages and currents with opposite phases are present at both ends of the line (Fig. 5.45). That is, at the line ends, the currents and voltages are equal to zero on the whole.

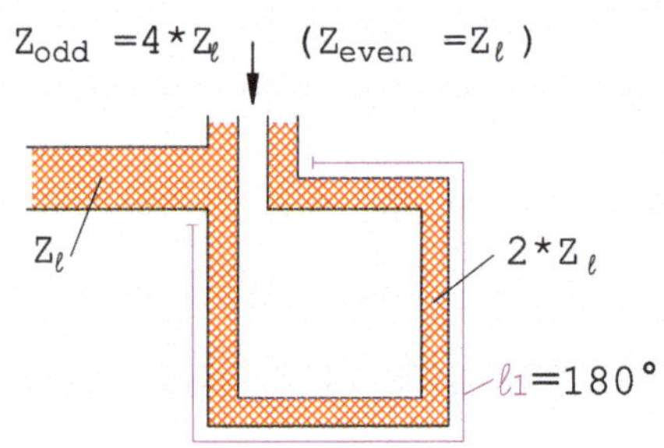

Fig. 5.45: Balun as $\lambda/2$ bypass in microstrip conductor technology (top view)

As the two halves of the line already have a characteristic impedance of $2 \cdot Z_\ell$, the characteristic impedance of $4 \cdot Z_\ell$ is set for the symmetrical line system.

This fixed impedance transformation as well as the narrow-band design are disadvantages of this otherwise very simple construction form of a balun.

——— *Exercise: M-Parameter of the λ/2 Offset-Line* ———

Given: *The sketch of the $\lambda/2$ offset line as well as the port impedances are given in the Fig. 5.45.*

Find: *Specify the general M-matrix for a $\lambda/2$ offset line for each of the two frequency points f_1 at $\lambda/2$ and f_2 at λ.*

Solution: *For the frequency point f_1 at $\lambda/2$, the signal is split into two equal parts at the unsymmetrical input port 1 into the following lines with the characteristic impedance $2Z_\ell$. In the upper branch, the phase shift is 0° and in the lower branch 180°. Consequently, the entire energy is converted to push-pull mode. This results in the adjustment for M_{11}^{0} and M_{22}^{-} as well as the transmission factor of 1 for M_{21}^{-0}.*

Due to the given reciprocity, M_{12}^{0-} is also the value 1. Because there is no loss between the unbalanced mode and the push-pull mode, the conversion variables are 0. A common mode wave, which arrives from MM-port 2 and is fed equally into the two lines, arrives at port 1 with 0° and 180° phase rotation. These two parts form a short circuit so that $M_{22}^{+} = -1$ applies.

The complete M matrix for the input frequency point f_1 at $\lambda/2$ is as follows:

$$\left[\mathbf{M}_{\lambda/2}\right] = \begin{bmatrix} 0 & 1 & 0 \\ 1 & 0 & 0 \\ 0 & 0 & -1 \end{bmatrix} \quad . \tag{5.73}$$

For the frequency point f_2 at λ, the signal at the asymmetrical input port 1 is divided into two equal parts into the following lines. In the upper branch, the phase shift is 0° and in the lower branch 360°. Consequently, all energy is converted into the common-mode. This results in the matching for M_{11}^{0} and M_{22}^{+} and the transmission factor of 1 for M_{21}^{+0}. Due to reciprocity, $M_{11}^{0+} = 1$. The conversion factors are again 0 due to the lossless transmission. A push-pull wave that is incident from the MM-port 2 and is fed into the two lines arrives at port 1 with a phase rotation of 0° and 360° respectively. These two components form a short circuit due to their opposite phase, so that $M_{22}^{-} = -1$ applies.

The complete M-matrix of the balun for the frequency point f_2 at λ is as follows:

$$\left[\mathbf{M}_{\lambda}\right] = \begin{bmatrix} 0 & 0 & 1 \\ 0 & -1 & 0 \\ 1 & 0 & 0 \end{bmatrix} \quad . \tag{5.74}$$

——— *Exercise: M-Parameter of the $\lambda/2$ Offset-Line* ———

The number of known balancing elements with lines is large, [48]. However, these all have the disadvantage that their structures are too large for modern RF electronic circuits. In the following, only the concepts of interest for highly integrated circuits will be presented.

Phase Shifter Balun

±90°-Baluns are particularly interesting for relatively narrow-band circuits and systems such as WLAN. If you need balun elements for broadband applications, for example, you should divide the problem into two steps:

1. Broadband design of ±90° phase shifter,
2. Combine phase shifter to balun.

The balun with $\lambda/2$ bypass showed that a 0° phase shifter and a 180° phase shifter could also be used instead of ±90° phase shifters. As with the bypass, these often have slightly different losses and therefore do not offer as good conditions as a -90° and a +90° phase shifter.

One way of increasing the bandwidths of phase shifters is specified in [103]. A new method for the development of high-precision and broadband phase shifters is shown in the next section.

If you want to combine two phase shifters to form a balun, you can save components by providing the same shunt components at the input of both phase shifters. Otherwise, attention must of course be paid to the impedance ratios.

±90°-Balun

A balun that splits a signal into two equal parts using high-pass/low-pass filtering will be presented as a ±90°-balun. This ±90°-balun is now the most commonly used balun in RF electronics, as it allows any impedance transformation in addition to the low component complexity.

The circuit topology for a setup with concentrated components is illustrated in Fig. 5.46.

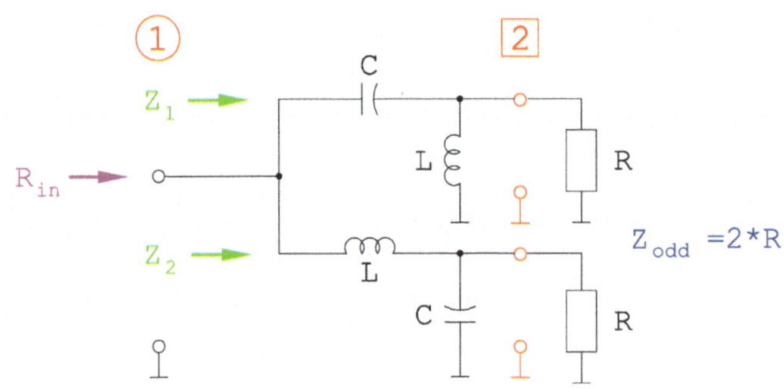

Fig. 5.46: ±90°-Balun realized with concentrated components

Both the low-pass filter and the high-pass filter transforming terminating resistor R to the input resistors $R_{in} = Z_1//Z_2$. The common mode and differential mode resistances can be determined from the resistance value R on the symmetrical port:

$$R = Z_{odd}/2 = Z_-/2 = 2\,Z_{even} = 2\,Z_+ \quad . \tag{5.75}$$

By selecting the same components for both branches and complying with the resonance condition $\omega_0^2 = \frac{1}{L\,C}$, a phase rotation of the low-pass branch of -90° is achieved. The high-pass branch, on the other hand, rotates the transmission signal by +90°. This results in a push-pull signal rotated by 180° at the balanced output.

To analyze and understand the circuit, the two phase shifters can also be interpreted as two series resonant circuits in different sequences of the two reactive elements L and C with load tapping. The dimensioning is thus for the general case:

$$L = 2\,C\,R\,R_{in} \qquad \text{with} \qquad C = \frac{1}{\omega_0\,\sqrt{2\,R\,R_{in}}} \quad . \tag{5.76}$$

As can be seen in further examples, the ±90°-balun also works very well with terminations $Z_{even} \neq R\,/\,2$.

With an exact design, the input and output are perfectly matched and the mode conversion to common mode is zero:

$$M_{11}^{0} = M_{22}^{-} = M_{21}^{+0} = M_{22}^{+-} = 0 \quad .$$

For use in more complex circuits, the ±90°-balun also has the excellent property that it reflects the common mode at the symmetrical port as an open circuit:

$$M_{21}^{-0} = M_{22}^{+} = 1 \quad .$$

——— *Exercise: Derivation of the Equations in (5.76)* ———

Using the circuit shown in Fig. 5.46 with the auxiliary variables Z_1 and Z_2, derive the results shown in (5.76) for the calculation of L and C.

——— *Exercise: Derivation of the Equations in (5.76)* ———

If you want to use this balun for two frequency bands (dual-band), you only need to replace the components with parallel resonant circuits. For the design, first calculate the two circuits for the two frequencies according to the equations in (5.76). Then you can use the calculation equations in the chapter 3.4.

These baluns can also be used to create an unbalanced signal divider. The input port always remains matched. The values of the components for 1 GHz and the corresponding S-parameters are printed in the table 5.1.

Tab. 5.1: Component and S-parameter values of an unbalanced signal splitter at 1 GHz

L/nH	C/pF	S21/dB	S22/dB	S31/dB	S33/dB
11.3 nH	2.3 pF	-3.0	-6.0	-3.0	-6.0
9.4 nH	1.9 pF	-1.7	-9.8	-4.9	-3.4
8.0 nH	1.6 pF	-1.0	-13.7	-6.8	-2.0
7.0 nH	1.4 pF	-0.6	-17.6	-8.8	-1.2
6.3 nH	1.3 pF	-0.4	-21.2	-10.6	-0.8
5.6 nH	1.1 pF	-0.3	-24.6	-12.3	-0.5
5.1 nH	1.0 pF	-0.2	-27.8	-13.9	-0.4
4.7 nH	0.9 pF	-0.1	-30.7	-15.3	-0.3
3.8 nH	0.8 pF	-0.1	-38.3	-19.1	-0.1

If higher bandwidths are to be achieved with this balun technique, the number of elements of the two phase shifters must be increased or other phase shifters with a wider bandwidth must be used, which will be discussed below.

Winding Transformer with Center Tap

The winding transformer with center tap is the most important component in RF measurement technology and especially for analytical investigations in circuit simulation. Although state-of-the-art network analyzers can already display M-parameters, circuit simulators do not yet provide any information about the multi-mode parameters.

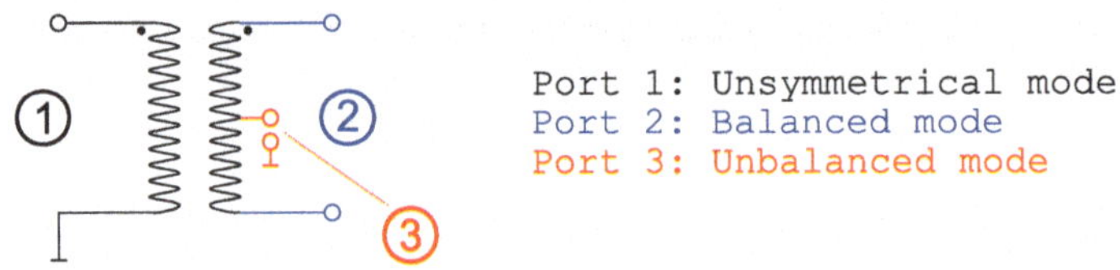

Fig. 5.47: Winding transformer with center tap as balun

Although the equations can be entered into the circuit simulator, this is very cumbersome. A simple remedy is to use ideal winding transformers with a center tap, which are available in every circuit simulator. One such balun is illustrated in Fig. 5.47.

Another example of the use of a winding transformer in this analysis is given in the Fig. 5.48. There, a ±90° -balun for Bluetooth applications is calculated with a purely symmetrical output port (ports 1 and 2) and an output port with a DC and push-pull wave termination. The latter port was realized with the help of a transformer. The 25 Ω resistor represents the impedance of the common port[7].

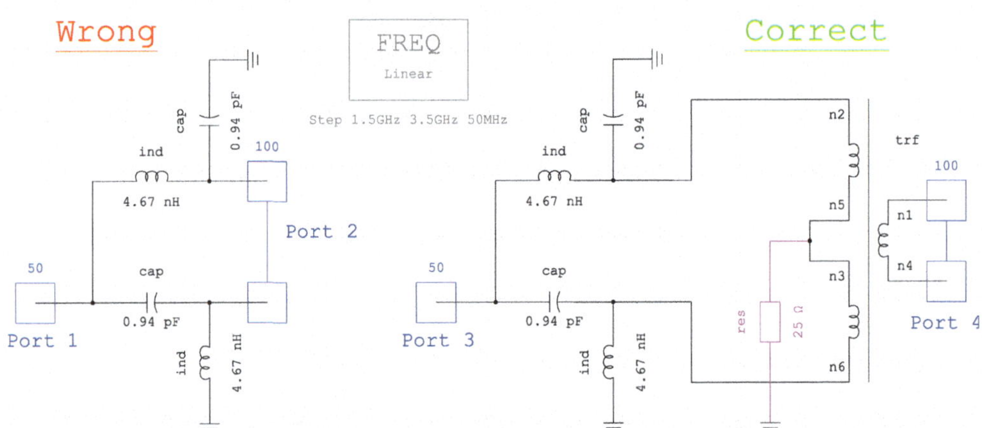

Fig. 5.48: Simulation of a ±90°-balun with purely symmetrical output port (left) and with DC and push-pull output port (right)

These two simulations produce two very different results, which also illustrate how necessary it is to take into account the existing common mode terminations. Although S_{33} is significantly better matched than S_{11}, S_{43} $(=M_{43}^{0-})$ is more narrowband than S_{21}![8].

In the left-hand simulation, the common mode wave is completed with $Z_{even} = \infty\,\Omega$.

With this procedure, the common-mode resistor always corresponds to the resistor that is connected to the middle attack and contacted to ground. With a transmission ratio of 1:1[9], the push-pull resistance corresponds to the resistance of the symmetrical port.

[7]In the simulation performed, the 25 Ω resistor was set to 30 Ω so that S_{33} is visible.

[8]For R=25 Ω one would even have broadband perfect matching.

[9]! 0.5 must be entered for Serenade, as only the turns ratio of a pair is considered.

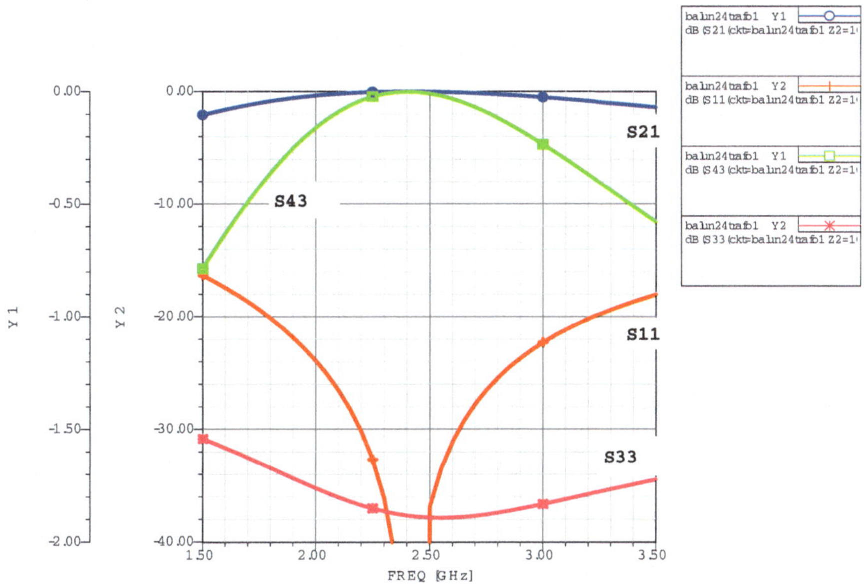

Fig. 5.49: Results of the ±90° -balun with purely symmetrical output port (S_{11} and S_{21}) and with DC and push-pull output port (S_{33} and S_{43})

With this procedure, you can also attach a port to the center tap and also obtain the common mode conversion parameters and the common mode parameters.

The fact that these winding transformers are only used as a component in RF measurement technology is due to the disadvantages of transformers, which have already been explained in the section 3.2.2. The advantage of transformers is that they can be realized very broadband with some effort.

Economical Balun

Many older books and Wikipedia present numerous balun solutions with magnetic cores in transformer style. These are of course not small, handy, can be fitted automatically and are inexpensive. That is why this chapter does not deal with them at all.

Finally, a very simple new circuit will be introduced under the name economical balun, which does not provide perfect balun functionality.

Many RF electronic products are used in the consumer market, where cost pressure is very high. In the meantime, the best semiconductor materials are no longer used for RF circuits, as the functionality of other materials is often sufficient and an overall solution for the circuit, which also includes a computer part, can then be produced more cheaply as a one-chip solution on CMOS.

You can use the economical balun with the same intention. The electrical behavior of a circuit does not have to be perfect. If the behavior is good enough and it fulfills the purpose, then the only selection criterion left to consider is the production price.

The structure of the economical balun consists of just two components, as shown in Fig. 5.50.

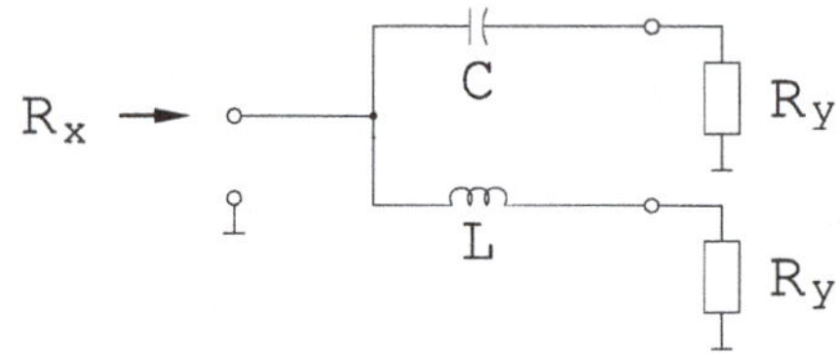

Fig. 5.50: Economical balun realized with only two concentrated elements

We know that a coil only turns a signal by -90° if the inductance is so high that almost no energy is transmitted. The situation is similar with capacitance. As a result, these two simplest phase shifters cannot perform the desired ±90° -phase rotations with the desired input matching.

If the resonance condition $\omega_0^2 = \frac{1}{L\,C}$ is specified for the design, the result is obtained for the calculation of a perfect input matching:

$$L = \frac{1}{\omega_0} \sqrt{2\,R_x\,R_y - R_y^2} \quad . \tag{5.77}$$

Eq. (5.77) can only be fulfilled for the condition $R_x > R_y$! The following relationship applies to the characteristic impedances of conventional symmetrical ground-coupled lines

$$R_y = Z_{odd}/2 = Z_-/2 = 2\,Z_{even} = 2\,Z_+ \quad . \tag{5.78}$$

A closer examination reveals that the circuit works perfectly for a (hypothetical) line with $Z_+ = 0\,\Omega$.

An example of a design at 2 GHz is intended to illustrate the functionality of the economical balun. The impedance and component values can be found in the Fig. 5.51.

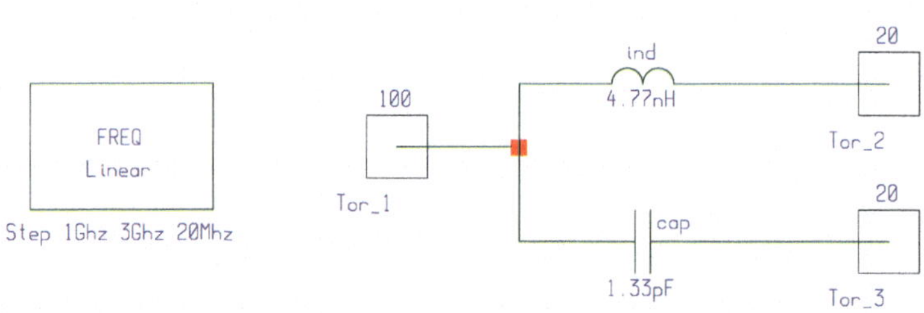

Fig. 5.51: An economical balun for the center frequency of 2 GHz implemented in the circuit simulation

The analysis shows that the input matching for the unbalanced mode is perfect, $M_{11}^0 = 0$. The matching M_{22}^- of the push-pull wave is more than sufficient in practice at -20 dB.

If you examine the S-parameters of this circuit, you will see that the signal distribution (amplitude condition of the balun) is perfect. However, as expected, the transmission phases are not exactly 180° apart. Fig. 5.52 shows the phase responses over the frequency.

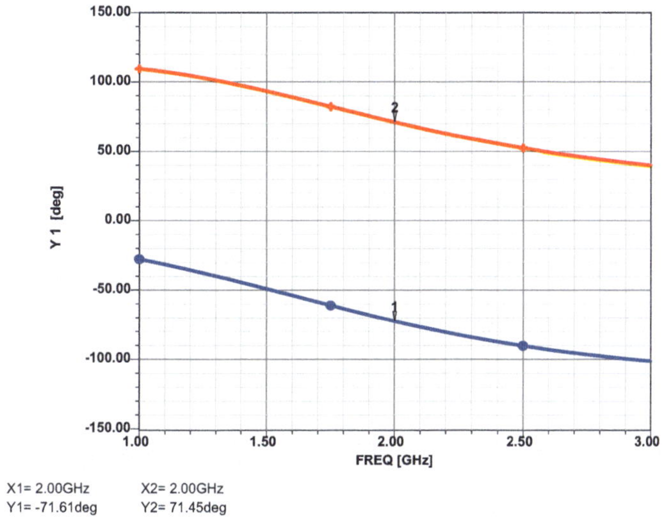

Fig. 5.52: Phase responses of S_{21} (inductive coupling, bottom) and S_{31} (capacitive coupling, top) of the calculated economical balun

It can be seen that the phase values at 2 GHz are around 71°. How this error affects the transmission loss and the mode conversions is shown in the Fig. 5.53.

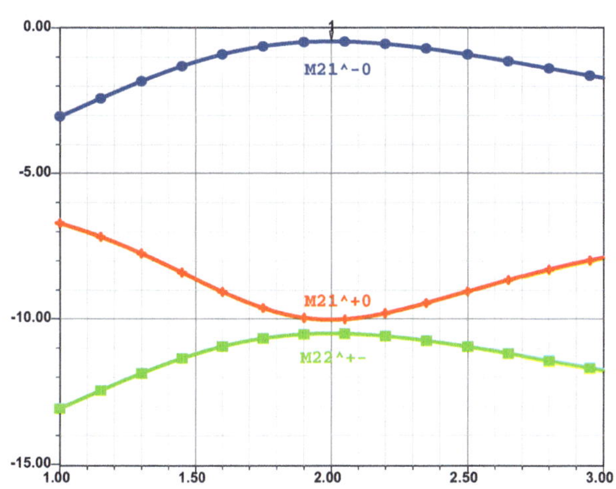

Fig. 5.53: M-parameters for the transmission and mode conversion of the calculated economical balun

This balun also has losses of almost 0.5 dB for this example, even for lossless components. In practice, however, the component losses of further components are not taken into account. This compensates for about half of the additional transmission losses.

The conversion values into the common-mode mode from Fig. 5.53 of around -10 dB show that the transmission losses for M_{21}^{-0} in the conversion into the common-mode mode is justified.

This economical balun can also be used as a **perfect signal divider** or as a **crossover**, which will also be shown in other parts of this book.

5.4 Circuit Synthesis of Symmetrical Networks

In this section, the introduced differential and common-mode quantities are to be used to analyze linear two- and multi-port components.

In other words, in practice these components are not operated with these modes. Only when calculating the design of the two- or multi-ports is it assumed that an port-pair or several port-pairs are subject to common and differential waves.

A theory for circuit synthesis of symmetrical networks called ASTRID (A Synthesis Technique for symmetRIcal Devices) was presented [57]. ASTRID is particularly suitable for the synthesis of filters, phase shifters, all-pass filters, impedance transformers, couplers, etc. ASTRID enables exact analytical solutions to be obtained using short calculations. Simple mono-mode networks (two-port) as well as mixed-mode networks can be synthesized.

First, the procedure for synthesizing two-ports is presented and illustrated using three examples. Then it is explained how to perform a synthesis with mixed-mode circuits.

A general symmetrical two-port network is shown in Fig. 5.54 as a normal two-port and as a one-port pair (for excitations with the common and push-pull mode).

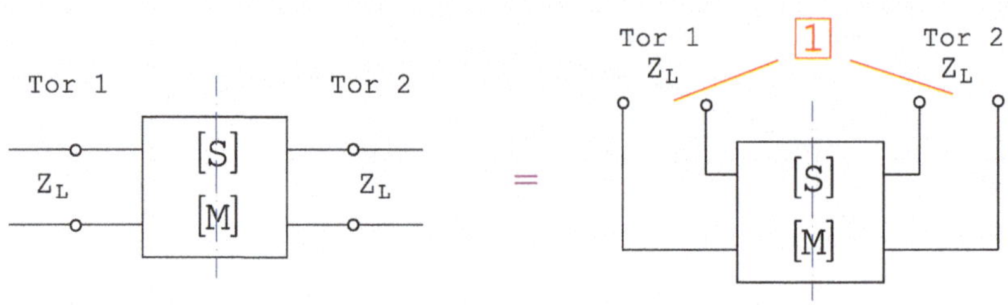

Fig. 5.54: General symmetrical network in S- or M-parameters with two ports (on the right: a port-pair)

Even if symmetrical two-ports contain active components, the scattering parameters $S_{11} = S_{22}$ and $S_{21} = S_{12}$ always apply. This means that only two conditions can be specified, such as the adjustment and a phase shift. For the direct use of the S-parameters in a general network, the procedure was presented in chapter 2.

Using ASTRID, the network is analyzed for desired properties for excitation using both unbalanced and balanced waves. This has the advantage that you only have to analyze two half networks instead of the full network. The example of calculating an attenuator in particular shows how short this makes the solution paths.

Matching

Any symmetrical network is to be adapted. An ideal open circuit (OC) in the symmetry plane is generated with a common-mode excitation at port 1 and port 2, and an ideal short-circuit (SC) in the symmetry plane is generated with the differential-mode excitation at port 1 and port 2.

The OC and SC levels that have been introduced mean that there is no longer any transmission, leaving two identical ports remain. The single port at port 1 has the input reflection coefficient for the open circuit $r_+ \sim M_{11}^+ = r_{even}$ and the input impedance $Z_+ \sim Z_{even}$ and for the short circuit the reflection factor short circuit the reflection coefficient $r_- \sim M_{11}^- = r_{odd}$ and the impedance $Z_- \sim Z_{odd}$ for the short circuit, Fig. 5.55.

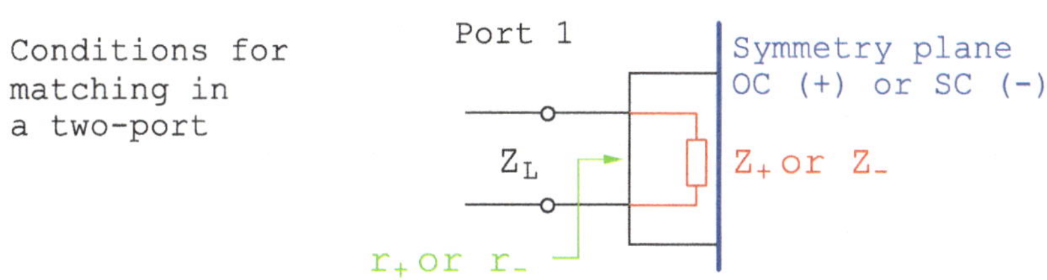

Fig. 5.55: Reduced symmetrical two-port for the OC and SC case

As the incoming waves can only be reflected in the plane of symmetry, the difference between the reflection factor of the common mode wave r_+ and the reflection factor of the push-pull wave r_- must be exactly 180°. I.e.:

$$\boxed{r_+ = -r_-} \quad . \tag{5.79}$$

To fulfill this condition, it is irrelevant whether the network is lossy[10] and what absolute electrical lengths between the ports and the plane of symmetry are present.

[10] As a rule, other circuit synthesis techniques are only capable of designing lossless circuits.

Since the left and right parts of the network are the same, a signal that is incident at port 1 and is only reflected in the middle with a reflection factor of 1 experiences the same attenuation and phase shift as a signal that is transmitted. The following therefore applies:

$$S_{21} = r_+ \quad . \tag{5.80}$$

In circuit synthesis, therefore, only half the network needs to be considered. Furthermore, it makes sense to work with the impedances Z_+ and Z_- rather than the reflection coefficients.

If the equation from (5.79) is used in the known relationship between reflection factor and characteristic impedance with the reference impedance Z_L

$$Z_x = \frac{1+r_x}{1-r_x} \cdot Z_L \qquad \text{and therefore} \qquad Z_+ = \frac{1+r_+}{1-r_+} \cdot Z_L \tag{5.81}$$

and multiplies the two equations together

$$Z_+ \cdot Z_- = \frac{1+r_+}{1-r_+} \cdot Z_L \cdot \frac{1-r_+}{1+r_+} \cdot Z_L \quad , \tag{5.82}$$

is obtained after reduction:

$$\boxed{Z_+ \cdot Z_- = Z_L^2} \quad . \tag{5.83}$$

Eqn. (5.83) is, as will be shown below, the interesting synthesis equation for filter synthesis. If you want to develop customized networks that also have other properties, such as a given signal attenuation or a phase shift, then based on the Eq. (5.81), synthesis equations arise that enable a very simple circuit design.

Matching and Attenuation

If a network is to have a certain attenuation $\alpha = |S_{21}| = S_{21}$ without additional phase shifting, then the conditions for the common-mode and differential-mode characteristic impedance can be calculated directly from Eqs. (5.80) and (5.81):

$$\boxed{Z_+ = \frac{1+\alpha}{1-\alpha} Z_L \ , \quad Z_- = \frac{1-\alpha}{1+\alpha} Z_L} \quad \text{resp.} \quad r_+ = \alpha \quad , \quad r_- = -\alpha \ . \tag{5.84}$$

At the end of this section, the simple application of these synthesis equations for a T attenuator is shown.

Matching and Lossless Phase Shifting

A phase shifter is adapted and has a known transmission phase $\varphi = \beta\ell$. For a lossless phase shifter and a general termination Z_A in the symmetry plane, the input resistance (e.g. from Eq. (4.99)) is known to apply at half the line length:

$$Z_{in} = \frac{Z_A + jZ_L \tan\varphi/2}{1 + j\frac{Z_A}{Z_L}\tan\varphi/2} \quad . \tag{5.85}$$

Es gilt

$$Z_+ = Z_{in}\,|_{Z_A=\infty\,\Omega} \quad , \quad Z_- = Z_{ein}\,|_{Z_A=0\,\Omega} \quad . \tag{5.86}$$

and thus results in

$$\boxed{Z_+ = -j\,\frac{1}{\tan(\varphi/2)}\,Z_L \quad , \quad Z_- = j\,\tan(\varphi/2)\,Z_L} \quad . \tag{5.87}$$

Alternatively, based on Eq. (5.80), the phase shift $S_{21} = r_+ = e^{-j\,\varphi}$ can be used directly in the Eq. (5.81) for calculating the necessary in-phase and opposite-phase resistance. The general synthesis equations are obtained for any phase shift:

$$\text{resp.} \qquad r_+ = e^{-j\,\varphi} \quad , \quad r_- = -e^{-j\,\varphi} \quad . \tag{5.88}$$

Matching, ±90°-Phase Shift and Impedance Transformation

For common-mode excitation, the termination is open-circuited ($Z_A = \infty$) and for push-pull excitation, the termination is short-circuited ($Z_A = 0$). If you insert these two results into (5.85) or the phase values into the Eq. (5.87), you get the following two conditions for a -90° phase shifter:

$$\boxed{Z_+ = -j\,Z_L \ , \quad Z_- = j\,Z_L} \qquad \text{bzw.} \qquad r_+ = -j \quad , \quad r_- = j \quad . \tag{5.89}$$

If you look at a 90° phase shifter, the signs change. If you compare this solution with the calculation of an LC phase shifter in chapter 3.2.3, you will see that the solutions are already given here.

It has already been shown in chapter 4.3.2 that a 90° long line can be used as an impedance transformer if Z_L is chosen appropriately. The same applies to the ±90° phase shifters presented here.. If you look at a 90° phase shifter, the signs change. If you compare this solution with the calculation of an LC phase shifter in chapter 3.2.3, you will see that the solutions are already given here.

Matching and 180° Phase Shifting

If we insert $\varphi = \pi$ into the equations under (5.87) and (5.88), we obtain the two conditions in each case

$$\boxed{z_+ = 0 \ , \quad z_- = \infty} \qquad \text{resp.} \qquad r_+ = -1 \quad , \quad r_- = 1 \tag{5.90}$$

for a lossless 180° phase shifter.

Matching and 360° Phase Shifting

If we insert $\varphi = 2\,\pi$ into the equations under (5.87) and (5.88), we obtain the conditions

$$\boxed{z_+ = \infty\,, \quad z_- = 0} \qquad \text{resp.} \qquad r_+ = 1 \;,\; r_- = -1 \tag{5.91}$$

for a 360° phase shifter without losses.

Table 5.2 clearly shows the results for the most important phase shifters. The phase shifts shown are the values of $\angle S_{21} = -\varphi$.

Tab. 5.2: Synthesis values for the calculation of phase shifters

Phase shift	Z_+	Z_-	r_+	r_-
−90°	$-j\,Z_L$	$j\,Z_L$	$-j$	j
90°	$j\,Z_L$	$-j\,Z_L$	j	$-j$
180°	$0\,\Omega$	$\infty\,\Omega$	-1	1
360°	$\infty\,\Omega$	$0\,\Omega$	1	-1

Isolation

If a network is to have perfect isolation at a certain frequency, there will be no difference between the reflection factors and the characteristic impedances of the common and differential mode waves. The following therefore applies:

$$\boxed{Z_+ = Z_-} \quad , \quad \boxed{r_+ = r_-} \quad . \tag{5.92}$$

Here too, freedom from loss is not assumed. This simple condition makes it much easier to specify the transmission zeros during filter synthesis.

Exercise: Design of a T-Attenuator

Given: *Calculate the resistances R_1 and R_2 of a 3 dB attenuator with a T-structure according to Fig. 5.56.*

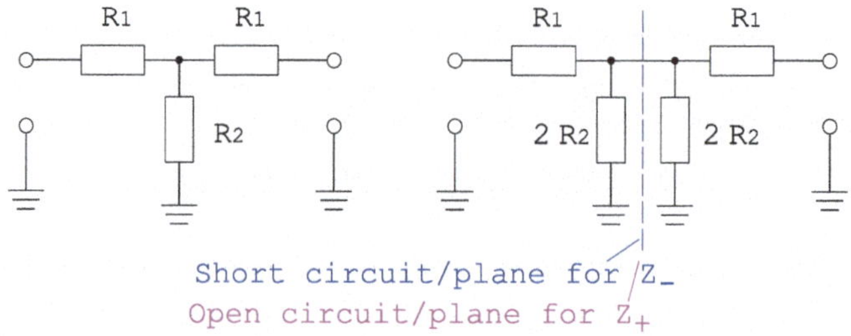

Fig. 5.56: Attenuator with T-structure (right: redrawn for ASTRID synthesis)

Solution: *An attenuation of 3 dB corresponds to a transmission value* $\alpha = 0.708$.

The two resistances result directly from the two equations under (5.84):

$$Z_+ = 292.5\,\Omega \quad \text{und} \quad Z_- = R_1 = 8.55\,\Omega \quad ,$$

with the specification that the push-pull resistor Z_- *corresponds directly to the required resistor* R_1. *The calculation of the common mode resistance* Z_+ *of the network according to Fig. 5.56 directly provides the second resistor sought:*

$$Z_+ = R_1 + 2\,R_2 \quad \Rightarrow \quad R_2 = \frac{Z_+ - R_1}{2} \quad , \quad R_2 = 142.0\,\Omega \quad .$$

——— *Exercise: Design of a T-Attenuator* ———

If you compare this simple exercise with the calculation of the Π attenuator, which was carried out in chapter 2, you can see the efficiency of this ASTRID method.

Filter Synthesis

Using the example of coupled resonator filters, which are very important in practice, the chapter 5 also shows the advantages of synthesis with ASTRID compared to the conventional synthesis technique ([48]). For example, it is shown that an optimal synthesis of resonator filters with any resonator quality is possible using ASTRID.

Areas of Application for ASTRID Synthesis Technology

The number of components used in a circuit plays a very important role, particularly in the high-volume communications market. To reduce the number of components, the functionalities of classic components are combined.

The multi-function components have phase-shift and low-pass characteristics, for example. A large number of new components are required in this area, not least due to cost pressure.

Another area in which ASTRID synthesis technology can be used is broadband technology. The trend in modern communication systems is towards larger bandwidths. If several matching points are distributed over a larger frequency range using ASTRID theory, it is possible to meet even the most demanding bandwidth requirements.

Due to the use of several services in one device, several frequency bands are often required. Here too, novel multi-band components can be derived using ASTRID synthesis.

In the following, the synthesis of a broadband 180° phase shifter will serve as a further example.

——— *Exercise: Design of a 180° Phase Shifter* ———

Given and find: *Calculate the 2 elements C and L (in this order) of a* symmetrical *180° phase shifter according to Fig. 5.57.* L_k *is given as 1 nH. The 180° phase shift is to take place in the ISM band at 2.4 GHz. The phase shifter is perfectly matched by definition!*

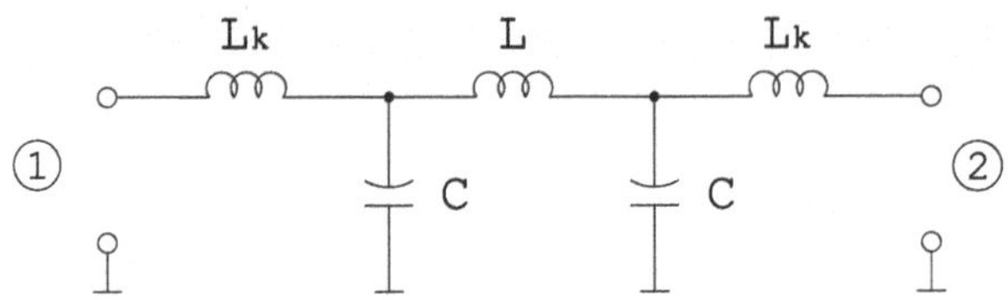

Fig. 5.57: Circuit diagram of the 180° phase shifter to be calculated

Solution:

1.: Common Mode Analysis

$$Z_+ = 0\,\Omega \quad \rightarrow \quad 0\,\Omega = j\,\omega\,L_k + \frac{1}{j\,\omega\,C} \quad \Rightarrow \quad C = \frac{1}{\omega^2\,L_k} = 4.40\,pF \tag{5.93}$$

2.: Differential Mode Analysis

$$Z_- = \infty\,\Omega \quad \rightarrow \quad Y_- = 0\,\tfrac{1}{\Omega} \quad mit \quad Z_- = j\,\omega\,L_k + \tfrac{j\,\omega\,L}{2-\omega^2\,L\,C} \tag{5.94}$$

$$Y_- = 0\,\tfrac{1}{\Omega} \quad \Rightarrow \quad 0\,\tfrac{1}{\Omega} = \tfrac{2-\omega^2\,L\,C}{Denominator} \quad \Rightarrow \quad L = \tfrac{2}{\omega^2\,C} = 2.00\,nH \tag{5.95}$$

——— *Exercise: Design of a 180° Phase Shifter* ———

5.4.1 Analysis of Symmetrical Two-Port Networks

The previous sections showed how easy it is to perform a synthesis using a balanced and unbalanced examination. Once this has been done, you generally want to examine the transfer function. This can also be advantageously specified directly using the balanced and unbalanced conductances, [110]:

$$S_{21} = \frac{z_- \; - \; z_+}{(1+z_-)(1+z_+)} \quad . \tag{5.96}$$

If a network cannot be found in the auxiliary sheet A.2, an analysis using the Eq. (5.96) is the simplest way.

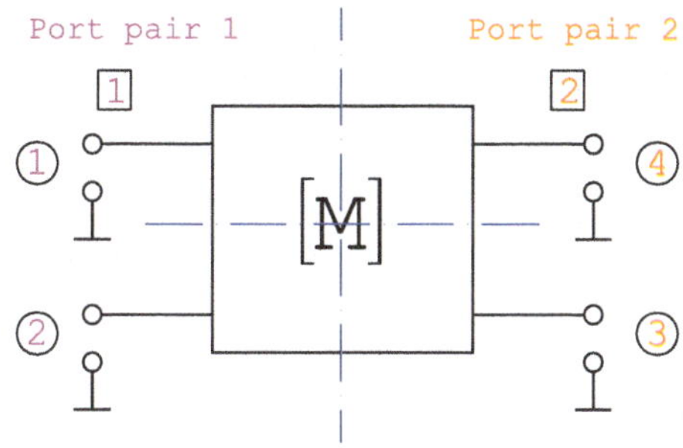

Fig. 5.58: Mixed-mode network with double symmetry

5.4.2 Synthesis of Symmetrical Mixed-Mode Networks

A symmetrical mixed-mode network should be defined in such a way that there is an axis of symmetry between the port pairs and that there is also symmetry in the division in an upper and lower section. This is illustrated in Fig. 5.58 by the indication of two symmetry axes and is thus referred to as `double symmetry`.

A symmetrical mixed-mode network has no mode conversion and can be reduced to just four M-parameters:

$$\begin{pmatrix} b_1^- \\ b_2^- \\ b_1^+ \\ b_2^+ \end{pmatrix} = \begin{bmatrix} M_{11}^- & M_{21}^- & 0 & 0 \\ M_{21}^- & M_{11}^- & 0 & 0 \\ 0 & 0 & M_{11}^+ & M_{21}^+ \\ 0 & 0 & M_{21}^+ & M_{11}^+ \end{bmatrix} \begin{pmatrix} a_1^- \\ a_2^- \\ a_1^+ \\ a_2^+ \end{pmatrix} \quad . \tag{5.97}$$

Whereas only half of the two-port network had to be calculated twice, with this mixed-mode network only a quarter has to be calculated four times. As a result, the number of equations increases, but the complexity does not increase.

The synthesis of such networks is carried out in such a way that in a first step only the push-pull network with the specification of M_{11}^- and M_{21}^- is synthesized. To do this short-circuits must be introduced in the horizontal symmetry line. What remains is a symmetrical two-port network with input port 1 and output port 4. This network can now be treated in the same way as the ASTRID method was introduced. In this case, $S_{11} = M_{11}^-$ and $S_{21} = M_{21}^-$.

In a separate second step, the common mode network is specified with the parameters M_{11}^+ and M_{21}^+. To do this, you must introduce empty runs in the horizontal symmetry line. This again leaves a symmetrical two-port network with input port 1 and output port 4 with the conditions $S_{11} = M_{11}^+$ and $S_{21} = M_{21}^+$ for the ASTRID method.

This separate approach allows you to achieve two completely different characteristics for the common and differential mode behavior of the circuits if you choose a suitable network. It is difficult to make specifications for the choice of network. It is helpful for the push-pull behavior to start from circuits in unbalanced conductor technology and to mirror them on the horizontal axis. To further influence the common mode behavior, shunt elements can be converted into cross elements. After the conversion, these elements no longer exist for the common mode signal.

5.5 Couplers and the Basics of Coupler Synthesis

Apart from the measuring bridge, there is no circuit in the electronics that comes close to a so-called RF coupler. A coupler has four matched ports and couples a signal that is incident on one port to two other ports. The fourth port is decoupled. The corresponding symbol is shown in Fig. 5.59. The S-matrix of an ideal (at least simply symmetrical)

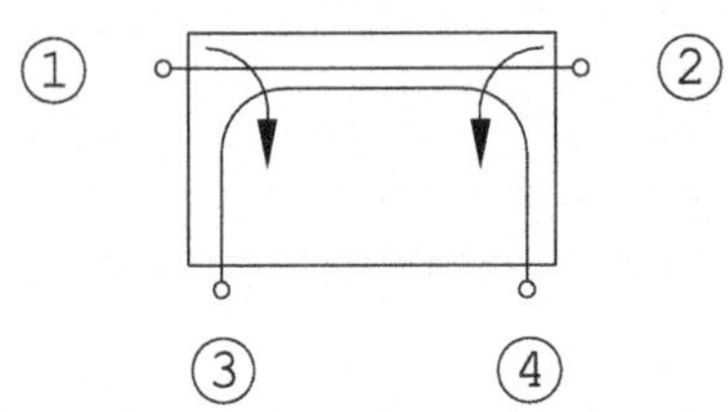

Fig. 5.59: Symbol of an RF coupler in single-pole representation

coupler is shown in (5.98).

$$[\mathbf{S_{kopp}}] = \begin{bmatrix} 0 & \tau & \kappa & 0 \\ \tau & 0 & 0 & \kappa \\ \kappa & 0 & 0 & \tau \\ 0 & \kappa & \tau & 0 \end{bmatrix} \tag{5.98}$$

The variable τ stands for the transmission factor (main path) and κ for the coupling transmission factor (coupling factor or value for short). The following applies to this ideal lossless coupler:

$$|\tau|^2 = 1 - |\kappa|^2 \quad . \tag{5.99}$$

In practice, the coupling values are between -40 dB and -3 dB. The table in the appendix (1) shows the solution of Eq. (5.99) for the most important coupling values. Here, S_{11} must be replaced by τ and S_{21} by κ. This table is helpful in practice to specify the transmission losses in the main path.

In practice, not only is the matching finite, but so is the isolation ν to the fourth port. For the latter case, the scattering matrix is obtained (with perfect matching values)

$$[\mathbf{S_{kopp}}] = \begin{bmatrix} 0 & \tau & \kappa & \nu \\ \tau & 0 & \nu & \kappa \\ \kappa & \nu & 0 & \tau \\ \nu & \kappa & \tau & 0 \end{bmatrix} \quad . \tag{5.100}$$

For the application, the so-called directivity

$$\delta^{dB} \quad = \quad - \left(\; \nu^{dB} \quad - \quad \kappa^{dB} \; \right) \tag{5.101}$$

of interest. This directivity reflects the signal-to-noise ratio of the measured variable. It is therefore an essential quality feature of a coupler.

In a coupler, two times two ports (here ports 1 and 4 as well as 2 and 3) shall be isolated from each other. Thus, the condition (5.92) must hold. Furthermore, a coupler shall be matched at all four ports. Thus, the condition (5.83) must hold.

The following therefore applies at the port pair 1: $M_{11}^{-} = M_{11}^{+} = 0$, which is synonymous with: $S_{11} = S_{12} = S_{21} = S_{22} = 0$. These two conditions are only applicable in the case of

$$\boxed{Z_+ \; = \; Z_L \; , \quad Z_- \; = \; Z_L} \qquad \text{resp.} \qquad \boxed{r_+ \; = \; 0 \quad , \quad r_- \; = \; 0} \tag{5.102}$$

This principle has been known for some time ([125]):

> Two ports are matched and decoupled from each other when both a differential-mode wave and a common-mode wave, which each enter the network via these two ports (this port pair), are matched.

The following couplers will show how this principle is applied.

5.5.1 Wilkinson Coupler

Wilkinson Coupler as Signal Divider

Other names for this Wilkinson coupler are 0°-coupler, signal divider, adapted three-port and Wilkinson divider. The variety of names for the component shown in Fig. 5.60 reflects its importance in practice.

With the Wilkinson coupler, one port is usually terminated with a resistor. Therefore, this coupler is only shown as a three-port device.

For this coupler, what was presented in the section 3.3.4 as an ideal signal divider and specified with Eq. (3.27) applies:

$$[\mathbf{S_{ST}}] \;\; = \;\; -j \begin{pmatrix} 0 & 0.7 & 0.7 \\ 0.7 & 0 & 0 \\ 0.7 & 0 & 0 \end{pmatrix} \quad . \tag{5.103}$$

Since this signal divider contains a resistor, it does not contradict the principle that there is no universally adapted lossless three-port network.

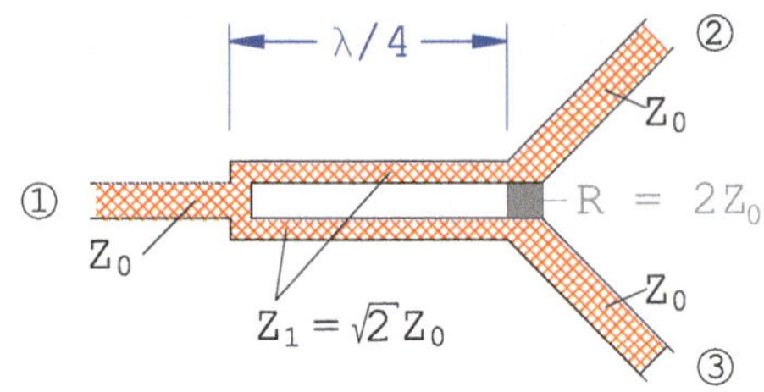

Fig. 5.60: Three-port Wilkinson coupler as a planar view on a microstrip line for use as a matched signal divider

To analyze this coupler, we first consider an common mode wave that propagates from port pair 2, which consists of port 2 and port 3, towards port 1. For this in-phase wave, we have a characteristic impedance at port pair 2 of $Z_{even} = Z_0/2$. This must be transformed up to Z_0 at port 1.

Therefore, a common-mode wave impedance of $Z_0/\sqrt{2}$ is required for the $\lambda/4$ line. Since Z_1 is the wave impedance of the unsymmetrical line and this corresponds to double the value of the common-mode wave, $Z_1 = Z_0 \cdot \sqrt{2}$. The resistor R has no influence on the common-mode wave, as no current flows through it in in-phase operation.

For the push-pull case, only the half of the coupler lying above the symmetry plane needs to be considered, as the potential in the middle is zero and can therefore be set to ground according to Fig. 5.61.

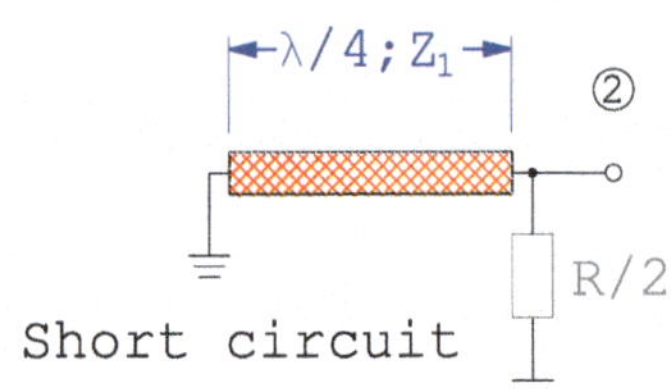

Fig. 5.61: Substitute circuit for a push-pull wave at the Wilkinson coupler

The short circuit at the end of the $\lambda/4$ line is transformed into an open circuit and the remaining resistance $R/2$ must correspond to the characteristic impedance Z_0.

This coupler is only suitable for a bandwidth of about 20 % due to the $\lambda/4$ lines it contains. The bandwidth can be significantly increased using multi-stage transformation lines. Fig. 5.62 shows the solution for a two-stage transformation.

Wilkinson-Coupler as Mode Splitter

If all four ports of the Wilkinson coupler are used, different modes occur. If the two ports 3 and 4 (Fig. 5.63) are combined into a port pair, the common-mode wave fed into this port pair is only transmitted to port 1 and the differential-mode wave fed in is transmitted to port 2.

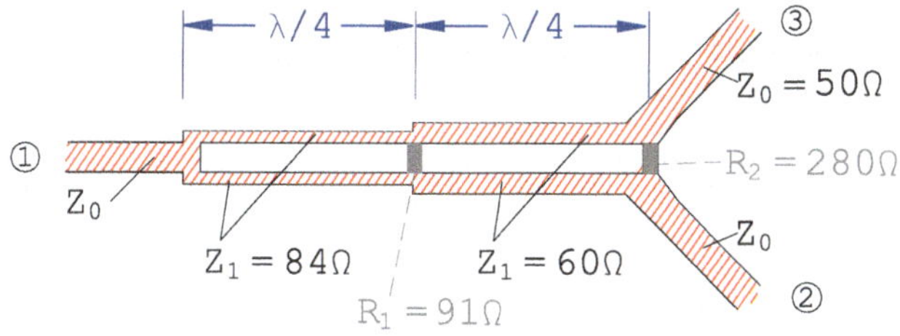

Fig. 5.62: Two-stage Wilkinson coupler as top view of a microstrip line for broadband applications

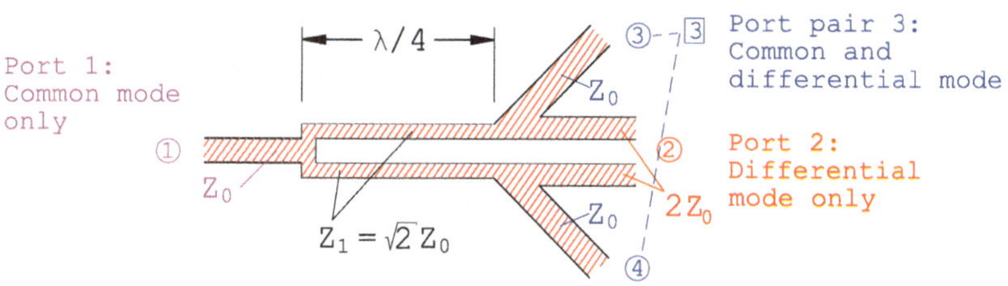

Fig. 5.63: Four-port Wilkinson coupler as a top view of a microstrip line for use as a mode switch

5.5.2 Line Coupler

The other couplers presented, such as the line coupler, which is very important in measurement technology, can be advantageously designed using the common and differential mode analysis.

The line coupler is based on the coupling of two TEM or quasi-TEM lines over the length $\beta\ell$. One possible form of implementation is shown in Fig. 5.64.

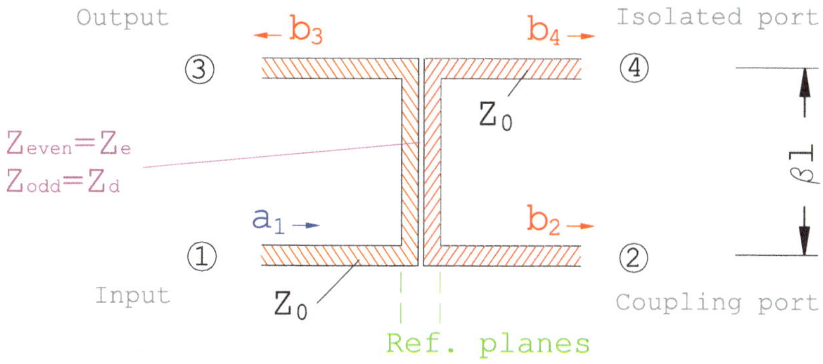

Fig. 5.64: Top view of a microstrip line coupler

The characteristic impedances of the common and push-pull are only used in a normalized way in the following. $z_e = Z_e/Z_0$ and $z_d = Z_d/Z_0$.

The mathematical description is based on the introduction to the coupler chapter (page 177). This four-port network is decomposed into two two-ports using common-mode and differential-mode excitation.

For one half of the line, the following applies in the coupling range (within the reference planes) using the auxiliary sheet A3 (page 442) for the normalized A-matrix:

$$\begin{pmatrix} A^+ & B^+ \\ C^+ & D^+ \end{pmatrix} = \begin{pmatrix} \cos(\beta\ell) & j \cdot z_e \sin(\beta\ell) \\ j \cdot \frac{\sin(\beta\ell)}{z_e} & \cos(\beta\ell) \end{pmatrix} \quad . \tag{5.104}$$

The corresponding scattering parameters for reflection and transmission can be calculated using the auxiliary sheet A.6:

$$M_{11}^+ = \frac{j \cdot (z_e - 1/z_e) \sin(\beta\ell)}{2 \cos(\beta\ell) + j \cdot (z_e + 1/z_e) \sin(\beta\ell)} \quad , \tag{5.105}$$

$$M_{21}^+ = \frac{2}{2 \cos(\beta\ell) + j \cdot (z_e + 1/z_e) \sin(\beta\ell)} \quad . \tag{5.106}$$

This two-port is also reciprocal and reflection-symmetric($M_{21}^+ = M_{12}^+$ and $M_{11}^+ = M_{22}^+$). The line half of the coupler can be calculated in the same way for the push-pull case:

$$\begin{pmatrix} A^- & B^- \\ C^- & D^- \end{pmatrix} = \begin{pmatrix} \cos(\beta\ell) & j \cdot z_d \sin(\beta\ell) \\ j \cdot \frac{\sin(\beta\ell)}{z_d} & \cos(\beta\ell) \end{pmatrix} \quad , \tag{5.107}$$

$$M_{11}^- = \frac{j \cdot (z_d - 1/z_d) \sin(\beta\ell)}{2 \cos(\beta\ell) + j \cdot (z_d + 1/z_d) \sin(\beta\ell)} \quad , \tag{5.108}$$

$$M_{21}^- = \frac{2}{2 \cos(\beta\ell) + j \cdot (z_d + 1/z_d) \sin(\beta\ell)} \quad \text{as well as} \quad M_{21}^- = M_{12}^- \text{ and } M_{11}^- = M_{22}^- \; . \tag{5.109}$$

The following values should be synthesized as broadly as possible for the four-port line coupler:

$$S_{ii} = 0 \; ; \quad S_{41} = 0 \; ; \quad S_{32} = 0 \quad . \tag{5.110}$$

The condition from Eq. (5.83) must be met in order to fulfill the matching requirement: $M_{11}^+ + M_{11}^- = 0$.

To fulfill the requirement of isolation, the condition from Eq. (5.92) must be met: $M_{21}^+ - M_{21}^- = 0$.

The corresponding solution of these two conditions for the design of a line coupler is:

$$\boxed{z_e \cdot z_d = 1} \quad . \tag{5.111}$$

Another variable that can be calculated directly from the M-parameters is the transmission factor:

$$\tau = S_{31} = \frac{1}{2}(M_{21}^{+} + M_{21}^{-}) \;=\; \frac{2}{2\cos(\beta\ell) + j\cdot(z_d + 1/z_d)\sin(\beta\ell)} \quad . \tag{5.112}$$

The complex value for the coupling factor can also be calculated directly from the M-parameters:

$$\kappa = S_{21} = \frac{1}{2}(M_{11}^{+} - M_{11}^{-}) \;=\; \frac{j\cdot(z_d - 1/z_d)\sin(\beta\ell)}{2\cos(\beta\ell) + j\cdot(z_d + 1/z_d)\sin(\beta\ell)} \quad . \tag{5.113}$$

It can be seen that τ and κ differ by 90° independently of $\beta\ell$! Because this coupler divides the two output signals over a wide bandwidth with this phase offset of 90° , the line coupler is also assigned to the class of 90°-couplers.

The realization of a line coupler is shown in the Fig. 5.65. In this two-stage coupler, which is described in more detail in [91], among other places, the bandwidth is higher than in the single-stage solution shown.

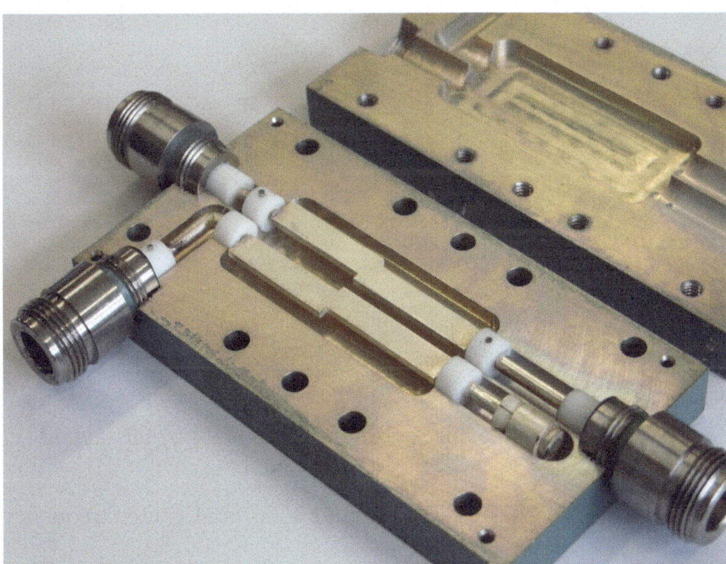

Fig. 5.65: Two-stage line coupler implemented in TEM line technology

The most complex construction in the coupler in Fig. 5.65 is the resistor to ground, which is located at the bottom right of a metal tube. By using a field simulator (see Chapter 2 in [50]), such a coupler can be optimally optimized and then manufactured with almost the same microwave performance.

The couplers presented below are also best developed by first calculating them using analytical solutions and then optimizing them using a simulator.

If you are working on planar substrates, a circuit simulator is often sufficient. A field simulator is generally better. The enclosing (often metallic) housing is also not unimportant for these longer cable constructions. The field simulator also calculates this influence correctly.

5.5.3 Hybrid Couplers

The two frequently used hybrid couplers (Fig. 5.66) are not only used as couplers, but also as 3 dB signal dividers and as phase shifters.

The 180°-hybrid[11], also known as a "rat-race" coupler, is used to generate push-pull signals from a single-ended signal.

The 90°-hybrid, also known as a "branch line" coupler, is used to generate *complex* signals.

Both are often sub-circuits of more complex components such as balanced mixers. The simple dimensioning rules for the two couplers are shown in Fig. 5.66.

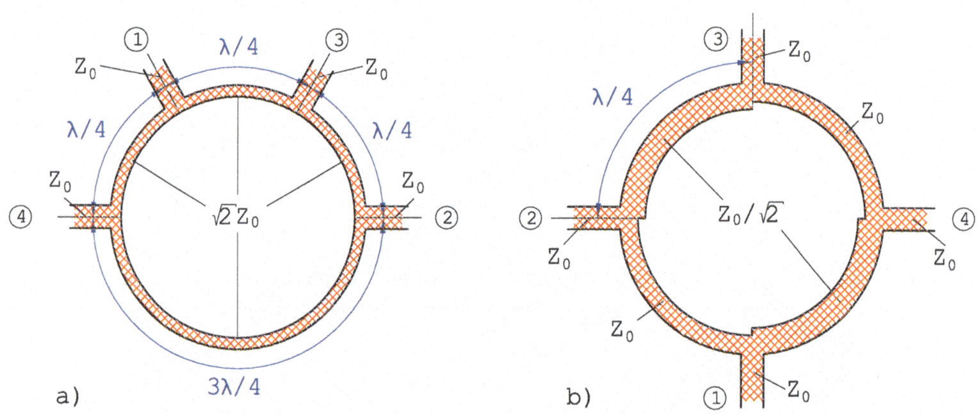

Fig. 5.66: Hybrid couplers: a) 180°-hybrid; b) 90°-hybrid

The scattering matrix of the 180°-hybrid looks as follows at center frequency:

$$\left[\mathbf{S}^{\mathbf{180^\circ-Hyb}}\right] = -j \begin{pmatrix} 0 & 0 & 0.707 & 0.707 \\ 0 & 0 & 0.707 & -0.707 \\ 0.707 & 0.707 & 0 & 0 \\ 0.707 & -0.707 & 0 & 0 \end{pmatrix} \quad . \tag{5.114}$$

The scattering matrix of the 90°-hybrid at center frequency is as follows:

[11]The specification in degrees always refers to the differential phase between the outputs.

$$\left[\mathbf{S}^{90^\circ-\mathbf{Hyb}}\right] = -\begin{pmatrix} 0 & 0 & 0.707 & j\,0.707 \\ 0 & 0 & j\,0.707 & 0.707 \\ 0.707 & j\,0.707 & 0 & 0 \\ j\,0.707 & 0.707 & 0 & 0 \end{pmatrix} \quad . \tag{5.115}$$

Both couplers can be calculated using excitations with common-mode and differential-mode signals. For both hybrid couplers, the following applies when exciting at port 1 and port 2.

$$M_{11}^{+} = M_{11}^{-} = 0 \tag{5.116}$$

and consequently matching at port 1 as well as port 2 and decoupling between port 1 and port 2.

Exercise: Designing a 90° hybrid coupler

Excite ports 1 and 3 with in-phase and opposite-phase waves and evaluate the resulting chain circuits according to Fig. 5.67, so that decoupling between ports 1 and 3 and all-round matching result from the conditions according to Eq. (5.116).

Furthermore, a 3 dB coupling (or signal splitting) should be implemented.

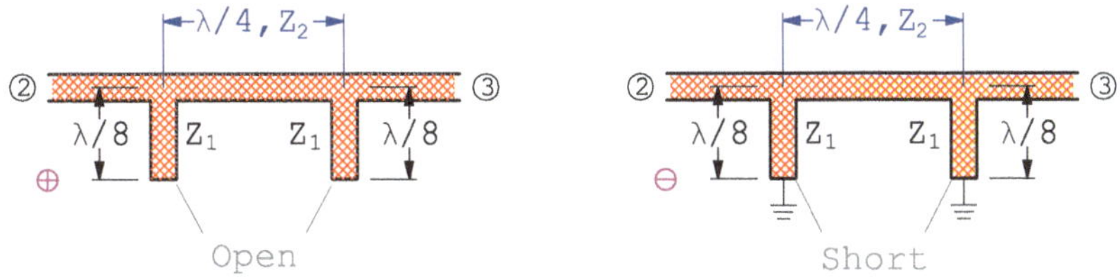

Fig. 5.67: Equivalent circuit diagram of the 90° hybrid coupler for common mode and push-pull excitation

Exercise: Designing a 90° hybrid coupler

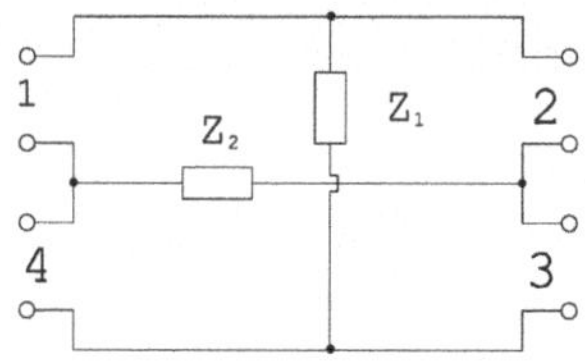

Fig. 5.68: Basic structure of the resistive coupler

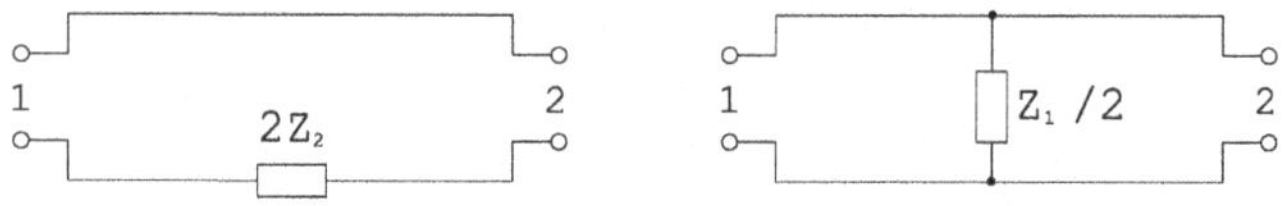

Fig. 5.69: Equivalent circuits for the common and differential mode excitation of the resistive coupler

5.5.4 Resistive Coupler and ±90°-LC-Coupler

The so-called resistive couplers, Bild 5.68, which are also shown in [125], deserve special mention.

The method of calculating this coupler corresponds to that of the line coupler. With a unbalanced and balanced excitation at the two ports 1 and 4, the two simple equivalent circuits shown in Fig. 5.69.

If this coupler is used according to [125] for broadband applications in measurement technology, it is advantageous to use two resistors ($R_1 = Z_1$ and $R_2 = Z_2$) that fulfill the condition:

$$Z_0^2 = Z_1 Z_2 \quad . \tag{5.117}$$

For $Z_1 = Z_2 = Z_0$ this results in a uniform signal division with 6 dB attenuation. In general, the attenuations of both signal paths are calculated via

$$S_{21} = \frac{Z_0}{Z_0 + Z_2} \quad \text{and} \quad S_{41} = \frac{Z_2}{Z_0 + Z_2} \quad . \tag{5.118}$$

This coupler, which can be built with only two components, is particularly interesting for mixed-mode applications. It is largely unknown that this coupler works as a 90-°-coupler when using only one capacitor and one inductor, [61]. This 90°-coupler shown in Fig. 5.70 can only be connected by one port in an GND-symmetrical way. The other ports are then GND-unsymmetrical.

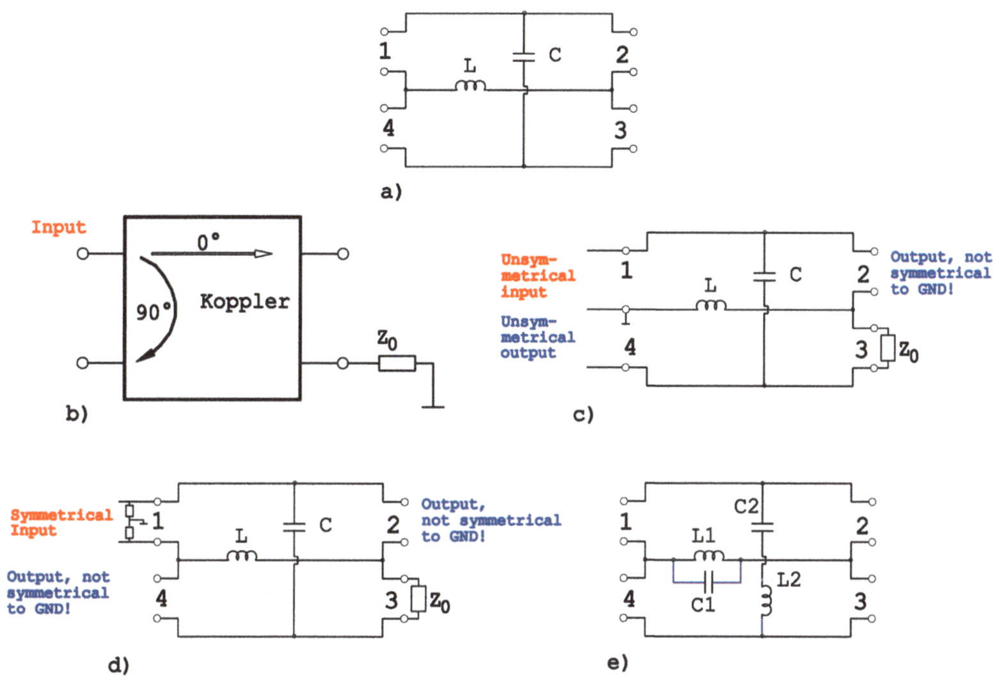

Fig. 5.70: Basic structure of the a) 90°-LC-couplers, b) its operating principle, c) in an asymmetrical circuit, d) in ground-symmetrical configuration, e) as a dual-band coupler

If this coupler is used in unsymmetrical setups, only two ports allow unsymmetrical connections, Fig. 5.70. These disadvantages have no effect on the wiring of symmetrical antennas, for example, [61]. In this application, this simple coupler assumes the functionality of a balun, a signal splitter and an isolator[12]!

For the -90°-coupler, only L and C need to be swapped, which also corresponds to the basic principle of the dual-band coupler. These ±90°-couplers each have only a 3 dB attenuation if $Z_0 = \omega L = \frac{1}{\omega C}$.

[12] Isolators are introduced in chapter 10.

Chapter 6

Resonators and Filters

Circuits that have a strongly frequency-dependent transmission behavior are referred to as filter circuits or filters for short. These filter circuits are used very frequently in applications. The components used are either capacitors and coils and/or RF-lines and/or so-called resonators, which are manufactured using a wide variety of technologies. As the synthesis of filters is a very broad field, this chapter can only provide an insight.

In principle, a distinction can be made between two classes of filters: On the one hand the low and high-passes and on the other hand the band-passes and band-stops.

Low and high-pass filters are usually synthesized from concentrated coils and capacitors. Quasi-concentrated components can also be used for realization. Filter design methods with Butterworth or Chebyshev standard low-pass filters are used to calculate them.

Band-pass and band-stop filters are often made from resonators. The simplest resonators can consist of a series or parallel L-C combination. Other resonators already presented are the $\lambda/4$ and $\lambda/2$ lines. Quartz crystals are important resonators in the lower frequency range and dielectric resonators in the upper frequency range. In practice, band-stop filters are less important than band-pass filters.

The most common ways to design band-passes are

1. Synthesis from Butterworth or Chebyshev standard low-passes.
2. Calculation as elliptical filters.
3. Working with Kuroda identities and unit lines.
4. Design of narrowband filters with resonators and inverters.
5. Coupled resonator filters.

The synthesis of low-pass, high-pass and band-pass filters from Butterworth or Chebyshev standard low-pass filters is discussed below as well as in many high-frequency works (e.g. [8, 91]). The disadvantage of these filters is that they have no transmission zeros in the stopband.

H. Heuermann, *Radio Frequency Technologies*,
https://doi.org/10.1007/978-3-658-50170-9_6

In practice, elliptical filters are used as improved filters if required. Fig. 6.1 shows the typical transfer behavior of an elliptical filter and a dual-circuit resonator filter (a Butterworth filter is usually significantly flatter in the stopband than the resonator filter).

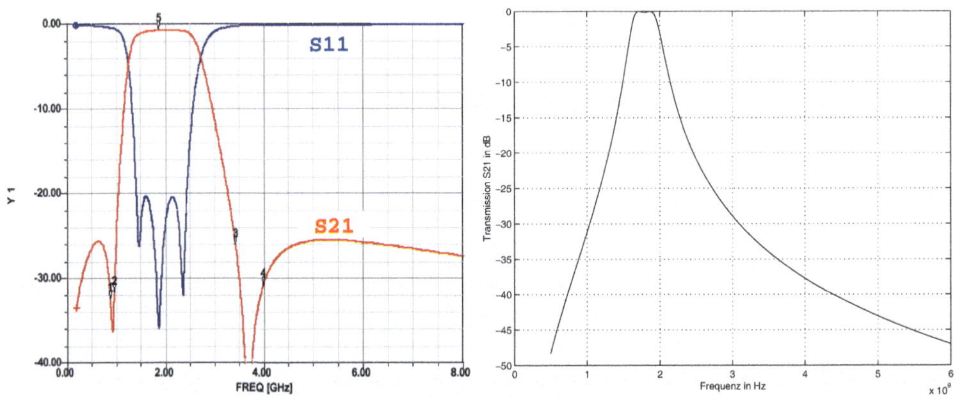

Fig. 6.1: Transfer functions of bandpass with (left:) elliptical filter and (right:) filter with two coupled resonators

Elliptical filters (also known as Cauer filters) can be used to realize the same stop-band characteristics across a wide band, which is their great advantage. A very good filter manual for the design of elliptical filters is [120].

How to synthesize microwave filters with distributed components from standard filters using the so-called Kuroda identities and unit lines can be read in [8] and the standard work on high-frequency and RF filter synthesis [91].

In filter synthesis for RF technology, the central research topic in recent decades has been design methods for narrow-band filters with resonators and inverters. To successfully implement these theories, resonators of the highest quality (= very low losses) are required. Therefore, these theories are not applicable to the majority of applications. A very good German-language introduction to one of these theories can be found in [8] and [73].

The theory of coupled band-pass filters was established many decades ago [32, 48]. As coupled band-pass filters have the widest range of applications, the theory for dual-circuit filters will be explained in detail below. In contrast to the solutions shown in [32, 48], these are exact closed equations that are very easy to apply.

Before we can discuss this theory, however, we need to understand the properties of resonators. This resonator theory is also very important for the design of oscillators. Furthermore, a new method for increasing the loaded Q-factor of resonators is presented.

6.1 Synthesis from Butterworth or Chebyshev Standard Low-Pass Filters

The following values are specified for low-pass filters (LP-filters), high-pass filters (HP-filters) and bandpass filters (BP):

A_d: Insertion loss in dB,

A_c: Maximum ripple in the pass-band in dB,

A_s: Minimum stop-band attenuation in dB,

f_d: Cutoff frequency,

f_c: Corner frequency (at 3 dB),

f_s: Stop-band edge frequency,

f_{max}: Maximum guaranteed frequency for A_s,

n: Filter order.

These variables are shown in Fig. 6.2 for an LP.

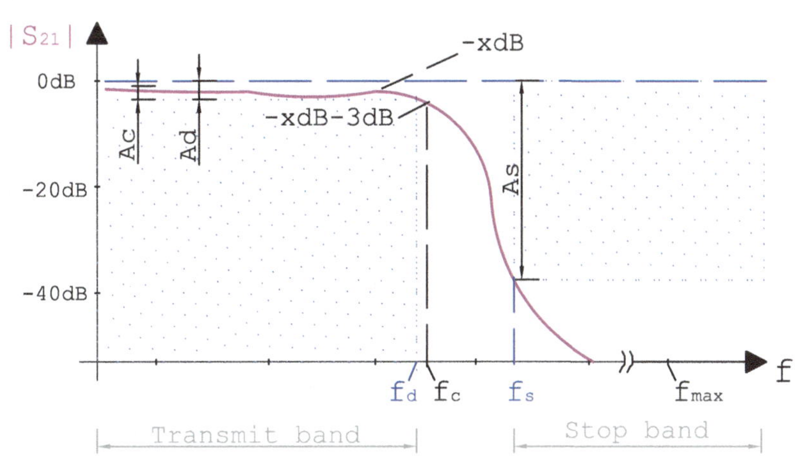

Fig. 6.2: Transfer function and specifications of a low-pass filter

In the filter design, the components are initially assumed to be loss-free. This means that A_d and A_c are identical for the first design step.

The transfer function of these filters is derived from the mathematical Butterworth and Chebychev polynomials.

Transfer Function of the Butterworth LP-Filters

The Butterworth filter has no ripple at all. However, A_s is lower than with the Chebyshev filter for the same component cost.

Mathematically, the transfer function is very easy to calculate:

$$|S_{21}|^2 = \frac{1}{1 + k_d^2 \cdot f_n^{2n}} \tag{6.1}$$

with the normalized frequency $f_n = f/f_c$ and the insertion loss value k_d and the filter order n. The associated transfer functions up to the 5th order are shown in Fig. 6.3.

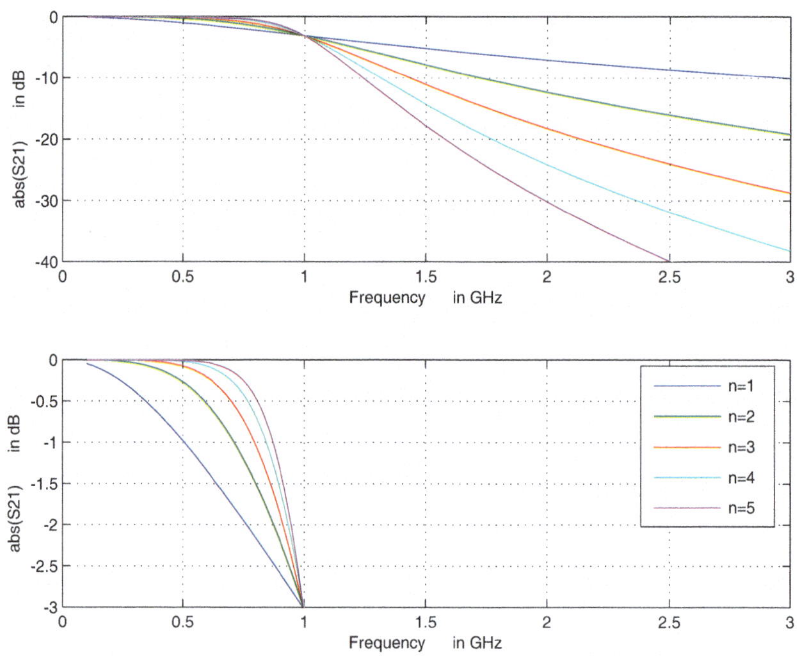

Fig. 6.3: Transfer functions over frequency of Butterworth low-pass filters (k_d=1)

If distortions (constant group delay times) are important, this Butterworth filter is preferred. If only low throughput losses with the highest possible rejection suppression are important, a Chebychev filter is preferred.

Transfer Function of the Chebychev LP-Filters

The ripple A_c of a Chebychev filter is calculated from the specification of the insertion loss value k_d:

$$A_c = 10 \cdot \log\left(1 + k_d^2\right) dB \quad . \tag{6.2}$$

The greater the ripple in the pass-band, the higher the blocking suppression values in the blocking range.

The transfer function of a Chebychev filter can be calculated using

$$|S_{21}|^2 = \frac{1}{1 + k_d^2 \cdot T_n^2} \tag{6.3}$$

with

$$T_1 = f_n; \qquad T_2 = 2f_n^2 - 1; \qquad T_3 = 4f_n^3 - 3f_n; \qquad T_n = 2f_n T_{n-1} - T_{n-2} \quad . \tag{6.4}$$

The associated transfer functions up to the 4th order are shown in Fig. 6.4.

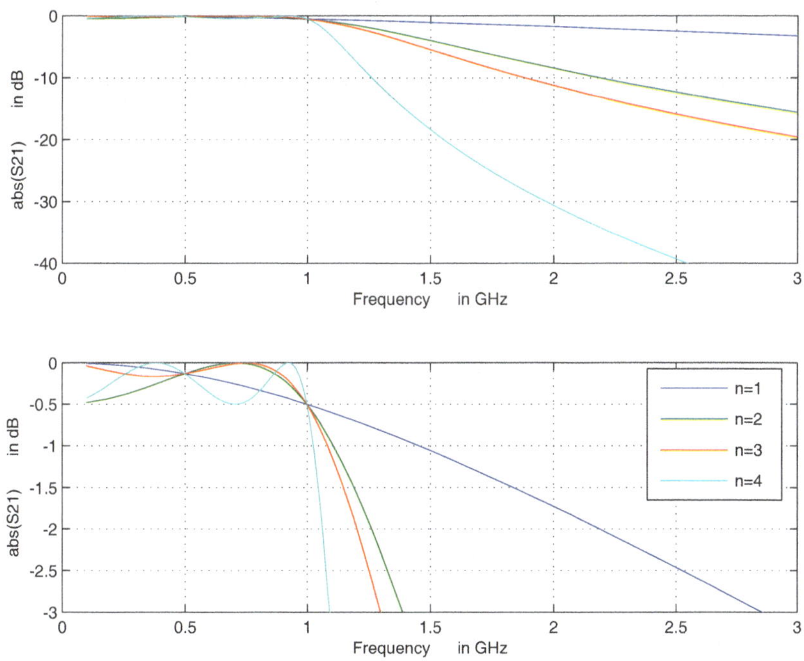

Fig. 6.4: Transfer functions over frequency of Chebychev low-passes (A_c=0.5 dB, k_d=0.349)

6.1.1 Circuit Design of Standard Filters

In practice, the circuit design for Butterworth and Chebychev filters (standard filters for short) is based on the low-pass filter and the use of tables. An LP can use a low-impedance input (capacitor to ground) or a high-impedance input (series coil) to generate the increasing isolation at high frequencies.

Circuit Design of LP-Filters

Two setups are therefore possible for the circuit design of a low-pass filter. Fig. 6.5 shows these two LPs for working with a table.

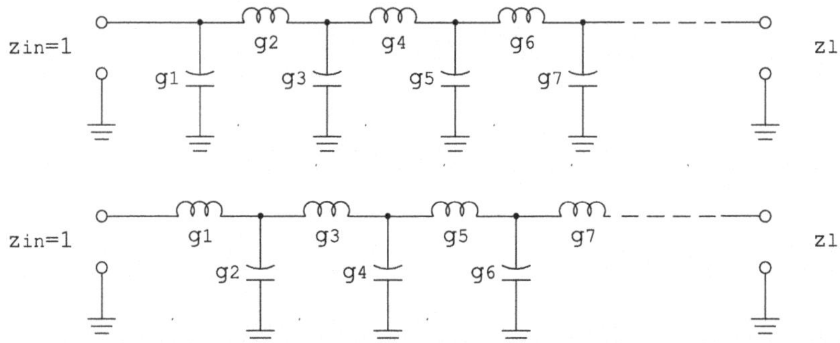

Fig. 6.5: Standard-LP with standardized impedance values for the filter design

The elements of a real low-pass filter can be easily calculated via

$$L_i = \frac{g_i Z_0}{\omega_d} \quad \text{and} \quad C_i = \frac{g_i}{Z_0 \omega_d} \tag{6.5}$$

with the values:

g_i: Element value of the normalized filter (from the table),

Z_0: System impedance (at the input; usually 50 Ω),

ω_d: Frequency at the pass-band limit: $\omega_d = 2\pi f_d$.

The element values g_i are calculated for a Butterworth filter with k_d=1 using the simple equation:

$$g_i = 2 \sin\left(\frac{\pi(2i-1)}{2n}\right) \quad \text{with} \quad i = 1, 2, \dots n \quad . \tag{6.6}$$

The values up to the 5th order are shown in the table.

Tab. 6.1: Element values g_i for a normalized Butterworth low-pass filter (A_c=3 dB, k_d=1)

n	g_1	g_2	g_3	g_4	g_5	z_l
1	2					1
2	1.414	1.414				1
3	1	2	1			1
4	0.7654	1.848	1.848	0.7654		1
5	0.6180	1.618	2	1.618	0.6180	1

For a Chebychev-LP, it is no longer possible to specify a simple equation for calculating the elements. The three most important tables for circuit design are shown below.

It should be noted that the asymmetrically constructed filters perform an impedance transformation and therefore cannot be used in a 50 Ω system.

Tab. 6.2: Element values g_i for a normalized Chebychev low-pass filter (A_c=0.2 dB, k_d=0.2171)

n	g_1	g_2	g_3	g_4	g_5	z_l
1	0.4342					1
2	1.038	0.6745				0.6499
3	1.228	1.153	1.228			1
4	1.303	1.284	1.976	0.8468		0.6499
5	1.339	1.337	2.166	1.337	1.339	1

Tab. 6.3: Element values g_i for a normalized Chebychev low-pass filter (A_c=0.5 dB, k_d=0.3493)

n	g_1	g_2	g_3	g_4	g_5	z_l
1	0.6986					1
2	1.403	0.7071				0.5040
3	1.596	1.097	1.596			1
4	1.760	1.193	2.366	0.8419		0.5040
5	1.706	1.230	2.541	1.230	1.706	1

Tab. 6.4: Element values g_i for a normalized Chebychev low-pass filter (A_c=1.0 dB, k_d=0.5089)

n	g_1	g_2	g_3	g_4	g_5	z_l
1	1.018					1
2	1.822	0.6850				0.3760
3	2.024	0.9941	2.024			1
4	2.099	1.064	2.831	0.7892		0.3760
5	2.135	1.091	3.001	1.091	2.135	1

Circuit Design Examples of LP-Filters

Entering the table values for a 5th order filter, a pass-band frequency of 1 GHz and a system impedance of 50 Ω results in the following values:

Butterworth:	L1=L5= 4.9 nH;	C2=C4= 5.2 pF;	L3= 15.9 nH
Chebychev (A_c=0.5 dB):	L1=L5= 13.6 nH;	C2=C4= 3.9 pF;	L3= 20.2 nH

These values were entered in the ADS circuit simulator: Fig. 6.6 shows the scattering parameters and the group delay values of these two filters.

It can be seen that the Butterworth filter is very well matched at low frequencies and only poorly matched at medium frequencies. The Chebychev filter, on the other hand, is moderately well matched over the entire passband. In terms of blocking behavior, it is superior to the Butterworth filter.

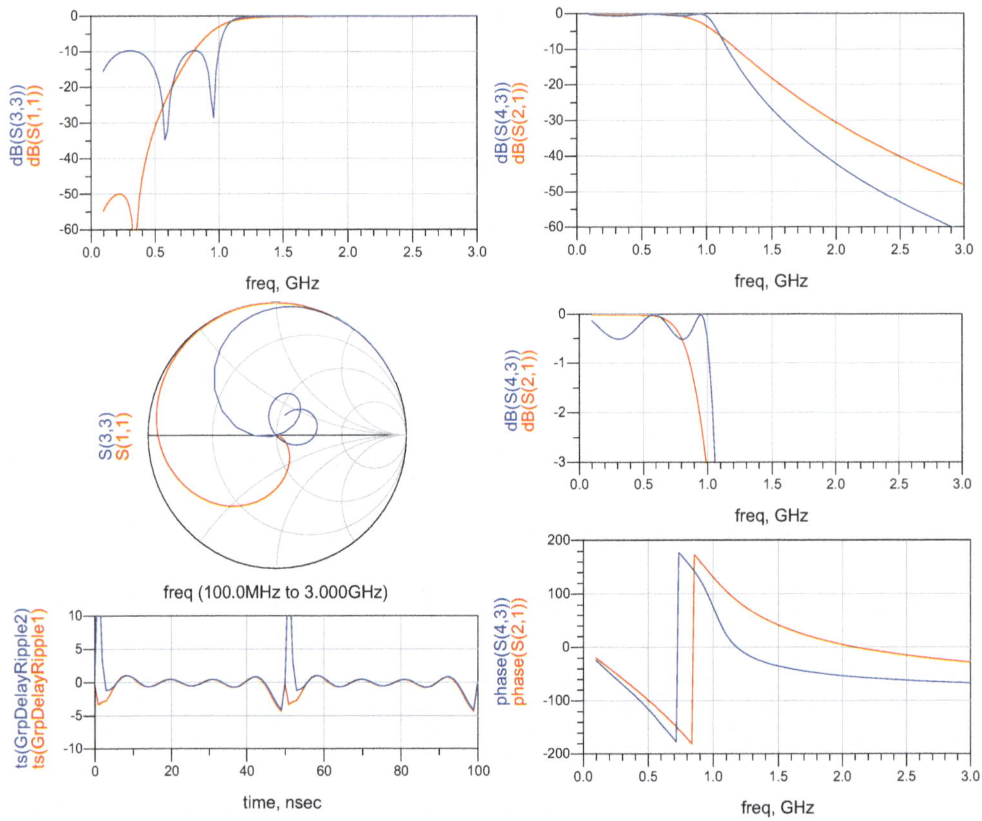

Fig. 6.6: Simulation results of the two 5th order filters (red: Butterworth, blue: Chebychev)

The Butterworth filter has less dispersive phase response in the transition range and a lower group delay.

Conversion in HP-Filters

Since high-passes block when low-passes let through and vice versa, a HP can be designed with the LP theory presented by introducing a new normalized frequency $f_n^{HP} = 1/f_n$. As a result, the element values of the unit filters must be inverted, see Fig. 6.7.

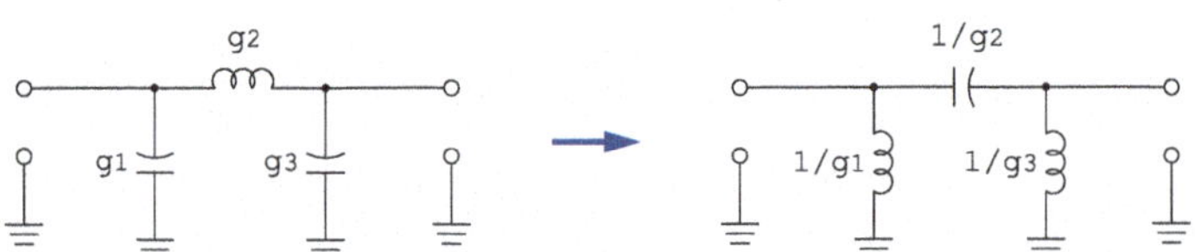

Fig. 6.7: Conversion of a standard low-pass to a high-pass

Conversion in BP-Filters

It is also possible to convert the standard LP into band-passes. However, a significantly greater effort is required to ensure compliance with the lower and upper band limits, which is described in detail in the specialist literature on filter synthesis (e.g. [120, 91]). The basic procedure is to transform a capacitor into a parallel resonant circuit and an inductor into a series resonant circuit, Fig. 6.8.

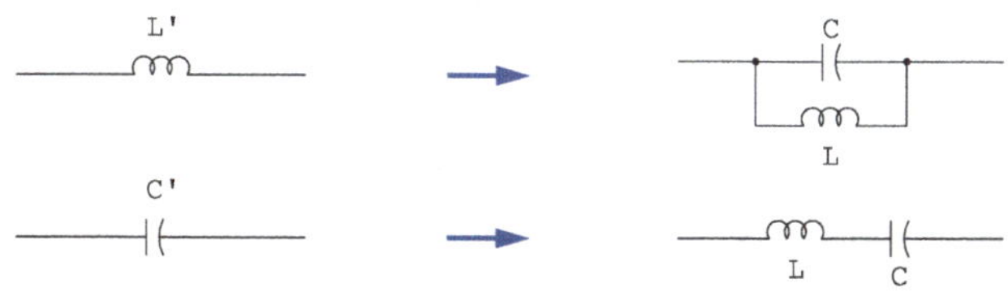

Fig. 6.8: Conversion path of a standard low-pass to a band-pass

This band-pass synthesis is complex and often cannot be implemented, as the components would have to be virtually lossless. Therefore, the synthesis for BP-filters presented below is preferred in practice for simple band-pass filters.

6.2 Synthesis of Special Filters

Synthesis of Low-Pass and High-Pass Filters from the Line Equivalent Circuit Diagram

This simple procedure, as shown in the example of the transmission line (see page 73), now enables the design of low-pass filters based on the specification of the corner frequency f, the line length ℓ and the capacitance C' and the simple equation:

$$L' = Z_L^2 C' \quad . \tag{6.7}$$

The number of series L's and shunt C's is derived from the 10°-rule. This theory therefore directly provides filters with a large number of elements.

The smaller C' and the longer the line is selected, the steeper the filter becomes. If the slope is increased by C', the matching in the passband continues to decrease. On the other hand, the matching remains unchanged if the cable is extended. However, the disadvantage of extending the line is that the component cost increases.

The HP design is carried out again, as shown in Fig. 6.7.

The RF Standard Bandpass Filter

Without going into the theory, the simplest microstrip line bandpass filter ([91], page 422), which is made with coupled $\lambda/4$ lines, will be presented here. With coupled lines, a common mode (for short: even) and a push-pull mode (for short: odd) can propagate[1].

[1]This is discussed in detail in chapter 6.

The associated characteristic impedances for the common-mode case Z_{even} and the differential-mode case Z_{odd} can be calculated for various conductor systems using software tools (e.g. Serenade: Transmission Line Tool).

The filter is characterized by the fact that it is very compact, can handle high power levels and can be manufactured well and inexpensively. The RF standard bandpass filter, which is sometimes also used only for ESD protection or for DC decoupling, requires the normalized characteristic impedances: $z_{even} = 3 \quad , \quad z_{odd} = 1 \quad .$

For a realization using microstrip lines, this filter is shown in Fig. 6.9.

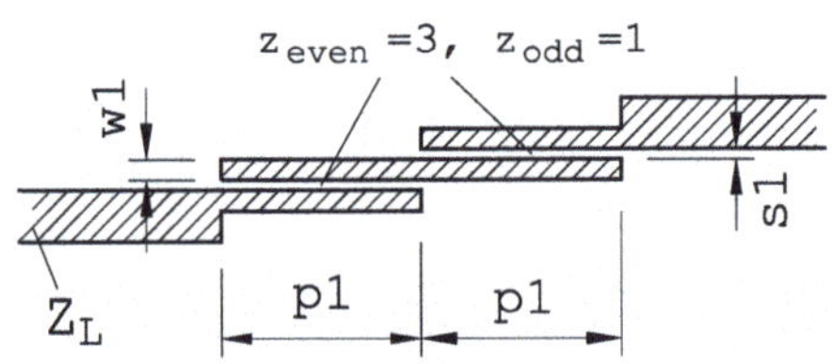

Fig. 6.9: RF standard bandpass filter using microstrip lines (p1: $\lambda/4$)

For the WLAN system at 5 GHz, the following results for a microstrip line with the following properties

$$h = 0.4\,\text{mm} \ , \ \epsilon_r = 4.5 \ , \ \tan(\delta) = 0.001 \ , \ \text{Met.: Cu mit t} = 0.17 \ \mu\text{m}$$

with the following data of the coupled lines:

$$w_1 = 0.153\,\text{mm} \ , \ s_1 = 0.042\,\text{mm} \ , \ p_1 = 8.66\,\text{mm} \qquad .$$

The reflection and transmission amounts of this filter are shown in the Fig. 6.10.

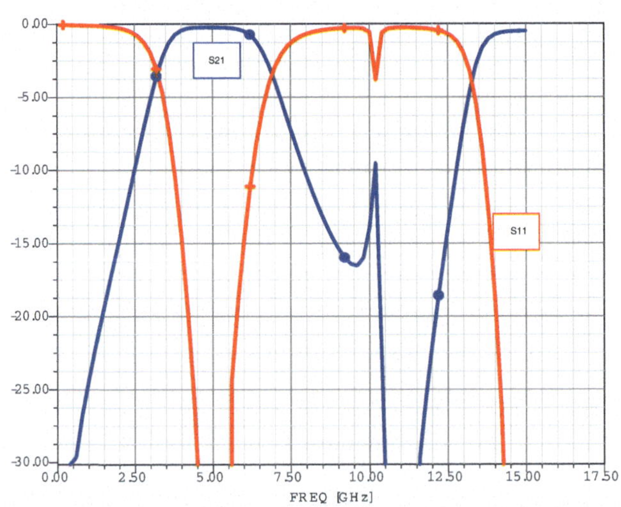

Fig. 6.10: WLAN example for the RF standard bandpass in microstrip technology

6.3 Fundamentals of Resonators

Resonators are of the utmost importance as elements in resonator filters and in RF frequency oscillators. It is extremely important for an RF engineer to have a basic understanding of how they work!

The simplest way to generate a MHz or GHz signal for a short time is shown in Fig. 6.11. In the left switch position, the capacitor is charged. After switching, this LC resonant circuit (resonator) oscillates briefly at the resonant frequency.

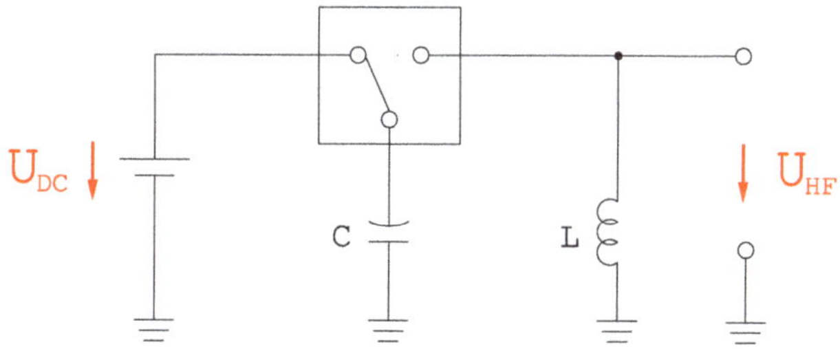

Fig. 6.11: Most simple setup for generating a sinusoidal HF or RF signal

A large part of the development time of a voltage controlled oscillator (VCO) is spent on optimizing the dimensions of the tuneable resonator.

The so-called Q-factors of the resonators are very decisive for the phase noise of an oscillator as well as for the band-pass or band-stop properties of resonator filters.

Fig. 6.12 provides an overview of the available qualities of resonators and elements manufactured in various technologies as a function of frequency. It is now possible to manufacture LC elements with a Q-factor of 10 at 10 GHz. Very robust BAW (Bulk Acoustic Wave) filters have recently been competing with SAW (Surface Acoustic Wave) filters.

The development of SAW resonators with Q-values that previously only featured quartz crystals made mobile communication technology possible in the first place.

SAW filters use surface acoustic waves on piezoelectric materials. BAW filters utilize silicon semiconductor technology and use acoustic bulk modes that move in the substrate and are reflected inside.

With regard to the internal structure of the quartz resonators important for digital and communications technology and the SAW resonators, please refer to [48]. These resonators and associated filters can be treated with the following theory.

A waveguide resonator is generally a $\lambda/2$ long waveguide line, which is presented in [48, 91, 131], for example. Thus, the waveguide resonators belong to the line resonators, which are discussed in general terms below.

The dielectric resonators, which are very important for the frequency range from 0.5 GHz to 20 GHz, are presented at the end of the next but one section.

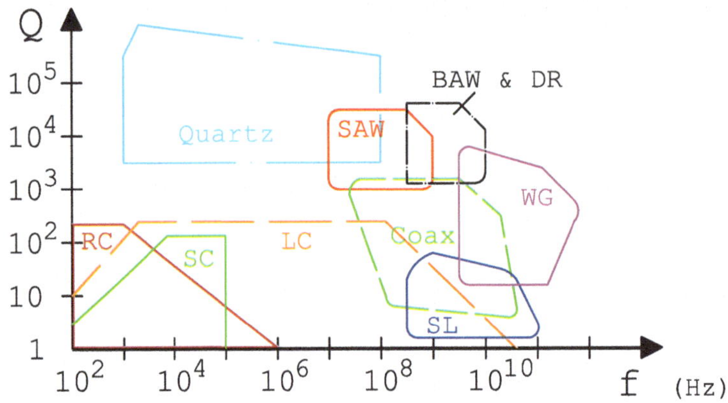

Fig. 6.12: Application area with realizable quality of the elements of different technologies

Characteristics of Resonators

In the resonant frequency range, every resonator behaves like a series or parallel resonant circuit. It is therefore sufficient to limit the presentation of the characteristics of resonators to these two simple special cases. Fig. 6.13 shows both lossy resonant circuits.

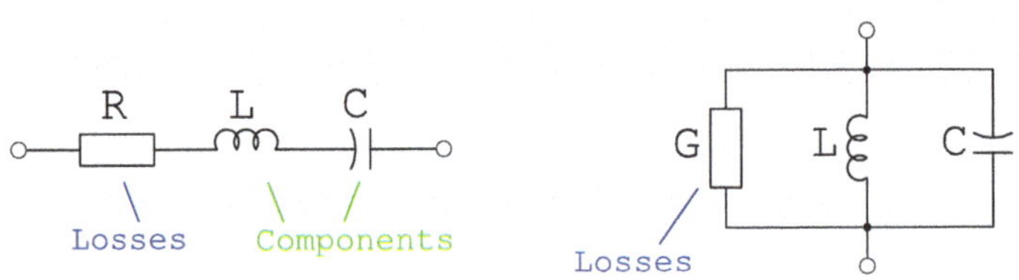

Fig. 6.13: Series and parallel circuit with RLC elements

The following parameters apply to these two resonators:

Series Circuit | Parallel Circuit

Impedance/Admittance:

$$Z_S = R + j\,(\omega L - \frac{1}{\omega C}) \qquad\qquad Y_P = G + j\,(\omega C - \frac{1}{\omega L}) \tag{6.8}$$

Both Circuits

Resonant frequency:

$$\omega_r = 2\pi\ f_r = \frac{1}{\sqrt{LC}} \tag{6.9}$$

Characteristic impedance and admittance, characteristic impedance and admittance:

$$X_K = \omega_r L = \frac{1}{B_K} = \frac{1}{\omega_r C} = \sqrt{\frac{L}{C}} \tag{6.10}$$

Series Circuit | Parallel Circuit

Both Circuits

Loss factor of the resonator:

$$\tan\delta_r = \frac{R}{\omega_r L} \qquad\qquad \tan\delta_r = \frac{G}{\omega_r C} \tag{6.11}$$

Resonator Q-factor:

$$Q_0 = \frac{1}{\tan\delta_r} \tag{6.12}$$

$$Q_0 = \frac{\omega_r\, L}{R} \qquad\qquad Q_0 = \frac{\omega_r\, C}{G} \tag{6.13}$$

Detuning:

$$\nu = \frac{\omega}{\omega_r} - \frac{\omega_r}{\omega} = \frac{\Delta\omega + \omega_r}{\omega_r} - \frac{\omega_r}{\Delta\omega + \omega_r} \tag{6.14}$$

Approximation for $\Delta\omega = \omega - \omega_r \ll \omega_r$:

$$\nu \approx \frac{2\ \Delta\omega}{\omega_r} \tag{6.15}$$

Normalized frequency:

$$\Omega = Q_0\, \nu \tag{6.16}$$

Resolution bandwidth:

$$B_{res} = \frac{2\ \Delta\omega_{3dB}}{\omega_r} = \frac{2\ \Delta f_{3dB}}{f_r} \approx \frac{1}{Q_0} \tag{6.17}$$

Impedance/admittance:

$$Z_S = R + j\omega_r L\nu \qquad\qquad Y_P = G + j\omega_r C\nu \tag{6.18}$$

$$Z_S = R\,(1 + j\Omega) \qquad\qquad Y_P = G\,(1 + j\Omega) \tag{6.19}$$

——— *Exercise: Derivation of the Equation (6.18)* ———

From the Eqs. (6.8), (6.9),(6.14) and (6.16) derive the equation for the impedance of a series resonant circuit according to Eq. (6.18).

——— *Exercise: Derivation of the Equation (6.18)* ———

Unloaded Resonators

At low frequencies in the electronics sector, it is possible to measure circuits with measuring devices (LCR meters, oscilloscopes) that have such a high input resistance that it is negligible. With such a measuring device, the resonator can be measured without load.

The `Q-factor` or the already presented resonator quality Q_0 of such an unloaded resonant circuit can be generalized by:

$$Q_0 = 2\pi \frac{\text{Stored energy}}{\text{Energy loss per period}} \quad . \tag{6.20}$$

This results for a series resonant circuit:

$$Q_0 = 2\pi \frac{\frac{1}{2} L \hat{I}^2}{\frac{1}{2} R \hat{I}^2 T_r} = \frac{\omega_r L}{R} \quad . \tag{6.21}$$

The following applies to the oscillation period: $T_r = 1/f_r$.

Eq. (6.21) verifies the values specified in Eq. (6.12) calculation rules for Q_0.

Eq. (6.12) serves as a dimensioning rule for the resonant circuits to achieve maximum resonator quality: If the components L and C have the same component qualities, the following applies:

- For a series resonant circuit, the inductance must be as large as possible and the capacitance as small as possible.
- For a parallel resonant circuit, the inductance must be as small as possible and the capacitance as large as possible.

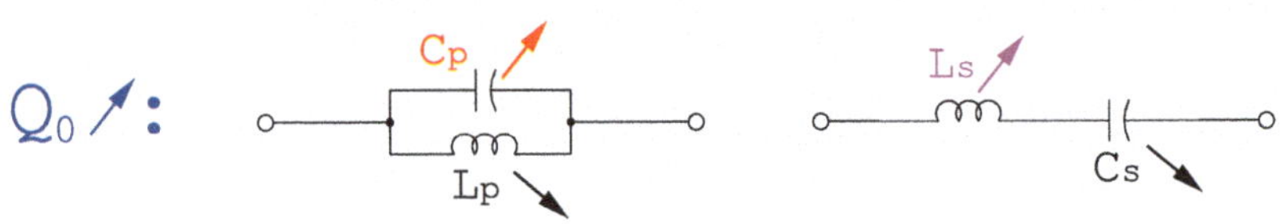

Fig. 6.14: Visualization of the statements for the increase in quality for resonators

With SMD or on-wafer circuits, the parasitic components limit the value range.

In the following, the impedance Z_S and the admittance Y_S over the normalized frequency Ω will be examined for such an unloaded series resonant circuit. Eq. (6.19) applies to the series circuit:

$$Z_S = R\,(1 + j\Omega) \quad , \qquad Y_S = \frac{1}{Z_S} = \frac{1}{R}\,\frac{1}{1 + j\Omega} \quad . \tag{6.22}$$

The normalized frequency Ω is proportional to the detuning ν and therefore zero at the resonance frequency f_r. It is known that the series resistance R can be determined from the resistance measurement of Z_S at the resonant frequency ($\Omega = 0$).

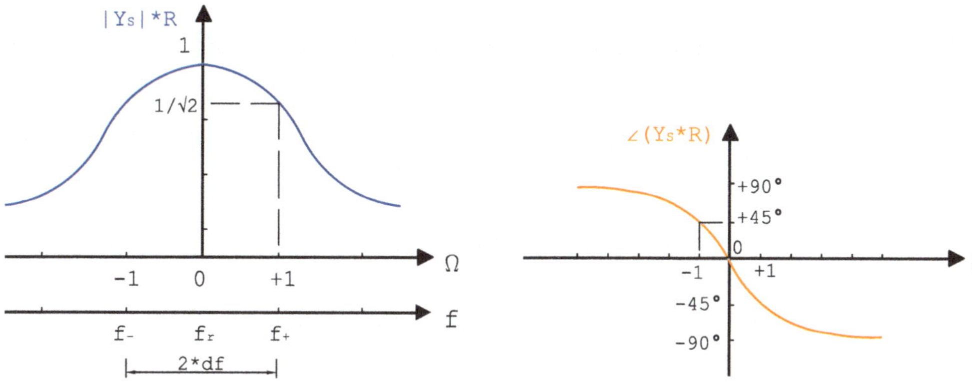

Fig. 6.15: Magnitude and phase of the admittance normalized to R plotted over the normalized frequency Ω

Below f_r the normalized frequency Ω is negative and above Ω is positive. If the product $Y_S\,R$ is plotted over the normalized frequency, the result is the Fig. 6.15 ($df = \Delta f$).

Near the resonance frequency, the following applies to the normalized frequency with (6.15) and (6.16)

$$\Omega \approx Q_0\,\frac{2\,\Delta f}{f_r} \quad . \tag{6.23}$$

If the two frequencies f_- and f_+ and thus the value $2\Delta f$ are determined via a magnitude or phase measurement of the admittance, the unloaded quality factor (Q_0) can be determined via the correlation that Ω has a magnitude of 1 at the associated frequencies.

$$Q_0 \approx \frac{f_r}{2\,\Delta f_{3dB}} \tag{6.24}$$

The same procedure can also be used for a parallel resonant circuit. In this case, it is advantageous to work with the normalized impedance Z_p/R. This gives the same result for the unloaded quality factor.

In the RF case, a resonator is always loaded both during a measurement and when used in a circuit.

Loaded Resonators

So far, only the actual (unloaded) resonator has been considered. In real RF circuits, a resonator must be coupled to the rest of the circuit. Each coupling network loads the resonator. The additional losses are taken into account by means of an additional resistance or conductance. The parallel resonator shown in Fig. 6.16 should serve as an example.

The conductance G_e represents the equivalent circuit for the losses of the coupling network. If the resonator is measured using a network analyzer in the 50 Ω system, G_e is calculated via $1/50\,\Omega$.

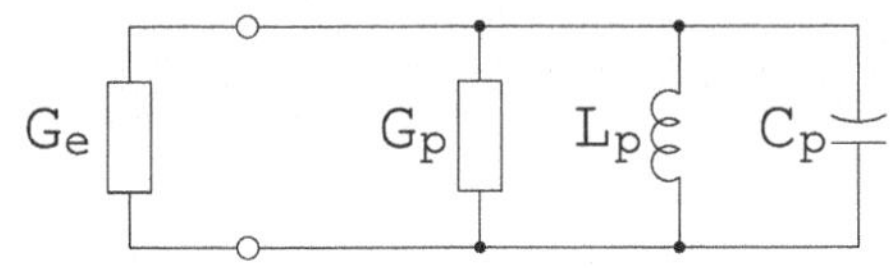

Fig. 6.16: Equivalent circuit diagram of a loaded resonator

Just like the resonator Q-factor Q_0, the so-called external Q-factor Q_{ext} is defined as the quotient of the characteristic value and the conductance G_e:

$$Q_{ext} = \frac{1}{\tan \delta_{ext}} = \frac{B_K}{G_e} = \frac{\omega_r C_p}{G_e} \quad . \tag{6.25}$$

The following still applies to the resonator Q-factor:

$$Q_0 = \frac{1}{\tan \delta_r} = \frac{B_K}{G_p} \quad . \tag{6.26}$$

The so-called "loaded Q-factor" Q_L is defined as follows:

$$Q_L = \frac{1}{\tan \delta_L} = \frac{B_K}{G_p + G_e} \quad . \tag{6.27}$$

Consequently, the following also applies:

$$\tan \delta_L = \tan \delta_r + \tan \delta_{ext} \quad ; \quad \frac{1}{Q_L} = \frac{1}{Q_0} + \frac{1}{Q_{ext}} \quad . \tag{6.28}$$

The size of the loaded Q-factor is ultimately the decisive value in a circuit realization so that a filter has as little transmission loss as possible with maximum stopband attenuation or an oscillator has the lowest possible phase noise.

Instead of the external Q-factor Q_{ext}, the *coupling factor* k is often used in practice. The real coupling factor is defined via

$$k = \frac{Q_0}{Q_{ext}} \quad . \tag{6.29}$$

This results in the following for the loaded quality using Eq. (6.28)

$$Q_L = \frac{Q_0}{1+k} \quad . \tag{6.30}$$

It can be seen that for a very weak coupling (k approaches zero), the loaded Q-factor is almost equal to the unloaded Q-factor. This is the case when the aim is to achieve an optimal resonator filter for oscillators with the lowest possible Δf value.

The couplings are assigned to three characteristic cases:

Weak or sub-critical coupling	:	$k < 1$	$\Leftarrow$ Oscillators
Critical coupling	:	$k = 1$	
Hard or super-critical coupling	:	$k > 1$	$\Leftarrow$ Filters

In practice, the coupling values for bandpass filters containing several resonators are in the super-critical coupling range. Forms of realization of couplings will be discussed in the presentation of embodiments and resonator circuits. First, the pure transmission line resonators without coupling losses are presented.

The following example, which was created using the freeware SynRF [133], represents a bandpass for 1 GHz consisting of the components $C = 5\,\mathrm{pF}$; $L = 5\,\mathrm{nH}$; $R = 1\,\Omega$; $Z_0 = 50\,\Omega$. The two ports load the resonator with 100 Ω. In contrast, the loss resistance

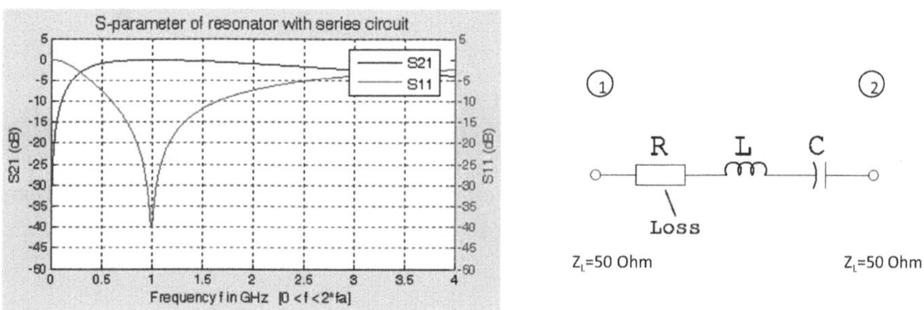

Fig. 6.17: RF behavior of a series resonant circuit used as a bandpass filter in a 50 Ω environment

is only 1 Ω. The total loss of 101 Ω in the series circuit results in a wide distribution of the filter behavior, as shown in Fig. 6.17.

If, on the other hand, the port impedances are only 1 Ω each, the filter has a total series resistance of only 3 Ω, which is a very narrow behavior, as shown in Fig. 6.18.

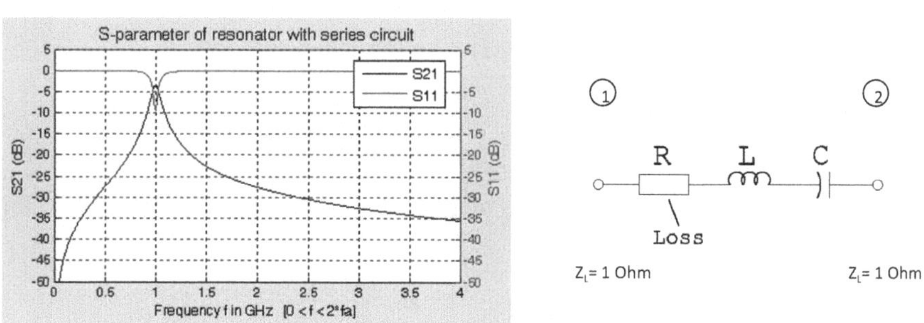

Fig. 6.18: RF behavior of a series resonant circuit used as a bandpass filter in a 1 Ω environment

6.3.1 Theory of the $\lambda/4$ Transmission Line Resonators

The properties of TEM and quasi-TEM lines were presented in chapter 4. The $\lambda/4$ lines have already been introduced as impedance transformers. As $\lambda/4$ lines are only half the length of $\lambda/2$ lines, the $\lambda/4$ lines are preferably used as resonators.

We start with a description of an open-circuited $\lambda/4$ line according to Fig. 6.19 and will transfer these results to a short-circuited $\lambda/4$ line. The electrical properties of a line

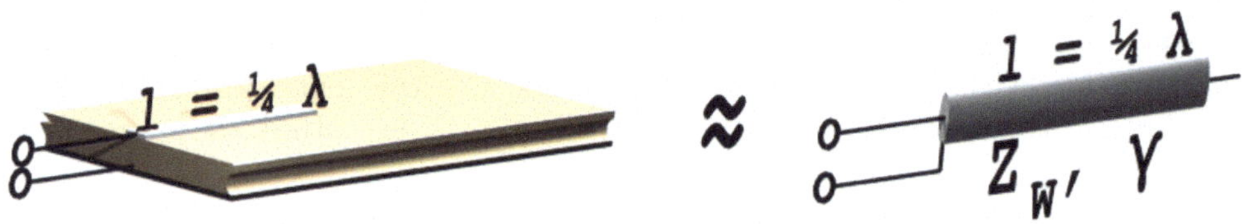

Fig. 6.19: Open-circuit $\lambda/4$ microstrip line ($Z_w = Z_L$)

can be fully described by specifying the characteristic impedance Z_L, the propagation constant γ, which is made up of the damping constant α and the phase constant β, and the mechanical length l.

The aim of this brief derivation is to show that, when the length $l = \lambda/4$, an open-circuit line behaves like a series resonant circuit and a short-circuited line behaves like a parallel resonant circuit.

At the end of the derivation, equations are available that can be used to determine the wave impedance of the $\lambda/4$ lines from a given reference filter[2].

Just as the input resistance Z_E for the lossless case could be derived from the line equations (Eq. 4.98), the input resistance of a line for the lossy case can be calculated from the general line Eq. (4.29).

$$Z_E = Z_A \frac{1 + \frac{Z_L}{Z_A} \cdot \tanh(\gamma l)}{1 + \frac{Z_A}{Z_L} \cdot \tanh(\gamma l)} = \frac{1 + \frac{Z_L}{Z_A} \cdot \tanh(\gamma l)}{\frac{1}{Z_A} + \frac{1}{Z_L} \cdot \tanh(\gamma l)} \quad . \tag{6.31}$$

If the line is terminated with an open circuit ($Z_A = \infty$) at the end, Eq. (6.31) simplifies to

$$Z_E = Z_L \frac{1}{\tanh(\gamma l)} = Z_L \frac{\cosh(\gamma l)}{\sinh(\gamma l)} \quad . \tag{6.32}$$

With the help of the complex function theory [14], $\gamma = \alpha + j\beta$ leads to

$$Z_E = Z_L \frac{\cosh(\alpha l) \cdot \cos(\beta l) + j \cdot \sinh(\alpha l) \cdot \sin(\beta l)}{\sinh(\alpha l) \cdot \cos(\beta l) + j \cdot \cosh(\alpha l) \cdot \sin(\beta l)} \quad . \tag{6.33}$$

[2] A reference filter is an LC filter that has been determined using any filter synthesis method.

The phase constant β can be expressed by the ratio of the angular frequency ω to the phase velocity v_p:

$$\beta = \frac{\omega}{v_p} \quad . \tag{6.34}$$

We would like to find a replacement circuit for the open-circuit line for the frequency range $\Delta\omega$ at the resonance circuit frequency ω_r and therefore introduce:

$$\omega = \omega_r + \Delta\omega \quad . \tag{6.35}$$

Using equations (6.34) and (6.35) and the length $l = \lambda/4$, as well as the previously derived relationship $v_p = \lambda f$ and $\lambda = \lambda_r$, we obtain the following for the argument

$$\beta l = \frac{\omega_r}{v_p}\frac{\lambda}{4} + \frac{\Delta\omega}{v_p}\frac{\lambda}{4} = \frac{\omega_r}{\lambda f_r}\frac{\lambda}{4} + \frac{\Delta\omega}{v_p}\frac{\lambda}{4} = \frac{\pi}{2} + \frac{\Delta\omega}{v_p}\frac{\lambda}{4} \quad . \tag{6.36}$$

If this result is used in the arguments, the following approximations can be made:

$$\cos\left(\frac{\pi}{2} + \frac{\Delta\omega}{v_p}\frac{\lambda}{4}\right) \approx -\frac{\Delta\omega}{v_p}\frac{\lambda}{4} \quad , \tag{6.37}$$

$$\sin\left(\frac{\pi}{2} + \frac{\Delta\omega}{v_p}\frac{\lambda}{4}\right) \approx 1 \quad . \tag{6.38}$$

Furthermore, the following applies approximately for $\alpha l = \alpha\frac{\lambda}{4} \ll 1$

$$\cosh(\alpha l) \approx 1 \; ; \qquad \sinh(\alpha l) \approx \alpha l \quad . \tag{6.39}$$

If these approximations of the Eqs. (6.37)-(6.39) are inserted into Eq. (6.33), the following applies:

$$Z_E \approx Z_L \frac{-\frac{\Delta\omega}{v_p}\, l + j\,\alpha\, l}{-\alpha\, l\,\frac{\Delta\omega}{v_p}\, l + j} \quad . \tag{6.40}$$

With $\alpha\, l\,\frac{\Delta\omega}{v_p}\, l \ll 1$ follows:

$$Z_E \approx Z_L\,(\alpha\, l + j\frac{\Delta\omega}{v_p}\, l) = Z_L\,(\alpha\,\frac{\lambda}{4} + j\frac{\Delta\omega}{v_p}\,\frac{\lambda}{4}) \quad . \tag{6.41}$$

Let us compare this result of the input resistance of an open-circuit $\lambda/4$ line with the input resistance of a series resonant circuit according to Eq. (6.18)

$$Z = R + j\omega_r L\nu \quad , \tag{6.42}$$

the losses of the cable can be converted directly into an ohmic load of the resonant circuit:

$$R \approx Z_L \, \alpha \, l \quad . \tag{6.43}$$

For the design of filters with line resonators from reference filters with oscillating circuits, the relationship between the wave impedance of the line and the characteristic impedance (or inductance) of the reference filter is sought. This relationship can be obtained by comparing the imaginary parts of Eqs. (6.41) and (6.42) and additionally introducing the following Eq. (6.15):

$$\frac{\Delta\omega \, \lambda}{4 \, v_p} = \frac{\Delta\omega \, \lambda}{4 \, \lambda \, f} = \frac{\Delta\omega \, 2\pi}{4 \, (\omega_r + \Delta\omega)} = \frac{\pi \, \Delta\omega}{2 \, (\omega_r + \Delta\omega)} \approx \frac{\pi \, \nu}{4 + 2\nu} \tag{6.44}$$

$$\omega_r L \, \nu \approx Z_L \frac{\pi \, \nu}{4 + 2\nu} \quad \Longrightarrow \quad L \approx Z_L \frac{\pi}{\omega_r \, (4 + 2\nu)} \tag{6.45}$$

To design the resonator, consider Eq. (6.45) at the resonance frequency ($\nu = 0$) and obtain a very simple relationship between the inductance of the series resonant circuit (known from the reference filter) and the desired characteristic impedance of the resonator:

$$Z_L \approx \frac{4}{\pi} \, \omega_r \, L \quad . \tag{6.46}$$

This means that, given a reference filter implemented as a (series) resonant circuit, the wave impedance of the resonator can be calculated and then the Q-factor and relative bandwidth of this resonator can be determined using Eq. (6.43) if the attenuation α of the line is known.

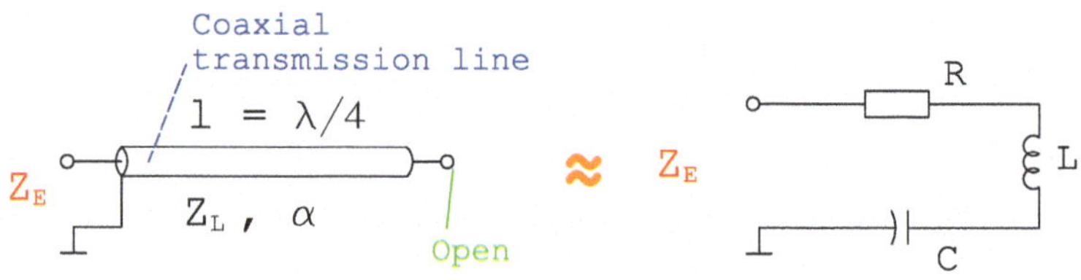

Fig. 6.20: Replacement circuit for an open-circuit $\lambda/4$ resonator in the resonance frequency range

If the results for L and R are used to calculate the resonator quality according to Eqs. (6.12), the following applies

$$Q_0 = \omega_r \frac{L}{R} \approx \omega_r \frac{Z_L \, \pi}{4 \, \omega_r \, Z_L \, \alpha \, l} = \frac{\pi}{4 \, \alpha \, l} \quad . \tag{6.47}$$

Using $l = \lambda/4$, the resonator quality can be calculated directly from the line attenuation:

$$Q_0 \approx \frac{\pi}{\alpha \, \lambda} = \frac{\pi \, f \, \sqrt{\epsilon_r}}{\alpha \, c_0} \quad . \tag{6.48}$$

The relationship between quality, wavelength (or line length) and damping constant α that can be derived from Eq. (6.48) is of great importance for passive elements and components in RF technology:

The losses of a component are inversely proportional to the volume.

Fig. 6.21 shows that the approximations introduced in the resonant frequency range are very accurate.

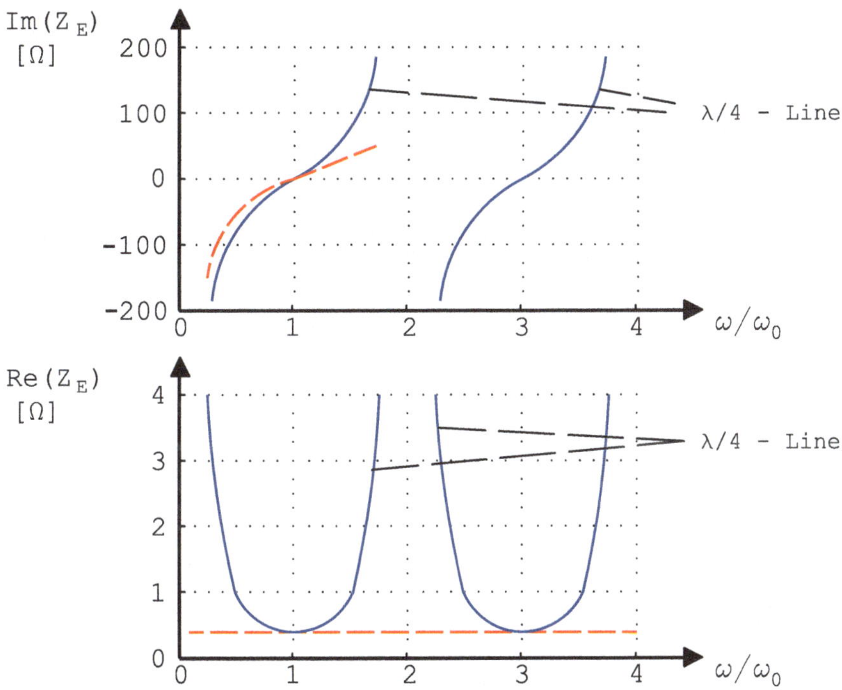

Fig. 6.21: Comparison of the impedances of an open-circuit $\lambda/4$ line and an RLC series resonant circuit (shown in red) over the frequency ($Z_L = 50\,\Omega,\ Q = 100$)

$\lambda/4$ Resonator with Short Circuit

The characteristic impedance of a $\lambda/4$ line resonator (Fig. 6.22) with a short circuit is derived from the parameters of a parallel circuit in the same way as for the open-circuit line resonator.

Fig. 6.22: Equivalent circuit of a $\lambda/4$ resonator short-circuited at the end at resonant frequency

For the Q-factor Q_0 this again results in Eq. (6.48). In this case, the characteristic impedance is advantageously calculated from the capacitance C:

$$Z_L \approx \frac{\pi}{4} \frac{1}{\omega_r C} \quad . \tag{6.49}$$

The following now applies to the loss calculation:

$$R \approx \frac{Z_L}{\alpha \, l} \quad . \tag{6.50}$$

Choice of Transmission Line Systems for $\lambda/4$ Resonators

If you want to replace parallel or series resonant circuits that are connected to ground, use the transmission line system in which the circuit is preferably laid out.

However, if you want to replace resonant circuits that are connected in series, you must use a balanced line system. Fig. 6.23 shows an example.

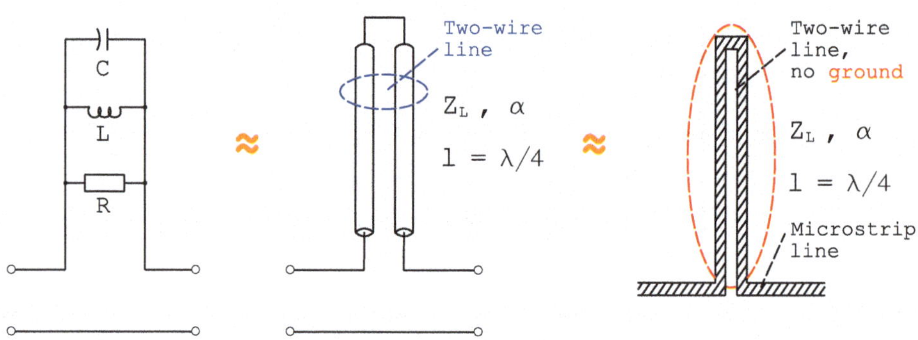

Fig. 6.23: Equivalent circuit and $\lambda/4$ resonator in symmetrical conductor technology of its parallel resonant circuit, which is connected in series

6.4 Circuit-Connected Resonators

So far, mainly the unloaded resonators have been considered. Both for measurement technology and for use as a circuit component, the resonator can no longer be considered unloaded.

A resonator can only be easily used as a component if it has a circuit in the form of a so-called coupling network at the input or output that optimizes it for a 50 Ω system.

In addition to the wiring, the installation position and the choice of resonant circuit play a decisive role in the functionality of a wired resonator. A distinction is made between three cases:

- Transmission resonator: The resonator is connected into the circuit as a feed-through element (two-port). The transmitted power is maximum in the resonance case, far from ω_r the incoming power is largely reflected at the input. The arrangement acts as *band-pass filter.*

- Reflexion resonator: The resonator is connected via a coupling network as a terminating element (single port) in a circuit. Due to its purely ohmic character, the resonator consumes maximum active power in the event of resonance.

- Reaction resonator: The resonator is coupled as a line discontinuity (two-port) to a continuous line. The transmitted power at resonance is minimal. Away from the resonance, most of the incoming power is transmitted. The arrangement represents an *band-stop filter.*

These three types of wired resonators are considered below.

6.4.1 Transmission Resonators

Resonators or the equivalent resonant circuits have either a very low or a very high input resistance at the resonant frequency.

In simple filter circuits, resonators are often directly electrically connected to the 50 Ω system. However, two identical resonators are never directly coupled to each other. This is shown in the section on dual-circuit resonator filters (Chap. 6.6).

For the realization of oscillators, on the other hand, one only wants to use the resonance properties of a resonator. The simplest transmission resonator consists of the direct coupling of the resonator to the line system with the characteristic impedance Z_L, as shown in Fig. 6.24.

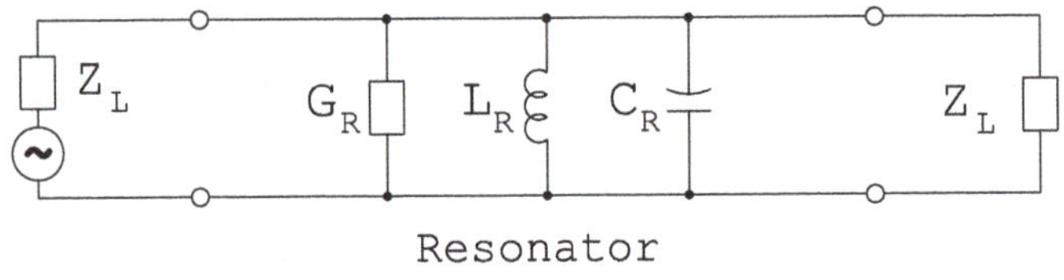

Fig. 6.24: Realization of a transmission resonator with parallel resonant circuit as the simplest band-pass resonator filter

As already introduced in the discussion of unloaded resonators, the resonator based on a parallel resonant circuit has the best quality (Q-factor) characteristics if the inductance is as small as possible. Component Q-value for coil and capacitor of 40 is a good average value for semiconductors and SMD components. The following Fig. 6.25 shows the loaded Q-factor Q_L and the transmission losses $|S_{21}|$ of the simplest transmission resonator shown in Fig. 6.24. It can be seen that the loaded quality for "component stops" of 1 nH or

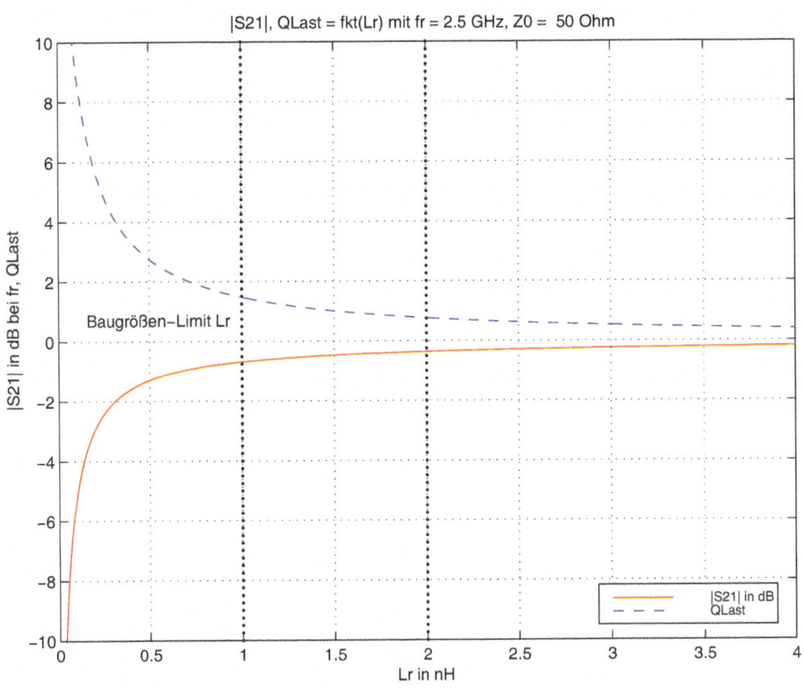

Fig. 6.25: Loaded quality (Q-factor) of a parallel resonant circuit plotted against the inductance in the 50 Ω system for 2.5 GHz and component Q-value of 40 and marked limits of the smallest L_r of the component series

2 nH are far below 2.

The less the resonator is loaded, the better its resonance properties. This is why loosely coupled resonators are preferred for oscillators.

This loose or weak coupling always involves an impedance transformation. If you want to implement a transmission resonator or a simple bandpass resonator filter in the 50 Ω system using a resonator, you can

1. connect the resonator (e.g. $\lambda/4$ line with short circuit to ground) directly.

2. connect the resonator via an impedance transformation network.

For example, with a $\lambda/4$ line with a short circuit to ground, whose input resistance is high at resonance and low in the blocking range, the impedance transformation would be used to transform the 50 Ω system up to around 200 Ω and then back down again (Fig. 6.26 with the transformation ratio n).

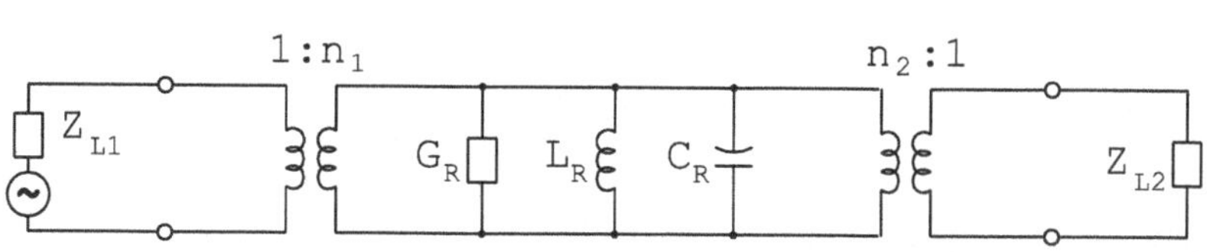

Fig. 6.26: Realization of a transmission resonator as a simple band-pass resonator filter with impedance transformers

Although this increases the losses in the pass-band, the blocking attenuation is significantly increased compared to a direct coupling. The increased internal impedance level increases the mismatch in the blocking range.

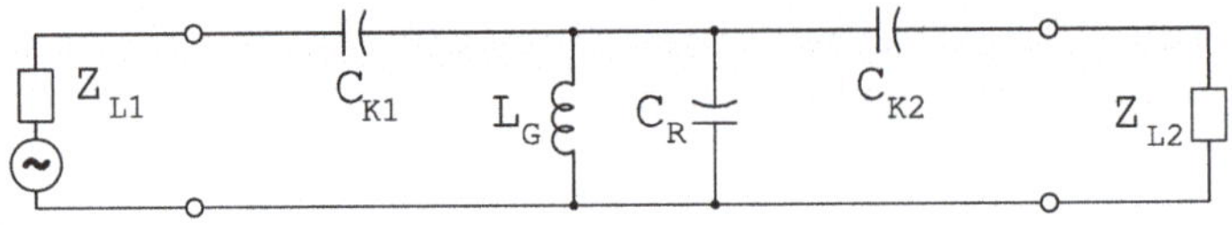

Fig. 6.28: Hardware setup of a transmission resonator as a band-pass resonator filter with Γ transformers

An RF impedance transformer can be realized using an inductor and a capacitor (see Γ transformer). With clever partitioning, only two additional components are required for a discrete realization. With distributed components, only one additional element is usually (optically) recognized at the input and output. Fig. 6.27 illustrates the most commonly used circuit with coupling capacitors at the input and output. The coupling inductors are not found in the circuit realization. They reduce the effective inductance of the resonator element L_R, which is generally referred to as detuning of the resonator due to the coupling.

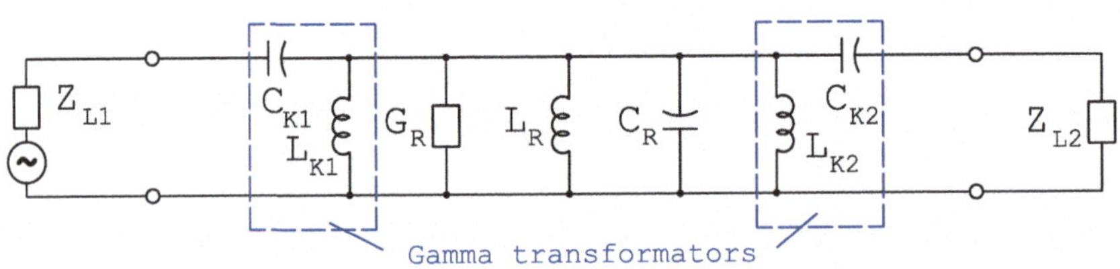

Fig. 6.27: Realization of a transmission resonator as a band-pass resonator filter with Γ transformers

Such a resonator realized by means of small coupling capacitors consists of only 4 components when assembled using SMD components, Fig. 6.28.

The following correlation applies to the inductances:

$$\frac{1}{L_G} = \frac{1}{L_R} + \frac{1}{L_{K1}} + \frac{1}{L_{K2}} \qquad \text{with} \qquad \omega_r^2 = \frac{1}{L_R C_R} \quad . \tag{6.51}$$

Again for component Q-factors of 40 and a frequency of 2.5 GHz, the loaded Q-factors and associated transmission losses of the resonator filter, now improved by the Γ transformation, were calculated according to Fig. 6.27 and the results printed in Fig. 6.29 for a 50 Ω system. With minimum component values of 1 nH and 0.5 pF, the improved loaded quality of around 7[3].

[3]The internal impedances were specified with R_2.

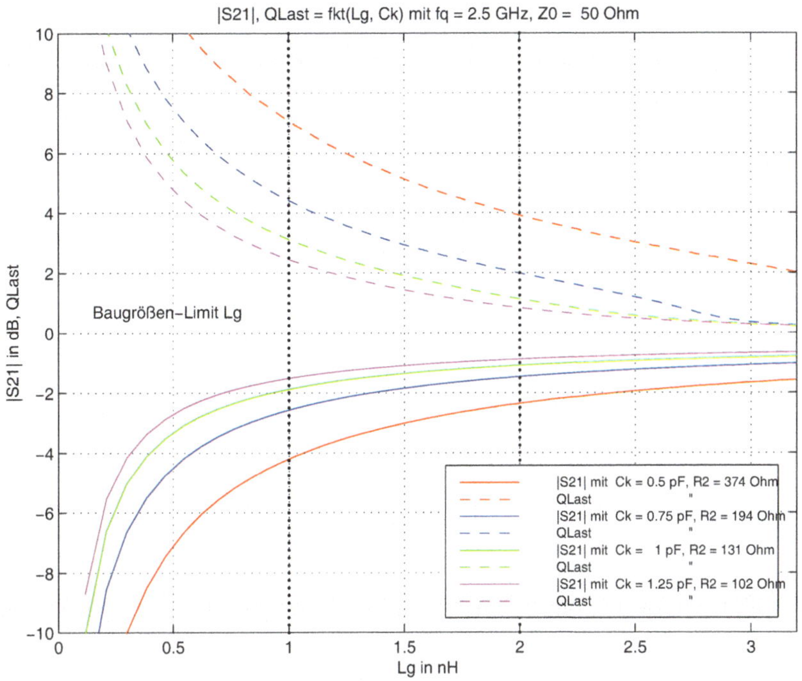

Fig. 6.29: Loaded quality Q_{load} of a parallel resonant circuit with Γ transformation plotted over the inductance L_g and as a function of the coupling capacitance C_k and with the limits of L_g (dotted lines)

The scattering parameters of the transmission resonator can be calculated for the series connection of the two transformers and the resonant circuit as shown in Fig. 6.26 as well as a series resonant circuit from the chain circuit and values of the partial circuits from the auxiliary sheets:

$$[S] = \begin{pmatrix} -\frac{1-k_1+k_2+j\Omega}{1+k_1+k_2+j\Omega} & \frac{2\sqrt{k_1k_2}}{1+k_1+k_2+j\Omega} \\ \frac{2\sqrt{k_1k_2}}{1+k_1+k_2+j\Omega} & -\frac{1+k_1-k_2+j\Omega}{1+k_1+k_2+j\Omega} \end{pmatrix} \quad . \tag{6.52}$$

For a series resonant circuit, however, the minus signs for S_{11} and S_{22} are omitted! In the Eq. (6.52), the real coupling factors result for a parallel resonant circuit

$$k_i = \frac{1}{n_i^2 \, G_R \, Z_{Li}} \qquad \text{with} \qquad i = 1, 2 \tag{6.53}$$

and for a series resonant circuit the coupling factors

$$k_i = \frac{n_i^2 \, Z_{Li}}{R_R} \qquad \text{with} \qquad i = 1, 2 \quad . \tag{6.54}$$

The transfer factor n_i can be found in the Fig. 6.26. As can be derived from the transmission equation for S_{21}, the loaded quality factor, which can now be calculated via

$$Q_L = \frac{Q_0}{1 + k_1 + k_2} \qquad \text{or} \tag{6.55}$$

can be determined from the relative 3 dB bandwidth of S_{21} using Fig. 6.30:

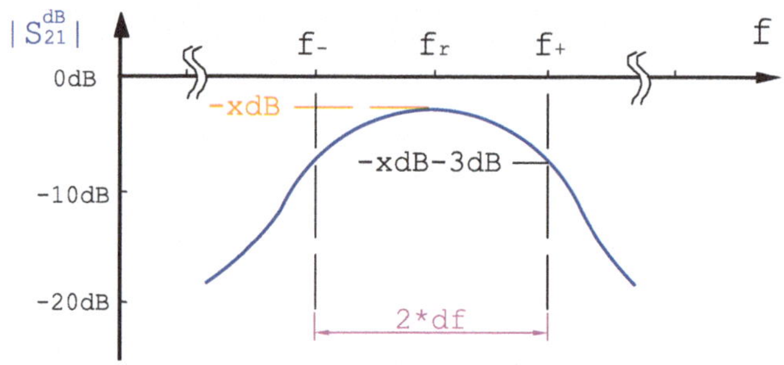

Fig. 6.30: Diagram showing the calculation of Q_L from the transmission measurement

$$k_1 = \frac{1 - S_{11}|_{f_r}}{S_{11}|_{f_r} + S_{22}|_{f_r}} \quad , \quad k_2 = \frac{1 - S_{22}|_{f_r}}{S_{11}|_{f_r} + S_{22}|_{f_r}} \quad . \tag{6.57}$$

The ring resonator in microstrip conductor technology according to Fig. 6.31 is a possible design with good quality characteristics.

Fig. 6.31: Ring resonator with capacitive coupling and decoupling (Umfang: engl. scope)

6.4.2 Reflection Resonators

A realization example for a loosely or weakly coupled resonator, which is very interesting for an oscillator, for example, is shown in Fig. 6.32 for a microstrip line circuit is shown.

A pure series capacitor or (generalized) the equivalent circuit according to Fig. 6.33 can be used for such a coupling.

In the resonator in Fig. 6.33, only the coupling capacitor is visible as an additional element. However, an impedance transformation is carried out using a Γ transformer. The parallel connection of the transformer inductance L_K and the resonator inductance results in the only available equivalent inductance of the line. The "effective" inductance of the resonator thus increases, which causes the resonant frequency to become a lower frequency *detuned*. If you want to design distributed circuits as shown in Fig. 6.32, you have to use the use a so-called field simulator to calculate the coupling networks.

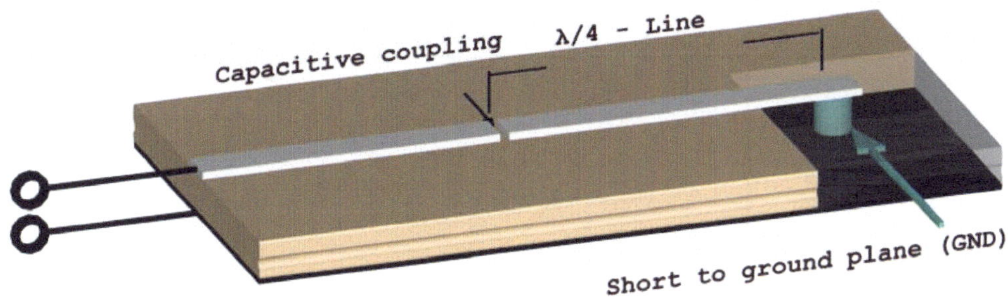

Fig. 6.32: Short-circuited $\lambda/4$ resonator in microstrip conductor technology with loose capacitive coupling

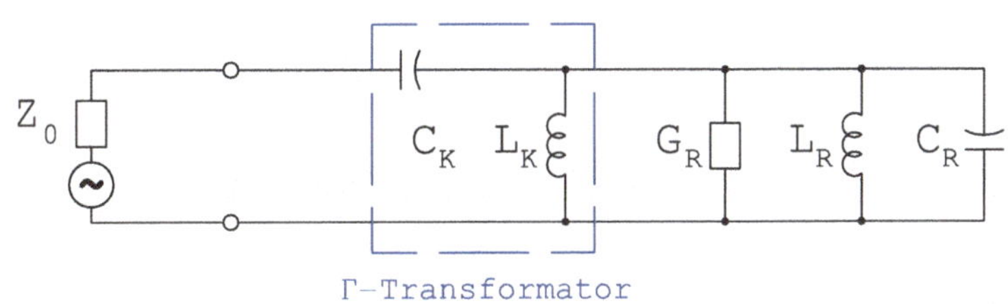

Fig. 6.33: Equivalent circuit of a resonator with coupling via a impedance transformation by means of a Γ-transformer

For the general calculation of a reflection resonator in which a series or parallel resonant circuit is connected via an impedance transformer with the transformation ratio n, one works with the input reflection value, which in turn can be calculated from the coupling factor k and the normalized frequency Ω:

$$S_{11} = \frac{1 - k + j\Omega}{1 + k + j\Omega} \quad . \tag{6.58}$$

The coupling factors for series and parallel resonant circuits are again calculated using the Eqs. (6.53) and (6.54).

By converting the Eq. (6.58), the real coupling factor can be calculated from the measurement:

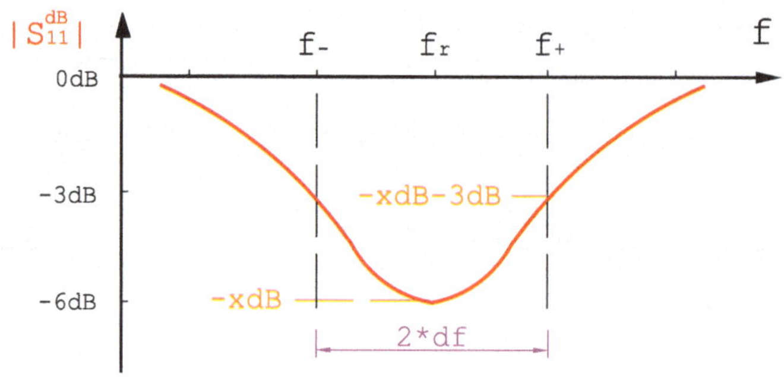

Fig. 6.34: Diagram showing the calculation of Q_L from the reflection measurement

$$k = \frac{1 - S_{11}|_{f_r}}{1 + S_{11}|_{f_r}} \quad . \tag{6.59}$$

For illustration purposes, $k = G_{in}/G_r$, where G_{in} is the conductance at the output of the Γ transformer. As the transformation ratio increases, the matching at the resonant frequency improves[4]. Again, it should be possible to determine the Q-factor of the connected resonator from the scattering parameters and the 3 dB drop according to Fig. 6.34.

In the vicinity of the resonant frequency, the following applies to the input impedance of the reflection resonator

$$Z_E \approx \frac{R_R}{n^2} \left[1 + j\, Q_0 \frac{2\, df}{f_r} \right] \quad . \tag{6.60}$$

By means of the ratio

$$F = \left(\frac{\frac{R_R}{n^2} + Z_L}{Z_E + Z_L} \right)^2 \tag{6.61}$$

the power dissipated in R_R or G_R far outside the resonance[5], and the power dissipated in R_R at resonance, the loaded Q-factor can be determined from the reflection measurement:

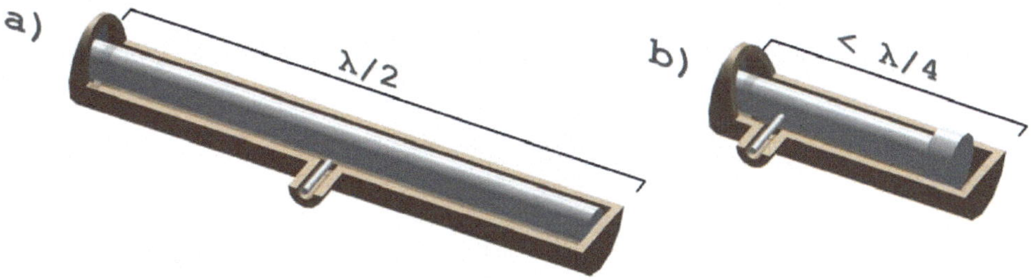

Fig. 6.35: Coaxial line resonators with a) capacitive coupling and b) galvanic coupling

$$Q_L = \sqrt{\frac{1-F}{F}} \frac{f_r}{2\,df} \quad . \tag{6.62}$$

Reflection resonators with very high qualities can be manufactured from coaxial cables, provided no dielectric is used. To ensure that the inner conductor of such coaxial conductors is supported, two resonators are connected in parallel (Fig. 6.35 a)). If a resonator with a short design is required, a capacitive termination can also be used (Fig. 6.35 b)).

A precise calculation of the coupling and decoupling networks can only be carried out with field simulators for all line systems.

6.4.3 Dielectric Resonator and Reaction Resonators

Dielectric resonators are used for applications in high Q-factor multi-circuit filters and in particular as a selective element in oscillators. The mostly cylindrical dielectric resonator consists of a low-loss, temperature-stable dielectric with a high permittivity ϵ_r. Thanks to its low price, high Q-factor and small dimensions, the dielectric resonator has also established itself in consumer products.

The electromagnetic fields that build up at resonance are shown in Fig. 6.36.

Details about the TE_{01} fundamental mode and the higher modes that develop can be found in the literature [48, 86]. When designing, it is important to consider that the dielectric resonator, like any line resonator, has further resonances at higher frequencies. Low-pass structures must be included so that an oscillator does not oscillate at higher resonances.

Based on the permittivity, the radius a and the height h, the resonant frequency of the fundamental mode of a cylindrical resonator can be calculated approximately [81]:

$$f_r = \frac{34}{a\sqrt{\epsilon_r}}\left(\frac{a}{h} + 3.45\right) \quad \text{in GHz resp. mm} \quad . \tag{6.63}$$

The ratio h/a should be in the range of 0.8 so that the distance to the next higher mode is large.

[4]In oscillators, the fact that the phase response also becomes increasingly steeper is exploited.

[5]Dissipated means that the power is converted into heat.

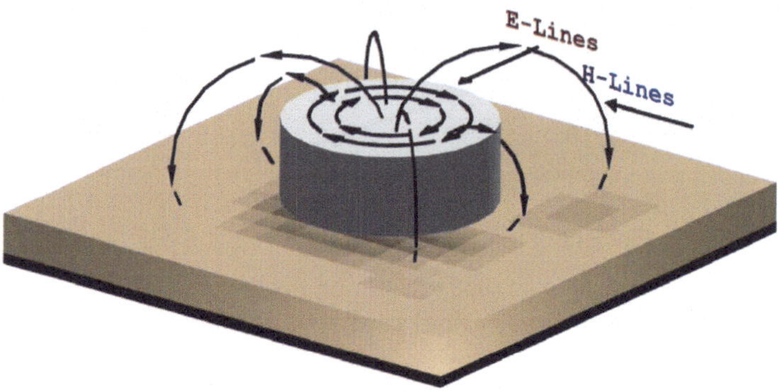

Fig. 6.36: Field distribution of the TE01 fundamental mode in the dielectric resonator

Barium titanate ($BaTi_4O_9$) with various additives (Zr, Sn, Zn, etc.) is used as the preferred dielectric. The properties of these dielectrics are at 10 GHz:

Q-factor: 3000 - 20000 ,
Temperature coefficient: ± 10 ppm/°C ,
Permittivity: 20 - 200 .

Dielectric resonators can be easily coupled to microstrip lines.

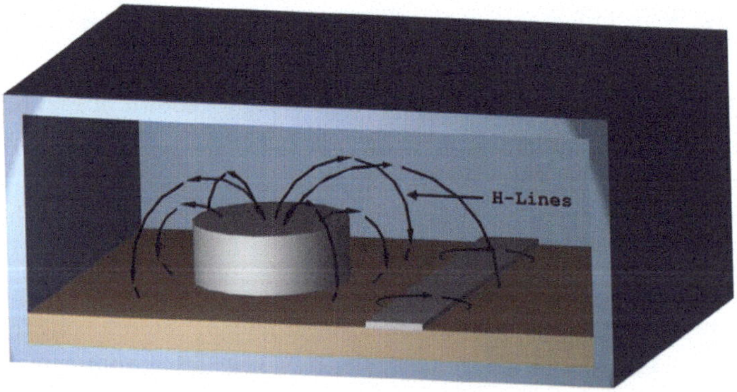

Fig. 6.37: Dielectric resonator in a shielded housing coupled to a microstrip line

The requirements for setting the resonance frequency are often so high that the resonators themselves are individually trimmed in consumer products using an adjusting screw located in the upper shielding area.

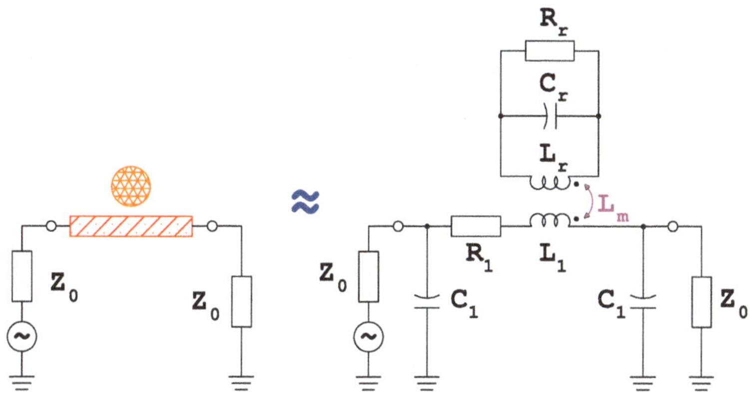

Fig. 6.38: Left: Top view of a dielectric resonator and a microstrip line; right: Equivalent circuit for a dielectric resonator and in a microstrip line

An equivalent circuit for a microstrip resonator is shown in Fig. 6.38.

In practice, the developer is well advised to precisely measure a dielectric resonator according to Fig. 6.38 in several positions using a vector network analyzer (see chapter 10) and then to fit the data using the equivalent circuit shown (see chapter 9).

These dielectric resonators are the most important representatives of reaction resonators, which are also referred to as *resonator as conduction discontinuity*. These are band-stop filters in transmission.

Similar to the reflection resonator, the phase for the reflected signal, which becomes steeper with increasing Q-factor, is also utilized here. In addition, the magnitude of the reflection has the same bandpass characteristic as the magnitude of a transmission resonator.

Reaction resonators can be realized either by a transformer connected in series and terminated by a parallel resonant circuit, as shown in Fig. 6.38, or by a transformer connected to ground and terminated by a series resonant circuit.

The scattering matrix for such a reaction resonator with shunt transformer and series resonant circuit is as follows:

$$[S] = \begin{pmatrix} \frac{-k}{1+k+j\Omega} & \frac{1+j\Omega}{1+k+j\Omega} \\ \\ \frac{1+j\Omega}{1+k+j\Omega} & \frac{-k}{1+k+j\Omega} \end{pmatrix} \quad . \tag{6.64}$$

For the resonator given in Fig. 6.38, however, only positive signs are to be used for S_{11} and S_{22}. A photo of the practical realization was shown in chapter 1 under Fig. 1.1.

Using the Eq. (6.64), the coupling factor results for the case of a parallel resonant circuit

$$k = \frac{1}{2\,n^2\,G_R\,Z_L} \tag{6.65}$$

and, in the case of a series resonant circuit, the coupling factor

$$k = \frac{n^2\,Z_L}{2\,R_R} \quad . \tag{6.66}$$

From the isolation of the bandstop behavior of the transmission at resonant frequency, the coupling factor can be calculated via

$$k = \frac{1}{S_{21}|_{f_r}} - 1 \quad . \tag{6.67}$$

For this resonator system, the unloaded Q-factor can be calculated directly from the 3 dB bandwidth of the transmission via

$$Q_0 = \frac{f_r}{2\,df} \quad . \tag{6.68}$$

The loaded quality can be determined from the unloaded quality using Eq. (6.30).

6.5 Dual-Mode Resonators

At the beginning of the section 6.3, the large number of technologies in which resonators are manufactured today was discussed in detail.

These different technologies were mainly developed in order to be able to bring resonators with improved Q-factors onto the market. Mainly, only the unloaded Q-factors can be increased by improving the technology.

For individual resonator circuits as well as for filters, the loaded Q-factor of a resonator is crucial. The following shows a recent invention that contains a concept for increasing the loaded Q-factors of resonators, [62].

As has already been shown, an impedance transformation can significantly improve the loaded Q-factor of a resonator.

The new dual-mode resonators are resonator circuits that combine an impedance transformation and the use of differential and common mode. The basic idea of mixed-mode resonators is as follows:

- The energy passes through a coupling network as mode 1.
- A narrow-band mode converter (with resonator element) generates mode 2.
- Mode 2 is reflected by the coupling network.
- The narrow-band mode converter generates mode 1 from mode 2.
- Mode 1 can pass through the coupling network.

The basic systems and some of the resulting circuits are presented below.

Dual-Mode Transmission Resonators

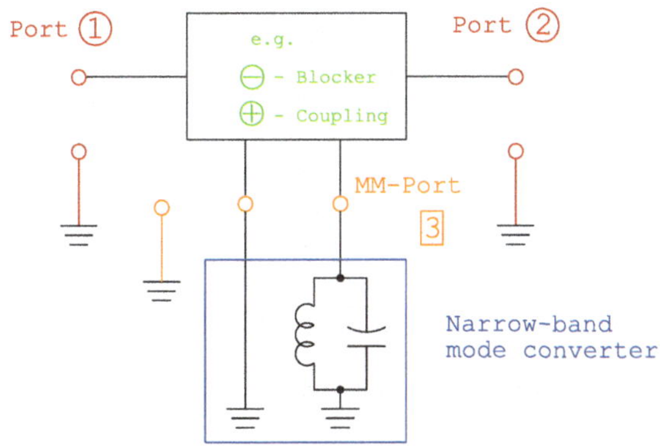

Fig. 6.39: Illustration of the concept for a dual-mode transmission resonator with a coupling network that blocks the differential mode

A possible concept for a dual-mode transmission resonator is shown in Fig. 6.39. Here, the resonator is part of a mode converter.

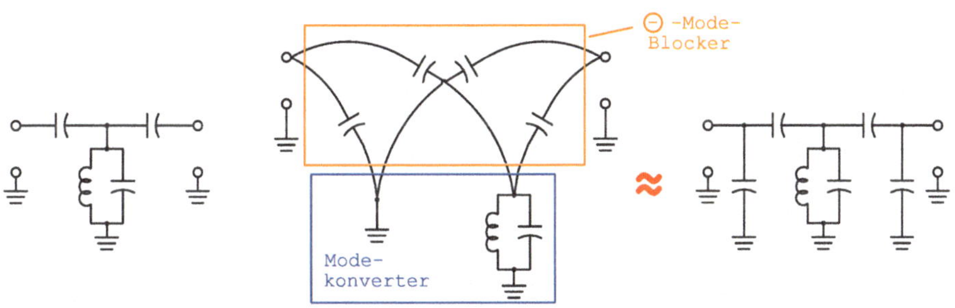

Fig. 6.40: Left: classic transmission resonator; center and right: dual-mode transmission resonator

With purely unbalanced signals, a signal is coupled in at port 1 and decoupled again at port 2. The coupled signal is split by the coupling network and fed to the frequency-selective reflection mode converter (section 5.2.5) as a common mode wave via MM port 3.

There it is converted into the push-pull mode at the resonant frequency. This differential mode signal is reflected by the coupling network due to the integrated mode blocker functionality and fed back to the mode converter. This energy, now converted into a common mode signal, can be decoupled through port 2.

This concept seems quite complex. However, the section 5.2.3 shows how easy it is to implement mode blockers. How simple the solution to the given problem is, is illustrated by the Fig. 6.40.

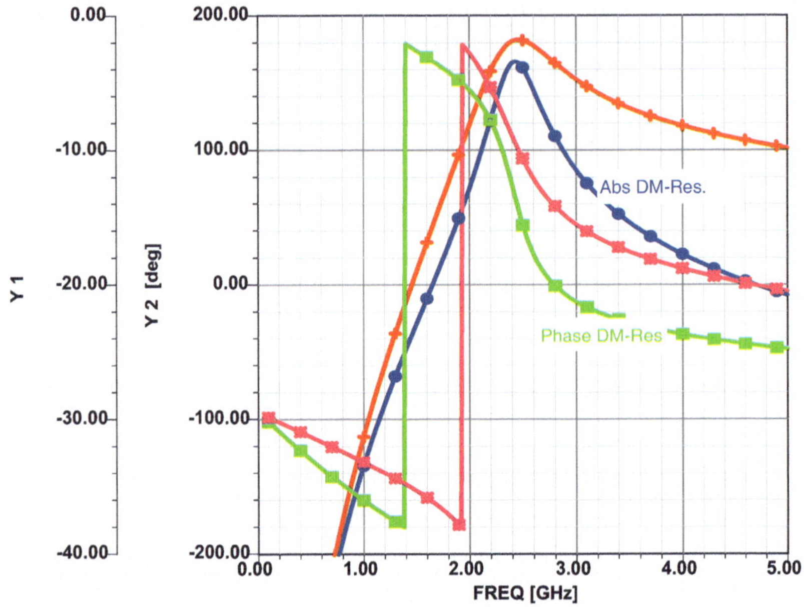

Fig. 6.41: Magnitude (Y1 in dB) and phase (Y2 in degrees) of S_{21} of a Γ transformation resonator and a dual-mode transmission resonator (short DM-Resonator)

Using a circuit simulation in particular, it is possible to very easily compare the properties of the resonator filter with Γ transformation and with dual-mode concept, see Fig. 6.41. The following conditions were used: L_{min} =1 nH, C_{min} =1 pF, $Q_L = Q_C$ =40, f=2.5 GHz. It can be seen in Fig. 6.41 that the filter characteristics of the dual-mode concept and also the phase steepness are significantly better at 2.5 GHz.

The Fig. 6.42 (components: $Q_L = Q_C = 40$, frequency: 2.5 GHz) shows an increase in Q-factor due to the mixed-mode concept. It can be seen that the loaded Q-factor at 2.5 GHz could be increased from around 7 to around 9.

Fig. 6.43 illustrates that the percentage effect is much more pronounced at higher frequencies. The boundary condition for the components was again: $Q_L = Q_C$ =40, L_{min}: 1 nH, C_{min}: 0.5 pF.

In general, however, it can be seen that the loaded Q-factor is still very dependent on the unloaded Q-factor. This concept can be developed even further. You can select other coupling networks with mode blocker functionality and also other mode converters. Similarly, the common mode can also be blocked instead of the differential mode. However, this requires ports 1 and 2 to be MM ports. Such an example will be presented for the dual-mode reflection resonator.

Dual-Mode Reflection Resonators

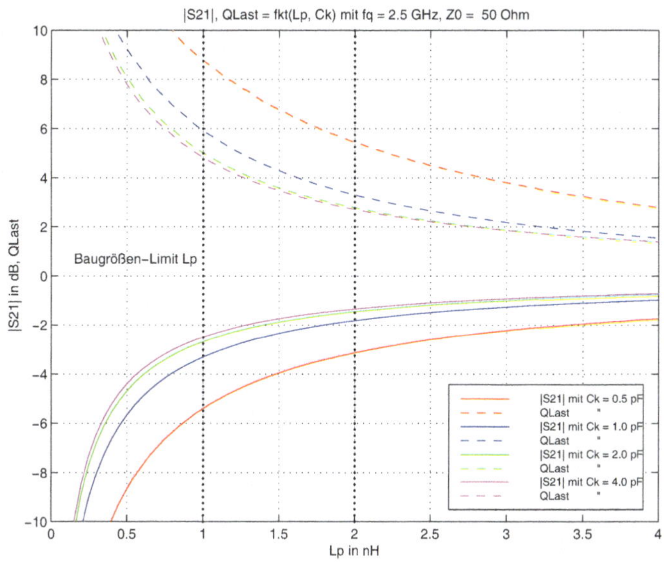

Fig. 6.42: Loaded Q-factor of a parallel resonant circuit in MM circuit

A concept for a reflection mode converter is shown in Fig. 6.44 is shown.

An oscillator, [58], was developed based on this concept, which is currently the most frequently used push-push, differential or cross-coupled oscillator in semiconductor technology.

This concept differs in the final realization only by an additional short circuit!

It could be shown both in the simulation and in the measurement that by using this the phase noise by a good 6 dB at 2.5 GHz can be reduced by using this dual-mode concept.

Fig. 6.45 shows the change of the previous standard resonator with connected upstream small capacitors for the implemented Γ-transformation and the new network with the DM Technology. Only a short circuit to ground needs to be introduced!

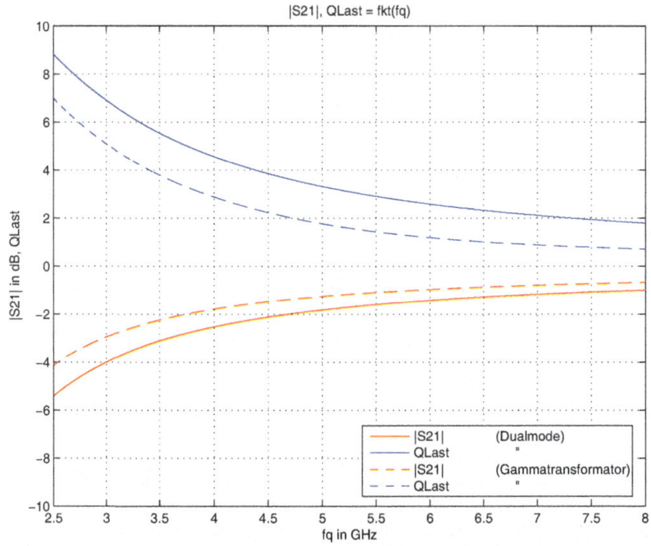

Fig. 6.43: Loaded Q-factor of a parallel resonant circuit with coupling and in MM circuit

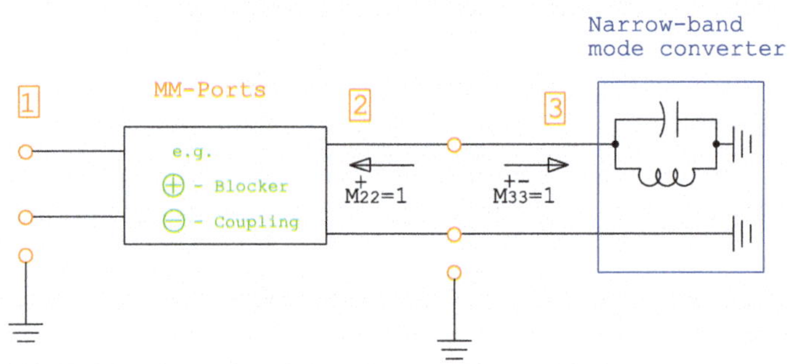

Fig. 6.44: Illustration of the concept for a dual-mode reflection resonator with a coupling network that blocks the common mode

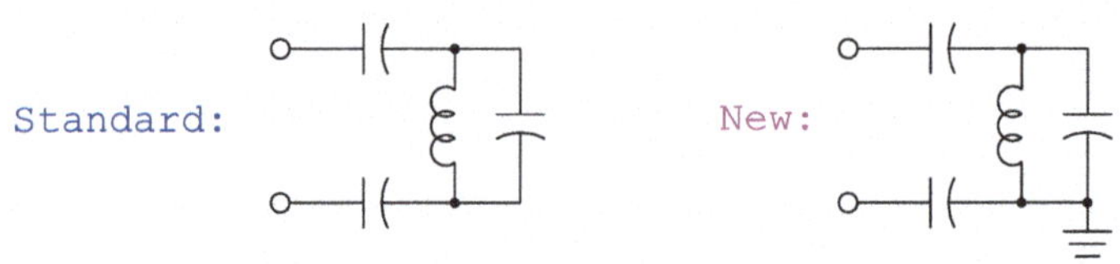

Fig. 6.45: Change in the reflection resonator after the introduction of DM technology

6.6 Coupled Dual-Circuit Resonator Filters of any Q-Factor

It has already been noted in the introduction to this chapter that, in addition to LC chain filters synthesized from standard low-pass filters, coupled resonator filters are very important for the design of band-pass filters. design of band-pass filters. Among for RF filters, these coupled resonator filters are by far the most widely used.

Dual-circuit filters are the most commonly used coupled resonator filters. The only exceptions are BAW and SAW filters as well as professional waveguide filters for base, satellite and similar stations.

High Q-factor single resonators are not suitable as filters for commercial frequency bands as they have significantly lower pass-band widths. Single resonators of medium and low Q-factors are not suitable as their slope is usually too low.

For two-circuit filters of medium Q-factor, the slope of the filter edges of the individual resonators is doubled from typically 20 dB/decade to 40 dB/decade, [48].

For two-circuit filters with medium Q-factor resonators, the approximate derivation of the literature sources [21, 32, 150, 48] was considered to be the best theory for practical use, providing quite useful starting values. For filters with resonators that have a very high Q-factor, useful theories can be found in [8, 73, 89, 91].

In 2004, an alternative and exact theory was introduced - based on the ASTRID method for symmetrical networks - which provides optimal results for both low-quality and high-quality resonators, [57].

ASTRID can also be used to design resonator filters with more than two resonators and a wide variety of configurations. However, the two-circuit resonator filter already shows all effects and can still be displayed very clearly.

ASTRID also enables the synthesis of lossy filters. However, since lossless filters offer sufficiently good starting values for realization in many cases, we will limit ourselves to these in the following.

6.6.1 Overview: Dual-Circuit Resonator Filters

Coupled resonator filters can be used to realize band-pass filters, band-stop filters and combinations of both. In practice, a bandpass filter is usually required, which may have very good selective harmonic suppression.

Just as with single resonators, there are two types of construction for realizing band-passes and band-stops from series and parallel resonators. These four cases are shown in Fig. 6.46.

The resonators can be realized in a wide variety of ways. The LC resonator can therefore also represent the equivalent circuit. The coupling networks with the coupling factors k can in turn be realized in a wide variety of techniques. Energy only needs to be coupled from one resonator to the other.

Examples of couplings are

> Voltage and/or current coupling and/or
>
> magnetic and/or electric field coupling or
>
> acoustic or optical coupling.

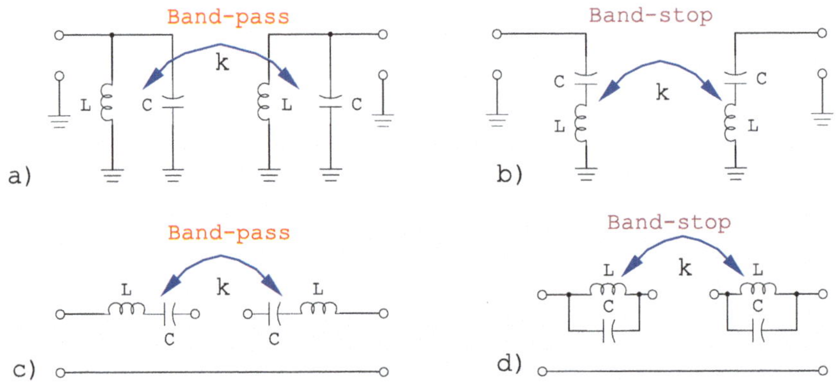

Fig. 6.46: Representation of all configurations of filters with two coupled resonators

Coupling networks can consist of only two-terminal elements or also of complex distributed networks. For filter synthesis, it is advantageous if the coupling network can in turn be modeled with concentrated equivalent circuits.

In the following, we will limit ourselves to the special case of a concentrated coupling network with two terminals. Examples of such networks are:

- One capacitor or one coil,
- A series or parallel resonant circuit,
- Any combination of capacitors and coils.

Furthermore, only the coupled parallel resonant circuit filter, which is very interesting in practice (filter a, Fig. 6.46), which is very interesting in practice, will be examined and analyzed in more detail.

6.6.2 Synthesis of Coupled Parallel Resonant Circuits with Voltage Coupling

Various coupling circuits are shown in Fig. 6.47 for a coupled resonator filter with parallel resonant circuits.

If highly integrated circuits with concentrated and/or quasi-concentrated inductors are realized, the magnetic coupling according to c) and d) is very advantageous, since in this case two inductors are realized on almost the surface of one inductor.

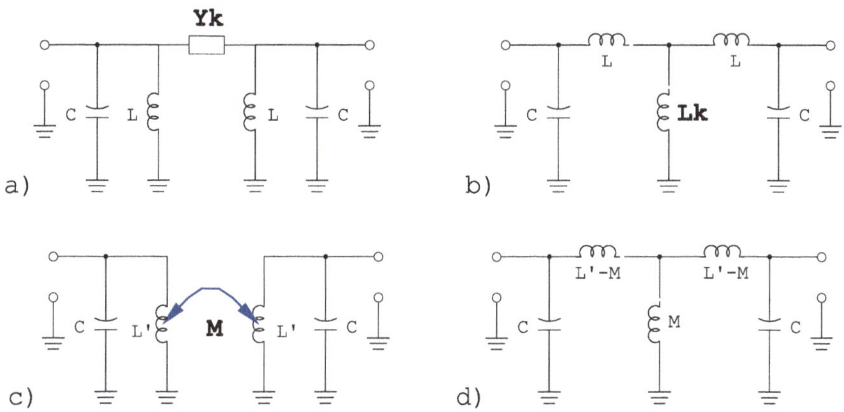

Fig. 6.47: Representation of coupled parallel resonant circuit filters with a) general voltage coupling, b) inductive current coupling, c) magnetic coupling and d) equivalent circuit diagram for c)

Current coupling requires that either the coils are on a common coupling coil (Fig. 6.47 b) and d)) or the capacitances are on a common capacitance, [48].

Voltage coupling is a more general case. Any terminal element[6] as shown in a) or even any two-port.

The further mathematical derivation should be based on a) and the use of a general terminal element.

Furthermore, according to Fig. 6.48, the parallel resonant circuit is to be described with the reactive conductance jB and the coupling element with the reactive conductance jB_k.

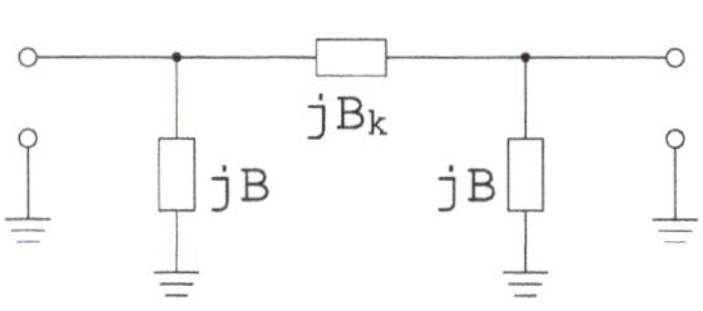

Fig. 6.48: Equivalent circuit of dual-circuit voltage-coupled resonator filters based on parallel resonant circuits

In practice, the coupling element $Z_k = 1/Y_k$ is significantly higher impedance than the resonant circuit elements. Each resonant circuit has the resonant circuit frequency ω_r.

The two resonant circuits are slightly detuned in the same way by the coupling element, as the circuit is symmetrical. If the coupling element only consists of a capacitor C_k, the resonant circuits are capacitively loaded.

A frequency ω_1 that is lower than ω_r is set via the path C_k and C, and a frequency ω_2 that is between ω_1 and ω_r is set via the path C_k and L.

[6] A very interesting special case are resonators as coupling elements, as the filter Q is then almost equal to the resonator Q-factor.

The filter has a perfect match at the frequencies ω_1 and ω_2. With this knowledge, you can start the filter synthesis with the ASTRID method. The corresponding frequency positions are shown in Fig. 6.49.

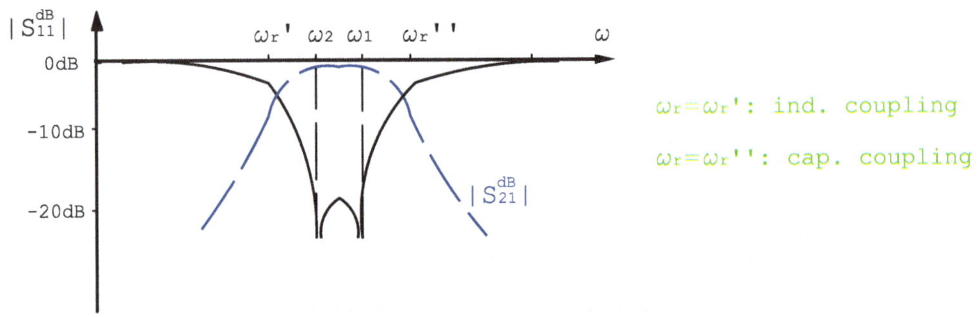

Fig. 6.49: Frequency points for a coupled resonator filter

The analytical calculation has the following starting points:

Given: Topology: Fig. 6.48, ω_1, ω_2 and ω_r (Specifications for the synthesis)

Find: L, C, B_k

For a common-mode excitation at both ports, there is a magnetic wall in the symmetry plane and thus an open circuit. In this case, the input admittance remains:

$$Y_{even} = j\,B \quad . \tag{6.69}$$

For push-pull excitation at both ports, an electrical wall is created in the symmetry plane In this case, the input admittance remains:

$$Y_{odd} = j\,(B + 2B_k) \quad . \tag{6.70}$$

The condition for the adjustment is as follows (5.83):

$$Y_{odd}\,Y_{even} = \frac{1}{Z_L^2} \quad . \tag{6.71}$$

Consequently, with Eq. (6.69) and Eq. (6.70):

$$-\,B^2 - 2\,B_k\,B = \frac{1}{Z_L^2} \quad . \tag{6.72}$$

Eq. (6.72) applies at the angular frequencies ω_1 and ω_2.

This results in the two equations

$$-B_1^2 - 2\,B_{k1}\,B_1 = \frac{1}{Z_L^2} \quad , \tag{6.73}$$

$$-B_2^2 - 2\,B_{k2}\,B_2 = \frac{1}{Z_L^2} \quad , \tag{6.74}$$

with the following conductance values for the parallel resonant circuit:

$$B_1 = \omega_1\,C - \frac{1}{\omega_1\,L} = \nu_1\,\omega_r\,C \quad , \tag{6.75}$$

$$B_2 = \omega_2\,C - \frac{1}{\omega_2\,L} = \nu_2\,\omega_r\,C \quad , \tag{6.76}$$

$$\text{with} \quad \nu_i = \left(\frac{\omega_i}{\omega_r} - \frac{\omega_r}{\omega_i}\right) \quad \text{and} \quad \omega_r^2 = \frac{1}{L\,C} \quad . \tag{6.77}$$

For the rest of the calculation, we need to choose the topology of the coupling network. In the following, we will first select a simple capacitor and then a resonator in the form of a third parallel resonant circuit.

6.6.2.1 The Capacitively Coupled Resonator Filter

A purely capacitive coupling (Fig. 6.50) has the advantage that it is often very easy to fine-tune, which is an important aspect of coupled resonator filters in practice.

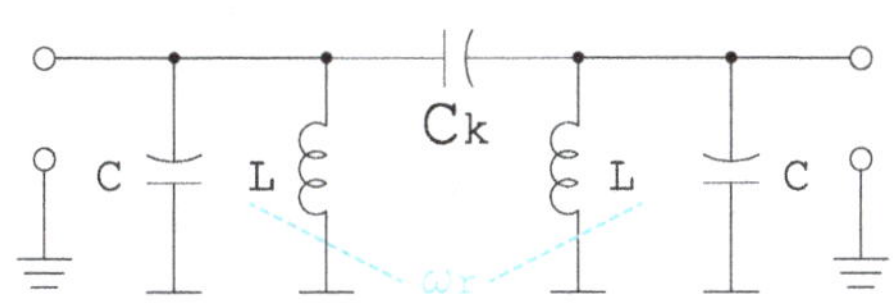

Fig. 6.50: LC circuit for a voltage-coupled resonator filter with coupling capacitor

When selecting a capacitor for B_k, the following applies

$$B_{k1} = \omega_1\,C_k \quad \text{and} \quad B_{k2} = \omega_2\,C_k \quad . \tag{6.78}$$

If the reactive conductance values of Eqs. (6.75), (6.76) and (6.78) are used in Eqs. (6.73) and (6.74), the result is

$$-\nu_1^2\,\omega_r^2\,C^2 - 2\,\omega_1\,C_k\,\nu_1\,\omega_r\,C = \frac{1}{Z_L^2} \qquad // \cdot \nu_2\,\omega_2 \quad , \tag{6.79}$$

$$-\nu_2^2\,\omega_r^2\,C^2 - 2\,\omega_2\,C_k\,\nu_2\,\omega_r\,C = \frac{1}{Z_L^2} \qquad // \cdot \nu_1\,\omega_1 \quad . \tag{6.80}$$

The next steps are the multiplication with the specified ν_i and ω_i variables and a subtraction of both equations:

$$\left(\nu_1\,\nu_2^2\,\omega_1\,\omega_r^2\;-\;\nu_1^2\,\nu_2\,\omega_2\,\omega_r^2\right)\,C^2 \;=\; \frac{\nu_2\,\omega_2}{Z_L^2}\;-\;\frac{\nu_1\,\omega_1}{Z_L^2} \quad , \tag{6.81}$$

$$\Longrightarrow \quad C \;=\; \sqrt{\frac{\nu_2\,\omega_2\;-\;\nu_1\,\omega_1}{Z_L^2\,\omega_r^2\,\nu_1\,\nu_2(\nu_2\,\omega_1\;-\;\nu_1\,\omega_2)}} \quad . \tag{6.82}$$

The inductance L is obtained from Eq. (6.77) and the coupling capacitance via Eq. (6.79) from

$$C_k \;=\; -\;\frac{\frac{1}{Z_L^2}\;+\;\nu_1^2\,\omega_r^2\,C^2}{2\,\omega_1\,\nu_1\,\omega_r\,C} \quad . \tag{6.83}$$

An example with the frequencies f_1=0.95 GHz, f_2=1.05 GHz and f_r=1.2 GHz results in the values: C=6.21 pF, L=2.83 nH and C_k=3.37 pF. The resulting reflection and transmission properties of the filter are shown in Fig. 6.51.

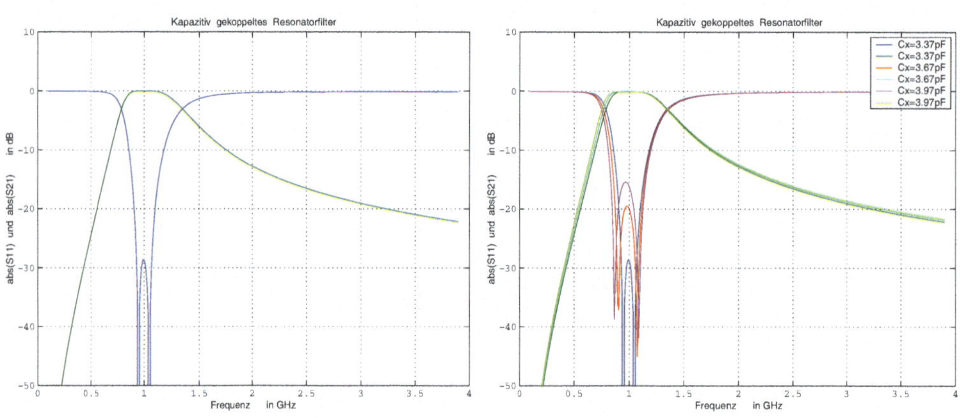

Fig. 6.51: Results of a synthesized capacitively coupled filter, (right with faulty coupling capacitances up to 20 %)

The Fig. 6.51 clearly shows the two frequency points with the perfect match.

Furthermore, the figure on the right shows that the sensitivity to deviations from the ideal value of the coupling capacitance is within normal limits with this design.

The suppression of high frequencies is relatively poor with this filter. A filter with an inductive coupling has a significantly better behavior in this respect, but has poor suppression at low frequencies.

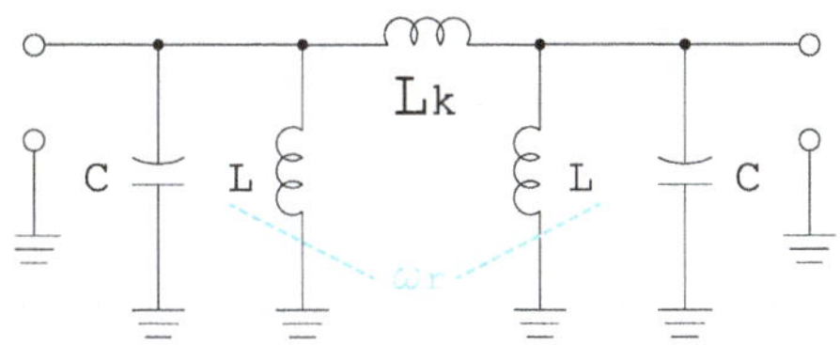

Fig. 6.52: LC circuit for a voltage-coupled resonator filter with coupling coil

This can be remedied by using a resonator as a coupling element.

6.6.2.2 The Inductively Coupled Resonator Filter

The inductively coupled resonator filter differs from the capacitively coupled filter only in the coupling element, 6.52.

The derivation corresponds to that of the capacitively coupled filter and the results differ only slightly:

$$C = \sqrt{\frac{\nu_2\,\omega_1 \;-\; \nu_1\,\omega_2}{Z_L^2\,\omega_r^2\,\nu_1\,\nu_2(\nu_2\,\omega_2 \;-\; \nu_1\,\omega_1)}} \quad , \tag{6.84}$$

$$L_k \;=\; \frac{1}{\omega_1}\,\frac{2\,\nu_1\,\omega_r\,C}{\frac{1}{Z_L^2} \,+\, \nu_1^2\,\omega_r^2\,C^2} \quad . \tag{6.85}$$

6.6.2.3 Resonators as Coupling Elements

In filter synthesis for coupled resonator filters, resonators were generally not permitted as coupling elements (e.g. [8, 48]).

However, the use of resonators is helpful, as some filters - such as the acoustic filters[7] - do not allow any other elements.

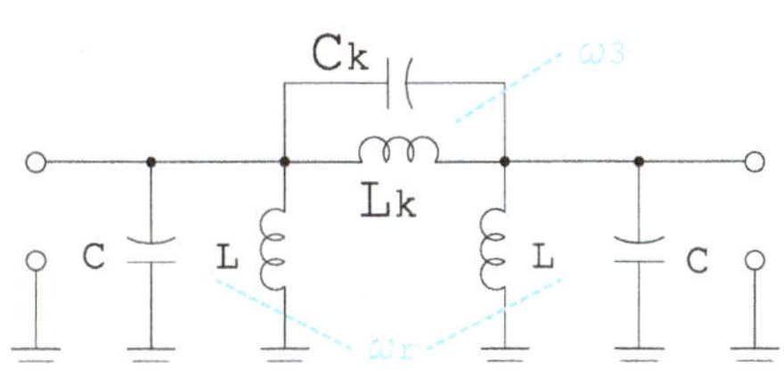

Fig. 6.53: LC circuit for a voltage-coupled resonator filter with coupling resonator

With a filter of this type (Fig. 6.53), there is a further degree of freedom with the additional component Lk. It makes sense to use this degree of freedom for the effective suppression of a harmonic. Although this condition can be easily taken into account using ASTRID theory, in this case the parallel resonance of L_x and C_x at ω_3 is the direct solution.

[7] SAW: Surface Acoustic Wave or BAW: Bulk Acoustic Wave.

The following applies to the reactive conductance values of the coupling resonator:

$$B_{k1} = \omega_1 C_k - \frac{1}{\omega_1 L_k} = \nu_{31} \omega_3 C_k \quad , \tag{6.86}$$

$$B_{k2} = \omega_2 C_k - \frac{1}{\omega_2 L_k} = \nu_{32} \omega_3 C_k \quad , \tag{6.87}$$

$$\text{mit} \quad \nu_{3i} = \left(\frac{\omega_i}{\omega_3} - \frac{\omega_3}{\omega_i} \right) \quad \text{and} \quad \omega_3^2 = \frac{1}{L_k C_k} \quad . \tag{6.88}$$

If these reactive conductance values of Eqs. (6.75), (6.76), (6.86) and (6.87) are substituted into Eqs. (6.73) and (6.74), the result is

$$-\nu_1^2 \omega_r^2 C^2 - 2 \nu_{31} \omega_3 C_k \nu_1 \omega_r C = \frac{1}{Z_L^2} \quad // \cdot \nu_2 \nu_{32} \quad , \tag{6.89}$$

$$-\nu_2^2 \omega_r^2 C^2 - 2 \nu_{32} \omega_3 C_k \nu_2 \omega_r C = \frac{1}{Z_L^2} \quad // \cdot \nu_1 \nu_{31} \quad . \tag{6.90}$$

The next steps are again the multiplication with the specified ν_i and ν_{3i} variables and a subtraction of both equations:

$$\left(\nu_1 \nu_2^2 \nu_{31} \omega_r^2 - \nu_1^2 \nu_2 \nu_{32} \omega_r^2 \right) C^2 = \frac{\nu_2 \nu_{32}}{Z_L^2} - \frac{\nu_1 \nu_{31}}{Z_L^2} \quad , \tag{6.91}$$

$$\Longrightarrow \quad C = \sqrt{\frac{\nu_2 \nu_{32} - \nu_1 \nu_{31}}{Z_L^2 \omega_r^2 \nu_1 \nu_2 (\nu_2 \nu_{31} - \nu_1 \nu_{32})}} \quad . \tag{6.92}$$

The inductance L is obtained from Eq. (6.77) and the coupling capacitance via Eq. (6.89) from

$$C_k = -\frac{\frac{1}{Z_L^2} + \nu_1^2 \omega_r^2 C^2}{2 \nu_{31} \omega_3 \nu_1 \omega_r C} \quad . \tag{6.93}$$

Finally, the inductance L_k can be calculated using the resonance condition from Eq. (6.88).

The example with the frequencies f_1=0.95 GHz, f_2=1.05 GHz and f_r=0.85 GHz[8] results in the values: C=9.93 pF, L=3.53 nH, L_k=6.60 nH and C_k=0.43 pF. The resulting reflection and transmission properties of the filter with now different slopes are shown on the right-hand side of the Fig. 6.54.

On the right-hand side of the Fig. 6.54, the resonant frequency f_r was increased from 0.85 GHz to f_r=0.93 GHz for the two frequencies f_1=0.96 GHz and f_2=1.04 GHz.

You can see that the steepness of the filter increases sharply. The steeper the filter is designed, the more sensitive it becomes to losses. This means that only resonators of the highest Q-factor can be considered for a steep-edged filter.

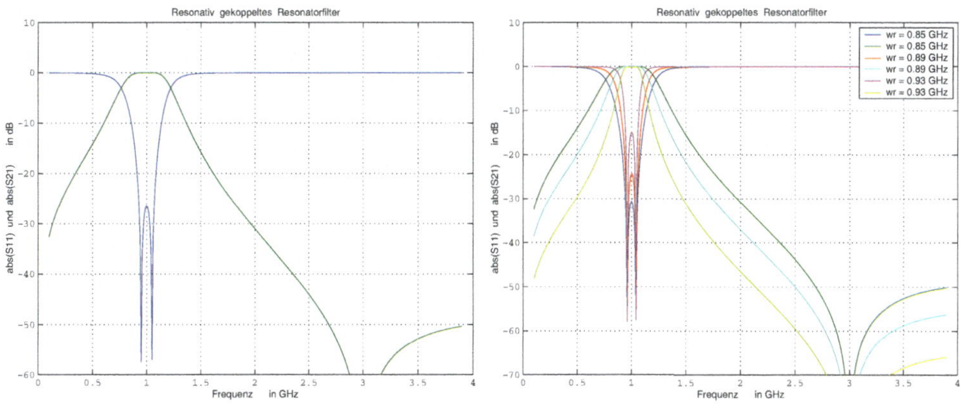

Fig. 6.54: Results of a synthesized coupled resonator filter, (right with increased ω_r)

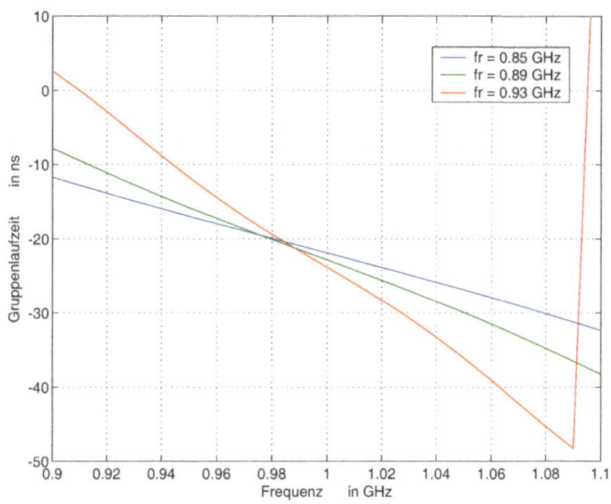

Fig. 6.55: Group delay times of the resonator filter with coupling resonator of different slopes

In [57], an SMD implementation was used to show that this filter synthesis, in contrast to other to other methods is also suitable for components with low Q-factors. Another advantage of this filter synthesis is the good group delay behavior. Fig. 6.55 shows the group delay times with very little ripple for the three different slopes according to Fig. 6.54.

This very low ripple of the group delay shows that these filters cause very little dispersive distortion.

6.6.2.4 Implemented Impedance Transformation

Elliptical filters ([120]) are closely related to the two-circuit filters presented here, which are coupled by means of resonators[9]. Elliptical filters with impedance transformation can only be obtained with asymmetrical topologies, ([120], p. 58 ff).

[8]Note, inductive coupling in the band is advantageous here.

[9]The (symmetrical) elliptical filters can also be derived using the ASTRID method.

This also generally applies to coupled resonator filters[10]! To integrate an impedance transformation in the filters derived in the previous chapter, it is advantageous to use a Γ impedance transformer.

The transformer with series capacitor introduced in the chapter 3.2.2 is the best choice for this case. This transformer according to Fig. 6.56 (left) is suitable because it reduces the size of an inductor and only requires an additional capacitor, which in practice is cheaper and of better quality than an inductor.

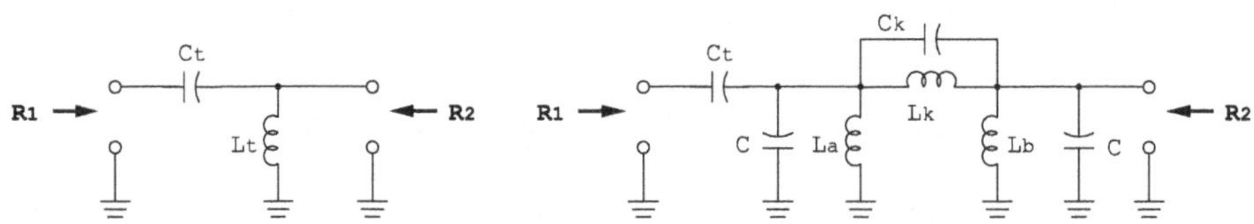

Fig. 6.56: Left: Used Γ impedance transformer; right: Modified resonator filter with impedance transformation

The following applies to this impedance transformer: $R_1 < R_2$.

The results are obtained using the center frequency $\omega_m = \frac{\omega_2 - \omega 1}{2}$:

$$C_t = \frac{1}{\omega_m} \sqrt{\frac{1 \,/\, R_1}{R_2 - R_1}} \quad , \tag{6.94}$$

$$L_t = R_1 \, R_2 \, C_t \quad , \qquad L_a = \frac{1}{\frac{1}{L} + \frac{1}{L_t}} \quad . \tag{6.95}$$

For an impedance transformation of $R_1 = 20\,\Omega$ and $R_2 = 50\,\Omega$ and the other data from the last example, the new components $C_t = 6.50\,\mathrm{pF}$ and $L_a = 2.29\,\mathrm{nH}$ result. The filter function was shown in the first edition and corresponds exactly to the left part of Fig. 6.54.

Double-Sided Impedance Transformation

If you want to produce very steep-edged filters, you are limited, as with resonators, by a "component stop" in the form of minimum inductances and minimum capacitances, particularly in semiconductor, LTCC and SMD technology.

[10] In contrast to what is specified in [48].

There is the same remedy here as with the resonators. If the filter is designed for a higher impedance level, a higher slope can be achieved with increased transmission losses. In this case, the impedance transformation shown above must be carried out on both sides and therefore two coupling capacitors must be used.

6.7 Crossovers

Crossovers have the task of connecting a port with at least two other ports, whereby the pass-through frequencies (individual frequency bands) always have different values. If more than two frequency bands are involved, this is referred to as a **multiplexer**, Fig. 6.57.

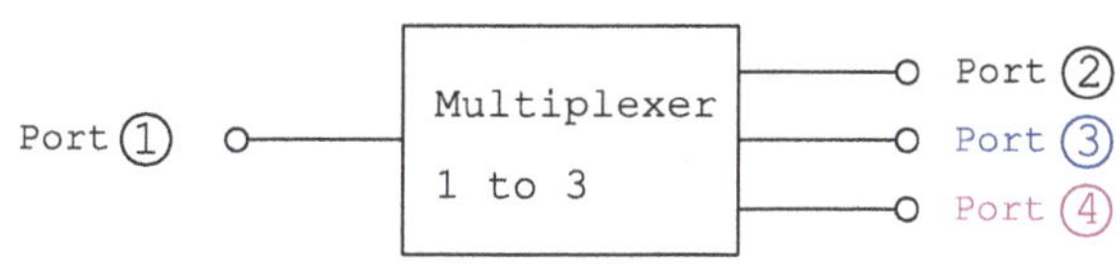

Fig. 6.57: General representation of a multiplexer with a broadband input port 1 and 3 narrowband output ports in 3 different bands

If there are only two frequency bands, this is called a diplexer if the bands are relatively far apart, or a duplexer if the bands are very close together. If the bandwidth between the two bands is equal to or less than the bandwidth of the individual bands, it is a duplexer. Duplexers are often also referred to as RX/TX switches.

The crossover has parallels to the crossover for modes. With the crossover, it must be ensured that the input resistance of a bandpass in the working band of another filter is high impedance (ideally corresponds to an open circuit), Fig. 6.58.

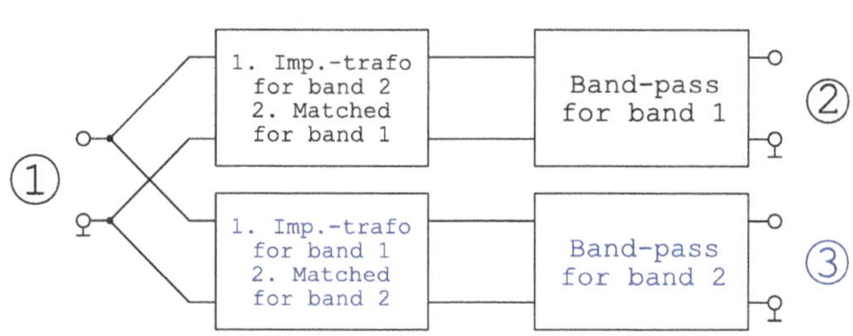

Fig. 6.58: Division of a duplexer or diplexer into bandpass filters and impedance transformers

This method of decoupling these two branches is not the only one, as the example of the ±90° coupler has already shown. This ±90° coupler represents a broadband solution of a TP/HP combination with impedance transformation. In [120] another way of a TP/HP combination using standard filters is shown. The most extensive literature on crossovers can be found in [91]. Another way of combining two bandpass filters will be presented in the next exercise.

——— *Exercise: Diplexer from Resonant Circuits* ———

Given: *A diplexer according to Fig. 6.59 should have a lossless transmission* ($|S_{21}| = 1$) *for* $f_1 = 1\,GHz$ *and a lossless transmission* ($|S_{31}| = 1$) *for* $f_2 = 2\,GHz$. *The filter is perfectly matched to the antenna port for both frequencies* ($S_{11} = 0$). *The two component values* $C_1 = 10\,pF$ *and* $L_1 = 3\,nH$ *are specified.*

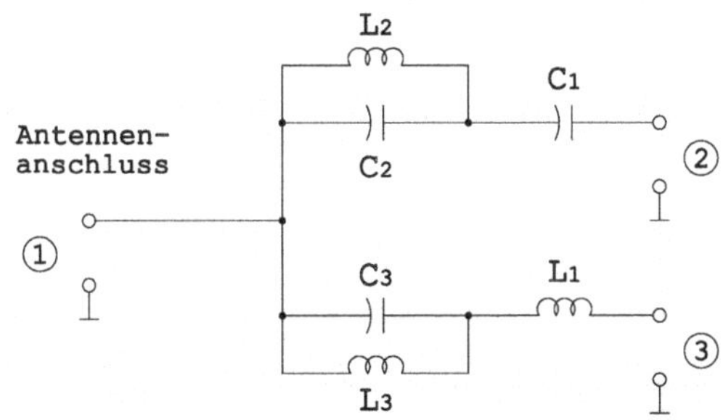

Fig. 6.59: Construction of a diplexer with series and parallel resonant circuits with the antenna port 1

Find: *Calculate the values of the components* C_2, C_3, L_2 *and* L_3.

Solution: *The two parallel resonant circuits each isolate one path:*

$$L_2//C_2 \text{ blocks at } 2\,\text{GHz} \sim \omega_2\,, \qquad L_3//C_3 \text{ blocks at } 1\,\text{GHz} \sim \omega_1\,.$$

$$\Rightarrow \quad \omega_2^2 = \frac{1}{L_2\,C_2}\,; \qquad \omega_1^2 = \frac{1}{L_3\,C_3} \tag{6.96}$$

The following must also apply to both series resonant circuits:

$$\omega_1^2 = \frac{1}{L_x\,C_1}\,; \qquad \omega_2^2 = \frac{1}{L_1\,C_x}\,. \tag{6.97}$$

These two simple conditions deliver directly: $L_x = 2.53\,nH$ *and* $C_x = 2.11\,pF$.

Using the results from Eqs. (3.36) to (3.38), L_2 *and* C_3 *can be calculated:*

$$\text{with} \quad \nu_1 = \frac{\omega_1}{\omega_2} - \frac{\omega_2}{\omega_1} = -\frac{3}{2} \quad \text{and} \quad \nu_2 = \frac{\omega_2}{\omega_1} - \frac{\omega_1}{\omega_2} = \frac{3}{2} \quad \text{applies} \tag{6.98}$$

$$L_2 = -\,L_x \cdot \frac{\nu_1\,\omega_1}{\omega_2} = 1.90\,nH \quad \text{and} \quad C_3 = C_x \cdot \frac{\omega_2}{\nu_2\,\omega_1} = 2.81\,pF\,. \tag{6.99}$$

The two remaining component values can be calculated using the conditions according to (6.96): $L_3 = 9.01\,nH$ *and* $C_2 = 3.33\,pF$.

——— *Exercise: Diplexer from Resonant Circuits* ———

6.7.1 Matched Filters

Filters often cause interference in the stopband due to strong reflections. For this reason, research and development has recently focused on the concept and product development of filters that are matched to a wide bandwidth and have the desired low-loss transfer function in the passband. The calculation of such bandpass filters follows the same path as the Astrid method and can be found in [97]. Highly integrated purchasable filters are available on the market.

Self-development is quite simple for low and high-pass filters. The core filter (LP or HP) can be connected to two diplexers, each of which is terminated with 50 Ω for the output in the stopband. The diplexers were introduced in the previous section. Alternatively to the diplexer, the double Γ transfomators presented in the chapter 3.3.4 can also be used before and after the filters.

In practice, it is very often sufficient to use a filter that is only matched on one side. For example, a first-order high-pass filter (coupling capacitor) can be adapted for low frequencies so that the 1/f noise is absorbed or the strong oscillation tendency at low frequencies is suppressed.

Such a filter can be calculated directly from the economical balun above the Eq. (5.77) and the resonant circuit resonance condition.

For the most common case where there is no impedance transformation, the simple calculation equations for the 3 dB corner frequency at ω_0:

$$Z_0 = R_x = R_y \quad , \quad L = Z_0/\omega_0 \quad \text{and} \quad C = \frac{1}{Z_0\,\omega_0} \quad . \tag{6.100}$$

The corresponding circuit for a one-sided matched coupling capacitor with the 3 dB corner frequency of 1 GHz and the associated component values is shown in Fig. 6.60.

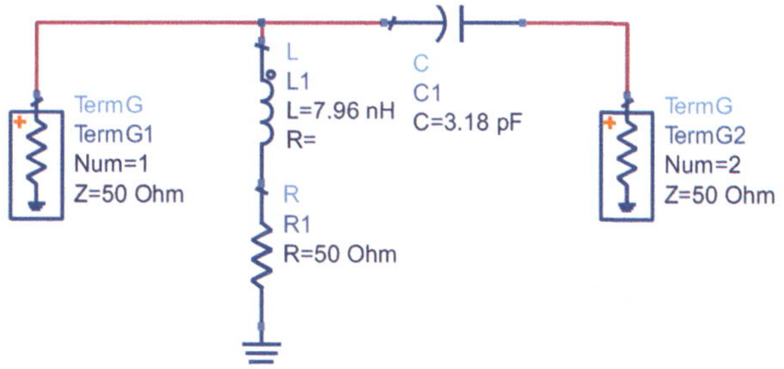

Fig. 6.60: Circuit of a coupling capacitor and a circuit extension for absorbing the energy incident from the left at low frequencies

The corresponding simulation results are shown in the following figure and show that the filter is perfectly matched across virtually the entire bandwidth.

No unilaterally matched filters have yet been introduced in the entire literature or in the reference book [98]. These one-sided matched filters reduce the number of components.

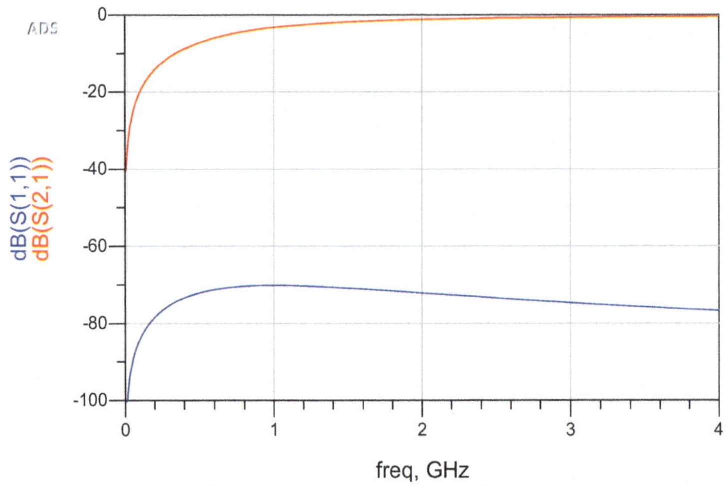

Fig. 6.61: Simulation results of a coupling capacitor with the circuit extension for absorbing the energy incident from the link

The simplest first-order filter published there is shown in Fig. 6.62, where L and C are swapped, as the standard low-pass filter is always introduced in filter theory and the high-pass filter is derived from it.

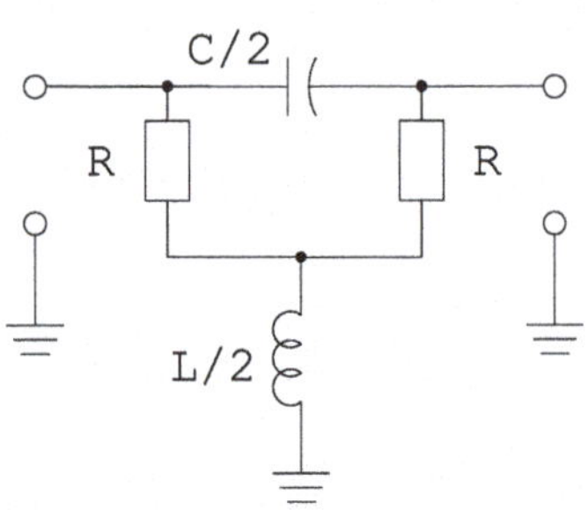

Fig. 6.62: Illustration of the first-order filter matched on both sides

The design of this filter is based directly on the results of Eq. (6.100), as it is is only the mirrored series connection of the filter from Fig. 6.60. The reflection behavior of the first-order filter matched on both sides is quite close to the values shown in Fig. 6.61. In the transmission behavior, the losses are somewhat higher, as expected.

Chapter 7

RF Switches

Switches are used in RF technology to route the output signal of a power amplifier to different paths and to connect an antenna to different transmit (TX) and receive (RX) branches. Switches are also found in modulators and switchable phase shifters. In measurement and production technology, RF switches are used to measure a measurement object using various devices.

This chapter presents the most frequently used RF switches. The electromechanical switches are presented first. The electronic switches are described in more detail below.

7.1 Coaxiale Relay

Electromechanical coaxial switches are preferably used in measurement technology, where the cost of the switch and the switching times play a subordinate role.

The most important specifications of a coaxial relay are:

- Frequency range,
- Input power,
- Transmission losses,
- Isolation,
- Input matching,
- Repeatability and
- Service life.

Frequency range: Coaxial relays are available from 0 Hz to 40 GHz.

Input power: A distinction must be made here between two operating modes: "Hot-Switching" and "Cold-Switching". Furthermore, a distinction must be made as to whether a power is permanently present (continuous wave signal) or whether the power is only present for a short time (pulsed signal). A pulsed signal that is switched in the time slots in which no power is present (cold switching) can generally be greater than 100 W.

H. Heuermann, *Radio Frequency Technologies*,
https://doi.org/10.1007/978-3-658-50170-9_7

Transmission losses: These losses increase due to the skin depth δ over the frequency. Typical values for a good switch are 0.2 dB at 4 GHz, 0.5 dB at 18 GHz and 1.0 dB at 40 GHz.

Isolation: The insulation values of coaxial relays are >90 dB at 18 GHz and >50 dB at 40 GHz.

Repeatability: The repeatability is specified by the manufacturers with a value of 0.03 dB after 5 million switching operations. This means that the difference between the transmission values is less than -60 dB (calculation: $\Delta S_{21}^{dB} = 20 \text{ lg} \left(|S_{21}' - S_{21}''|\right)$). Due to these excellent properties, relays are also used in precision measuring devices.

Lifetime: The service life is specified by the manufacturers as over 5 million switching operations, but in practice (e.g. semiconductor measurement technology) it is significantly higher.

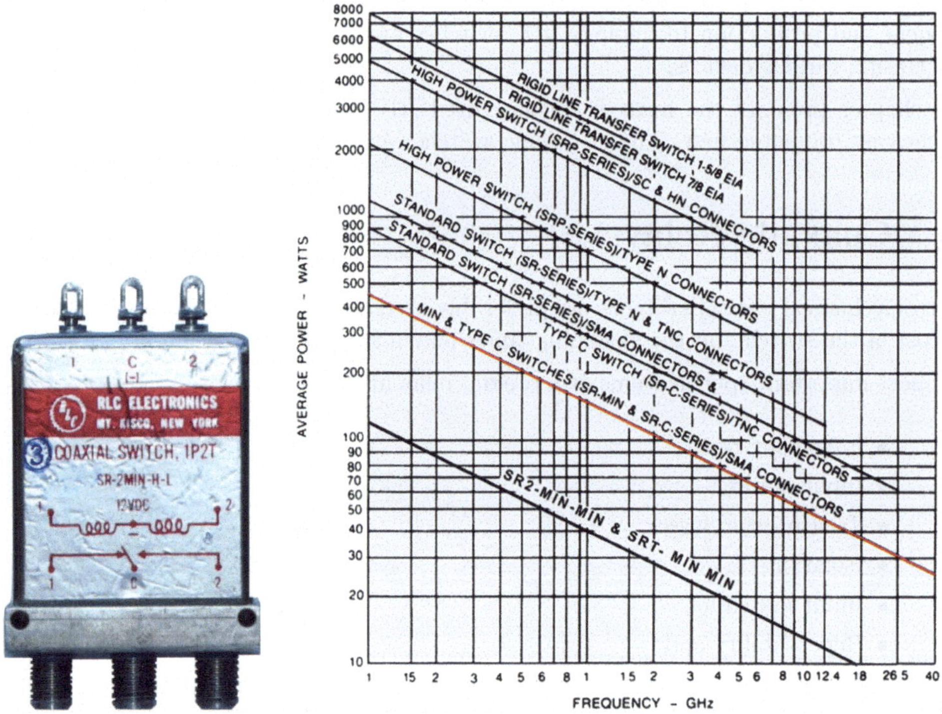

Fig. 7.1: Left: RF relays with coaxial connectors; right: frequency-dependent losses of relays (source: online catalog of the company RLC Electronics)

Fig. 7.1 shows a photo of a coaxial relay on the one hand and illustrates the power-frequency diagram on the other. This shows that a coaxial relay can handle a lot of RF power in the MHz range. The higher the frequency, the lower the power compatibility of the relay. The increasing losses due to the decreasing skin depth are a significant factor here.

In practice, so-called *latching relays* are also very interesting. These bistable relays have the property that they can assume two different stable switching states when de-energized.

7.1.1 Reed-Relais

The so-called reed relay works with spring contacts that are located in an airless chamber or a chamber filled with inert gas (glass tube) and are actuated by a magnetic field. In contrast to normal relays, which are actuated purely electromechanically, reed relays have a much longer service life of 1 million switching cycles and more.

The two reed switch contacts are spring-mounted and made of precious metal-coated ferromagnetic material (soft iron). There is a gap between the two contacts in the idle state.

When a magnetic current is applied from outside, the two spring contacts are closed. This closing process is carried out by means of an electromagnet located around the glass body. Fig. 7.2 shows a photo of the inner reed contact.

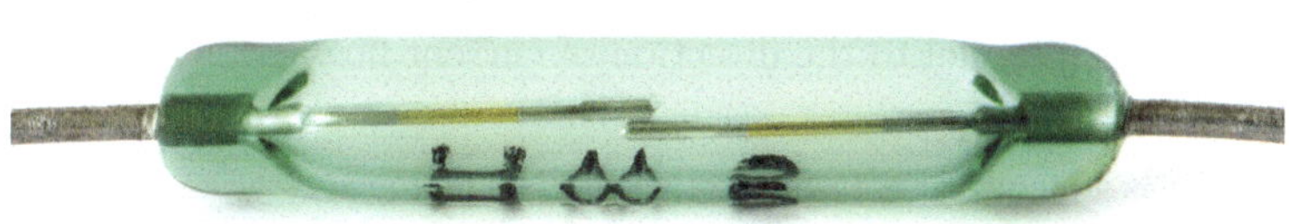

Fig. 7.2: Photo of a reed contact in the closed state (source: wikipedia.de)

Fig. 7.3 illustrates how a coaxial relay is created from a reed contact. The spring contacts in the glass tube are provided with a shield that acts as a ground plane, as is the case with a coaxial line.

In addition, the spring contacts are dimensioned in such a way that the criterion of a 50 Ω characteristic impedance is fulfilled. A coil is wound around the shield, which causes a magnetic field to close the spring contacts and does not affect the RF signal.

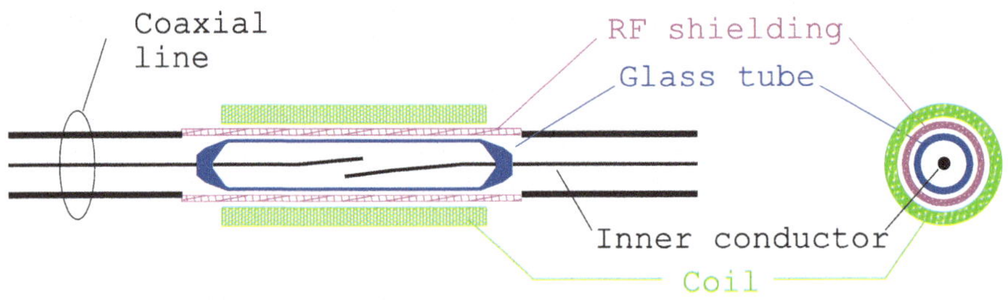

Fig. 7.3: Illustration of the structure of a reed relay

The reed contact was patented back in 1936 by W. B. Elwood. This contact is also used in electronics, including in SMD relays. The advantages of this contact are Durability, currentless stationary operation, explosion-proof and higher current compatibility. The disadvantage compared to standard relays is the higher cost.

7.1.2 Freewheeling Diode

The diode described below has many other names besides freewheeling diode. These names are flyback diode, snubber diode, commutating diode, flywheel diode, suppressor diode and clamp diode.

The flow of current creates a magnetic field in a coil. If the current flow is interrupted, the coil tries to maintain its "old" state. Since the current no longer flows after disconnection, but at the same time the charged energy W cannot simply escape, the voltage of the coil quickly increases to relatively high values, as shown in Fig. 7.4.

High voltages often interfere with other circuits and can destroy electronic components. To ensure that the current can continue to flow through the inductive load (motor, relay coil, etc.) after the voltage is switched off and to avoid voltage peaks due to induction, a diode is connected antiparallel to the direction of current flow through the load as shown in Fig. 7.5.

When the switch is closed, the freewheeling diode D has no effect as it blocks. The current increases according to an exponential function $i = \frac{U}{R_{Cu}}\left(1 - e^{-t/\tau}\right)$ with the time constant $\tau = \frac{L}{R_{Cu}}$ up to a final value that is only determined by the driving voltage and the DC resistance of the circuit.

When the switch is opened, the coil acts as a generator and drives the current through the conducting freewheeling diode, initially at the same strength. The magnetic energy is converted into heat.

The current decays with $i = \frac{U}{R_{Cu}}\left(e^{-t/\tau}\right)$ with the time constant $\tau = \frac{L}{R_{Cu}}$ as well as the induced voltage U_L, see Fig. 7.4.

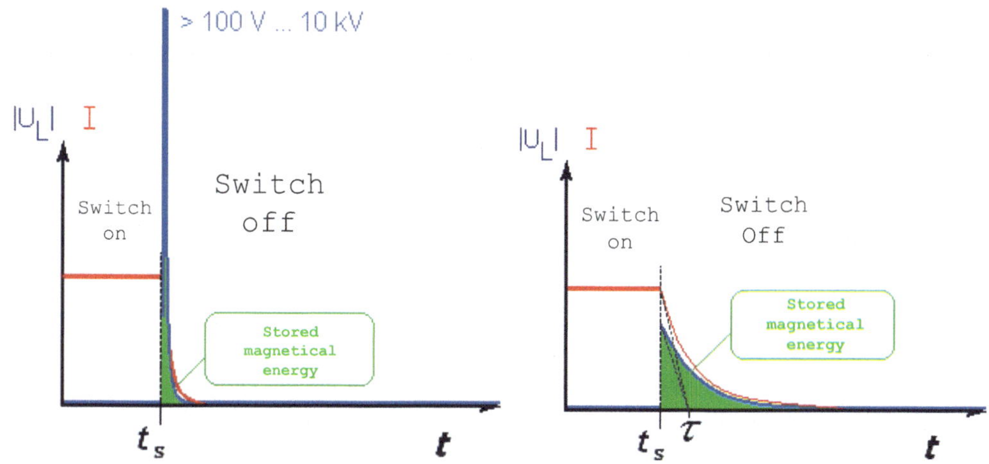

Fig. 7.4: Voltage at relay contact, left: without freewheeling diode; right: with freewheeling diode

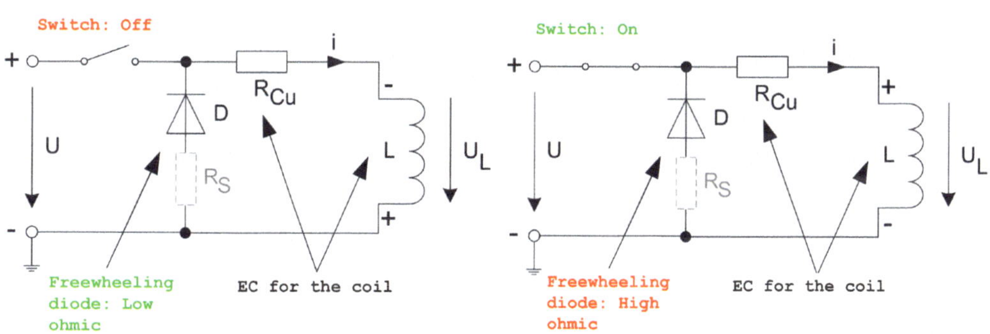

Fig. 7.5: Freewheeling diode on the coil with the equivalent circuit (EC) for the coil

By converting the induced voltage formula U_L, the discharge time $dt = -L\frac{di}{U_L}$ can be determined. When the switch is open, the forward voltage of the diode drops across the coil and causes a time constant that is in the millisecond range.

As a result, the stored magnetic energy is released (constantly) over a longer period of time. With a series resistor R_S in the diode branch, the time constant τ can be reduced and thus the switching of a relay accelerated.

The mathematical relationship is as follows:

$$\tau = \frac{L}{R_{Cu}+R_S} \quad .$$

The wiring of the coil through the freewheeling diode and R_S has a noticeable effect on the switching time. With optimum wiring, you can achieve less than 0.2 ms.

7.2 MEMS

Electromechanical switches that are produced using semiconductor manufacturing technology are called Micro Electro-Magnetic Switches or MEMS for short.

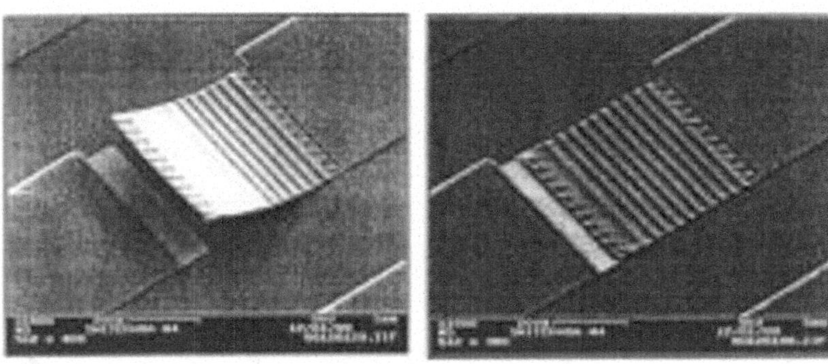

Fig. 7.6: Photo of a MEMS switch in open and closed state

With MEMS, the conductor of a RF line is switched via an electrostatic force. The disadvantage of MEMS is that the voltage for this powerless switching must generally be greater than 20 V.

MEMS are divided into two categories:

1. Ohmic switches,
2. Capacitive switches.

Like the coaxial relays, the resistive switch consists of a metallic contact structure. However, as the switch forces in MEMS are very small, resistive switches fail due to "sticking".

In a capacitive switch, there is a thin dielectric layer between the two conductor contacts. This switch therefore only switches a capacitance from a large value (e.g. 20 pF, continuity) to a small value (e.g. 0.05 pF, insulation). These switches have a significantly longer service life than ohmic switches, but can only be used in the RF range.

Due to their reliability problems, the relatively slow switching times (typically 5 ms) and the high switching voltage required, MEMS have not yet become established in mass products, but are being strongly promoted due to their advantages over electronic switches. The advantages are:

- broadband and high cut-off frequency,
- low transmission losses and high isolation,
- linearity,
- ESD resistance,
- power-free control, and
- no 1/f noise.

The disadvantages of MEMS technology are the high control voltages, large sizes, high costs, complex housing technology and poor reproducibility.

7.3 PIN Diode Functionality and Switch

A large number of diodes are still used in RF technology. A very good description of the most important diodes can be found in [145]. Only one type of diode construction is suitable for switching an RF signal.

These so-called PIN diodes consist of three layers: Highly doped p^+ contact zone, intrinsic i-center zone and highly doped n^+ contact zone. The name of this special diode is derived from this three-layer structure.

PIN diodes have been the most commonly used RF switches in modern communication technology over the last century. They are characterized as RF switches by

- the low price,
- the great linearity,
- the high power stability,
- high ESD robustness,
- low losses and
- a small design.

The disadvantage of using PIN diodes as RF switches is that the control signal cannot be powerless when switched through. A continuously flowing direct current in the range of a few mA is required to switch the PIN diode through.

Fig. 7.7: PIN diode switch as series switching element

For a basic understanding of this RF switch, it is sufficient to consider the PIN diode as a controllable RF resistor. By means of a current control, the series resistor (PIN diode) in the RF path is set to high or low resistance via a DC path (coils with perfect RF isolation, e.g. 100 nH), Fig. 7.7.

The steady-state conditions of the PIN diodes in the reverse and flux ranges are described below. The switching behavior of the diode switches is then described. Finally, applications as RF switches are discussed.

7.3.1 Structure of a PIN Diode

A direct current can only flow across a diode if a positive voltage between the anode and cathode exceeds the value of the diffusion voltage U_D, Fig. 7.8.

Fig. 7.8: Circuit diagram and associated structure of a standard diode

The PIN diode differs from the PN diode in that it has an additional layer. A schematic section through a PIN diode with abrupt p^+/ν and ν/n^+ transitions and typical doping is shown in Fig. 7.9. The inner zone (center zone) with width w is very weakly and quite homogeneously doped.

An ideal intrinsic conduction cannot be set for a semiconductor in practice. There is always an impurity in every semiconductor, no matter how pure it is.

If weak (excess) doping with donators $N_D \leq 10^{13}\,\text{cm}^{-3}$ occurs, this is referred to as ν doping and the PνN structure (slightly n-conducting).

If weak (excess) doping occurs with acceptors $N_A \leq 10^{13}\,\text{cm}^{-3}$, this is referred to as π doping and the PπN structure (slightly p-conducting).

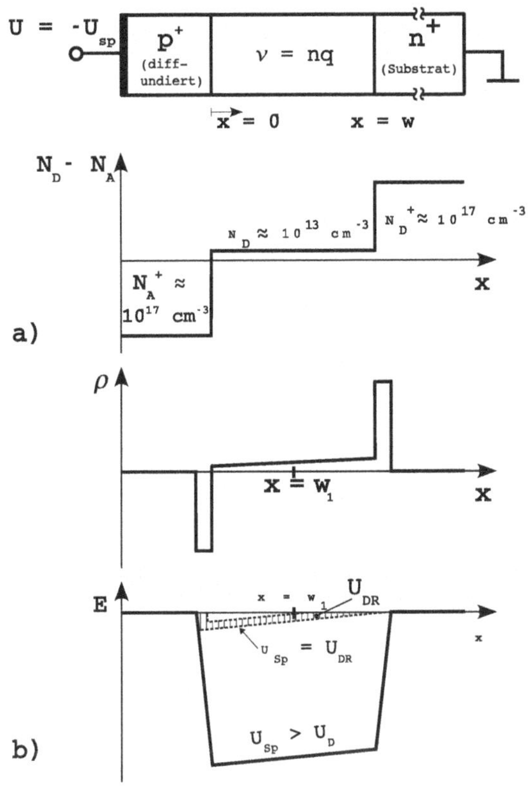

Fig. 7.9: a) Structure and doping profile of a PIN diode with pνn structure and *abrupt* transitions. b) Representation of the space charge area $\rho = q(N_D - N_A)$ and the field strength E with partial (dashed curves) and complete (dotted: $U_{sp} = U_{DR}$; solid: $U_{sp} > U_{DR}$) clearing of the ν zone

7.3.2 PIN Diode in the Stationary Blocking Area

The contact zones are very highly doped and therefore very good conductors. This is why the reverse voltage U_{sp} only drops in the i-zone.

As with abrupt p$^+$n diodes, for junction widths $w_1 < w$ the junction width w_1 follows from

$$U_D + U_{sp} = \frac{q}{\epsilon}\, N_D\, \frac{w_1^2}{2} \qquad \Rightarrow \qquad w_1 = \sqrt{\frac{2\,\epsilon}{q\,N_D}\,(U_D + U_{sp})} \tag{7.1}$$

with the elementary charge $q = 1.602 \cdot 10^{-19}$ As, the dielectric constant ϵ and the diffusion voltage

$$U_D = kT\,\ln(N_A^+\,N_D/n_i^2) \qquad \text{(p}\nu\text{n-structure)} \quad , \tag{7.2}$$

with the inversion density (also intrinsic line density) $n_i^2 = n \cdot p$.

The center zone or i-zone is completely cleared when $w_1 = w$. The related punch through voltage U_{DR} is

$$U_{DR} = \frac{q\, N_D\, w^2}{2\, \epsilon} - U_D \qquad (\text{p}\nu\text{n-structure}) \quad . \tag{7.3}$$

If larger reverse voltages are applied, the width of the space charge zone remains almost at w. However, the field strength E increases evenly, as shown in Fig. 7.9. The equivalent circuit according to Fig. 7.10 shows the conditions for operation in the reverse direction. The ohmic resistances of the contact zones are denoted by $r_b/2$ in each case. C_s is the

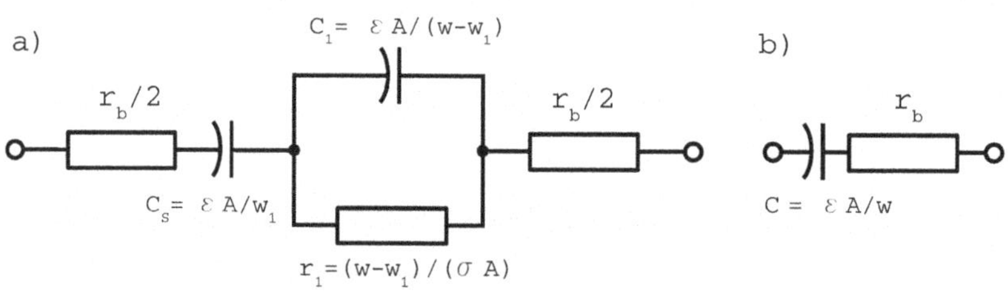

Fig. 7.10: a) Equivalent circuit of a PIN diode in blocking mode; b) Approximate equivalent circuit for higher frequencies

capacitance of the cleared i-zone.

The area $w - w_1$ of the remaining i-zone has a resistance r_1 with the conductivity $\sigma = q\mu N_D$ and a capacitance C_1, which takes into account the displacement current in the uncleared area is taken into account.

For example, for a doping of $N_D = 10^{12} - 10^{13}\,\text{cm}^{-3}$ and a mobility of $\mu_n = 1500\,\text{Vs/cm}^2$ (silicon), the dielectric relaxation frequency of the i-zone $2\pi f_d = \sigma/\epsilon$ is in the range 25-250 kHz.

The displacement current then predominates in the higher frequency range. The following applies $1/\omega C_1 > r_1$. The parallel circuit consisting of C_1 and r_1 can therefore be combined and simplified to a series element $z_x = z_1$:

$$z_x = \frac{1/(j\omega C_1)}{1 + 1/(j\omega C_1\, r_1)} \simeq \frac{1}{j\omega C_1} + \frac{1}{\omega^2\, C_1^2\, r_1} \quad . \tag{7.4}$$

The capacitive component of z_x becomes c with c_s:

$$\frac{1}{C} = \frac{1}{C_s} + \frac{1}{C_1} = \frac{w}{A\, \epsilon} \quad . \tag{7.5}$$

The real part of z_x with the specifications for c_1 and r_1 from Fig. 7.10

$$r_x = \frac{\sigma}{\omega^2\, A\, \epsilon^2}\,(w - w_1) \;\simeq 0 \tag{7.6}$$

is zero for $w = w_1$ and for high frequencies, as it is proportional to $1/\omega^2$.

This leaves only the series connection of the capacitance c and the series resistance as an alternative circuit for the blocking mode of the PIN diode r_b (0.5-5 Ω).

The series resistance is made up of the two resistances of the highly doped zones with the lengths l_{p+} and l_{n+}:

$$r_b = \frac{l_{p+}}{q\, N_A^+\, \mu_p\, A} + \frac{l_{n+}}{q\, N_D^+\, \mu_n\, A} \quad . \tag{7.7}$$

7.3.3 PIN Diode in the Stationary Flow Area

When operating in the flux direction, the applied flux voltage injects charge carriers into the i-zone. Due to the high doping of the contact zones, the current at the p^+i junction is almost exclusively a hole current and at the ni^+ junction an electron current.

The induced minority charge carriers have a density p or n that is much greater than the doping concentration in the high-purity i-zone. They therefore recombine preferentially with each other and have the same lifetime τ.

Because of $n \gg N_D$, $p \gg N_D$ (for a ν zone), the neutrality condition $n \simeq p$ follows. Consequently, the total diode current corresponds to the recombination current of the holes and electrons. The flux current results from the ratio of the stored charge Q of one sign and the recombination lifetime τ:

$$I_f = \frac{q\, n\, A\, w}{\tau} = \frac{q\, p\, A\, w}{\tau} \quad . \tag{7.8}$$

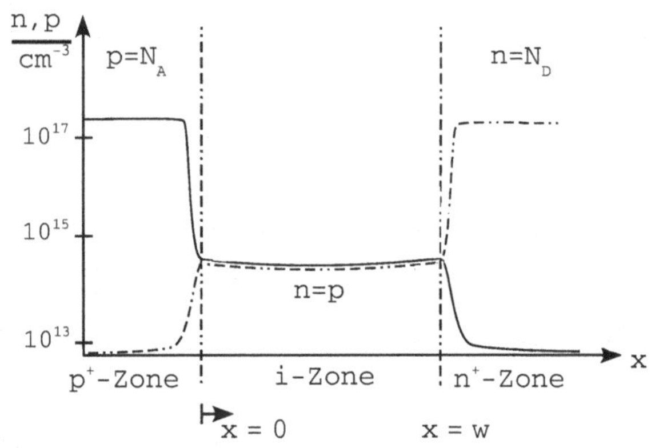

Fig. 7.11: Schematic distribution of the injected charge carriers for a PIN diode operated in the flux direction

Fig. 7.11 illustrates the distribution of the injected charge carriers for a PIN diode with high doping of the contact zones for a large ratio of diffusion length

$$L_{n,p} = \sqrt{D_{n,p}\, \tau} \tag{7.9}$$

and the width w of the i-zone. The diffusion coefficient D can also be expressed as a function of mobility using the Nernst-Einstein equation.

$$D_{n,p} = \frac{k\,T}{q}\;\mu_{n,p} \tag{7.10}$$

Because of

$$\tau_{n,p} = \frac{m_{n,p}\,\mu_{n,p}}{q} \qquad \text{applies} \qquad L_{n,p} = \sqrt{k\,T\,m_{n,p}\,\frac{\mu_{n,p}}{q}} \quad . \tag{7.11}$$

The mobility of the electrons of GaAs is much greater than that of silicon (8000 compared to 1350), but the mobility of the holes is significantly worse for GaAs (300 compared to 480).

For this reason, silicon has long been the only semiconductor considered for PIN diodes, especially as silicon technology is far better mastered and more cost-effective.

For the differential resistance r_f of the i-zone, the conductivity $\sigma = q(n\mu_n + p\mu_p)$ and $n = p$ with the Eq. (7.8):

$$r_f = \frac{w}{\sigma\,A} = \frac{w}{q\,n\,(\mu_n + \mu_p)\,A} = \frac{w^2}{\tau\,I_f\,(\mu_n + \mu_p)} \quad . \tag{7.12}$$

As the flux current I_f increases, the charge carriers that are injected into the i-zone increase proportionally and the flux resistance r_f decreases.

If the flow resistance is to be as low as possible, w should be reduced, as there is a quadratic relationship. On the other hand, the blocking capacitance C only increases linearly with $1/w$ (Eq. (7.5)).

Fig. 7.12 shows the AC equivalent circuit of a PIN diode for operation in the flux direction.

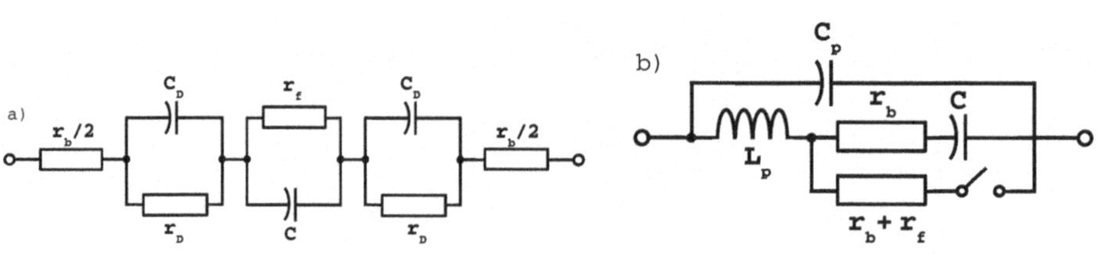

Fig. 7.12: a) Detailed equivalent circuit of a PIN diode in the flux range; b) Summarized equivalent circuit for blocking and flux range with ideal switch and case capacitance C_p as well as the parasitic inductance C_p.t L_p

Recombination losses and diffusion capacities occur at the contact zones. These were taken into account in the detailed equivalent circuit diagram (ECD) using r_D and c_D. The following applies as a guide value for these losses

$$1/r_D + j\omega C_D = \frac{q\,I_f}{kT}\;\sqrt{\omega\,\tau}\;(1+j)/2 \quad . \tag{7.13}$$

This conductance is so large in the RF case that it can be neglected.

In addition to the remaining three elements of the PIN diode (r_b, C and r_f), the ECD in the Fig. 7.12 b) also contains the two parasitic elements C_p and L_p, which are mainly caused by the housing and the bonding wires. In practice, C_p is often greater than the blocking capacitance C, and it is therefore important to choose the smallest possible housing or to work without a housing. For PIN diodes in SMD housings, shunt capacitors to ground must always be added to both connections.

General note: It is customary to label the equivalent quantities belonging to the semiconductors (here e.g. r_f) with small letters and the equivalent quantities belonging to the housings with capital letters. The "inner" semiconductor equivalent circuit is called the intrinsic model and the "outer" package equivalent circuit is called the extrinsic model (Fig. 7.12: L_P and C_P).

7.3.4 Switching Behavior of PIN Diodes

The switching processes are only presented here in qualitative terms. Due to their design, the switching times are significantly faster than with mechanical switches, but are already in a critical range when used as RX/TX switches and as switches in modulators.

Switching by Large Signals

If the level of an RF signal to be switched is high enough, then half a wave of this RF signal is sufficient to open and close the switch. To do this, all of the charge stored in the diode must be cleared. To ensure that this happens, the AC amplitude of the RF signal must be

$$\hat{I} > \frac{\omega}{2}\,\tau\, I_f \quad . \tag{7.14}$$

This means that for a diode with a recombination lifetime $\tau = 10\,\mu$s at a frequency of 1 GHz and a flux current of $I_f = 10\,$mA, the alternating current must reach amplitudes of more than 3 A[1].

Eq. (7.14) states that at low frequencies, even a small applied RF current is sufficient to switch through a blocked diode. For this reason, PIN diodes are not used below 1 MHz.

The Switch-On Process

Electrons and holes move due to an applied E-field of field strength E with the mean velocities

$$v_n = -\mu_n\, E \qquad \text{resp.} \qquad v_p = \mu_p\, E \quad . \tag{7.15}$$

In the i-zone, the field strength of the E-field is so great that this linear relationship no longer applies and the drift velocity of the electrons almost corresponds to the saturation velocity v_s. v_s corresponds to the thermal velocity of the charge carriers of 10^7 cm/s. In the following, it is roughly approximated that the holes also move at the saturation velocity.

[1]These are conditions that can already occur with GSM handsets.

At the switch-on moment at $t = 0$, charge carriers are injected from the highly doped contact zones into the i-zone at the saturation velocity v_s. This penetration of the i-zone is only complete when the charge carriers have reached the contact zone opposite contact zone has been reached. The time required is

$$t_1 \simeq w/v_s \quad . \tag{7.16}$$

Consequently, a fast switch-on time can be set via a short width in the i-zone. The switch-on process is illustrated in Fig. 7.13.

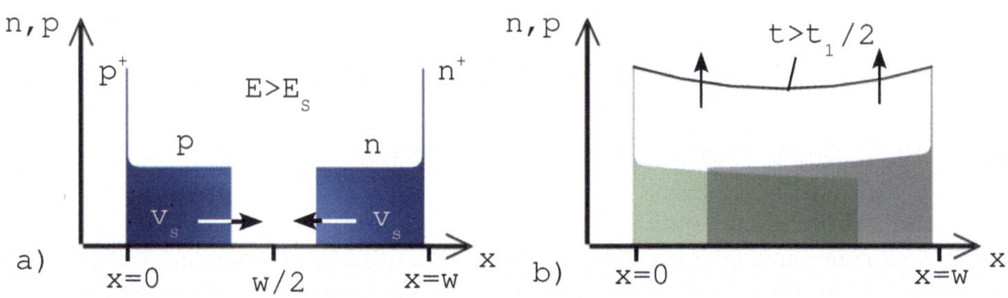

Fig. 7.13: Distribution of the injected charge carriers in the i-zone in the switch-on process for a) $t < w/(2v_s)$ and b) $w/(2v_s) < t < w/v_s$

While the switch-on time is usually acceptable, there are often problems with the switch-off time.

The Switch-Off Process

From the switch-off moment at $t = 0$, all charge carriers that are in the i-zone should be removed. In the best case, this is done by an imposed blocking current I_{sp}[2]. Fig. 7.14 shows the processes that take place when the PIN diode is switched off over time.

[2]This requires a negative voltage supply, which you may want to save.

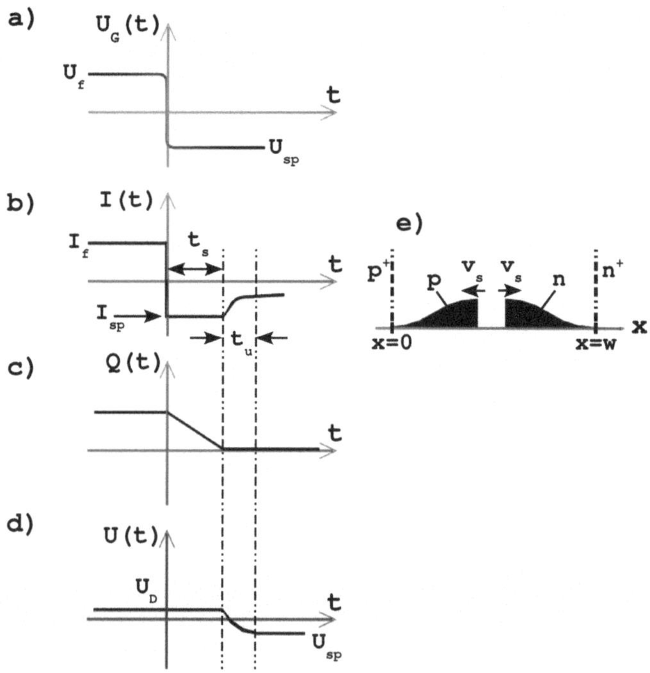

Fig. 7.14: Switch-off process of a PIN diode: a) Voltage curve of the control signal; b) Current curve of the control current; c) Charge in the i-zone d) Voltage across the diode; e) Distribution of the injected charge carriers

The storage time

$$t_s \simeq \tau \; I_f / I_{sp} \tag{7.17}$$

is required until the returning charge carrier clouds separate. Furthermore, the transitional period

$$t_u \simeq w/(2 v_s) \tag{7.18}$$

is required to completely clear the i-zone and switch off the diode:

$$t_2 = t_s + t_u \qquad . \tag{7.19}$$

The switching times of PIN diodes can reach into the μs range.

Especially if no negative reverse current is available, it should be ensured that the charge carriers have a "discharge path". Excessively large, charged coupling and block capacitors often delay the switching times in practical circuits.

7.3.5 Types of PIN Diodes

Various PIN diode designs are shown in the Figs. 7.15 and 7.16.

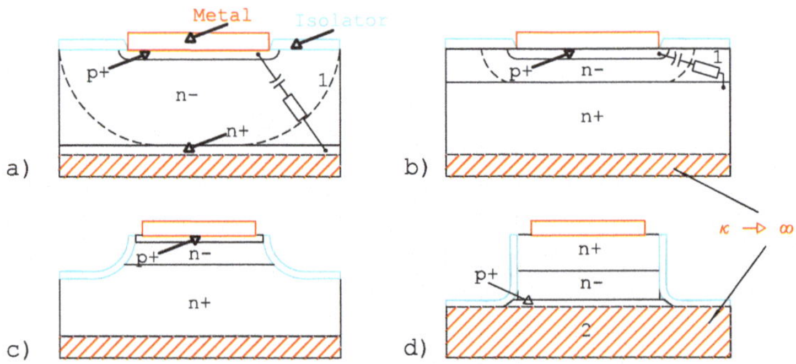

Fig. 7.15: Types of PIN diodes a) Planar diode with long central zone; 1. parasitic RC element in the edge area b) Planar diode with short center zone; c) Mesadiode d) "Upside-down"-Mesadiode; 2. Metal heat sink directly on the active layer

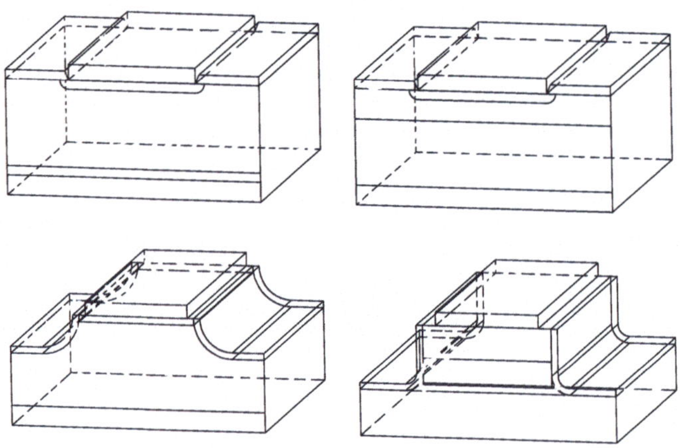

Fig. 7.16: 3-D representation of PIN diode designs from Fig. 7.15

In the past, the more elaborate mesa designs were preferred, as these have smaller parasitic elements. With increasing experience in silicon technology, however, it is now possible to produce PIN diodes with excellent properties using the less favorable planar design.

The advantage of the planar technology according to b) is its integrability in so-called MMICs (Monolithic Microwave Integrated Circuit) and the very favorable manufacturing price. Up to now, PIN diodes are not currently used in MMICs.

7.3.6 PIN Diodes Switch Arrangements

PIN diodes were the most frequently used switches in RF technology (see [8, 145]). Due to the disadvantages already mentioned, they are often combined with $\lambda/4$ lines. As PIN diodes are difficult to integrate with other semiconductor technologies, PIN diode switches are generally discrete structures.

Fig. 7.17 shows a single-pole and a double-pole switch for the case that positive and negative control signals (U_D) are available.

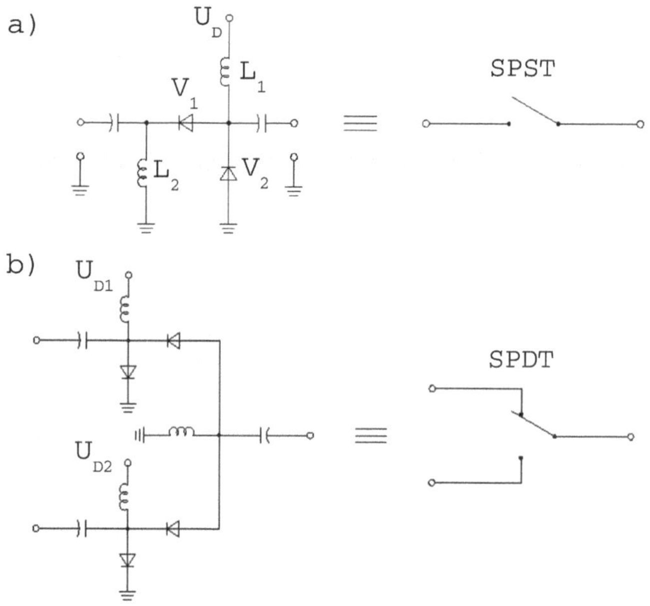

Fig. 7.17: PIN diode switch: a) SPST (Single Pole Single Thru) and b) SPDT (Single Pole Double Thru)

Such circuits can be realized in SMD technology up to 8 GHz and typically still have an isolation of 40 dB at 5 GHz with only 0.7 dB insertion loss and more than 12 dB reflection loss.

Often no negative control signal U_D is available. In this case, a distinction must be made as to whether switches are required for broadband applications or narrowband applications.

For broadband applications, V_2 (Fig. 7.17) may have to be omitted, and therefore only about half the isolation is achieved.

Most applications are narrowband and often PIN diode switches are required as transmitter-receiver switches for operation with only one antenna.

In this case, a $\lambda/4$ line can be inserted between L_1 and V_2 and the direction of V_2 can be rotated, as shown in Fig. 7.18.

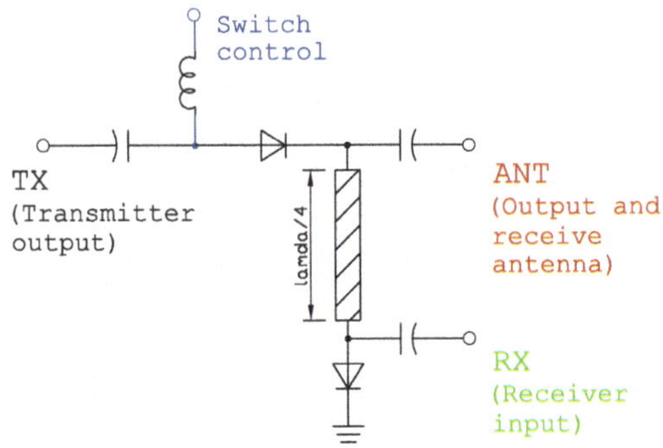

Fig. 7.18: PIN diode switch for TX/RX switching with only one positive control signal and no power consumption in the RX case

This switch arrangement has the two advantages that 1. no control current is required when receiving (RX) and a long so-called "standby" time can be achieved and 2. the integer harmonics of the amplifier are very well suppressed. This switch is included in almost every GSM device in this way.

In addition to pure switches, PIN diodes can be used to produce a range of variable components. Examples of variable attenuators and phase shifters are presented below.

Controllable Attenuator

An electronically variable attenuator should have a very high reflection attenuation over a wide frequency range and its insertion loss should be adjustable over a wide range. These requirements can be met with high precision by setting two shunt PIN diodes and one series PIN diode each via a computer-controlled current control (Fig. 7.19)[3].

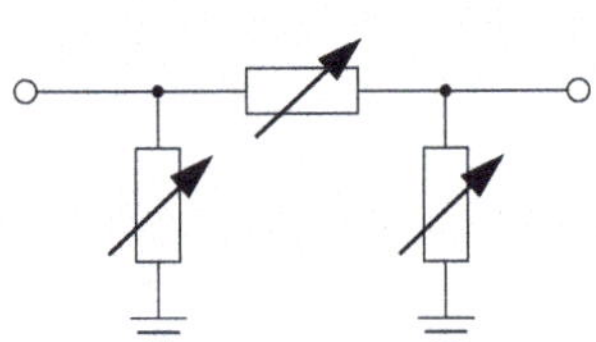

Fig. 7.19: Principle circuit diagram of a controllable attenuator that can be realized using two shunt diodes and a series PIN diode

Controllable attenuators are used in radio and television receivers (tuners), for example, to protect the receiver from excessive received signals. It must not be possible to adjust the attenuation precisely for this purpose. For cost reasons, computer control is often not necessary.

[3]The design of this π attenuator has already been completed in Chap. 2.3. The missing relationship between flux current $I_f = I_R$ and adjustable resistance can be found in the PIN diode data sheets.

In the TV, a peak value rectifier (detector) detects the input power, the detected signal is converted into a constant current by means of an operational amplifier circuit, and this control current controls a simple attenuator according to Fig. 7.20[4].

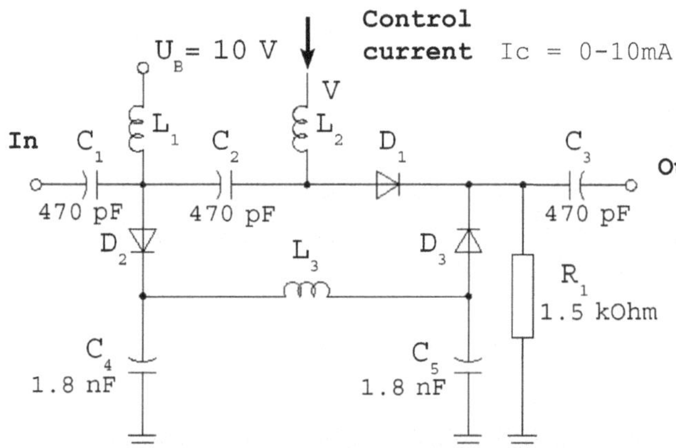

Fig. 7.20: PIN diode attenuator, as used in television tuners

R_1 is selected so that D_2 and D_3 for $I_R = 0$ are set to $75\,\Omega$ (system impedance of the antenna). The switch blocks and is matched. As the control current increases, the current through D_2 and D_3 decreases. If more than 9 V drops across R_1, D_1 must be fully controlled, as D_2 and D_3 are no longer conductive. This circuit can only be finely optimized by selecting the diodes D_1 and D_2/D_3.

This circuit achieves a reflection attenuation of 10 dB over the entire attenuation range from 0.5 dB to around 60 dB for radio and television tuners.

Switchable Phase Shifter

Switchable phase shifters are required if you want to modulate signals, generate complex-value (IQ) signals or control group antennas.

The transmission transit time phase shifter (Fig. 7.21) can be manufactured very precisely. The SPDT switches required for this are shown in Fig. 7.17.

Since a $l_2 = \mathrm{n}\cdot\lambda/2$ long line with any characteristic impedance connects two ports with the same input impedance, the phase shifter fails at these lengths (or the corresponding frequencies)!

[4]The capacitance of the capacitors is so large that they form short circuit for the RF signal, and the high inductance of the coils block the RF signal.

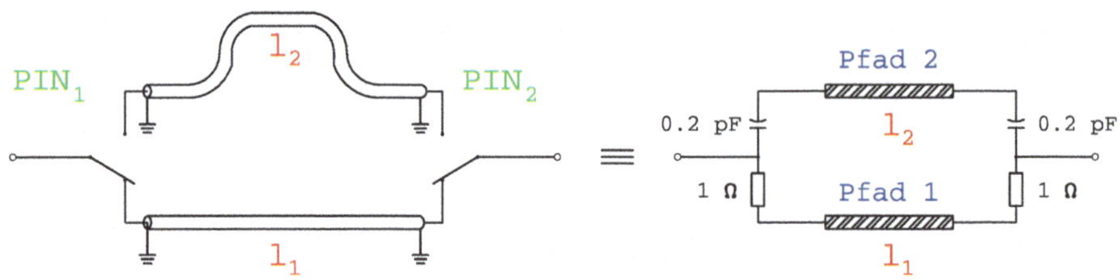

Fig. 7.21: Schematic diagram of a transmission transit time phase shifter with PIN diodes and associated ECD

If this phase shifter is used in the familiar BPSK modulation, for example, the modulation frequency may only be a fraction of the maximum possible switching frequency. Transistors offer significantly higher switching frequencies for such applications.

7.4 RF Transistor Switch

40 years ago, only field-effect transistors of the semiconductor type GaAs Mesfet switches were initially used as RF transistor switches for small signal applications, which were partially replaced at higher power levels by HEMT switches, which can also switch signals up to several watts. In the lower GHz range, the considerably cheaper CMOS switches are also currently entering the market.

The advantages of all these transistor switches over PIN diode switches are:

- faster design, integrable,
- faster shifting operations,
- virtually power-free operation.

The disadvantages of these transistor switches compared to PIN diode switches are:

- somewhat more expensive, more sensitive to ESD,
- lower upper operating frequency,
- a little more insertion loss,
- lower power compatibility.

Insulation values, space requirements and the number of components in the overall circuit are comparable.

However, these new FET, HEMT and CMOS switches are increasingly being used for mass applications in multiplex applications, as the circuit complexity for PIN diode switches is no longer acceptable given the number of channels. It is not unusual for a switch IC to link 2 antenna ports and 8 RX/TX ports.

Modern FET solutions can be used in broadband up to the double-digit GHz range. A Field Effect Transistor (FET) is a frequently used semiconductor component in electrical engineering. In the initial phase of GHz electronics, this type of transistor was used exclusively in this microwave range. The transistor has three connections: Port, source and drain.

Compared to bipolar transistors, this type of transistor only requires a control voltage to be applied between the gate and source in order to change the conductivity of the drain-source path. As the gate has a small capacity, there is a short-term current flow during recharging and therefore low losses. However, these are very low compared to the bipolar transistor, which is why the control can be regarded as virtually powerless.

Two different technologies for N-channel field-effect transistors are shown in 7.22. The BULK structure represents the conventional structure. The SOI structure is somewhat more complex, but has significant advantages, which will be shown below.

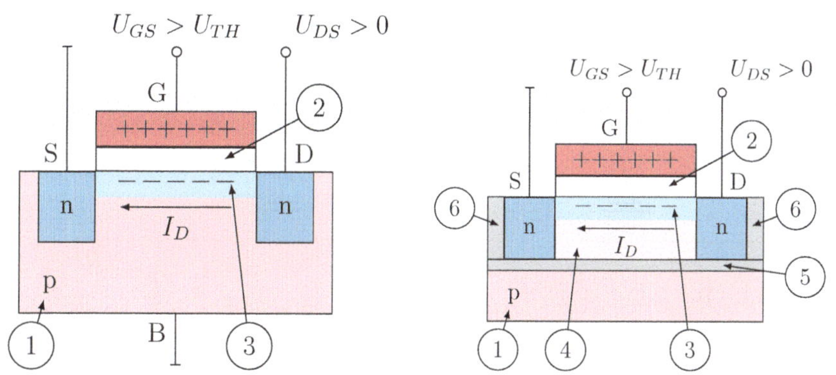

Fig. 7.22: Illustrations of N-channel MOSFET sections of the older design with body or bulk contact B (left) and the newer SOI design (right), **1**: Substrate (slightly p-doped), **2**: Silicon dioxide, **3**: Channel for electron current, **4**: intrinsische zone, **5**: Buried oxid, **6**: Trench insulation.

Both technologies are incorporated into a p-doped semiconductor substrate, the so-called wafer (1). Two n-doped regions are doped into this semiconductor substrate **D**rain and **S**ource). The semiconductor substrate is separated from the **G**ate by a thin oxide layer (2).

By applying a port-source voltage that is greater than the threshold voltage[5], the eponymous electron channel (n-channel) (3) is formed, which enables charge carrier transport, [138].

In SOI technology, this is already the entire structure of the transistor. Depending on the degree of integration, so-called shallow trench insulation (6) can also be used here. This ensures greater isolation of neighboring transistors from each other.

[5]The threshold voltage describes the port-source voltage required to form the charge carrier channel below the gate that is necessary for the current flow.

The problem with this technology is that the relatively wide channel between drain and source results in a relatively high capacitance between the two zones in the blocking state. As a result, the isolation decreases significantly with increasing frequency. In addition, a leakage current can flow from the drain to the source and bulk contact.

SOI technology (SOI: Silicon on Insulator) is used to reduce these problems. The boundaries of the electron channel are defined by inserting the so-called buried oxide (5). This layer thus forms a high-resistance insulation to the underlying substrate.

This can significantly reduce leakage currents. In addition, the parasitic capacitance of the blocking transistor is reduced as the channel becomes narrower. As a result, the insulation can be significantly improved. In addition, an attempt is made to create as intrinsically conductive an area as possible below the port.

In some cases, this area is slightly p-doped to increase the insulation effect. However, this has an influence on the conductivity of the drain-source path when the transistor is switched through. A trade-off must therefore be made here between insulation and conductivity [84, 139].

In addition to these two different technologies, there are numerous other substrate structures. In the RF sector, for example, silicon-on-sapphire is another widely used technology.

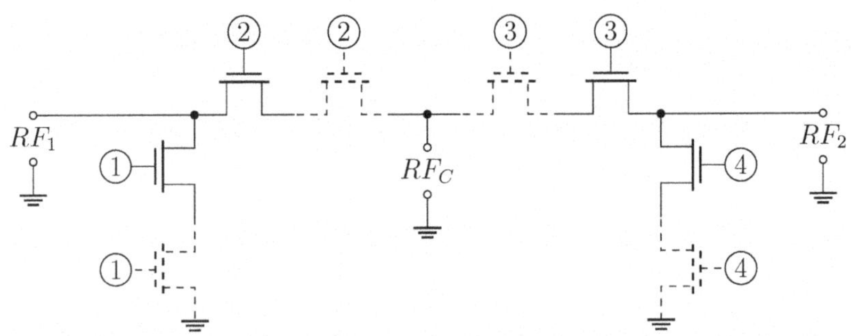

Fig. 7.23: Realization form of an SPDT switch using field-effect transistors

Fig. 7.23 shows the schematic structure of an SPDT switch with field-effect transistors. The control contacts (1) and (2) as well as (3) and (4) each form a switching unit. The center port (RF_C) corresponds to the common port of both switching paths. n order to realize a through-connection to port 1 (RF_1), the gates of the transistors (2) must be supplied in such a way that they enable a through-connection.

In addition, the signal path must be connected with high impedance to ground (1). Port 2 must be isolated so as not to disturb the signal path. This is done by blocking the transistors in the transmission path (3) and connecting the transistors to ground (4). If the signal path is now to be switched to port 2, all switching signals must be inverted, [147].

In a purchasable switch, the control lines of the individual transistors are interconnected by control electronics in such a way that the signal paths can be controlled via a single pin. For switches with a high number of switching contacts, the signal path can sometimes be selected using a bus.

The switch shown is a reflective switch. This means that the port, which is not part of the signal path, has a reflective effect. If a 50 Ω resistor were introduced into each of the two paths to ground, the switch would have an absorbing behavior.

Shaded additional transistors are shown in all paths. The reason for this is that, despite the reduced parasitic properties due to SOI technology, there is still quite a large capacitance (C_{OFF}) between drain and source in the blocking state. By connecting the transistors in series, the parasitic capacitance can be significantly reduced. As the transistors are manufactured in the same process, it can be assumed that the capacitances are approximately the same. This results in the following equation for the total capacitance for N transistors connected in series:

$$C_{Ges} = \frac{C_{Off}}{N} \quad . \tag{7.20}$$

In return, however, the total resistance of the transistors increases in the connected state. Again, it can be assumed that the resistance is the same due to the same manufacturing process. This results in a total resistance of

$$R_{Ges} = R_{On} \cdot N \quad , \tag{7.21}$$

where N again represents the number of linked transistors. The insulation of the switch can essentially be influenced by a lower capacitance. However, it also has an influence on the transmission of the switch, as the shunt transistors form a path to ground via this capacitance.

On the other hand, increased resistance in the signal path naturally also means increased transmission loss. When developing these switches, a compromise must therefore be made between capacitance in the blocking state and resistance in the conducting state. It should also be noted that both the drain and source of the transistor form an additional capacitance through the buried oxide. Although this is usually very low, with a high number of transistors per path it can also have a negative effect on the transmission properties [82, 139].

Compared to PIN diodes, these switches have the advantage that they can be controlled without power, which is a great advantage especially in portable applications. Furthermore, FET switches can usually be built more compactly as they require less external circuitry such as coils and capacitors. This makes it possible to provide a large number of signal paths in a small space. DP8T switches are therefore not uncommon in transceivers.

Furthermore, these switches do not require coupling capacitors and can therefore also transmit DC voltages. Depending on the internal design, there may still be a lower cut-off frequency. However, this is usually significantly lower than that of PIN diode switches [147].

In contrast to FET switches, PIN diodes have slower switching times due to storage effects in the space charge zone. On the other hand, however, they can usually transmit significantly higher power than FET switches and have better linearity.

A major advantage over MEMS switches is that they are more reliable and also work much more reproducibly. They are also characterized by better insulation.

A FET switch is therefore always recommended when a high level of insulation is required in conjunction with low-power control. The condition here is that the power to be transmitted is relatively low in order to avoid problems caused by the non-linear behavior of the field-effect transistors.

7.5 Switches for Differential Circuits

An ideal switch in electronics and RF technology has a very low on-resistance for the switched-through case and a very high insulation as well as high-impedance inputs for the blocking case. In RF technology, this means that the switch should have the lowest possible transmission losses and the best possible input matching when switched through. In the open state, a switch should preferably simulate an open circuit on both sides and have the lowest possible transmission.

The technological forms of realization of electronically controllable switches include contactors, relays, MEMS, transistor switches and, only in RF technology, PIN diode switches. All these forms of realization can be described in a first approximation by a series resistor for the connected state and a blocking capacitance in the open state.

However, since the smallest capacitances of only 200 fF already allow noticeable transmissions in RF circuits, the insulation properties of individual switches in particular are quite difficult to realize. For example, the PIN diode most frequently used in RF circuits in the design (e.g. BAR 88 from Infineon) only has an isolation of 23 dB at 2 GHz. At 10 GHz it is only 10 dB isolation. The individual switches listed for the various technologies are referred to below as switch elements.

RF transmitter/receiver systems (transceivers) are divided into a small signal transceiver IC and the so-called front end. The highly integrated transceiver IC provides frequency-stabilized signal conversion to the low-frequency range. Nowadays, these ICs are almost exclusively built using so-called symmetrical circuit technology. Front ends contain the components power amplifiers, filters, switches, low-noise receiving amplifiers and antennas.

These components are usually manufactured using different technologies. Consequently, frontends are discrete circuits. For antennas, amplifiers and filters, there are nowadays realization forms in balanced circuit technology that are significantly more advantageous than the unbalanced circuits. Since in principle any unbalanced circuit can be converted into a balanced circuit by mirroring (and thus almost double the component effort), solutions for switches are also known.

However, such solutions for front ends have not yet become established because, among other things, the complexity of the components is considered too great and therefore too expensive. Consequently, RF switches are manufactured and used almost exclusively in unbalanced circuit technology.

As switches with only one PIN diode often have insufficient insulation in practice, a feed-through switch is often realized by interconnecting two diodes, as shown in Fig. 7.17 a) as an SPST switch. The left-hand switch port has a high-impedance input resistance and the right-hand switch port has a low-impedance input resistance when blocked.

SPST switches are used as subcomponents in more complex switch configurations, such as an SPDT switch (Fig. 7.17 b)), which can be used as a changeover switch. Consequently, further consideration is limited to SPST switches, as this is the main sub-component of all more complex switches.

If possible, an RF switch should have the same behavior as you would expect from an ideal switch. However, the SPST only has one port as a PIN diode realization in the locked state, which has a high-impedance input resistance. The disadvantage of the low-impedance input resistor is that no switches can be interconnected here, for example to split the signal into different channels. Another disadvantage is that the SPST requires two polarities of a control voltage.

These two disadvantages could be eliminated by not using the shunt diode in the SPST. In this case, however, the isolation, which is now based solely on the blocking capacitance of the series diode, is insufficient. Another disadvantage of this design is that the entire transmission power flows through the series diode in the event of transmission and this diode attenuates the power due to its finite series resistance and must be designed for the transmission power.

If you want to produce front-end circuits in balanced circuit technology, which has proven advantages for amplifiers, filters and antennas, the component cost of an SPST circuit mirrored in the ground plane is quite high.

This section deals with two new architectures for SPST switches, [66], which are referred to below as `X-SPST-` and `Z-SPST switches`. Both can only be implemented using balanced circuit technology.

Both switches have the advantage that they have a higher (theoretically infinite) insulation than SPST switches in the locked state. Both switches exhibit almost perfect insulation if the components used are exactly the same. This means that the properties of the components do not have to be optimized. Equality of the components is particularly good in PCB production for line structures and in semiconductor production for elements on a wafer.

In their minimum configuration, both switches only require two switch elements (e.g. PIN diodes) and therefore have a component cost that is comparable with known unbalanced switches.

7.5.1 Description of the Symmetrical Switch Arrangements

Both new SPST architectures are based on work on so-called mode blockers, section 5.2.3. The cross blocker can be found in this section. The reflection factor r of the mode blocker can be determined via

$$r = \frac{1 - 2y}{1 + 2y} \tag{7.22}$$

with the normalized conductance $y = Z_0/Z$.

It was also pointed out that any two-port can be used instead of any impedance. No conditions are placed on this two-port and it can therefore also be non-reciprocal and active. As we will see, a low-loss cable is a very good choice. Isolation of the push-pull signal also occurs across a wide band when using a two-port of this type.

Description of the X-SPST Switch with Concentrated Elements

The general circuit of the X-SPST switch is shown in the implementation with concentrated elements in the left-hand part of Fig. 7.24.

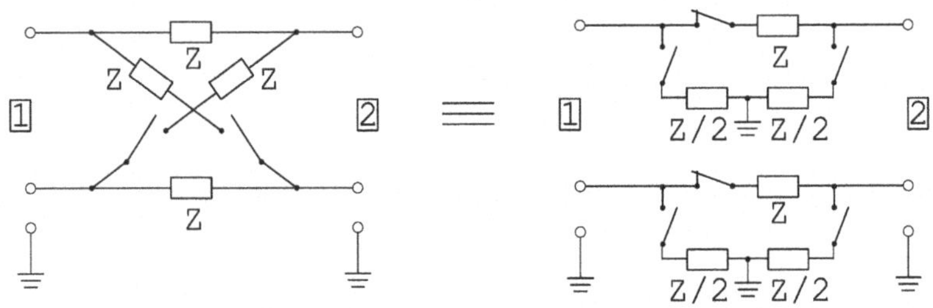

Fig. 7.24: General circuit diagram of the through-connected X-SPST switch with concentrated elements and associated Equivalent Circuit Diagram (ECD)

This switch simulates the cross blocker when the switch elements are closed. This means that when the switch elements of the X-SPST are closed, the series switches of the equivalent circuit diagram (ECD) are open and the shunt switches are closed. The switch elements of the X-SPST (as shown in Fig. 7.24) are open in the event of a pass-through. In this case, only the concentrated elements contained in the longitudinal branches are part of the current-carrying circuit, provided the switch elements have good isolation. The latter is already the case with standard PIN diodes. You are free to choose the component or the circuit of concentrated components to realize the impedance Z. An important criterion for the design is certainly the resulting reflection coefficient for the blocking case, which can be calculated using Eq. (7.22).

The circuitry is very simple if only a coupling capacitor is selected for Z. Even less effort is required for a very short cable section that can still be modeled sufficiently well using a concentrated inductance. In both cases, however, it must be taken into account that the input reflection factor r almost corresponds to a short circuit.

The concentrated impedance can also be a coupling element of a filter, whereby an otherwise series-connected filter merges with the switch. The impedance jumps in the filter allow even better isolation to be achieved.

Description of the X-SPST Switch with Distributed Elements

As shown in Fig 7.25, a two port with the general scattering matrix **[S]** can also be used instead of a terminal element Z. What is different is that a second switching element must be used for each cross-path in circuit implementations for the networks **[S]** that noticeably rotate the transmission phase (e.g. from 10°).

A very interesting two-port is a transmission line. However, close examination reveals that such switches can only be produced in very good quality up to a line length of $\lambda/2$. In the frequency range in which the line has a length of $\lambda/4$, the input reflection coefficients r are in the range of an often desired open circuit.

This topology is of course particularly interesting in the highest frequency range (e.g. for automotive radar systems at 77 GHz), as there are only a few concentrated components in front-end technology in this frequency range.

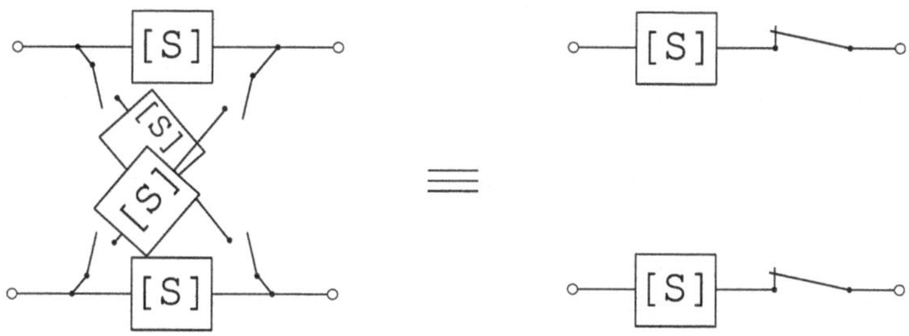

Fig. 7.25: General circuit diagram of the through-connected X-SPST switch with distributed elements and associated simplified ECD

Advantages of the X-SPST Switch

The advantage of the X-SPST switch is that the transmission power does not flow through the switch elements when switched through. This switch therefore has significantly lower transmission losses than an SPST switch. Furthermore, switch elements can be used whose maximum power compatibility is far below the applied power. This reduces the costs of large signaling systems immensely. In addition to the two switches, four other components are required. Capacitors, cables or other elements can be used here to improve functionality in terms of reflection behavior and insulation.

Description of the Z-SPST Switch

The Z-SPST switch (Fig. 7.26) also uses the special properties of the cross blocker.

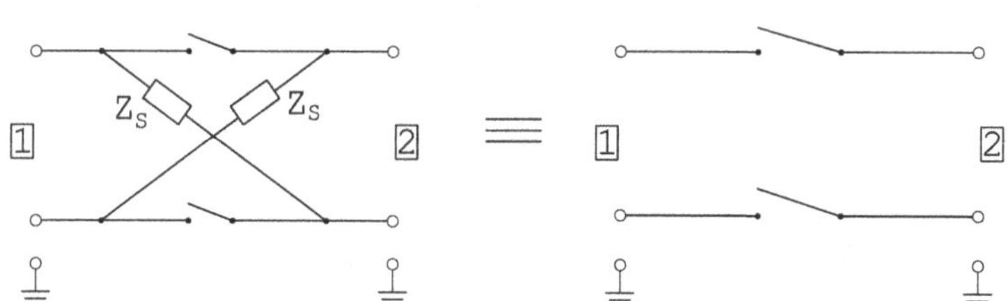

Fig. 7.26: Circuit diagram of the open Z-SPST switch and associated well approximated ECD

With this architecture, the switching elements are inserted in the longitudinal branches, as with traditional switches. In addition, there is an impedance element Z_S in each of the transverse branches, which simulates the properties of the open switching element. The use of a capacitor is a sufficiently good simulation for many switching elements. For more complex switching elements or over large bandwidths or at high frequencies, a switch element that is not switched through can also be used directly. This means that the switch corresponds to a cross blocker in the isolation state.

The mode suppression structure works alongside the isolation effect of the switches. This results in very good isolation even with component fluctuations. Another advantage of this switch is that the input resistance is broadband and quite high on both sides. In addition, some systems for the use of PIN diode switches only allow current consumption in transmission mode. This innovative switch also guarantees this.

Advantages of the Z-SPST Switch

The Z-SPST switch behaves in a broadband manner as you would expect from a very good switch. In the locked state, the inputs are high-impedance on both sides and the isolation is still very good even with large component dispersions. In terms of power compatibility, the Z-SPST switch behaves like a familiar SPST switch.

The Z-SPST switch architecture in particular also offers great potential for improvement for applications in low-frequency electronics and electrical engineering. The isolation of semiconductor switches can be significantly increased, which means that semiconductor switches are increasingly replacing mechanical switches.

Further Possible Applications for Both Balanced Switches

The new architecture of the balanced switches will boost the introduction of balanced switching technology for front-end circuits. This allows the other advantages of the amplifier, filter and antenna components to be utilized. Furthermore, the architecture of the X-SPST switch in particular, with its many degrees of freedom, allows the integration of very low-loss switches within other components such as filters. The design of such multifunctional components allows components and therefore space and costs to be saved.

Since the X-SPST switch is an extremely interesting component for power applications in particular, it is also attractive to use this balanced switch in unbalanced systems via two baluns, which are each realized by a 180° bypass, for example.

A large field of application is seen in particular with the introduction of broadband wired communication technology for cost-effective systems using "twisted pair" technology.

The switches can also be used for DC voltages, provided that a positive and negative voltage supply with the same amplitude is available.

Results from Practice

The excellent properties of these switches were verified in [28, 65] with many measurements. Isolations of more than 50 dB were achieved with simple PINs in 0402 housings and switches in the double-digit watt range could be realized with these extremely inexpensive diodes.

Chapter 8

Linear Amplifiers and Noise

The following introduction to the development of linear RF amplifiers only deals with amplifiers in which the voltage and current amplitudes of the signal variables to be processed are small compared to the values of the operating points. This theory can be used to design all RF amplifiers with output powers of up to 27 dBm and also power amplifiers for linear systems (e.g. 6G, WLAN). This covers the majority of amplifier applications. Only the power amplifiers cannot be described in their full breadth by the techniques presented.

Due to the rapid development of semiconductor technology, transistor amplifiers are also used in the highest frequency range. A few decades ago, the use of GaAs transistors dominated, some of which are still frequently used at frequencies above 4 GHz. SiGe transistors are now characterized by high operational reliability, low costs, long service life and very low noise. The following is an example of amplifier dimensioning using an amplifier with SiGe transistor for a center frequency of 0.9 GHz.

First, the characteristics of these amplifiers are described. Then a practical example is used to show how to optimize a small signal amplifier for a relatively narrow frequency band. The basic work steps, such as setting the operating point and MAG tests, are discussed. The results of a non-linear simulation are also briefly presented. Finally, the noise characteristics and the noise phenomena of RF circuits and simple noise figure calculations and noise figure measurements.

H. Heuermann, *Radio Frequency Technologies*,
https://doi.org/10.1007/978-3-658-50170-9_8

8.1 Characteristics of Small Signal Amplifiers

8.1.1 Power Amplification

A variety of definitions of power gain (power amplification or just amplification) can be found in the literature of an amplifier (seven in [48]), which can be calculated from the scattering parameters of the transistor and the reflection values of the matching networks. Only the most important definitions and parameters that are important for designing an amplifier are presented here without derivation.

The power amplification v_p is defined by:

$$v_p = \frac{\text{Active power delivered to the load}}{\text{Maximum available active power at the amplifier input}} = \frac{P_L}{P_{\max}} \quad . \tag{8.1}$$

The source that feeds the amplifier should have the reflection factor r_S. The input reflection factor of the load connected to the output of the amplifier should be denoted by r_L.

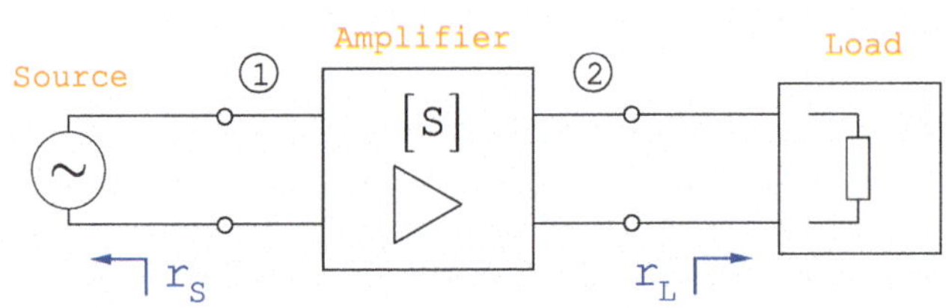

Fig. 8.1: Amplifier with generator and load circuit

Via a brief derivation (see Chap. 10 it can be shown that the power gain can be calculated exactly from the reflection coefficients and the scattering parameters S_{ij} of the amplifier:

$$v_p = \frac{|S_{21}|^2 \left(1 - |r_S|^2\right) \left(1 - |r_L|^2\right)}{\left|(1 - S_{11}\, r_S)\,(1 - S_{22}\, r_L) - S_{21}\, S_{12}\, r_S\, r_L\right|^2} \quad . \tag{8.2}$$

If ideal conditions are introduced with regard to the reverse isolation ($S_{12} = 0$) and the input and output matching ($S_{11}^* = r_S$, $S_{22}^* = r_L$), the maximum power amplification results:

$$v_{p\max} = \frac{|S_{21}|^2}{\left(1 - |S_{11}|^2\right)\left(1 - |S_{22}|^2\right)} \quad . \tag{8.3}$$

The maximum power gain indicates the gain in RF energy that can be achieved in the best case with a given transistor and is therefore very important for estimations. However, it must always be noted that the value is based on at least one approximation.

The approximation of perfect backward isolation ($S_{12} = 0$) would mean that a linear amplifier never becomes unstable. In practice, however, the optimizations with regard to stability take up a great deal of space.

8.1.2 Stability

The greatest danger when designing an amplifier is that the amplifier oscillates. To safely avoid oscillations, you need to know the stability condition of active two-port amplifiers.

A necessary condition to ensure that an amplifier does not oscillate is to check the forward and backward transmission values:

$$\left|S_{21}^{dB}\right| < \left|S_{12}^{dB}\right| \quad . \tag{8.4}$$

This simple equation explains the phenomenon of instability:
The backward isolation is less than the gain.

Here, the isolation on the transistor decreasing at higher frequencies can be just as much to blame as the often underestimated DC supply paths, Fig. 8.2.

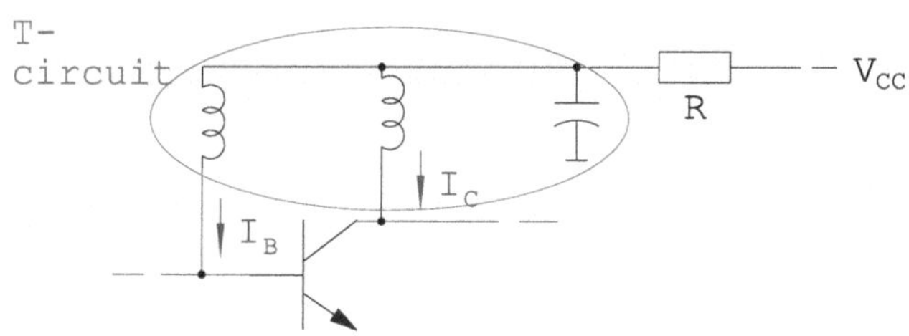

Fig. 8.2: DC-Supply lines for base and collector form via the PI circuit, e.g. 180°-phase shifter

The Fig. 8.2 highlights a T-circuit (series-L, shunt-C, series-L) in which the output is fed back to the input in the MHz range. The figure also illustrates that a phase condition is added to the amplitude condition of (8.4).

Applies with $n = 0, 1, 2, 3, \ldots$:

$$\angle S_{21} + \angle S_{12} = \mathrm{n} \cdot 2\,\pi \quad , \tag{8.5}$$

then `positive feedback` is present and the amplifier oscillates, provided that Eq. (8.4) is not fulfilled at the same time.

As there are often very large phase rotations above the frequency in the RF range, there is usually always a frequency value to fulfill the phase condition for the positive feedback.

However, this simple analysis is not sufficient in practice, as the external terminating impedances of the amplifier are significantly included in the stability analysis.

Stability Factor k

A large number of publications have shown how the **stability factor k** can be calculated using the S-parameters of a two-port (e.g. in [48]). At this point, we will only present the result of these derivations that is important in practice, which can be used to ensure so-called **absolute stability**:

$$\boxed{\mathrm{k} > 1} \qquad \text{with} \qquad k = \frac{1 + |\det S|^2 - |S_{11}|^2 - |S_{22}|^2}{2\,|S_{12}\,S_{21}|} \tag{8.6}$$

$$\text{whereby applies :} \qquad \det S = S_{11}\,S_{22} - S_{12}\,S_{21} \quad . \tag{8.7}$$

The following two conditions must also be met:

$$|S_{12}\,S_{21}| < 1 - |S_{11}|^2 \quad \text{and} \quad |S_{12}\,S_{21}| < 1 - |S_{22}|^2 \quad . \tag{8.8}$$

The conditions after (8.8) are not examined in practice. The circuit simulation provides the stability number k, which must be taken into account in all cases when designing the amplifier.

If all these criteria for absolute stability over the frequency are met, it is absolutely guaranteed that no possible passive reflection termination will cause this active two-port to resonate.

This simple stability test is applied to the entire amplifier. In practice, the circuit simulator is mainly used to check the k-factor. It is calculated at each optimization step in parallel with the other calculations.

Stability Control

For practical control of amplifiers, so-called **tuners** are connected at the input and output. Tuners are low-loss $50\,\Omega$ lines, often in the form of open stripe transmission lines, in which an adjustable capacitive load can be shifted using one or more micrometer screws. The input reflection coefficients of these tuners therefore always show circles in the Smith chart if these sliders are moved. Using these simple tools, all reflection values in the Smith chart can be quickly set using a finite discretization.

There are also automatic measuring stations (so-called **load-Pull measuring stations**) commercially available. However, their acquisition costs are in ranges that justify their purchase only for special applications.

Detailed Stability Analysis

A linear amplifier is described as *absolutely stable* if it always works stably with any generator and load with the reflexions coefficient of $|r_G| < 1$ and $|r_L| < 1$ at input and output port.

Conditionally stable means that an amplifier is only stable in certain areas of the passive terminations, but becomes unstable in the other areas. To illustrate these termination conditions, Fig. 8.3 of the amplifier serves as a general two-port with the generator impedance Z_G or the generator reflection factor r_G and the load impedance Z_L or the load reflection factor r_L.

The input reflection coefficient r_1 of the second port with termination is then

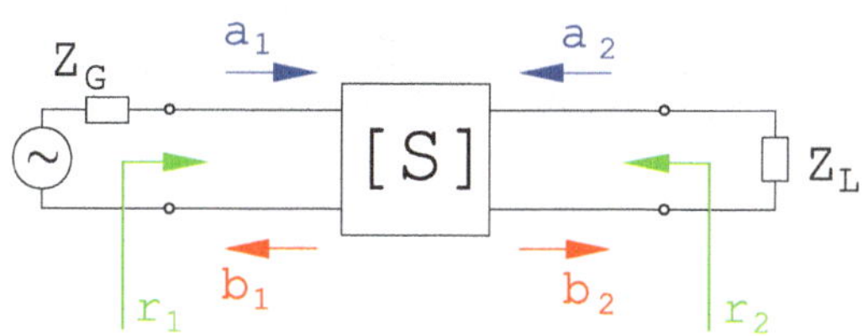

Fig. 8.3: Amplifier as a general two-port with generator and load circuit

$$r_1 = \frac{b_1}{a_1} = S_{11} + \frac{S_{21}\,S_{12}\,r_L}{1 - S_{22}\,r_L} \tag{8.9}$$

and the output reflection coefficient

$$r_2 = \frac{b_2}{a_2} = S_{22} + \frac{S_{21}\,S_{12}\,r_G}{1 - S_{11}\,r_G} \quad . \tag{8.10}$$

For absolute stability, $|r_1| < 1$ and $|r_2| < 1$ are required. Otherwise, if the magnitude of a reflection factor is $|r_i| > 1$, the associated input resistance has a negative real part, as can easily be seen from the definition for the reflection factor. If the real part of an input resistor is negative, it can attenuate a passive lossy network and generate oscillations. In the case of the general second port, therefore

$$\left| S_{11} + \frac{S_{21}\,S_{12}\,r_L}{1 - S_{22}\,r_L} \right| < 1 \quad \text{and} \quad \left| S_{22} + \frac{S_{21}\,S_{12}\,r_G}{1 - S_{11}\,r_G} \right| < 1 \tag{8.11}$$

must be fulfilled.

Only the conditions for r_1 are shown below. The same procedure applies to r_2 and the same statements apply.

The Eq. (8.9) is a fractional rational function (conformal mapping) that maps circles from the r_1-plane to circles in the r_L-plane. Center M_L and radius ρ_L of the image circles in the r_L plane result from (cf. [140])

$$M_L = \frac{(S_{22} - S_{11}^* \det S)^*}{|S_{22}|^2 - |\det S|^2} \quad \text{and} \quad \rho_L = \frac{S_{12}\,S_{21}}{|S_{22}|^2 - |\det S|^2} \tag{8.12}$$

with $\qquad \det S = S_{11}\,S_{22} - S_{12}\,S_{21} \quad .$

From the geometric observation (Fig. 8.4) of the mapping properties of the r_1 plane in the r_L plane, statements can now be made about the stability of the two-port. In the r_1 plane, reflection coefficients $|r_1| > 1$ result in unstable operating states (hatched). To determine which load resistances belong to these reflection factors, the unit circle $|r_1| = 1$ is mapped into the r_L plane.

The load resistor should be purely passive, so only the values within the input circuit are relevant in the r_L plane. If there are double-hatched areas within the unit circle via the r_1 transformation, the amplifier is unstable for r_L values that lie in these double-hatched areas. Stability applies to the white areas. Overall, it is only conditionally stable in this case.

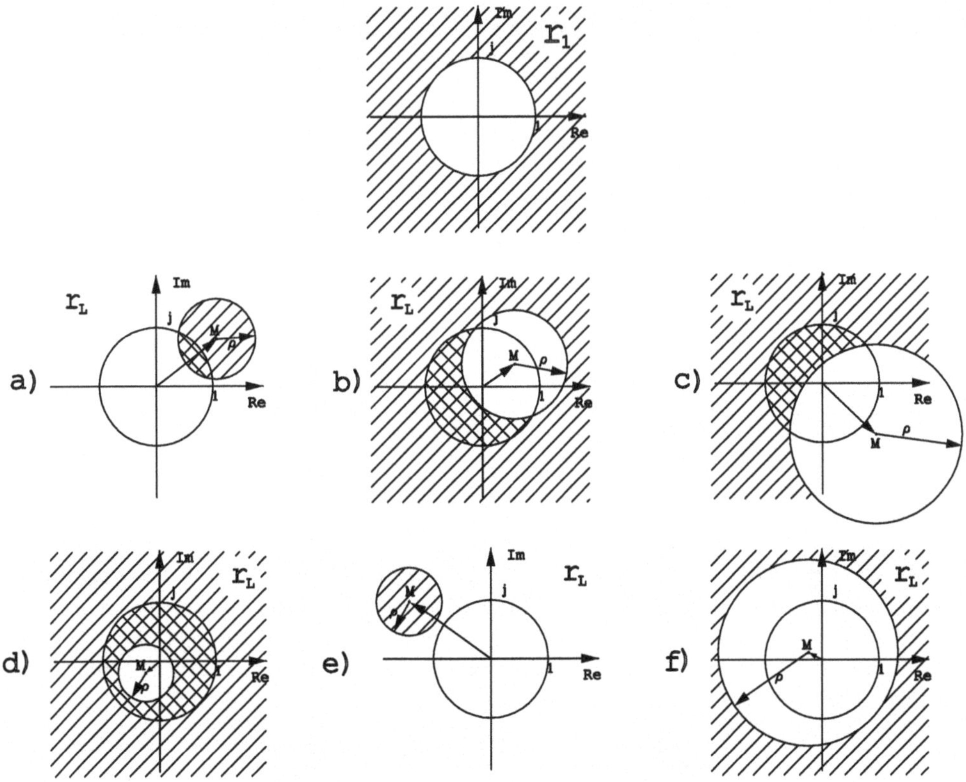

Fig. 8.4: Geometric stability analysis of an amplifier in the r_L-plane

Case a, b, c, and d show conditionally stable cases. Any passive termination can be connected in e and f (absolutely stable).

Similar considerations result from Eq. (8.10) for r_g. In fact, however, if the input reflection factor $|r_L| < 1$ is $|r_1| < 1$, the amplifier will be stable for every generator reflection factor r_G with a positive real part. A stability check with regard to $|r_2| < 1$ is therefore not necessary.

Based on these considerations, the already introduced k-factor, which greatly simplifies representability, was derived.

8.1.3 Maximum Power Gain and Maximum Stable Gain

In a system, an amplifier is expected to increase the signal power by a desired value. The required minimum gain often determines the choice of transistor type.

RF transistors have a strong frequency response with regard to the maximum possible power amplification. This frequency-dependent maximum available gain (MAG) can be calculated via the scattering parameters using the following relationship:

$$\mathrm{MAG} = \left| \frac{S_{21}}{S_{12}} \left(k - \sqrt{k^2 - 1} \right) \right| \quad . \tag{8.13}$$

This relationship only applies to $k \geq 1$! In practice, the MAG value should be at least 2 dB above the desired gain. In contrast to $v_{p\max}$, the MAG value S_{12} is not set to zero in the calculation.

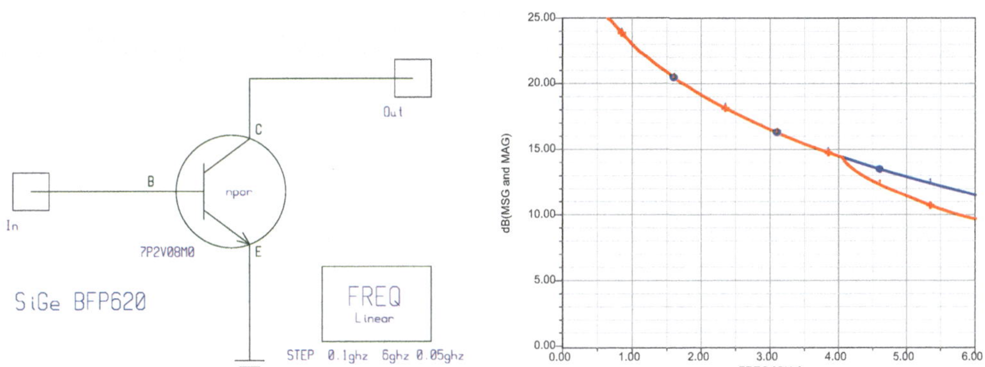

Fig. 8.5: Left: Simulation for MAG; right: MAG results in blue and MSG calculation in red for the BFP620 at the operating point of 2 V

In areas where stability is not given ($k < 1$), the MAG value can lead to incorrect results because the gain may become infinite with self-excitation. In practice, however, this only applies to more complex circuits. In the case of given transistor values as in Fig. 8.5, this case occurs $k < 1$ up to 4 GHz.

In data sheets, the maximum stable gain (MSG) is preferred. This is simply calculated as follows:

$$\mathrm{MSG} = \left| \frac{S_{21}}{S_{12}} \right| \quad . \tag{8.14}$$

It applies: $\quad \mathrm{MSG} = \mathrm{MAG} \quad$ at $\quad \mathrm{k} = 1 \quad .$

The advantage of the MSG over the MAG is that it is valid for every circuit (even for $k < 1$). However, in practice, the MSG is larger than the MAG for a stable circuit ($k > 1$), as can be easily derived from the Eq. (8.13).

8.2 Design of Narrowband RF-Amplifiers

The following example uses a narrowband 900 MHz small signal amplifier to illustrate the design process. This is a typical low noise amplifier (LNA) as used in receivers.

This design was carried out for a SiGe transistor from Infineon (BFP620), which was measured at an operating point of 2 V.

This transistor of the SIEGET family (SIEmens Grounded Emitter Transistor) is characterized by the fact that the collector connection is no longer on the back of the chip, which means that the emitter inductance could be reduced by a factor of 4, which has a noticeable effect on the transistor properties, [76].

To design a linear RF transistor amplifier, some data of the transistor type to be used must be known. In the simplest case, the S-parameters are available from the manufacturer. This is now the rule and should be assumed.

If the S-parameters are not given, they can also be measured using a network analyzer. Care must be taken to ensure that the operating point to be expected in the circuit to be designed is maintained during the measurements.

If it is not necessary to know the exact S-parameters for a specific operating point of an individual transistor, it will not be necessary to determine the S_{ij} by measurement and you can refer to the data sheet of the corresponding type. Here, the S-parameters are specified above the frequency and at a precisely defined operating point within the scope of the sample scattering.

These so-called small signal amplifiers are most frequently used in emitter circuits in RF applications. For the emitter circuit, the manufacturers provide scattering parameter measurement files for various operating points in the so-called S2P file format.

All of the following parameters and procedures will be explained in this chapter using a transistor as a practical example. A shortened specification of the linear parameters given by the manufacturer in the number of frequency points looks as follows:

```
!  INFINEON TECHNOLOGIES Small Signal Semiconductors
!  BFP620 , Si-NPN RF-Transistor in SOT343
!  Vce=2 V,   Ic=8 mA
!  Common Emitter S-Parameters:
#  GHz  S  MA  R  50
! f            S11              S21              S12              S22
! GHz       Mag    Ang       Mag    Ang       Mag    Ang       Mag    Ang
  0.200  0.8312  -21.8  21.109  163.1  0.0183   76.1  0.9477  -15.4
  0.400  0.7710  -42.9  19.447  147.1  0.0351   67.5  0.8729  -30.0
  0.600  0.6996  -62.3  17.279  133.5  0.0478   60.6  0.7748  -42.5
  0.800  0.6264  -79.1  15.163  122.4  0.0575   54.7  0.6721  -52.9
  1.000  0.5666  -93.5  13.241  113.6  0.0654   50.5  0.5828  -60.9
  1.200  0.5217 -107.3  11.680  106.0  0.0714   47.1  0.5049  -68.4
  1.400  0.4798 -119.5  10.348   99.5  0.0769   44.3  0.4438  -74.9
  1.600  0.4556 -131.0   9.297   93.7  0.0825   42.7  0.3878  -80.5
  1.800  0.4388 -141.3   8.419   88.6  0.0866   41.5  0.3404  -86.1
  2.000  0.4229 -150.0   7.670   84.1  0.0917   40.1  0.3021  -90.9
  3.000  0.4130  172.7   5.243   64.8  0.1147   34.7  0.1662 -120.6
  4.000  0.4749  144.9   3.914   48.5  0.1359   27.8  0.1538 -173.1
  5.000  0.5311  125.9   3.037   34.4  0.1524   21.3  0.1899  151.8
  6.000  0.5797  113.2   2.457   22.5  0.1699   15.8  0.2239  124.4
!
! f         Fmin      Gammaopt        rn/50
! GHz        dB       MAG      ANG      -
  0.900     0.64     0.22      25     0.12
  1.800     0.71     0.09      97     0.08
  2.400     0.75     0.06     139     0.09
  3.000     0.87     0.09    -175     0.10
  4.000     0.99     0.19    -147     0.08
  5.000     1.17     0.26    -125     0.11
  6.000     1.34     0.38    -101     0.17
```

The upper block shows the complex scattering parameters of the transistor and the lower block shows the parameters of the transistor required for the noise figure calculation. This specifies the minimum noise figure F_{min}, values for the noise adjustment Gammaopt, the equivalent noise resistance r_n normalized to $50\,\Omega$ and, if applicable, the termination conditions for maximum power amplification.

When designing a linear amplifier, the following properties must be taken into account:

- Operating voltage and power consumption
- Bandwidth and center frequency
- Gain
- Input and output reflection
- Output power
- Heat management
- Stability (s. section 8.1.2)
- Intermodulation(see section 8.3.1)
- Noise figure (s. section 8.6.3)

These properties of the amplifier can all be calculated in a linear and a non-linear circuit simulation and observed in the current design. This is important for the developer to maintain an overview.

8.2.1 Analysis of the Transistor at the Operating Point

The operating voltage and current consumption depend on the transistor size (gate and base width) and the required output power as well as the maximum current and voltage values that can be applied.

As a rule, a transistor is selected that theoretically provides 3-6 dB more gain than is required at an available operating voltage (e.g. 3 V). If the required amplification cannot be realized, a multi-stage amplifier must be built.

The power supply for the base and collector[1] is always via low-pass filters. As a rule, the filter starts with a high-impedance input from the RF signal (i.e. a coil). Particularly at the output, very small coil values are selected, as the series resistance is therefore also small and the efficiency is high. Furthermore, the transistor at the output has a low impedance, so that the low-pass functionality is already given with very low input impedances.

The operating point is set according to the criteria known from electronics. It is important that the operating point setting corresponds to the setting at which the scattering parameters of the transistor were measured!

Linear amplifiers are used in so-called `Class A` operation ($U_{CE} = V_{CC}/2$). The following is a brief description that is limited to pure linear RF amplifiers.

[1] The same always applies to the gate and drain of field-effect transistors.

Control Range

The maximum operating voltage is often limited by the transistor. In the smartphone sector, low voltages are used in order to consume less energy. In our example, a maximum voltage of 2 V should be selected for the BFP620, as its maximum value is limited to $U_{CE} \leq 2.3\,V$.

The collector resistor R_C, Fig. 8.6, serves as the current limiter for the current I_C on the collector side.

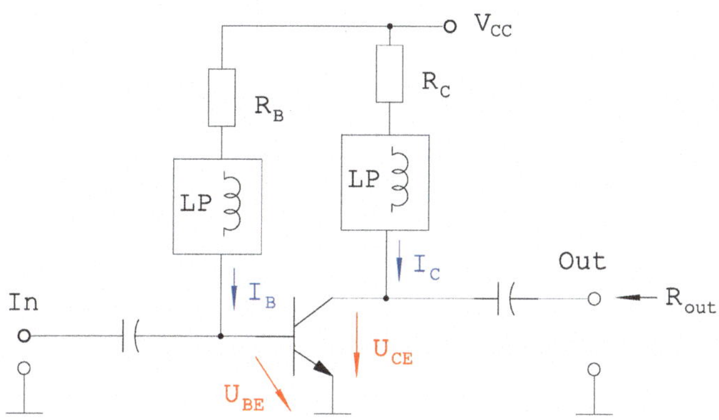

Fig. 8.6: Linear RF transistor amplifier with low-pass filters (LP) and resistors for operating point adjustment (with coupling capacitors and without RF matching elements)

In practice, around half the value of the operating voltage V_{CC} is selected as the operating point voltage at the collector U_{CE}.

$$R_C = \frac{V_{CC}}{2\,I_C} \quad . \tag{8.15}$$

For the selection of the collector current I_C, three different applications of the amplifier can be distinguished ($R_{out} \simeq R_C$):

a) Designed for maximum output power $\Longrightarrow$ $R_{out} < 50\,\Omega$

b) Designed for best broadband matching $\Longrightarrow$ $R_{out} = 50\,\Omega$

c) Designed for minimum power consumption $\Longrightarrow$ $R_{out} > 50\,\Omega$

These three cases are shown in the output characteristic 8.7 of the transistor.

I_B must now be selected via the calculation of $R_B = (V_{CC} - 0.7\,V)/I_B$ so that the desired collector current I_C is set via the current amplification at the output. As will be shown below, in addition to creating a linear simulation, it is also necessary to create a simple non-linear circuit simulation. This non-linear simulation can be used directly to design R_B without further analytical calculations.

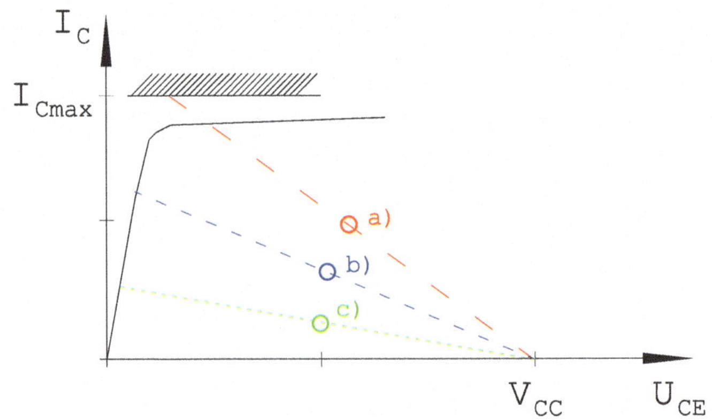

Fig. 8.7: Linear working range for the three cases of operating point adjustment

Is this "simple" operating point adjustment is not sufficient, for example because this circuit is used over large temperature ranges, integrated circuits such as the BCR400, which are known as operating point stabilizers, can be used.

S-Parameter Analysis

The next step is to analyze the achievable properties of a transistor with given scattering parameters for an operating point. This often only involves the achievable maximum available gain (MAG).

However, one should not make the mistake of neglecting the parasitic variables that exist in practice. Fig. 8.8 shows a *wrong* pure transistor analysis in the left part and correct analysis in the right part.

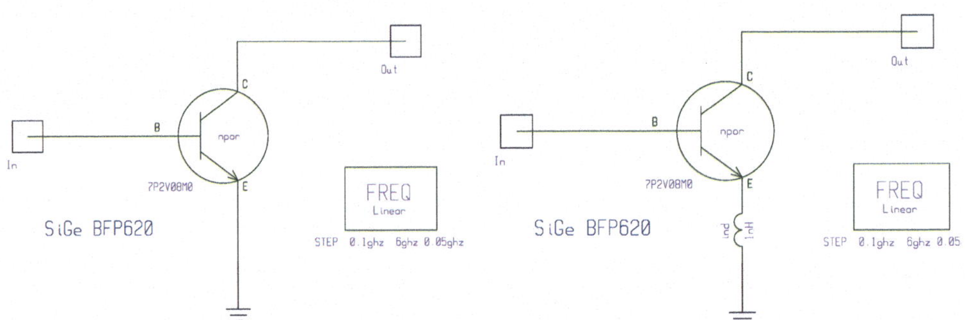

Fig. 8.8: Left: k-simulation model of the pure transistor; right: k-simulation model of the parasitically connected transistor

The simulation of the k-factor of the circuit in the left part of Fig. 8.8 corresponds exactly to the simulation in Fig. 8.5.

The fact that these two simulations of Fig. 8.8 provide very different information is shown by the display of the corresponding k values in Fig. 8.9.

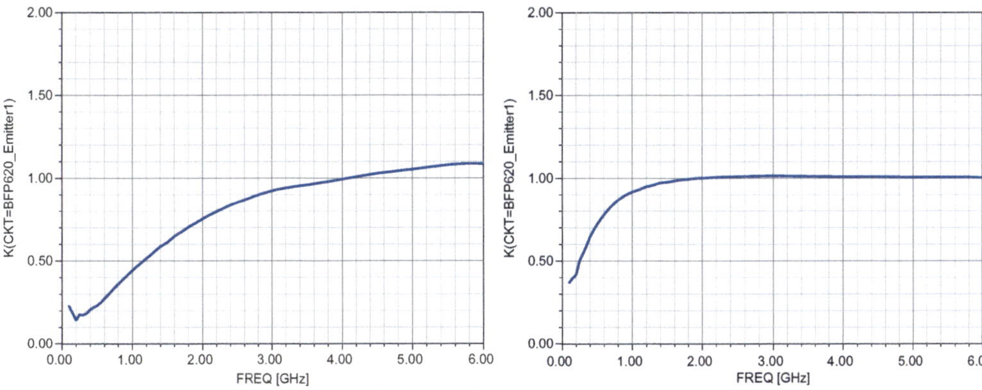

Fig. 8.9: Left: K-factor simulation results of the pure transistor; right: simulation result of the parasitically connected transistor

It can be seen that the stability of the parasitically connected transistor is significantly better. The reason for this is that the inductance[2] causes a current feedback.

The results of the k-factor of the wired transistor show that a MAG examination is only accurate from approx. 1.5 GHz. The resulting MAG values are printed in Fig. 8.10.

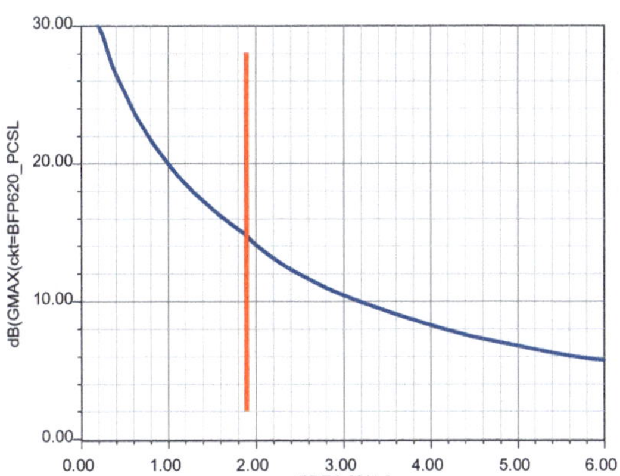

Fig. 8.10: MAG simulation result of the parasitically connected transistor

[2]In microstrip conductor circuits, the feedthroughs to ground behave approximately like an inductance.

Fig. 8.5 still showed a gain of around 23.5 dB at 900 MHz. In practice, the value of approx. 21 dB given in Fig. 8.10 at 900 MHz does not deviate much from the exact value.

8.2.2 Circuit Optimization of the Transistor Amplifier

Apart from measuring devices and optical transmission links for which broadband characteristics are required, RF amplifiers are generally used in narrowband applications. In the following, we will limit ourselves to narrowband applications. However, the procedures shown here can also be used for broadband applications with multi-stage impedance transformers.

For the following calculation of the transformation networks for input and output-side matching, it is first assumed that the transistor is reaction-free ($S_{12} = 0$).

This rough approximation leads to a sensible topology of the transformation network, which is precisely designed using an optimizer. Since the low-pass filters for the current feeds and operating point settings also have a noticeable influence on the electrical behavior of the transformation networks, a meaningful calculation can only be carried out with a circuit simulator.

The topology of the impedance transformation network can be easily obtained by using the so-called Smith chart tool. This is described in detail in the section 4.3.5 and Appendix A.11. Other possibilities of a direct synthesis for real impedances via closed equations are presented in section 3.2.2.

In most cases, the matching networks are designed for power matching (conjugate complex input resistors), but also for noise matching if necessary.

Even with the choice of transistors, it is very helpful to carry out stability tests throughout the design process.

The topologies of the input and output networks to be designed depend on two factors:

1. The base and collector must be connected to a DC voltage source via a coil.

2. The complex input and output reflection factors given at 900 MHz.

To 1.: The so-called "choke" coils can be part of the RF matching network. Subsequently, relatively large RF shunt capacitors (approx. 100 pF[3]) are used so that good RF short circuits can be realized. This means that the choke coil for the RF matching network only has the functionality of a shunt coil.

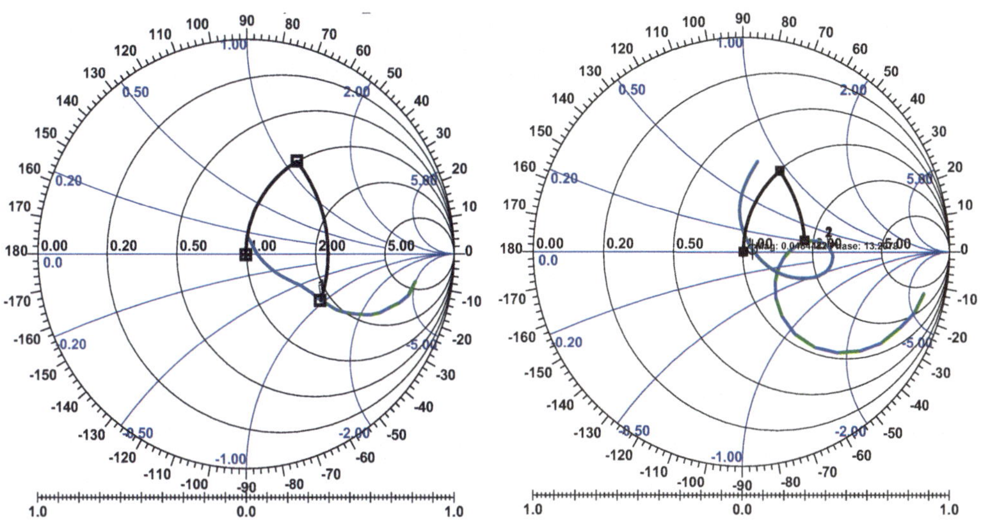

Fig. 8.11: Impedance transformations at 900 MHz at the input (left) and at the output (right)

To 2.: The technique to be used for the narrowband adjustment of any complex reflection coefficient was explained with the introduction of auxiliary sheet A.7 (rotation in the Smith chart). The transformation curves resulting from this theory in the Smith chart are shown in Fig. 8.11.

Both at the input and at the output, the matching network from the transistor was a coil to ground and a series capacitor. It was a boundary condition that the shunt coil should be one of the two network elements so that the number of components and the size of the inductance could be kept small. This means that you already have a coupling capacitor[4] and therefore save two additional components.

The resulting simple circuit including the serial parasitic inductors of the capacitors of L = 1 nH is shown in Fig. 8.12.

The resulting coil and capacitor values were taken from the standard series. Since stability for low frequencies could not be guaranteed with the given matching network, a rather high-impedance and therefore almost ineffective resistor had to be used for voltage feedback at 900 MHz. The component design therefore had to be carried out in two stages.

The resulting characteristic properties of this circuit are shown in the following figures. The Fig. 8.13 shows the adjustments at the input and output in the form of S_{11} and S_{22} as well as the transmission behavior in the form of S_{21}.

The isolation and stability are shown in Fig. 8.14 in the form of S_{12} and k.

Negative Feedback

[3]The size of the capacitor value determines the given technology of the capacitor. The larger, the better the functionality, provided that the parasitic values and losses remain unchanged.

[4]Coupling capacitors are used to decouple the DC voltage of two components connected in series.

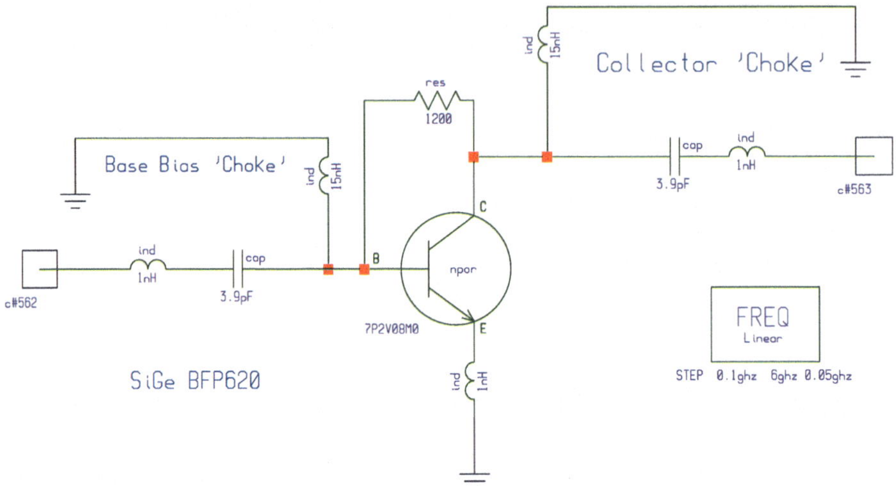

Fig. 8.12: Linear simulation model of the BFP620 transistor for a 900 MHz amplifier

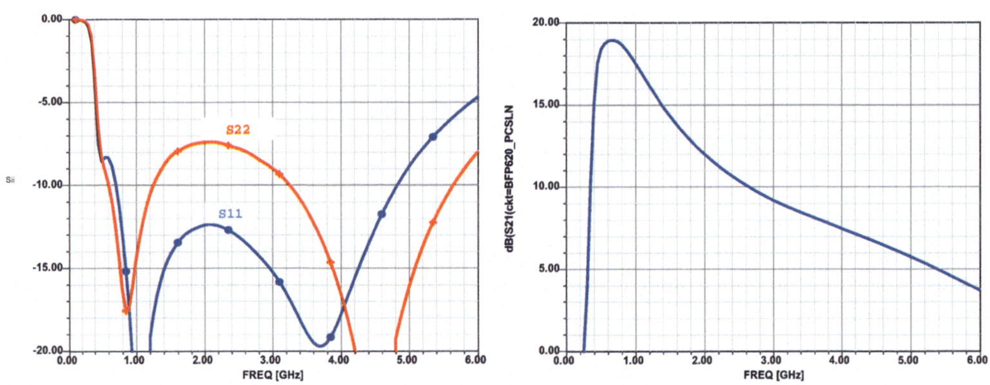

Fig. 8.13: Input and output-side adjustments and amplification of the 900 MHz amplifier

The two negative feedback measures of interest for linear RF amplifiers, which must be introduced if stability is not given, have already been introduced.

Current feedback: On the one hand, an impedance between the emitter and the ground causes an ever-increasing voltage to be applied to the emitter as the collector-emitter resistance decreases and thus as the collector current increases.

Consequently, U_{BE} and therefore I_B decreases. As a result, however, the collector current also decreases via the current amplification factor. This means that the collector current cannot increase arbitrarily or that the current gain decreases as the collector current increases.

Due to the finite contact lengths between the emitter and the earth, there is always a current feedback at high frequencies.

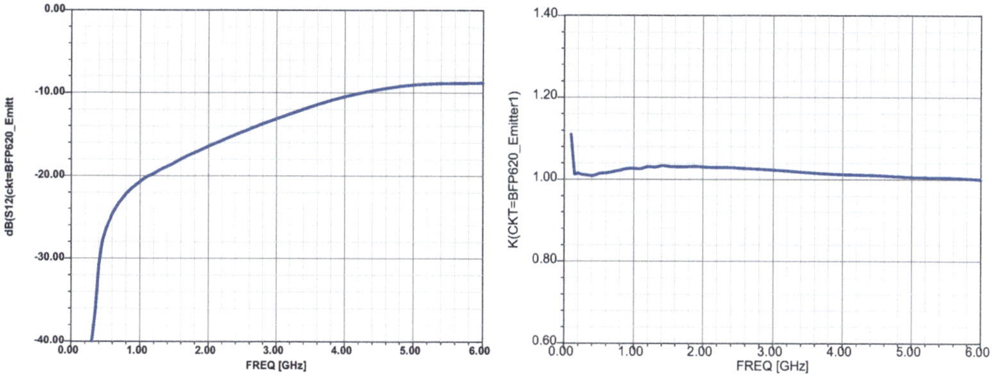

Fig. 8.14: Insulation and stability of the 900 MHz amplifier

Voltage feedback: During voltage feedback, no large electrical phase rotation may occur in the feedback network between the collector and base connection of the transistor. In the simplest case, a small part of the output voltage is fed back to the input voltage via a resistor. As the output voltage at the collector is 180° out of phase with the input voltage at the base, the feedback voltage reduces the input voltage and therefore also limits the amplification.

However, finite, non-negligible transmission line lengths must be taken into account for practical feedback. This feedback can therefore only be used for frequencies up to around 1 GHz in discrete SMD solutions. This is mainly used to limit the amplification and thus the oscillation tendencies for the lower frequency range.

8.3 Analysis of the Non-Linear Behavior of Amplifiers

For quite a few amplifier designs, it is far from sufficient to limit oneself to the analysis of linear behavior. This statement generally applies to power amplifiers, but also to small signal amplifiers if they are to be used in applications with high system requirements, as is the case with the 6G system, for example.

It can almost always be said that every semiconductor component exhibits non-linear behavior. For many components such as small signal amplifiers, these effects are very small, but must be investigated and, if necessary, optimized to meet the system specification.

In the following, a brief introduction to the basics of non-linear effects will help to better understand the non-linear simulation that the practitioner will then carry out.

8.3.1 Basics of Non-Linear Distortions

Linear amplifiers exhibit only very slight non-linear effects. Therefore, the weak effects around the operating point can be described with a polynomial representation [48]:

$$y_{(t)} - y_0 = a_1(x_{(t)} - x_0) + a_2(x_{(t)} - x_0)^2 + a_3(x_{(t)} - x_0)^3 + \dots \quad . \tag{8.16}$$

Here, $y_{(t)}$ is the output and $x_{(t)}$ the input variable, which are general substitute variables for the base voltage or base current and the collector voltage or collector current. x_0 and y_0 denote the operating point. The coefficients a_i are assumed to be time-independent, so that a quasi-static analysis is possible.

When excited with a mono-frequency signal

$$x_{(t)} = \hat{x}_1 \cos(\omega_1 t) \qquad \text{and} \quad x_0 = 0 \tag{8.17}$$

signal components arise with

$$\begin{array}{ccr} \omega = 0 & y_0 = -\frac{1}{2}a_2\hat{x}_1^2 & \text{DC component,} \\ \omega_1 & y_{1(t)} = \hat{x}_1(a_1 + \frac{3}{4}a_3\hat{x}_1^2) & \text{Fundamental wave } = \text{1. Harmonic,} \\ 2\omega_1 & y_{2(t)} = \frac{1}{2}a_2\hat{x}_1^2 & \text{2. Harmonic,} \\ 3\omega_1 & y_{3(t)} = \frac{1}{4}a_3\hat{x}_1^3 & \text{3. Harmonic.} \end{array}$$

On the one hand, a DC signal component can be recognized, which can noticeably change the operating point in power applications. On the other hand, you can see that the signal is not transmitted linearly at the operating frequency (fundamental wave).

Finally, you can recognize the emergence of harmonics. These are the easiest to measure in circuits using a spectrum analyzer and give the experienced engineer a quantitative measure of the non-linear behavior of the amplifier.

If a signal with two spectral components is now considered,

$$x(t) = \hat{x}_1 \cos(\omega_1 t) + \hat{x}_2 \cos(\omega_2 t) \quad , \tag{8.18}$$

a third degree polynomial representation results in the frequency components listed in Table 8.1, indicating the most important amplitudes.

Tab. 8.1: Frequency components when a cubic characteristic curve is driven by two signals with ω_1 and ω_2

	$\omega = 0$		$\omega_1 - \omega_2$
	$x_0 + \frac{1}{2}a_2(\hat{x}_1^2 + \hat{x}_2^2)$		$\hat{x}_1^2\hat{x}_2^2 a_2$
$2\omega_1 - \omega_2$	ω_1	ω_2	$2\omega_2 - \omega_1$
$\frac{3}{4}\hat{x}_1^2\hat{x}_2 a_3$	$\hat{x}_1\{a_1 + \frac{3}{4}a_3(\hat{x}_1^2 + 2\hat{x}_2^2)\}$		$\frac{3}{4}\hat{x}_1\hat{x}_2^2 a_3$
$\omega_1 + \omega_2$	$2\omega_1$	$2\omega_2$	$\omega_2 + \omega_1$
	$\frac{1}{2}\hat{x}_1^2 a_2$		
$2\omega_1 + \omega_2$	$3\omega_1$	$3\omega_2$	$2\omega_2 + \omega_1$
	$\frac{1}{4}\hat{x}_1^3 a_3$		

A polynomial representation of a higher degree would result in further harmonics. A parameter for the non-linear distortion, the distortion factor, is calculated from the amplitudes of the harmonics as follows

$$Kl = \sqrt{\frac{y_2^2 + y_3^2 + \ldots}{y_1^2 + y_2^2 + y_3^2 + \ldots}} \quad . \tag{8.19}$$

With the effective values, this distortion factor is defined as

$$Kl = \frac{\text{Effective value of all higher harmonics (no fundamental)}}{\text{Effective value of all harmonics}} \quad . \tag{8.20}$$

8.3.2 Non-linear Effects and Simulation of an RF Amplifier

A strong non-linear behavior is reflected in the fact that there is no linear relationship between the input and output power. Fig. 8.15 illustrates such a non-linear behavior by indicating the so-called IP_3 and P_{-1dB} points.

The P_{-1dB} point indicates the input power at which the gain has decreased by 1 dB compared to a linear gain.

The IP_3 point indicates the calculated input power at which the non-linear mixing product of the second and third harmonics (in the case of excitation of two closely neighboring signals in the fundamental wave range) corresponds to the input power of the fundamental wave.

At the so-called third-order intercept point (IP_3), the extrapolated useful power and the extrapolated interference power are therefore just equal.

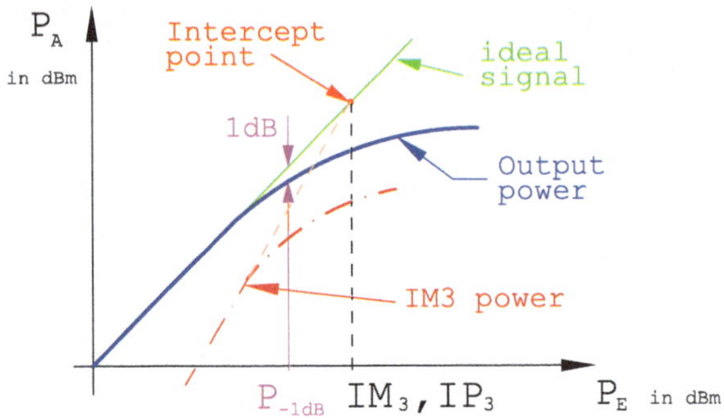

Fig. 8.15: IP_3- resp. IM_3- and P_{-1dB}-Points of an amplifier

Power amplifiers are operated in this power range of the P_{-1dB} point. Although small signal amplifiers are not used in this range, they can also be specified using these parameters.

When using small signal amplifiers, the variables of the harmonic generation, which in turn are very closely linked to the intercept point (IP_3), are of most interest with regard to the non-linear behavior. The higher the (IP_3) point, the more linear the amplifier behaves.

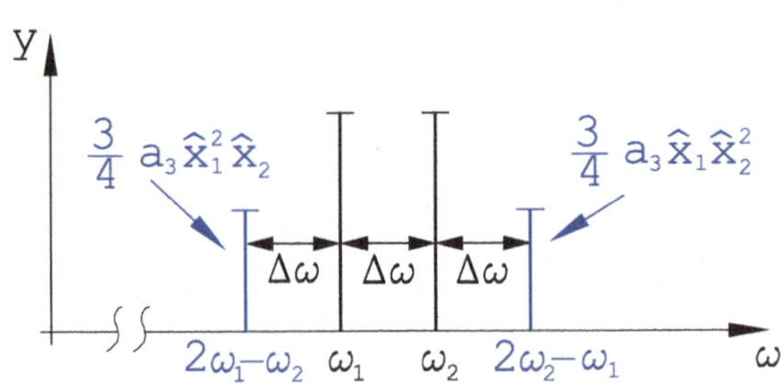

Fig. 8.16: Spectrum with modulation of a cubic characteristic curve with two closely adjacent frequencies ω_1 and $\omega_2 = \omega_1 + \Delta\omega$

If the continuity characteristic is now driven with two frequencies ω_1 and ω_2, additional spectral components result according to the above diagram (Table 8.1). If the two frequencies with

$$\omega_2 = \omega_1 + \Delta\omega \quad , \tag{8.21}$$

are close together, the non-linear mixed products are close to the fundamental frequencies, which is illustrated in Fig. 8.16. The components at $2\omega_1 - \omega_2$ and $2\omega_2 - \omega_1$ are referred to as third-order intermodulation products (IM3). In Fig. 8.16, the amplitudes of the wanted and disturbance signals are plotted logarithmically.

Where x_1 and x_2 are two amplitude-modulated signals ($\hat{x}_1(t) = \hat{x}_1 \cos(\Omega_1 t)$, $\hat{x}_2(t) = \hat{x}_2 \cos(\Omega_2 t)$ with $\Omega_i \ll \omega_i$), so can be recognized in the amplitude y_1 of the fundamental wave

$$y_1 = a_1\,\hat{x}_1 + \frac{3}{4}\,a_3\,\hat{x}_1(\hat{x}_1^2 + 2\,\hat{x}_2^2) \tag{8.22}$$

and the influence of the distorted modulation $\hat{x}_2$. The modulation of y_1 is formed by the modulation of $\hat{x}_1$ (desired signal to be amplified) and the distorted modulation of $\hat{x}_2$ (neighboring interference signal). This effect is referred to as *cross modulation*.

This effect is a very critical factor for linear amplifiers. Specifying the IM$_3$ point allows conclusions to be drawn about the magnitude of the cross-modulation and therefore helps with system design.

A simulator based on the "Harmonic Balance" method is used to simulate a non-linear amplifier. This already offers the older student version of the simulator "Serenade". The mathematical background to this process cannot be discussed in detail in this book. Harmonic Balance is in detail presented in [50].

The necessary non-linear transistor models generally correspond to the Spice models. These models describe the electrical properties via polynomials and fit a large number of measured values that were recorded at different operating points for the respective transistor. For example, the so-called "Gummel-Poon" model is used for the BFP 620 transistor. The corresponding values used in the following are

```
****************************************************************
* Infineon Technologies                                         *
* SPICE Library Version 4.1                      18.08.2025     *
* www.infineon.com/products/discrete/index.htm                  *
*                                                               *
****************************************************************
.MODEL BFP620 NPN(
+ IS = 3.543E-16        BF = 557.1          NF = 1.021
+ VAF = 1000            IKF = 2.262         ISE = 2.978E-12
+ NE = 3.355            BR = 100            NR = 1
+ VAR = 1.2             IKR = 0.00631       ISC = 1.923E-14
+ NC = 2.179            RB = 2.674          IRB = 1.8E-05
+ RBM = 2.506           RE = 0.472          RC = 2.105
+ XTB = -0.9            EG = 1.114          XTI = 3.43
+ CJE = 3.716E-13       VJE = 0.8986        MJE = 0.3152
+ TF = 1.306E-12        XTF = 2.71          VTF = 0.492
+ ITF = 2.444           PTF = 0             CJC = 2.256E-13
+ VJC = 0.7395          MJC = 0.3926        XCJC = 1
+ TR = 3.884E-10        CJS = 6E-14         VJS = 0.5
+ MJS = 0.5             FC = 0.8215)
****************************************************************
```

A non-linear simulation, which is to examine the circuit according to Fig. 8.12, is illustrated in the Fig. 8.17.

An RF source with a defined power and a DC voltage source with associated DC current paths were also implemented.

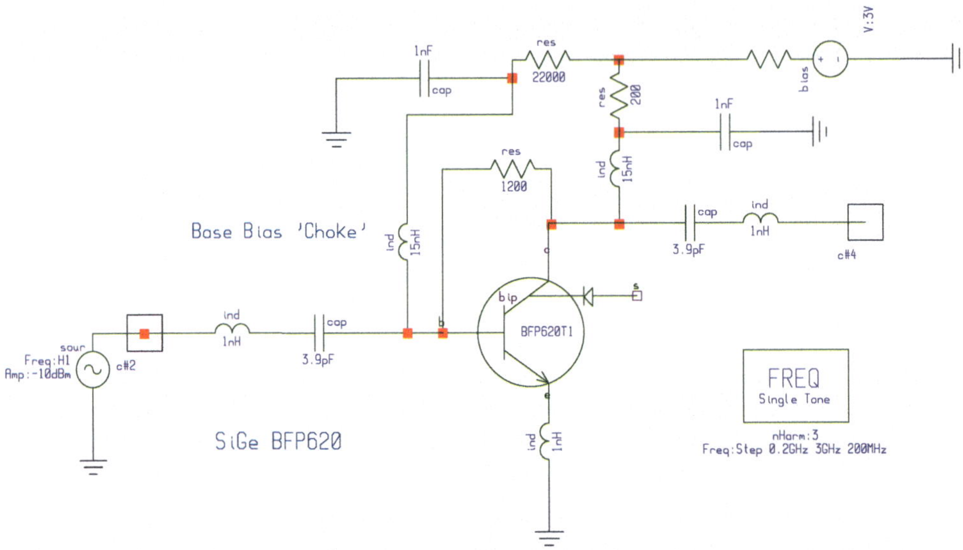

Fig. 8.17: Non-linear simulation model of the BFP620 transistor for a 900 MHz amplifier

Such a (numerical) simulation is no longer exact, but sufficient in practice, and no longer provides the full information of the complex scattering parameters and the variables that can be derived from them. Other variables can be calculated instead, such as the power of the harmonics and the temporal behavior of the signal. Such results are shown in Fig. 8.18.

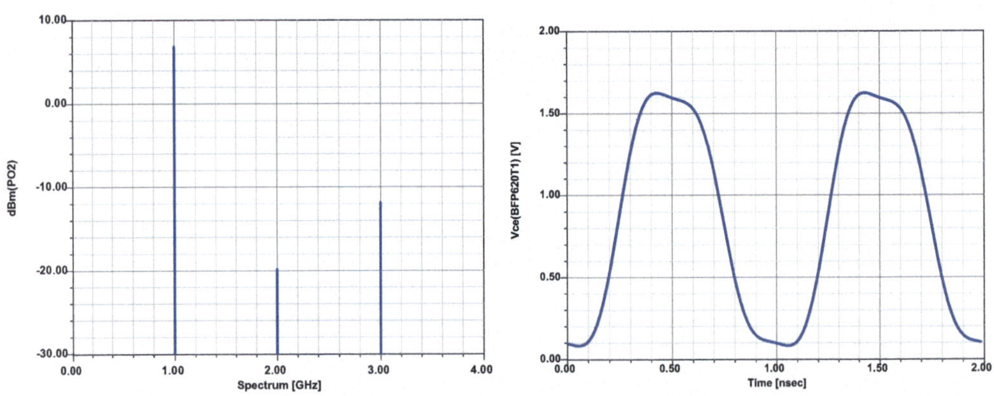

Fig. 8.18: Fundamental and harmonic waves as well as the temporal voltage behavior at the collector-emitter path of the 900 MHz amplifier

You can see a strong second harmonic and a noticeable deformation of the sinusoidal signal that was fed in at the input.

For the development of power amplifiers, a large parameter description in the form of mixed-frequency S-parameters [50] or X-parameters [50, 112] is also used.

8.4 Multi-Band LNAs for Software-Defined Radios

The introduction of the term Software-Defined Radio (SDR) was associated with a radio system that, in addition to the signal processors, should only consist of a D/A converter, a power amplifier and an A/D converter.

It is now clear that SDRs, which have software-based signal processing with coding, modulation, demodulation and decoding, place significantly more hardware demands on the modern front end.

Among other things, broadband LNAs are used that support bands with a typical bandwidth of more than 500 MHz. Alternatively, LNAs are used that cover several frequency bands, whereby the bands have increased in bandwidth.

This coverage of several frequency bands is achieved either by band switching, by tuning using varactor diodes and even by connecting several LNAs in parallel.

Furthermore, the amplification and output power must often be adjustable. Related circuit solutions and a lot of further literature on LNA for SDR can be found in [4].

In order to meet these conditions, much more development time must be invested in the LNA and other front-end components in the SDR.

8.5 Compensation of the Miller Effect

The Miller capacitance is the parasitic capacitance between the base and collector. As the frequency increases, more and more energy is coupled back from the output to the input via this capacitance.

This negative feedback is the main reason why the gain of transistors decreases with increasing frequency.

In [134] and [143] it is described how these miller-capacitances can be compensated for differential amplifier circuits. The results are shown in Fig. 8.19. By adding two cross-connected compensation capacitances C_k between base and collector as shown in Fig. 8.19, the parasitic base-collector capacitances C_p are neutralized if both values are equal. By neutralizing the miller-capacitance C_p, the gain-bandwidth product of the amplifier and thus the maximum stable gain increases.

This compensation circuit represents a cross-mode blocker (realized by the four identical capacitors or transistors) for the push-pull signal, which is connected in parallel to the two transistors.

Since the cross blocker perfectly suppresses the push-pull signal across the entire bandwidth, hardly any energy is coupled from the output to the input. Among other things, the backward isolation is also increased immensely. This measure can also increase the gain of a step without causing stability problems.

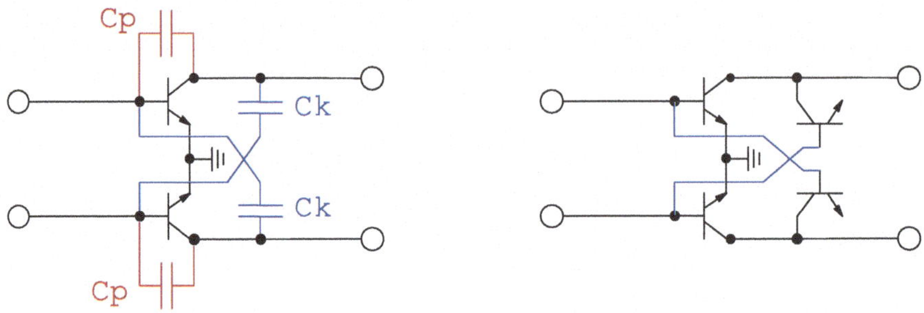

Fig. 8.19: Circuit diagrams of the compensated transmission amplifier: Left: Compensation of the capacitive component C_p by C_k and right: Compensation by identical inactive transistors

8.6 Basics of the Theory of Electrical Noise

Electrical circuits in every frequency range always have interference signals in addition to their actual useful signals.

In addition to possible deterministic interference such as IM3 products and radiation from other circuits, stochastic interference in the form of electrical noise `always` occurs. Electrical noise is always present in electronic circuits.

8.6.1 Basic Terms and Definitions

Statistical processes or signals whose course is unpredictable and which do not contain any usable information are referred to as `noise`.

A noise signal with power N is superimposed on each useful signal with power P. The ratio P/N is referred to as `signal-to-noise power ratio` or `signal-to-noise ratio` for short.

The term `noise matching` is used when the impedance transformation of one or more ports is not optimized for power matching, but is designed to achieve the best signal-to-noise ratio.

The `noise figure` F shows the ratio of the signal-to-noise ratio at the input of a two-port to the signal-to-noise ratio at the output.

8.6.2 Noise Sources

In practice, noise energy is generated by a large number of possible sources. Only the most important cases for modern RF technology are discussed below. The noise sources and phenomena are described in detail in [126].

Thermal Noise

Due to the random thermal movement of electrons and/or ions[5]. In metals, resistance materials or semiconductors, a so-called resistance noise or thermal noise is caused.

If any resistor or even a complex impedance is considered as a noise source and this is adjusted, the maximum available noise power is obtained:

$$N_{av} = k\,T\,\Delta f \quad . \tag{8.23}$$

The three contained terms mean:

k : Boltzmann constant: $1.38\,10^{-23}$ Ws/K

T : Absolute temperature in K

Δf: Bandwidth (The system is often limited by bandpass filters.)

It can be seen that the available noise power is independent of the size of the noisy resistor. On the other hand, N_{av} increases proportionally to the temperature and the bandwidth of the system.

The basic Eq. (8.23) is also used for the other noise sources and the noise at the output of two-port devices. In these cases, T is replaced by an effective noise temperature T_e.

Passive components such as resonators are optimized for the lowest possible losses in order to maintain the best possible signal-to-noise ratio. With increasing deviation of the effective resistance from the system resistance ($50\,\Omega$), less noise energy is coupled through the component.

Shot Noise

In addition to thermal noise, semiconductor components and electron tubes exhibit an additional noise phenomenon, shot noise. In semiconductors, the random charge movement through the potential threshold is responsible for this. In contrast to simple conductors and passive components, no charge transport takes place in semiconductors. Instead, energy is transported by quantized charge carriers, which have a discrete character. The shot noise was already predicted by Schottky in 1918 on the basis of theoretical considerations!

The calculation of the available noise power of a semiconductor component based on the shot noise must be carried out individually.

Example: PIN diode: It can be shown that the effective noise temperature T_e of a PIN diode is a good approximation of the absolute temperature T, [126]. Consequently, the shot noise has no major influence due to the large number of charge carriers. A PIN diode can be regarded as a complex impedance with two or more states.

[5] Ions are generally used to describe positive ions, i.e. atoms or molecules that are missing an electron. Cation is the exact term for these ions. Atoms or molecules that have an additional electron are referred to as anions.

Example: Schottky diode: A metal-semiconductor junction is used to produce diodes for RF and microwave applications that are based only on majority carrier effects and therefore have very good properties for use in detectors and mixers.

Schottky diodes of this type can be regarded as approximately lossless and therefore exhibit virtually no thermal noise. The main noise influence lies in the shot noise. The non-linear characteristic curve between current I and voltage U of a Schottky diode is characterized by

$$I = I_{ss}\left(e^{\left[\frac{q}{n\,k\,T}U\right]} - 1\right) . \tag{8.24}$$

In addition to the elementary charge q, the saturation current I_{ss} and the empirically determined ideality factor $n = 1.05...1.15$ are included. In the engineer's application, the temperature voltage is

$$U_T = \frac{n\,k\,T}{q} = 25.9\,\text{mV} \tag{8.25}$$

for the values $n = 1$ and $T = 300\text{K}$.

The diode is activated at the set small signal AC conductance value

$$G_s = \frac{I_0 + I_{ss}}{U_T} \tag{8.26}$$

and considered and adjusted for the operating point with the direct current I_0.

If we examine the exact analysis of the shot noise phenomenon for the Schottky diode, we arrive at the simple result for the effective noise temperature ([126])

$$T_e = \frac{n}{2}\,T \quad . \tag{8.27}$$

Consequently, a Schottky diode produces approx. 50 % less noise than a comparable resistor. In practice, this means that a receiver mixer constructed with a Schottky diode and adapted to $50\,\Omega$ is less noisy than many other mixer designs.

Further examples of various special diodes and the different transistors must be looked up in the special literature.

Pink Noise

In practice, thermal noise and Schottky noise can be regarded as approximately frequency-independent "white noise".

In the measurement technology of noise spectra, a logarithmically increasing value of the noise energy density can be recognized in current-carrying semiconductors below a cut-off frequency f_c. This phenomenon is referred to as pink noise, 1/f noise or flicker noise.

It cannot yet be fully explained scientifically. It occurs to different degrees in different semiconductor materials[6] and is also dependent on the design of the components. Bipolar transistors are less sensitive (f_c=100 Hz to 1 kHz). With field-effect transistors, values for the cut-off frequency of up to 10 MHz can be achieved depending on the design.

Unfortunately, this low-frequency noise is mixed into the higher-frequency signal bands by non-linear mixing processes and considerably worsens the signal-to-noise ratio there.

[6] It occurs much more strongly in GaAs than in Si.

One hypothesis for the 1/f noise is the simple assumption that a transistor amplifies at low frequencies with the 1/f characteristic. This assertion is supported by the results of oscillator development, [74]. These results, which are also shown in Fig. 8.20, demonstrate this potential influence by the fact that a negative feedback for low frequencies is deliberately implemented in the amplifier unit. This is intended to reduce the amplification in the lower frequency range (1-1000 kHz). The gain required for the capacitor required for DC voltage decoupling prevents use at lower frequencies.

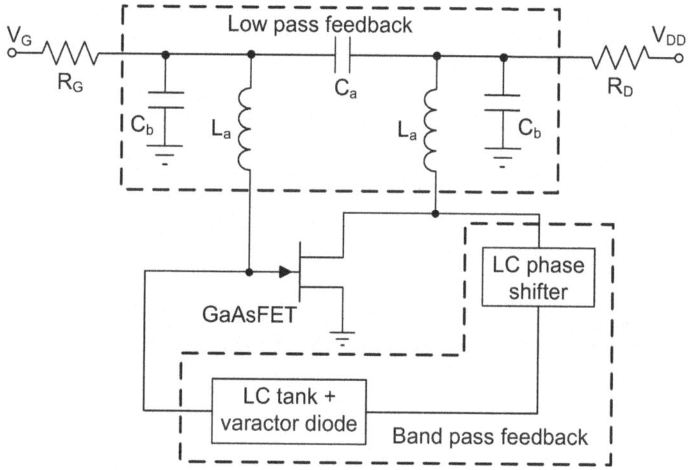

Fig. 8.20: Setup of an oscillator with 1/f noise suppression circuit

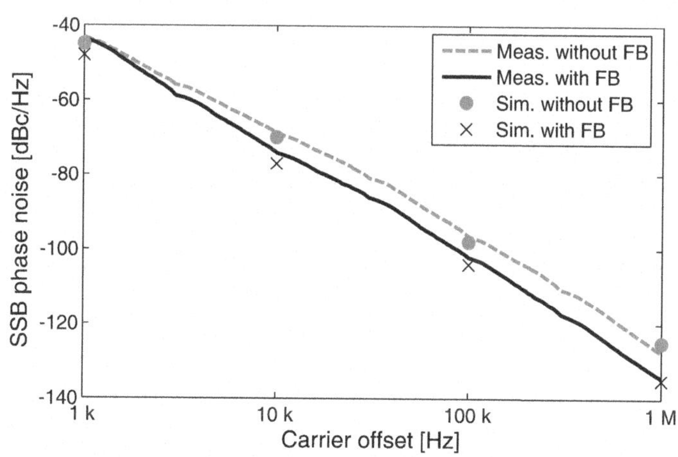

Fig. 8.21: Measurement results of an oscillator with 1/f noise suppression circuit

It can be seen that this measure is successful and noticeably improves the phase noise of the oscillator.

Other Sources of Noise

The most important noise sources for circuit technology were listed. Another source of noise in semiconductor technology is the so-called generation and recombination noise, [48].

Influence noise and current distribution noise play an important role in tube technology. In satellite communication, the angle-dependent antenna noise temperature must also be taken into account, which is dependent on a large number of noise sources such as the earth's atmosphere, the sun and other cosmic radio sources.

8.6.3 Noise Figure of a Two-Port and a Cascade

Every two-port or four-pole contained in a transmission chain worsens the signal-to-noise ratio, as there are no loss-free electrical networks. A data signal is referred to as a signal and is abbreviated to S. A noise signal is abbreviated as N (noise), Fig. 8.22.

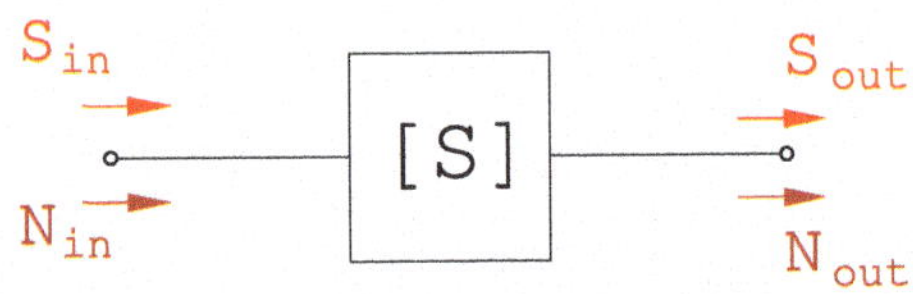

Fig. 8.22: Data and noise signals at the input and output of a two-port device

The noise figure F is a measure of how much the signal-to-noise ratio (SNR) is degraded by a two-port:

$$F = \frac{S_{in}/N_{in}}{S_{out}/N_{out}} \quad . \tag{8.28}$$

A lossless network therefore has the ideal noise figure: $F = 1$. The noise figure can only assume values greater than 1. The logarithmic value applies to all technical components:

$$F^{dB} > 0 \text{ dB} \qquad \text{with} \qquad F^{dB} = 10 \lg(F) \text{ dB} \quad . \tag{8.29}$$

The degree of degradation of the signal-to-noise ratio of a two-port can also be described alternatively by the effective noise temperature, which can be calculated directly from the noise figure:

$$T_e = T(F-1) \quad \text{resp.} \quad F = 1 + \frac{T_e}{T} \quad . \tag{8.30}$$

Before we can go into the calculation of the noise figure of passive two-port circuits and the calculation of the total noise figure of chain circuits, it is necessary to introduce the so-called available gain.

Available Gain

The available gain G_{av} of a two-port reflects the ratio between the available output power with power adjustment at port 2 (P_{2av}) and the maximum power that can be delivered by the source (generator) (P_g).

$$G_{av} = \frac{P_{2av}}{P_g} \qquad \text{resp.} \qquad G_{av}^{dB} = 10\ \lg(G_{av})\ \text{dB} \tag{8.31}$$

These conditions at the input and output do not have to correspond to the conditions with which the second port is operated. However, if everything is set up using the 50 Ω technique, the (logarithmic) value of the available gain can simply be calculated from the forward transmission:

$$50\,\Omega\text{system} \quad \Longrightarrow \quad G_{av}^{dB} = 20\ \lg(|S_{21}|)\ \text{dB} \quad . \tag{8.32}$$

The Noise Figure of a Passive Two-Port

The noise of passive two-port devices (which also do not contain diodes) is purely thermal. A short derivation leads to the simple general solution:

$$F = \frac{1}{G_{av}} \quad . \tag{8.33}$$

In the 50 Ω line technique, the noise figure can therefore be specified directly from the scattering parameters:

$$50\,\Omega\text{system} \quad \Longrightarrow \quad F^{dB} = -S_{21}^{dB} = 20\ \lg\left(\frac{1}{|S_{21}|}\right)\ \text{dB} \quad . \tag{8.34}$$

The Noise Figure of an RF Amplifier

Typical noise figures of low-noise amplifiers (LNA for short) for a frequency of 2 GHz are around 1 dB with noise matching. A power amplification of almost 15 dB is achieved.

In the past, amplifier noise was a dominant problem in receiver circuits. Nowadays, the losses of the passive components and the filter bandwidths predominate, as will be shown below. This is also shown in Fig. 8.23.

In practice, the losses of the antenna and the filters connected before the LNA have an even greater influence on the system noise figure than those of the LNA transistor, which is explained in the next section.

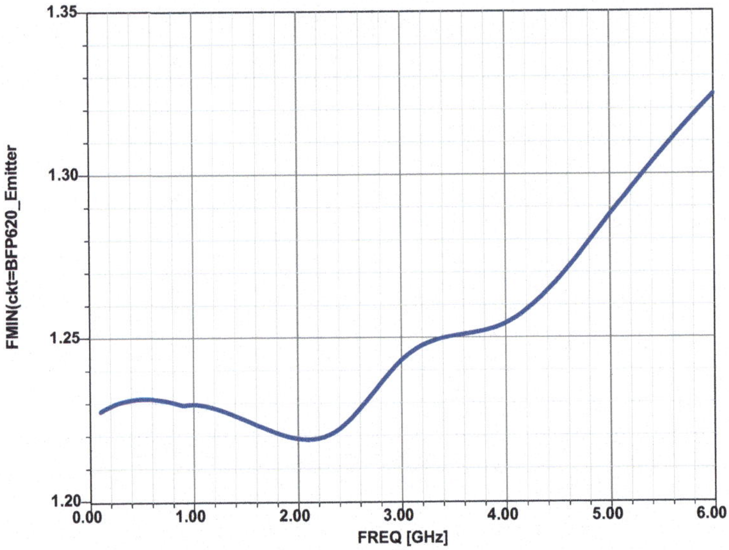

Fig. 8.23: Noise figure over the frequency of the BFP620 amplifier optimized for 900 MHz in SiGe technology (FMIN in dB)

Noise Figure of a Cascade of Two Ports

If several two-ports are connected in cascade (chain circuit) and either the noise figure F_i or the noise temperature T_{ie} and the available gain G_{iav} of each two-port $i = 1, 2, 3, ...$ are known, the total noise figure F or the total noise temperature T_e can be calculated via

$$F = F_1 + \frac{F_2 - 1}{G_{1av}} + \frac{F_3 - 1}{G_{1av} G_{2av}} + \frac{F_4 - 1}{G_{1av} G_{2av} G_{3av}} + ... \tag{8.35}$$

respectively

$$T_e = T_{1e} + \frac{T_{2e}}{G_{1av}} + \frac{T_{3e}}{G_{1av} G_{2av}} + \frac{T_{4e}}{G_{1av} G_{2av} G_{3av}} + ... \, . \tag{8.36}$$

This cascade formula (also known as Friis's formula) provides the design specification for a low-noise receiver:

1. The lowest possible noise amplifier (low F_i) with the highest possible gain (G_{iav}) is required as directly as possible at the input.
2. All the following components have only a minor influence on the total noise figure F.

Measuring the Noise Figure

The noise figures of modern amplifiers are so low that, unless you produce transistors yourself, you don't need to measure them, but can take them from the simulation.

The noise figure of the passive components corresponds to their attenuation and can therefore be traced back to a scattering parameter measurement.

In the past, the noise of active components was measured using either the *3 dB method* or the *Y-factor method*, [125]. Standard measuring devices are now used, [76].

8.6.4 Design of Low-Noise Receivers

When designing transmission systems, the calculation of the receive dynamics (signal-to-noise ratio at the receiver) plays a very important role. The transmission signal is heavily attenuated by free-space attenuation or transmission line attenuation. The receiver must be designed in such a way that the inherent noise (or receiver noise level) is lower than the received signal.

It is helpful to introduce the logarithmic value of the noise power introduced according to equation (8.23) for the bandwidth of $\Delta f = 1\,\text{Hz}$ and room temperature ($T = 290\,\text{K}$):

$$N_{1Hz}^{dBm} = 10\,\log\left(\frac{k\,T\,1\,\text{Hz}}{1\,\text{mW}}\right)\,\text{dBm} = -174\,\text{dBm} \quad . \tag{8.37}$$

The fundamental noise power N_{noise} at the receiver can be calculated for a given total noise figure F either via

$$N_{noise} = k\,T\,\Delta f\,F = N_{av}\,F \tag{8.38}$$

or as a logarithmized value

$$N_{noise}^{dBm} = N_{1Hz}^{dBm} + \Delta f^{dB} + F^{dB} \tag{8.39}$$

with

$$\Delta f^{dB} = 10\,\log\left(\frac{\Delta f}{1\,\text{Hz}}\right)\,\text{dB} \quad . \tag{8.40}$$

For example, with $N_{1Hz}^{dBm} = -174\,\text{dBm}$ and $\Delta f = 10\,\text{MHz}$ ($\Delta f^{dB} = 70\,\text{dB}$) and $F^{dB} = 3\,\text{dB}$ a basic noise level of $-101\,\text{dBm}$.

This practical example shows that the bandwidth of the receive filters has a dominant influence.

Fig 8.24 shows a possible structure of a receiver circuit consisting of various components. Their function and necessary properties for the development of a low-noise and robust RF receiver are explained below.

ANT: The noise power N_{ANT}^{dBm} and the signal power S_{ANT}^{dBm} can be specified behind the antenna (typical abbreviation ANT). It is often neglected that an antenna has transmission losses, which is particularly unacceptable for small compact antennas. The noise power N_{ANT}^{dBm} can be calculated from Eq. 8.38, whereby the reception bandwidth of the antenna must be taken into account.

BP1: This filter is becoming increasingly important, as the many interfering signals from neighboring receive bands can drive the LNA into compression on the one hand or interfere with the receive signal through IM products on the other. To reduce the total noise figure, this filter should be as low-loss as possible.

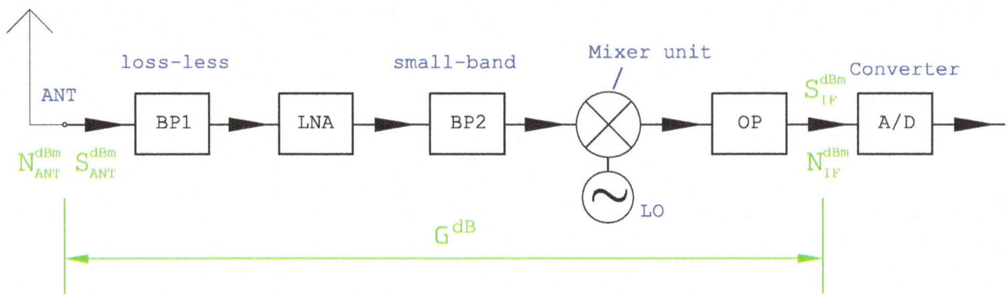

Fig. 8.24: Typical structure of an RF receiver circuit

LNA: In addition to the lowest possible noise figure, the LNA should also have the highest possible gain and the greatest possible suppression of interference signals from the DC supply line.

BP2: If BP1 does not fulfill the necessary bandwidth reduction, this filter BP2 must take over this function. BP2 no longer needs to be optimized with regard to losses.

LO: A local oscillator and a mixer stage convert the RF signal into the intermediate frequency range (typically in the lower MHz and upper kHz range). These two components and the correct information conversion are described in more detail in chapter 10.

OP: This operational amplifier amplifies the received signal. As the individual amplifiers (LNA and OP) operate in two different frequency ranges, the overall gain is significantly increased without the risk of oscillation of the receiving stage.

A/D: The received signal is digitized at the A/D converter and made available to a microprocessor.

The noise power at the A/D converter is calculated from the basic noise power (Eq. 8.39) and the gain of the receiver chain G^{dB}:

$$N_{ZF}^{dBm} = N_{1Hz}^{dBm} + \Delta f^{dB} + F^{dB} + G^{dB} \quad . \tag{8.41}$$

In contrast, the signal power at the A/D converter is simply calculated via

$$S_{ZF}^{dBm} = S_{ANT}^{dBm} + G^{dB} \quad . \tag{8.42}$$

The signal-to-noise ratio (SNR) at the A/D converter is calculated from the power ratio:

$$\text{SNR}_{\text{ZF}}^{\text{dB}} = S_{ZF}^{dBm} - N_{ZF}^{dBm} \quad . \tag{8.43}$$

The SNR at the A/D converter is always worse than at the antenna by the value of the noise figure of the receiver.

Two further variables on the receiver are the sensitivity and the dynamic range.

Sensitivity: The sensitivity at the receiver is the smallest dBm value that can be detected by the A/D converter.

Dynamic: The dynamic range at the receiver is the difference in dB between the smallest value that can be detected by the A/D converter and the maximum detectable A/D converter value.

Chapter 9

Modeling, Fitting and Coil Design

The current state of RF technology is that you can buy very good, inexpensive and excellently reproducible components or, in some cases, manufacture them yourself at even lower cost. Precise RF circuits can be developed with these very similar components, provided you have precise models of these components and the necessary circuit simulator.

Unfortunately, the models of components for the RF range are often not available at all or only inadequately. Component manufacturers do use test sockets to check the reproducibility of the components.

However, the manufacturers usually do not provide accurate models for the GHz range. As a result, good RF developers often have to characterize or model their components themselves.

Standard RF developers ignore this step and then only design circuits with moderate performance. By studying this chapter and applying it, you will bring your company's hardware forward!

This chapter uses the modeling of a semiconductor coil to demonstrate how the process of characterization based on given S-parameters takes place.

The coil is chosen because it is the component that RF developers most frequently design themselves. There are numerous examples in PCB and semiconductor technology.

Modeling can be divided into the two steps `Model determination` and `Adjustment` of the components of the model to the measured values, which is referred to below as *fitting*.

Simple and often sufficient models were presented in the Chap. 3.1.2 for various components and SMD as well as semiconductor implementations.

It goes without saying that different RF components sometimes require different models and that a large number of models are of technical interest in practice. Unfortunately, it is not possible to go into all these models of the various components in detail for all component types.

H. Heuermann, *Radio Frequency Technologies*,
https://doi.org/10.1007/978-3-658-50170-9_9

However, the example of the planar semiconductor coil is intended to illustrate the difficulties involved in creating models and fittings.

At the end of this chapter, the design of planar coils, which are typical for LTCC and semiconductor technology, is discussed in detail.

9.1 Models for Fitting a Coil

The Anglo-Saxon term *fitting* refers to the process of optimizing the variables of an equivalent circuit (EC, model) against one or more measurements in such a way that the electrical characteristic of the equivalent circuit corresponds as closely as possible to the characteristic of the measurement results.

How to carry out the associated precise measurements of the components is briefly explained in chapter 10.5.4.

Fitting does not only take place for measured objects. It is also often used to build a library based on a replacement model for a class of components using results from the field simulation.

In the following, a model of a 3.6 nH coil is used to show what a fitting process looks like in practice.

To this end, two models are first presented that provide good or very good results for the fitting and the physical description.

<u>The PI-Model</u>

The so-called PI model of a coil that is connected in series (Fig. 9.1) contains the elements that have the greatest influence on a design.

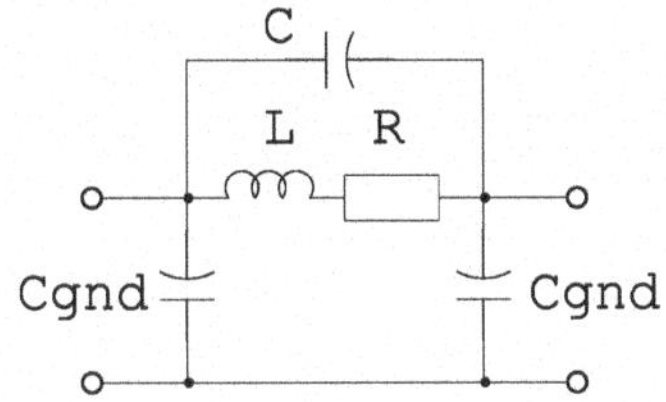

Fig. 9.1: PI equivalent circuit diagram of a coil

The most dominant element (= core element) is the series inductance L, which is made up of the actual coil and connecting elements such as bonding wires.

The resistance R takes into account `frequency-independent` the losses of the coil.

The capacitor C contains the parasites coupling capacitance between the input and output windings of the coil.

The two generally equal capacitors Cgnd take into account the parasitic capacitance between the mass and the windings of a coil.

Depending on the size of the series inductance L and the parasitic capacitors C and Cgnd, a parallel resonance results at frequencies between 0.01-100 GHz. For this resonance, the capacitors must be taken into account as follows:

$$Cr = \frac{C\,Cgnd}{2\,C + Cgnd} \quad . \tag{9.1}$$

If necessary, capacitors C or Cgnd can be omitted, which means that the remaining capacitors Cgnd or C are larger. The prerequisite is that one of the two capacitor values is noticeably larger than the other. In this case, a good fitting can still be carried out, but the capacitors can no longer be used as a substitute for the physical effects.

This PI model presented here is preferably used for SMD components. When working with coils manufactured using semiconductor technology, it is no longer possible to fit the substrate losses with the series resistance R. In this case, two resistors Rgnd, which are connected in parallel to the capacitors Cgnd, must also be used.

If you want to be able to separate the physical effects of a coil in order to optimize it further, you can also use the extended PI model according to Fig. 9.2. In this improved model, in addition to Cgnd, a parasitic resistance Rgnd is also taken into account.

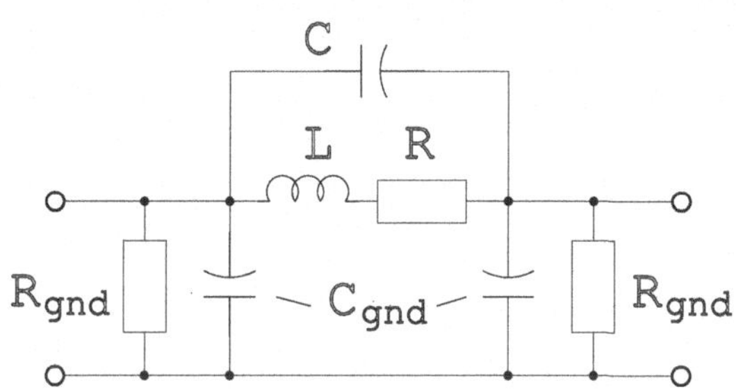

Fig. 9.2: PI equivalent circuit diagram of a coil for semiconductor applications

The reduced model can be used well if you are not interested in interpreting the physical effects of a coil.

Using this simple equivalent circuit, it is possible to calculate the electrical behavior of the coil in a component (e.g. a filter) very precisely. This model only offers satisfactory results when predicting losses. If a better prediction is required, the PISK model presented below must be used.

The PISK-Model

The largest loss mechanism in a coil at frequencies above 2 GHz is the skin depth δ. The resistance losses increase linearly to a good approximation at higher frequencies.

Therefore, the frequency-dependent losses can be modeled very well using a parallel connection of an inductor (Lsk) and a resistor (Rsk) (Fig. 9.3).

It should be noted that Lsk and Rsk are only an equivalent circuit to describe the skin depth and can no longer be explained by a physical sub-element of the coil.

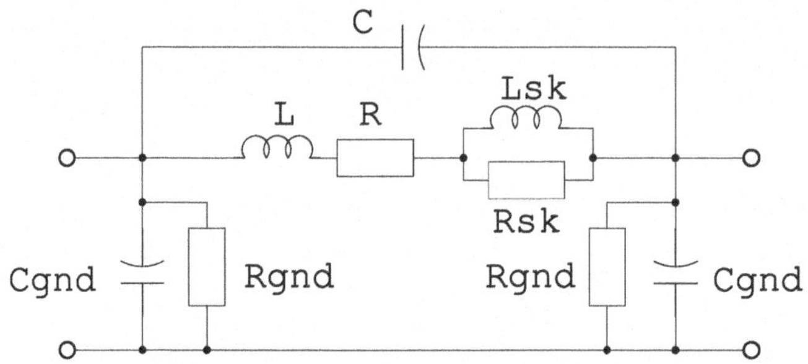

Fig. 9.3: PISK equivalent circuit diagram of a coil

The series resistance R now includes the ohmic losses of the coil at DC.

Furthermore, the PISK model includes the two resistors Rgnd, which in turn take into account the substrate losses that increase at lower frequencies. This effect is often barely recognizable.

The elements L, C and Cgnd have the same functionality as in the PI model.

9.2 Sensitization Structures for Optimal Fitting

When you start fitting, you often make the mistake of thinking that you have a good model because the curves of the complex S-parameters of the measurement and the model in the circuit simulator match.

A model for a measurement (or a field simulation) that is a great fit does not usually produce a good result!

The reason is that you have to make sure that the gradient of the individual variables (values of the elements) is sufficiently steep.

In other words, if you can vary the value of an element and there is no noticeable change in the electrical behavior of the component, you cannot perform a precise fitting.

This statement, which is very important in practice, will be illustrated by showing the fitting results of a coil without and with sensitization.

9.2.1 Fitting a Coil Without Sensitization

Fig. 9.4 shows the scattering parameters of the measurement and the equivalent circuit of a 3.6 nH coil superimposed on each other. As expected for a series coil, the matching (S_{11}) are small at low frequencies and large at high frequencies (i.e., strong reflection). They are shown in dark. On the other hand, it can be seen that the transmission value (S_{21}) is almost 1 at low frequencies (= 0 dB) and decreases sharply at high frequencies.

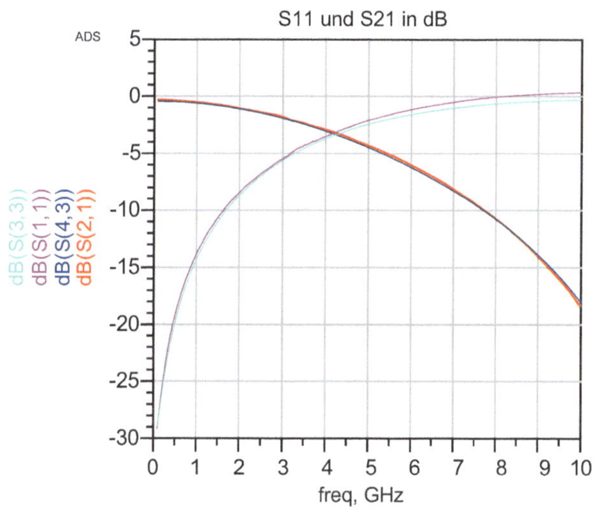

Fig. 9.4: Fitting results for the amounts of S_{11} and S_{21} of the PI equivalent circuit diagram and the measurement of a 3.6 nH RF coil

As the curves from 50 MHz to 10 GHz lie well *over each other*, one would think that the fitting was carried out very successfully.

A very decisive variable for the use of a coil is the series resistance R. The size of the series resistance often determines whether the losses of the entire component are too high or not[1].

In the following Fig. 9.5 you can see that the scattering parameters change only slightly if the values 3, 4 and 5 Ω are used for the series resistance.

If you want to be sure that you have determined the correct value for the series resistance, you must select a configuration that reacts sensitively to small deviations.

In the current case of the series coil, a large proportion of the energy is reflected. Only a small part is transmitted. This means that the value of the series reactance is between 40 and 200 Ω. Changes in the range of 1 Ω have almost no influence on the transmission value.

[1]The losses of capacitors are generally lower than those of coils.

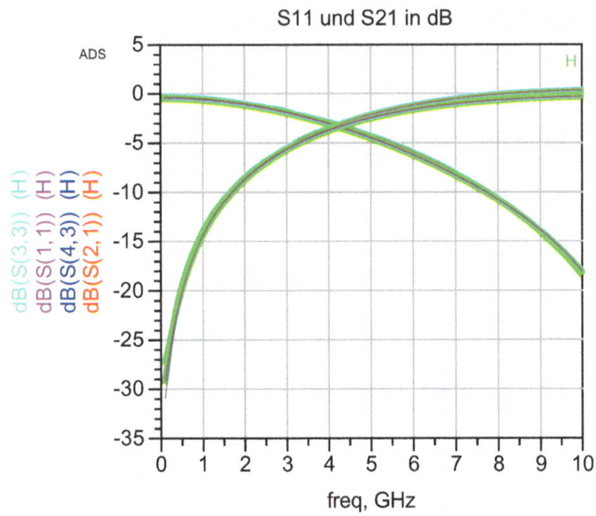

Fig. 9.5: Tuning results of the PI equivalent circuit diagram of a 3.6 nH RF coil with variation of the series resistance R with 3, 4 and 5 Ω

In the following, two methods are presented that clearly show changes in 1 Ω.

9.2.2 Fitting a Coil Using a Low-Pass Filter

A simple low-pass filter consists of a series coil and two capacitors connected to ground. A low-pass filter has very low losses in the passband. Consequently, S_{21} of a low-pass filter in the passband is very sensitive to a change in the series resistance.

If the same capacitors are used as ideal components (as contained in the circuit simulator) before and after the equivalent circuit and also before and after the measurement data, the resulting scattering parameters have changed, but S_{21} is very sensitive to a change in the series resistance of the coil.

Fig. 9.6 shows the PI equivalent circuit with two upstream 1 pF capacitors.

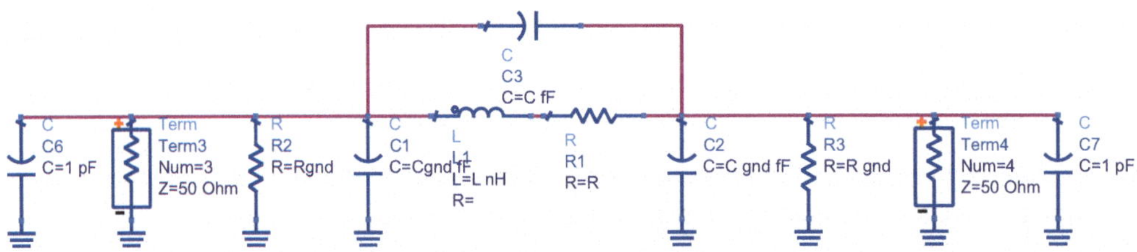

Fig. 9.6: PI-EC for a series coil with two capacitors extended to a low-pass filter

Fitting always requires two calculations that are fitted against each other. The measurement corresponding to Fig. 9.6 is shown in Fig. 9.7.

The differences between the PI and PISK models will now be discussed in more detail.

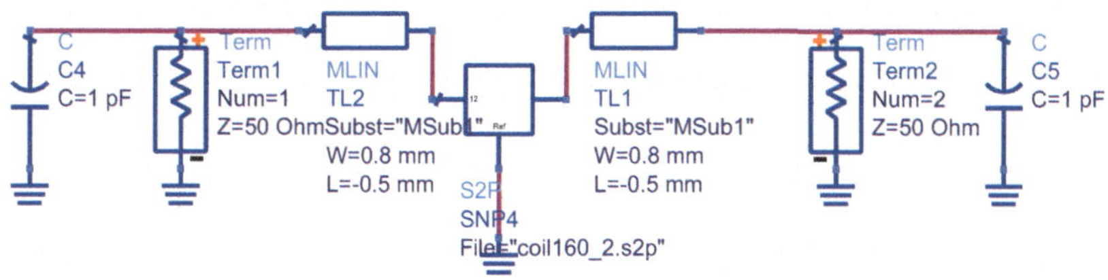

Fig. 9.7: Measurement file as a "black box" surrounded by two de-embedding lines of negative length and extended to a low-pass filter using two capacitors

Fig. 9.8 shows the transmission and reflection of the coil expanded to a low-pass filter (on a large scale).

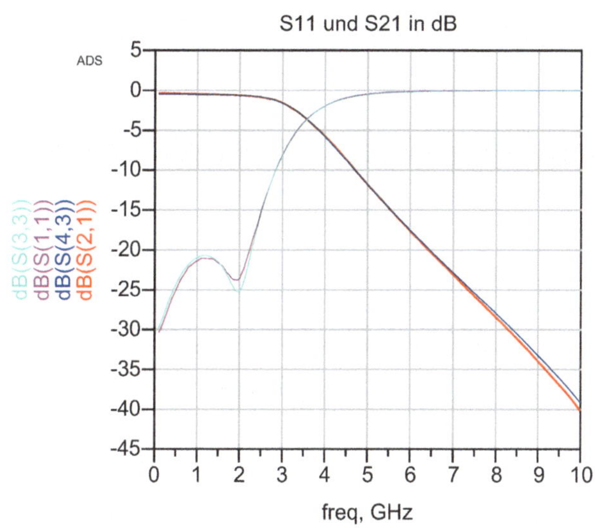

Fig. 9.8: Results of the PI equivalent circuit diagram of a 3.6 nH RF coil that has been extended to a low-pass filter

The Fig. 9.9 shows the tuning results with the PI model up to only 3 GHz. The values of the series resistance were tuned in 1 Ω steps. A clear "broadening" can already be seen here for the transmission. An improved resolution shows that the 3 Ω curve at 1.2 GHz and the 4 Ω curve from 2.2 GHz fit the measurement curve best. Overall, the losses varied by around 1.5 Ω to 3 GHz.

The dispersive behavior due to the finite penetration depth can be seen from the fact that the measured transmission values fit best at low frequencies with the 3 Ω-PI model and at high frequencies with the 4 Ω-PI model.

If a fitting is carried out with the PISK model, the values are obtained:

$$L=3.58\,\text{nH}\ ;\quad R=2.8\,\Omega\ ;\quad Rsk=1.1\,\Omega\ ;\quad Lsk=0.15\,\text{nH}\ ;$$
$$C=50\,\text{fF}\ ;\quad Cgnd=66\,\text{fF}\ ;\quad Rgnd=9700\,\Omega\ .$$

When fitting with the PISK model, it turns out that the resistor with around 3 Ω (exactly 2.8 Ω) provides the best match. Furthermore, it can be clearly seen that a fitting over the entire frequency range corresponds very well with the measured values (Fig. 9.10).

The procedure demonstrated here of initially fitting a coil without sensitization in a first step and then fitting the parameters for the losses and penetration depth by using a low-pass filter in a second step is referred to as **two-stage fitting**.

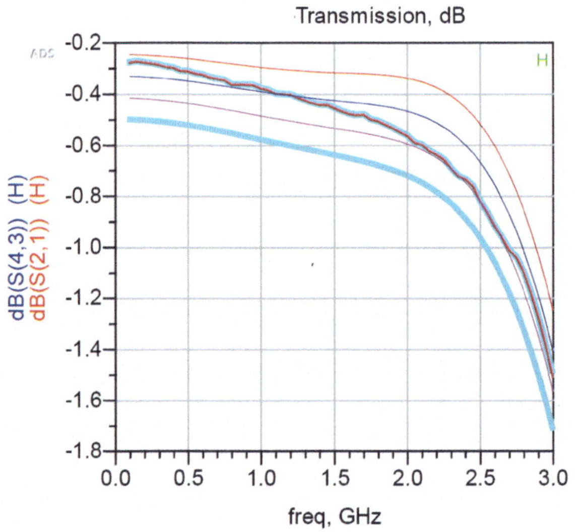

Fig. 9.9: High-resolution S_{21} tuning results of the PI equivalent circuit of a 3.6 nH RF coil, which was extended to a low-pass filter, when varying the series resistance with 3, 4 and 5 Ω in comparison with the measurement results

9.2.3 Evaluation of a Coil Using the MAG

The sensitization fitting of a coil with a low-pass filter is illustrative. However, the representation of the sensitization for the series resistance is limited in frequency. The greater the coil value, the greater the limitation of the frequency range.

An approach that is just as clear for experienced RF technicians is to no longer consider S_{21}, but the so-called MAG (Maximum Available Gain) (see also Chap. 8).

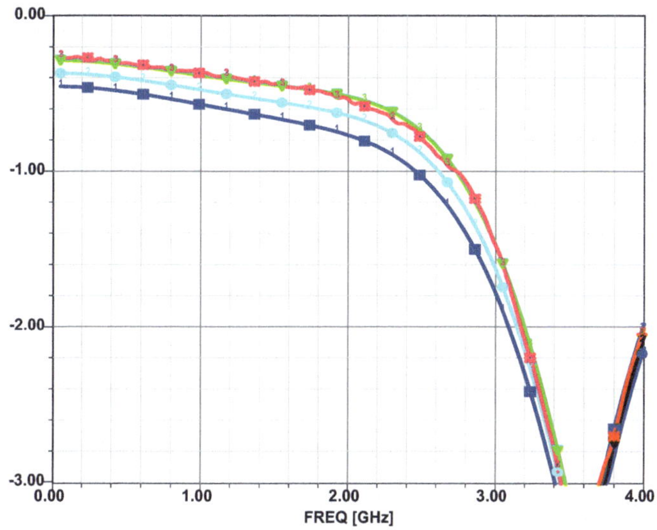

Fig. 9.10: High-resolution tuning results up to 4 GHz of the PISK replacement circuit of a 3.6 nH RF coil, which was extended to a low-pass filter, when varying the series resistance with 2.8, 3.8 and 4.8 Ω in comparison with the measurement results

The maximum available gain can be calculated from the scattering parameters of a two-port according to:

$$\text{MAG} = \left| \frac{S_{21}}{S_{12}} \left(k - \sqrt{k^2 - 1} \right) \right| \tag{9.2}$$

$$\text{with} \quad k = \frac{1 + |\det S|^2 - |S_{11}|^2 - |S_{22}|^2}{2\,|S_{12}\,S_{21}|} \tag{9.3}$$

$$\text{and} \quad \det S = S_{11} \cdot S_{22} - S_{12} \cdot S_{21} \quad . \tag{9.4}$$

This calculation mathematically completes the two-port in such a way that there is always an optimum match on the input and output sides.

In other words, the MAG indicates the maximum S_{21} value that can ideally be achieved with this component.[2]

Each RF circuit simulator can specify the MAG value (Ansoft/Serenade: GMAX) of a two-port directly based on the Eq. (2.36) and (2.37).

The great advantage of using the MAG value is that transmission sensitization is carried out here, which is frequency-independent!

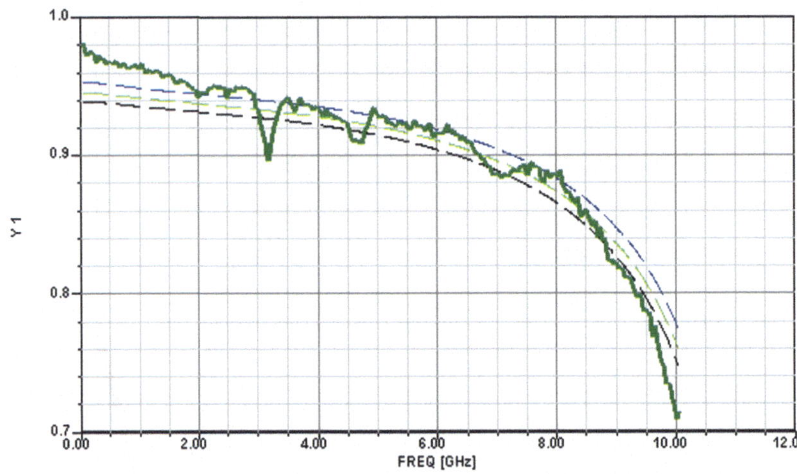

Fig. 9.11: Tuning results up to 10 GHz of the MAG values of the PISK equivalent circuit of a 3.6 nH RF coil when varying the series resistance with 2.8, 3.8 and 4.8 Ω in comparison with the measurement results

The MAG results when varying the series resistance are shown in Fig. 9.11.

The results in Fig. 9.11 show that the equivalent circuit of the PISK model delivers good results for loss assessments from 2 GHz to 8 GHz. It can be seen that measurement errors now have a significant influence.

9.3 The Practice of Modelings

For this chapter, the freely available program Serenade SV8.5 from Ansoft and ADS from Keysight were used for modeling.

The student version SV8.5 released at that time may also be used in industry and can still be found on the Internet. However, the PC date must be adjusted for the installation. Serenade SV8.5 is a student version based on the full version of the V8.5 stand. This program was still the latest version in 2001, is free to use and can still be installed on computers.

Structure and Sequence of a Coil Modeling Simulation

Like any modern program, Serenade has a project structure.

In the simplest sense, fitting is the alignment of two circuits (schematics).

[2]The use of MAG for passive components is not widespread. Originally, the MAG was used to evaluate transistors in order to recognize how much amplification can be achieved with the transistor in the "best case".

1. Realization of the Measured Value Schematic

To do this, you first need a circuit in which the measurement results of a so-called S2P file[3] are integrated (Fig. 9.12).

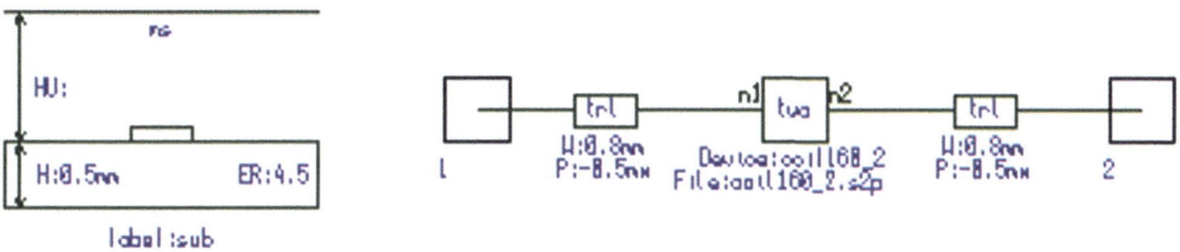

Fig. 9.12: Schematic for integrating the measured values (simulator: Serenade)

Fig. 9.12 contains lines with a negative length so that the measurement data is de-embedded in the reference plane of the coil connections (background is explained in Chap. 2).

To define this line, a through-connection line was measured beforehand and the losses and phase of a microstrip line were optimized so that the values match the through-connection.

This resulted in the `coil` series:

P=-0.5 mm; W=0.8 mm; H=0.5 mm; ER=4.5; Met1=au 1um; TAND=0.25 .

These values were entered in the left definition block for the microstrip line and in the trl elements of the line sections.

To integrate the measurement data (name e.g.: coil160_2.s2p), a two-port black box is used and the following two entries are made:

Device: coil160_2 File: coil160_2.s2p .

It is very important to create a symbol from the measured value schematic at the end. This command can be found under "Draw".

Next, you need the circuit in which the model is contained.

[3] S2P- and TOU- are the "classic" ASCII file formats for frequency-dependent n-port scattering parameters.

2. Realization of the EC Schematic

In addition, the main simulation must be drawn with the PI or PISK EC included. Fig. 9.13 shows an example of the PI model.

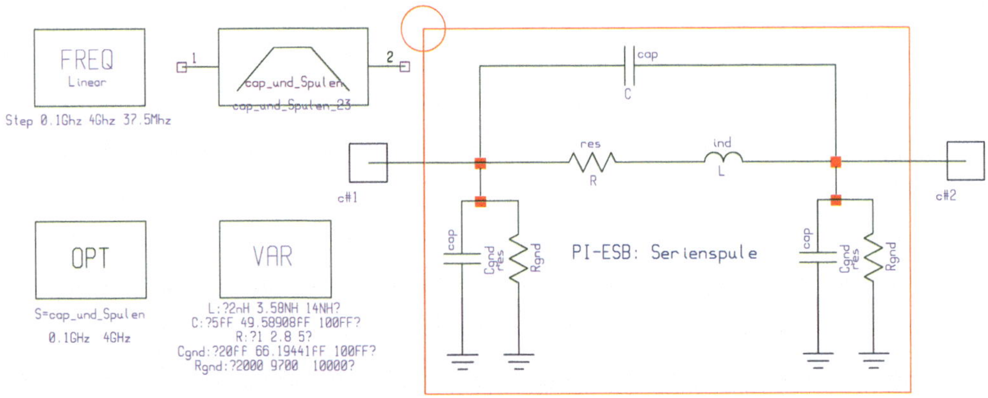

Fig. 9.13: Schematic for a fitting with the PI model (simulator: Serenade)

This schematic contains the following five blocks:

- Circuit: EC with elements that only have variable names.
- Frequency block (FREQ): Specifies linear sweep with start and stop frequencies as well as the step size.
- Optimization block (OPT): Specifies the frequency range for the fitting and the optimization targets (here: S=cap_und_Spulen).
- Variable box (VAR): Contains the variables with the upper and lower limit values as well as the start and optimized values.
- Integration of the symbol of the file to be optimized against (here: cap_und_Spulen.sym).

To check, simulate the circuit and look at the scattering parameter values of the measurement and the EC (with the start values).

3. Broadband Fitting of the Coil

First, the ESB of the coil is fitted against the measurement over the entire bandwidth (here 0.1-10 GHz). As the series coil blocks very well towards higher frequencies, this fitting is very sensitive to the parasitic variables C, Cgnd and Rgnd.

At this point, the PI model is already running into the existing restrictions of the SV version of Serenade: Only four variables are permitted for optimization.

For this reason, not all five variables can be set free and the series resistance R is set to a fixed (well-estimated) value. R is chosen because this value is not sensitive in this fitting step. R is set to a fixed value by no longer specifying upper and lower limits, Fig. 9.13.

Next, use the optimization function, start with the random method and use the gradient method if necessary.

After the optimization has been completed, the error F should be significantly smaller than 0.01. Fig. 9.13 shows the optimized values for this run of the 3.6 nH coil in the variable block.

After this first fitting step, the parasitic variables C, Cgnd and Rgnd are fixed. This means that the value obtained from the optimization remains in the variable block and the limits are eliminated. It is best to copy the variable block for backup and deactivate it using the right mouse button.

On the other hand, the value of the series resistance R is now enabled.

The second stage of the fitting is now carried out using the low-pass filter.

4. Fitting the Coil Extended to Low-Pass

In this second fitting step, the remaining parameters are to be calculated.

A simple low-pass filter is realized by connecting a capacitor to ground at both ports of both schematics (rough guideline: 2 nH: 0.5 pF; 4 nH: 1 pF; 8 nH: 2 pF).

If the adjustment is better than 10 dB, the low-pass filter is sufficiently sensitive. This is shown for the schematic of the PI equivalent circuit in Fig. 9.14.

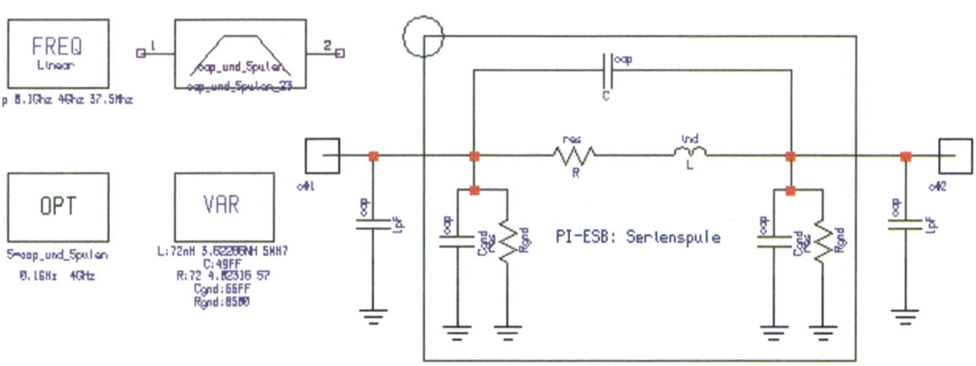

Fig. 9.14: Schematic for a fitting of the PI model in the low-pass extended stage

Attention: The upper frequency limit of the optimization frequency must also be reduced: A good choice is the 3 dB cut-off frequency (here just under 4 GHz).

Now run the optimization routines again and obtain the final results. As a check, keep looking at the scattering parameter values of the EC and the measurement (extended to low-pass).

Fig. 9.15 shows the result for the PI model after the 2nd fitting.

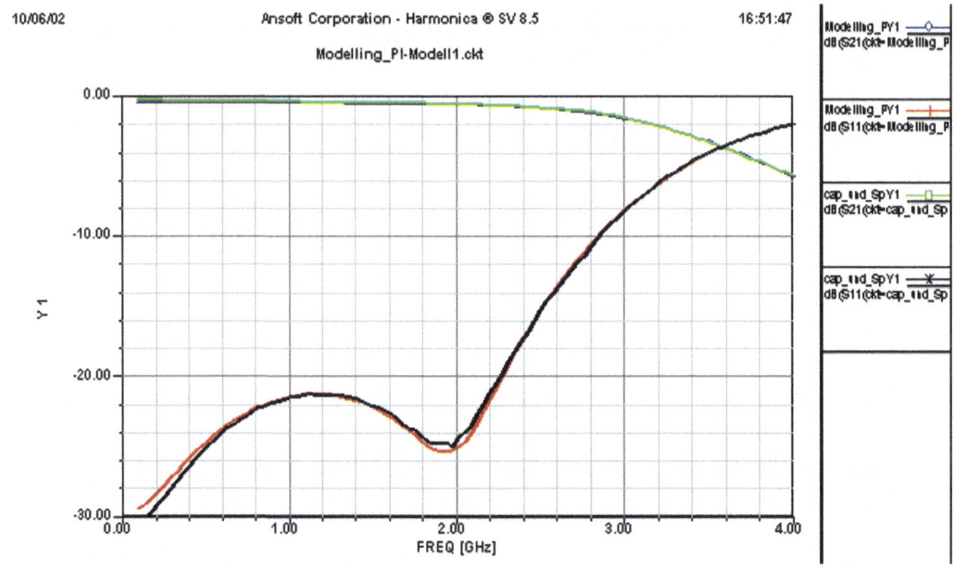

Fig. 9.15: Transmission and matching of a measurement of a 3.6 nH coil extended to a low-pass filter in comparison with the fitted values of the PI model

For the superficial observer of these fitting results, the fitting for the S_{11} and the S_{21} curves of Fig. 9.15 looks the same. However, the transmission values are very large numerical values that are almost 1 as an absolute value up to 2.5 GHz.

The reflection values of this well-matched low-pass filter are in the same frequency range at -15 dB and better (or lower). Consequently, the numerical values of the reflections are less than 0.18, with -30 dB even only 0.01.

The very good fitting of these reflection values is therefore significantly better than the results of the transmission values.

9.3.1 The Practice of Modeling with ADS

As an alternative to Serenade, the much more modern but not free ADS (Advanced Design System) software from Keysight can be used. In addition to a comprehensive circuit simulator, this software includes a 2.5D and a 3D field simulator and a layout tool for creating printed circuit boards.

This RF circuit simulator is the most widely used on the market. Particularly when using models from semiconductor manufacturers, ADS has advantages over the numerous other modern simulators available on the market due to its widespread use, see [70].

The circuit simulator contains the following solvers:

Linear scattering parameters,

Harmonic balance and

Various time domain calculations (similar to Spice)

and also supports the calculation of digital circuits. The software used for the book is a full version of ADS.

In terms of content, the structure of the circuits for fitting does not differ from that of Serenade. The values specified in the Serenade chapter are the same. These can also be found in the following Fig. (9.16).

The optimization in ADS is a little *more open* than in Serenade. The advantage is that the two circuits to be compared can be found in one simulation. In ADS, in contrast to Serenade, it is now necessary to set up equations for the optimization. These are specified in the *MeasEqn* block. As the optimization goals in ADS have to be optimized individually by amount and phase, it is also necessary to enter a corresponding weighting in the goals blocks.

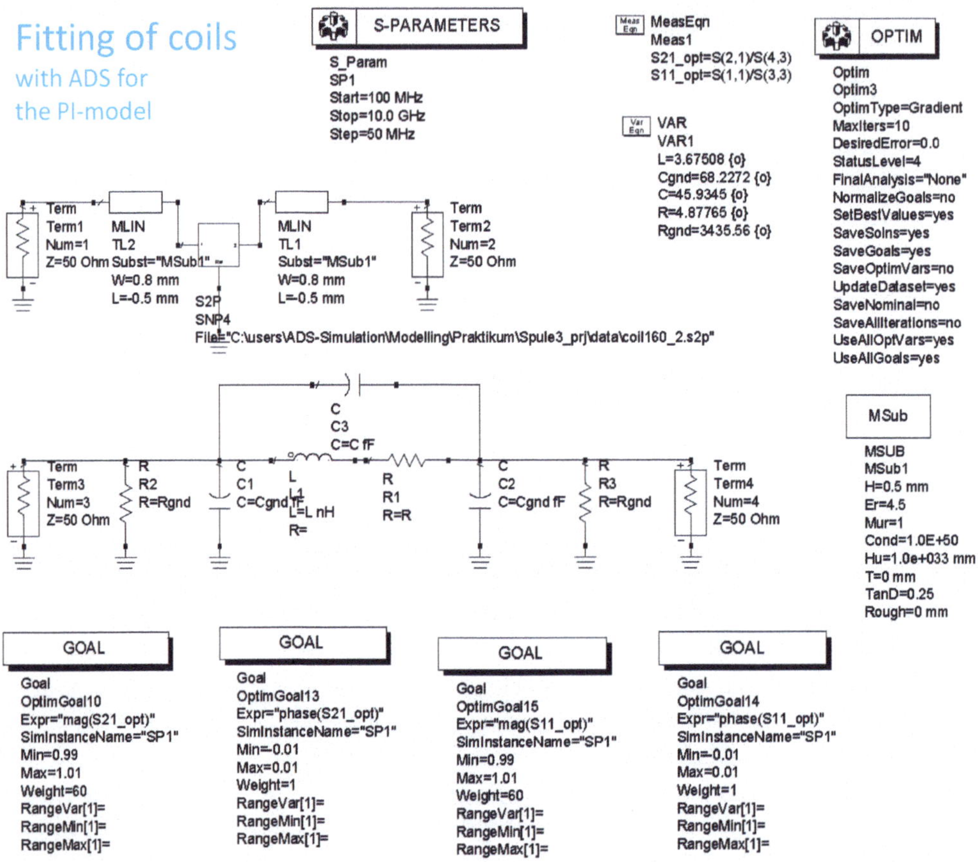

Fig. 9.16: ADS schematic for a fitting of the PI model

A frequent source of error when optimizing with ADS is that the new variables have to be updated after an optimization has been completed. For this it is necessary to push "Simulate" and them "Update Optimazation Variables".

If you want to display old and new simulations in one screen, you must select and "History" to activate the corresponding button.

The scattering parameters must be displayed as clearly as possible in ADS. Only the magnitudes and phases of the forward S-parameters are relevant for fitting. The results of the same fitting are printed in the Fig. (9.17).

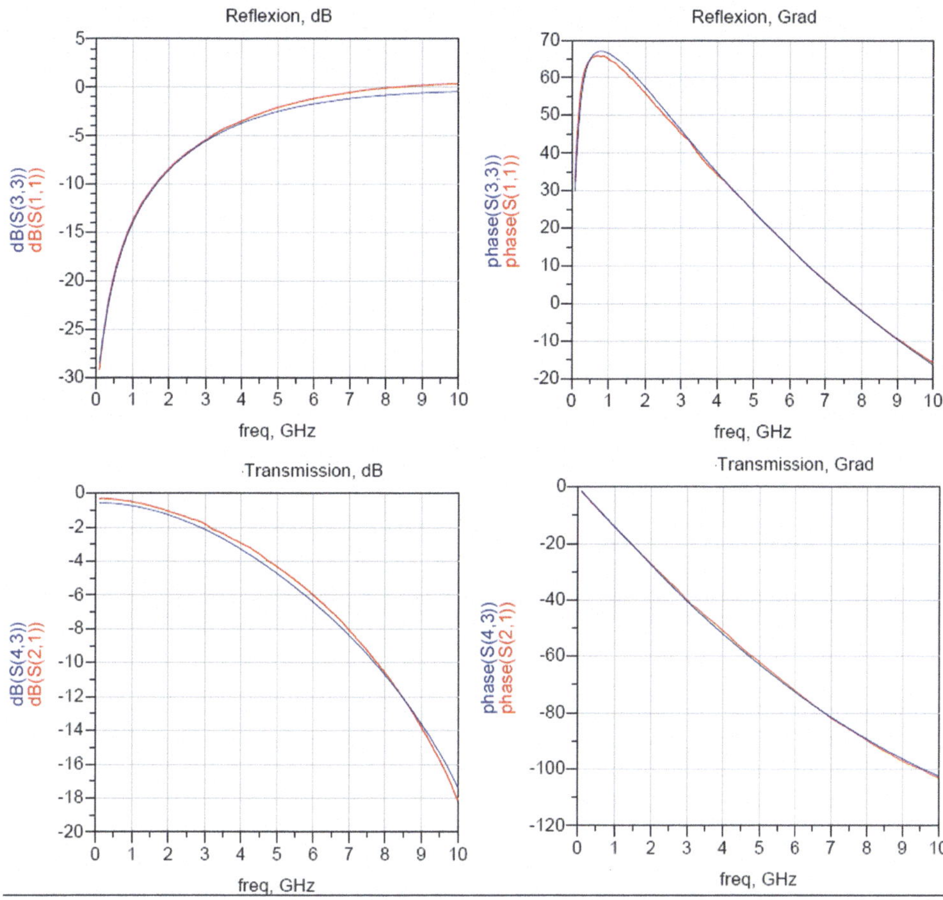

Fig. 9.17: ADS S-parameter results for a fitting of the coil with the PI model

When fitting, it is often not necessary to clearly assign the simulation curves to the axes. However, this is important for other simulations and can be carried out in ADS using the "Trace-Options" command.

9.4 Design and Optimization of Planar Coils

So far, the Chap. 9 has described the modeling and fitting of semiconductor coils based on given S-parameters, which can originate from both a field simulation and a measurement. In this section, essential properties of such semiconductor coils are presented, followed by design guidelines.

Finally, conceptual optimization options for the design of differential circuits are presented[4].

9.4.1 Characteristic Properties of Planar Coils

When designing (planar) integrated coils, the aim is to increase the effective Q-factor while maintaining or reducing the space requirement. At low frequencies (a few MHz), the ratio between the core inductance and the line losses can be used when considering the Q-factor.

The conductor losses R_{dc}, which arise due to the finite conductivity of the metallization used, can be taken into account.

$$R_{dc} = \frac{l_{sp}}{w\,t\,\sigma} \tag{9.5}$$

R_{dc} is calculated from the following variables: l_{sp} corresponds to the total conductor length of the coil, w to the width of the conductor tracks, t to the metallization height and σ to the electrical conductivity.

As the frequency increases, the consideration of current displacement (skin effect) leads to the addition of the metallization height t by the effective equivalent conductor thickness δ according to [6]

$$R_{RF} = \frac{l_{sp}}{w\delta(1 - e^{-t/\delta})\sigma} \quad . \tag{9.6}$$

At the same time, the interaction between the core inductance L and the parallel capacitance C (see Fig. 9.1) leads to a self-resonance of the coil at

$$f_r = \frac{1}{2\pi}\frac{1}{\sqrt{LC}} \quad . \tag{9.7}$$

The resulting effective impedance can be described by $Z_{eff} = j\omega L_{eff}$, where the effective inductance L_{eff} is a function of the natural resonance of the coil as follows

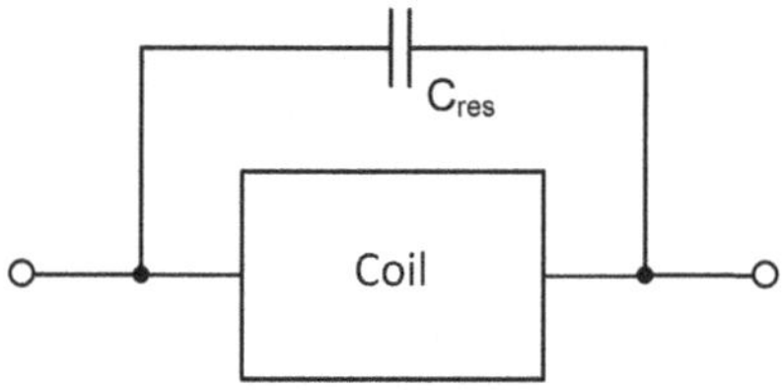

Fig. 9.18: Sensitization of a coil with regard to the operating frequency

$$L_{eff} = \frac{L_k}{1 - \omega^2 L_k C_p} \quad . \tag{9.8}$$

The relative bandwidth is small for the majority of available communication systems. A consideration of the coil quality around the operating frequency can lead to improved accuracy when determining the equivalent circuit parameters. For this purpose, the S-parameter set can be extended by an ideal capacitor C_{res}, [121] (see Fig. 9.18).

The quality of the coil Q_r can now be described using the 3 dB bandwidth Δf_{3dB} around the resonant frequency f_r:

$$Q_r = \frac{f_r}{\Delta f_{3dB}} \quad . \tag{9.9}$$

Alternatively, the quality Q_ϕ can be determined via the change in the phase response $d\phi$ at the resonant frequency can be determined [101]:

$$Q_\phi = \frac{\omega_0}{2} \left. \frac{d\phi}{d\omega} \right|_{\omega_0} \quad . \tag{9.10}$$

9.4.2 Dimensioning of Planar Coils

Integrated planar coils have been used for decades [16], [144], to realize compact and cost-effective circuits with good broadband properties. The relationship between the coil geometry and the resulting inductance has been frequently investigated over the years [43], [42], [105] and summarized in numerous design guidelines.

In this section, an approximation formula for the inductance of a planar coil is presented which, together with the subsequent recommendations on the essential geometry parameters, forms the basis for the design of planar coils with the highest quality.

[4]This section was fundamentally developed by Dr. A. Sadeghfam, whose doctoral thesis was carried out in the Department of High and Very High Frequency Technology at FH Aachen.

A planar coil is designed to realize the required inductance. The following formula takes into account the number of turns n and the coil geometry to provide a good approximation of this value [96]:

$$L = \frac{\mu_0 \, n^2 \, D_{av} \, c_1}{2} \left[ln \left(c_2 / \rho \right) + c_3 \rho + c_4 \, \rho^2 \right] \quad . \tag{9.11}$$

The parameters c_i depend on the coil geometry and can be taken from the table 9.1. Furthermore, D_{av} represents the average coil cross-section, while ρ is a measure of the coil area. The relationship between D_{av} and ρ and the inner and outer coil diameters, D_i and D_0, are

$$D_{av} = \frac{1}{2} \left(D_0 + D_i \right) \qquad \text{and} \qquad \rho = \frac{D_0 - D_i}{D_0 + D_i} \quad . \tag{9.12}$$

Coil geometry	c_1	c_2	c_3	c_4
square	1.27	2.07	0.18	0.13
hexagonal	1.09	2.23	0.00	0.17
octagonal	1.07	2.29	0.00	0.19
round	1.00	2.46	0.00	0.20

Tab. 9.1: Geometry-dependent coefficients for determining the inductance of single-layer coils according to [96]

General Recommendations for Dimensioning Integrated Coils with Optimized Quality

Technological changes in semiconductor processes in recent years have allowed these processes to become established in the mass production of low-cost RF circuits for applications up to several tens of gigahertz. Among the most important changes are the introduction of a high-impedance carrier and the definition of multiple "thick" metal layers.

However, the electrical properties of planar coils are essentially a function of the coil geometry. In a given manufacturing process, non-optimal dimensioning can lead to a significant impairment of the coil quality and thus to a deterioration in the electrical properties of the circuit.

There are a large number of recommendations for determining the geometric dimensions of an optimal planar coil (e.g. [6], [88], [90], [108]), which are summarized in this section.

Fig. 9.19 shows the layout of an integrated coil with the essential geometry parameters, namely the conductor width w, the distance s between the conductor tracks, the number of turns n, the inner coil diameter l_{in} and the outer side length of the coil l_{out} and forms the basis for the following dimensioning recommendations:

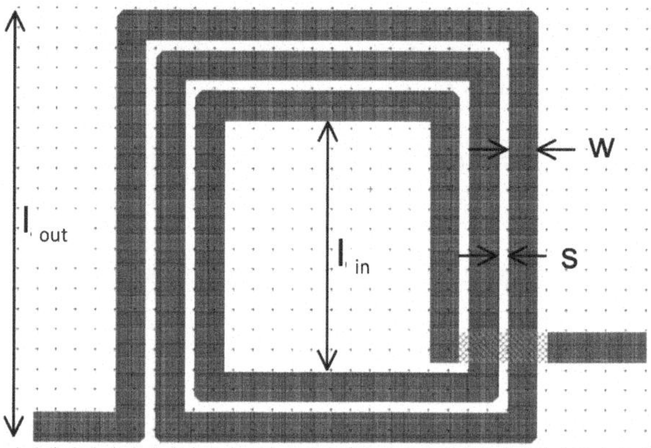

Fig. 9.19: The main geometric dimensions of a single-layer coil

- **The Conductor Width w:**

  The width of the conductor track is proportional to the cross-section and directly influences the metallic losses and thus the quality of the coil.

  However, wider conductor tracks lead to a longer circulation of the magnetic flux density around the conductor tracks and thus reduce the effective inductance per unit length.

  In order to achieve the desired inductance, this reduction in the effective inductance per unit length would have to be compensated for by increasing the conductor track length. An increase in conductor track length in turn leads to an increase in metallic losses.

  Furthermore, wider conductor tracks increase the space required for the coil, which increases the surface area and therefore the cost of the coil arrangement.

  The correct choice of conductor width should be determined for each application depending on the frequency. Experience values for conductor widths are between $10\,\mu m$ and $20\,\mu m$.

- **The Number of Windings n:**

  The inductance of a coil is ideally proportional to the square of the number of windings.

  However, the effective proportionality factor depends on the actual coupling between the coil windings and the magnetic resistance of the coil arrangement and, as will be shown below, can deviate greatly from this value.

- **The Distance Between the Conductor Tracks s:**

  Smaller gap widths lead to a shorter circulation of the magnetic flux density and thus increase the effective inductance per unit length and the coil quality. On the other hand, smaller gaps increase the capacitive coupling between the conductor tracks and lead to a lower (natural) resonance frequency.

In most cases, the minimum technologically feasible gap width corresponds to the metallization height and is a few micrometres. The resulting parasitic capacitance is in the order of a few tens of fF.

These values mean that for applications up to a few gigahertz and coil sizes of a few tens of nH, the smallest possible column width is preferable.

Smaller column widths also mean that less space is required.

- **The Inner Diameter of the Coil l_{in}:**

The smaller the inner diameter of the coil (or the coil eye), the higher the magnetic field density, which leads to an increase in magnetic resistance and eddy currents in the conductor paths around the coil eye.

On the other hand, small coil eyes are accompanied by a higher number of windings, which leads to a higher inductance coating.

Table 9.2 summarizes an investigation in which the relationship between the inner coil diameter l_{in}, the number of turns n and the coil quality Q_l was discussed, [90].

There is a clear tendency for smaller bobbin eyes to lead to a lower quality despite more windings.

Our own investigations confirm the guideline value of approx. 100 μm as a suitable diameter for the coil eye.

Number of turns n	Max. quality $Q_{l,max}$	Outer side length l_{out}	Coil eye l_{in}	Total length of the coil
3.5	5.8	255 μm	177 μm	3.02 mm
5.5	5.6	199 μm	75 μm	3.00 mm
7.5	5.0	190 μm	20 μm	3.12 mm

Tab. 9.2: Relationship between the inner coil diameter l_{in}, the number of turns n and the coil quality Q_l of a 5 nH coil with a conductor width w of 10 μm and a conductor gap s of 1.5 μm, [90]

- **Topology of the Coil:**

The shape of the coil influences the resulting Q-factor. Studies have shown that a round topology has up to ten percent better quality than a square coil.

- **Process Parameters:**

Technology-related parameters such as the metallization height t and the substrate losses can strongly influence the quality. In principle, processes with a greater metallization height and lower substrate loss are preferable.

The so-called "segment model" ([124]) can be used to determine the optimum cross-section (also for multi-layer manufacturing processes). This allows the different geometry parameters of integrated coils to be investigated in a time-optimized manner using a field simulator.

9.4.3 Conceptual Optimization of Differential Circuits

The use of differential components in modern telecommunications is state of the art.

The construction of a differential circuit based on an unbalanced circuit is usually done by doubling the unbalanced circuit and requires twice the number of elements and area.

The use of elements with mode-dependent properties, such as the so-called balanced coil, enables the use of new design methodology and leads to differential circuits with the same number of elements and area requirements as the unbalanced circuit.

This section presents such a design process and the elements required for it.

The Balanced and Differential Coil

A so-called balanced coil can be used to replace two serial coils in a differential circuit, [122]. Fig. 9.20 shows the basic structure.

The coupling of the two coils results in an increased magnetic coupling, comparable to a coil with twice the number of turns.

To replace two serial coils of the respective inductance L_0, a balanced coil consisting of two coils of inductance $L_k = L_0/(1+k)$ can be used, where k corresponds to the coupling factor ($0 \leq k \leq 1$) between the coils.

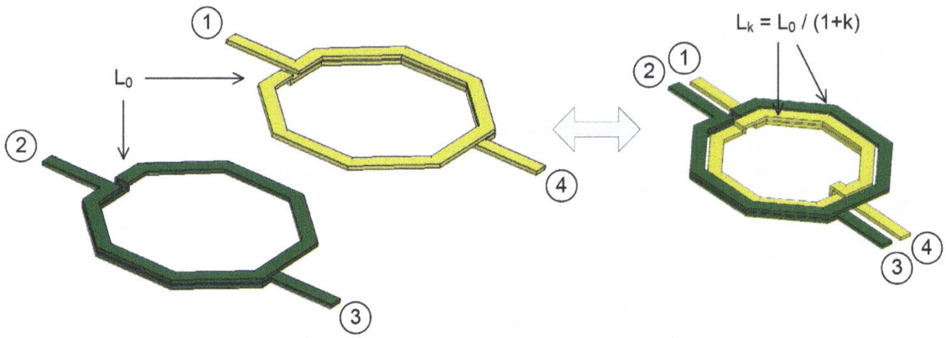

Fig. 9.20: Structure of the balanced coil: Two coupled coils of the inductance $L_k = L_0/(1+k)$ replace two serial coils of inductance L_0

In practice, it is also as shown in the Fig. 9.20: The balanced coil is the outer diameter of one of the two individual coils, which reduces the space requirement immensely.

The introduction of an effective inductance for the differential mode $L^-_{eff,dm}$ leads to the equivalent circuit diagram shown in Fig. 9.21. With an ideal coupling of 1, for example, two coils of inductance 2 nH can be coupled to replace two serial coils with 4 nH each.

Practical values for the coupling in semiconductor circuits are technology-dependent and lie between 0.6 and 0.9.

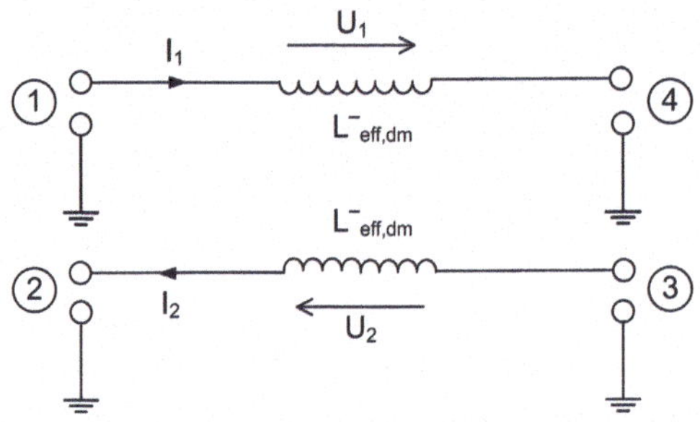

Fig. 9.21: Equivalent circuit diagram of the balanced coil with the effective inductance $L^-_{eff,dm}$

A so-called differential coil [18] can be used to replace two shunt coils in a differential or symmetrical circuit.

The structure of the differential coil corresponds to that of a balanced coil earthed on one side (e.g. ports 3 and 4 from the Fig. 9.20). Here too, the effective inductance increases by the factor $1 + k$.

Increasing the effective inductance of both balanced and differential coils means that smaller coils with lower losses[5] can replace larger, non-coupled coils.

This reduction in ohmic losses depends on the coupling between the coils and the number of windings. The more favorable ratio between (effective) inductance and losses leads to an increased quality of the new arrangement.

The improvement in quality is usually $\sqrt{1+k}$.

Design Method for Differential Circuits

The conventional design of differential circuits is based on doubling the unbalanced output circuit. This means that the electrical properties remain largely unchanged.

The introduction of the new mode-dependent elements enables the realization of differential circuits with the same space requirement and the same number of elements as the unbalanced circuit.

Table 9.3 shows the basic elements of unbalanced circuits and the corresponding elements for a differential circuit according to the conventional design or a design using the mode-dependent elements.

The optimized differential circuit is created by replacing the elements of the unbalanced circuit with the corresponding mode-dependent elements using this "transformation table". The resulting quality of the optimized differential circuit is better than the unbalanced circuit by a factor of $\sqrt{1+k}$.

By correctly dimensioning coupled coils, the equivalent circuit parameters can be determined in such a way that the manageable arrangement can be used as the core element of a more complex circuit.

An example of this is the so-called rondel coil [121], which consists of two coupled coils and can replace a transformer and up to six coils in a filter with integrated balun.

[5] The ohmic losses are proportional to the length of the cable used as a first approximation.

Asymmetrical circuit	Differential circuit	Multimode circuit (Push-pull mode)
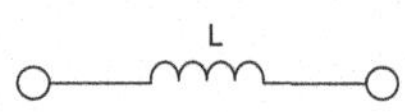 Serial coil	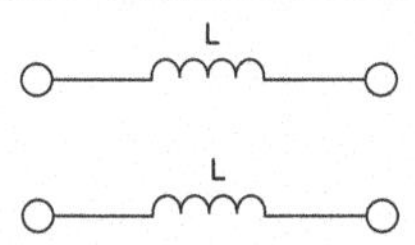 Two serial coils	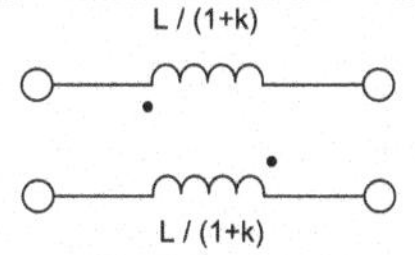 One balanced coil
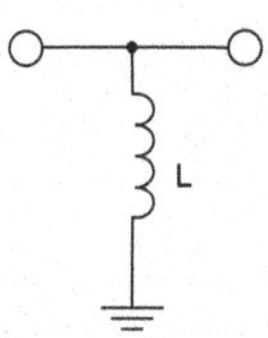 Shunt coil	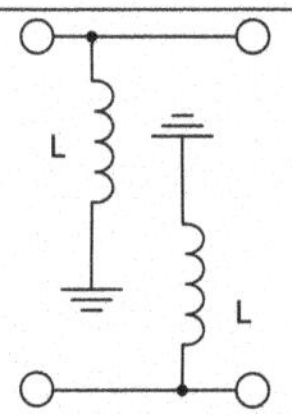 Two shunt coils	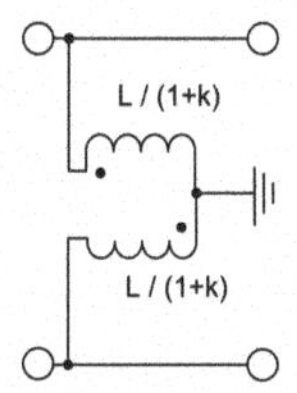 One differential coil
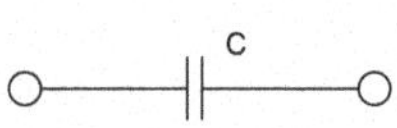 Serial capacitor	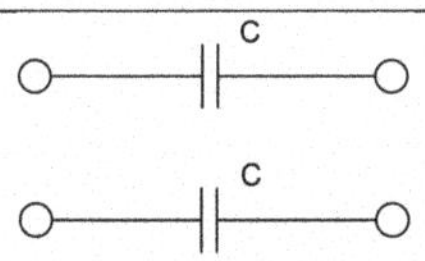 Two serial capacitors	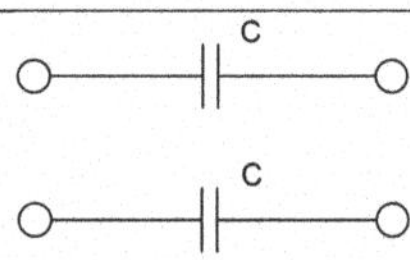 Two serial capacitors
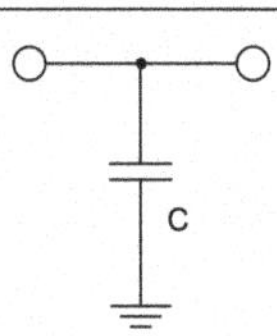 Shunt capacitor	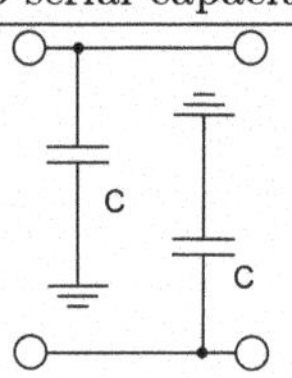 Two shunt capacitors	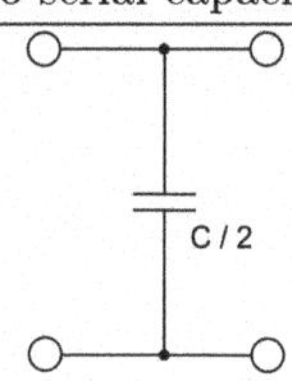 One shunt capacitor

Tab. 9.3: Transformation table with the comparison of the basic elements of an unbalanced circuit and the corresponding elements of a differential circuit according to the conventional and the new design method with multimode elements

Chapter 10

Fundamentals of the System Conception

RF components are created so that they can be used in systems and the latter have optimum functionality. The design of RF systems can be seen as a huge area, starting with simple transmission paths for scalar measurement technology and ending with highly complex modern communication systems in which digital modulation techniques play an important role.

This chapter first introduces the basic techniques and important parameters for calculating and describing simple transmission systems. The signal flow diagrams preferred in the American literature are then briefly discussed.

For a detailed understanding of common measurement and communication systems, it is essential to introduce other important RF components. Components such as detectors, mixers, IQ modulators, oscillators, synthesis generators and antennas are introduced. These will only be explained here in sufficient detail to enable the reader to integrate these components, which can generally be purchased, into their system in a qualified manner.

From the large number of possible systems, the scalar and vectorial scattering parameter measurement systems will be presented here. These are very clear and form the conceptual basis for many RF systems that are more complex in detail. Furthermore, many measurements of physical quantities and effects such as humidity, dust, temperature, lengths, distances, velocities and more can be carried out using scattering parameter measurements, so that the immediate field of application of these systems is large.

Furthermore, some basics for the transmission of so-called IQ signals are presented. Finally, a new radio system based on the use of two modes is presented as a system solution.

H. Heuermann, *Radio Frequency Technologies*,
https://doi.org/10.1007/978-3-658-50170-9_10

10.1 Design of Simple RF Transmission Paths

Whether in communication technology or in measurement technology, many RF systems can be described by a transmission path consisting of a chain circuit of many individual components.

Most of the two-port components of these transmission chains used in practice, such as amplifiers, filters, switches and cables, have already been introduced. Their interconnection can be calculated exactly for a transmission system with the transmission and chain parameters presented in section 2.5 (page 28).

In the following, this *known* inner two-port block will only be referred to as the transmission two port. The transmission path is reduced to just one transmitter, the transmission two-port and one load, as shown in Fig. 10.1.

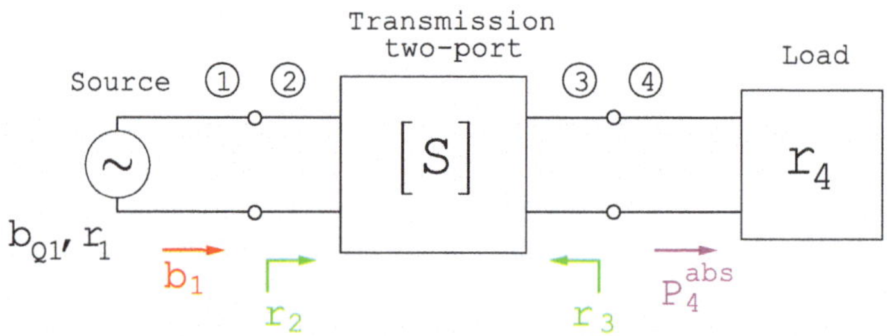

Fig. 10.1: Two-pole representation of a unidirectional transmission path

Furthermore, purely sinusoidal signals whose frequencies can be varied are used to analyze the circuits. The sinusoidal source is described electrically with an internal resistance Z_1 or a reflection factor r_1 and a source wave b_{Q1}. The source delivers the value b_{Q1} as an outgoing wave, provided it is terminated with the system impedance Z_0 or $r_2 = 0$. In this case, the power is

$$P_1^{Sy} = \frac{1}{2}\,|b_{Q1}|^2 \tag{10.1}$$

available from the source.

The numbers circled in Fig. 10.1 represent the numbers of the respective ports. Consequently, the input reflection coefficient of the transmission two port is S_{22} and the output reflection coefficient is S_{33}. On the other hand, the value r_2 indicates the input reflection at a termination with the load r_4. Consequently, r_2 and S_{22} do not have to match, as it is defined for S_{22} that $r_4 = 0$.

The wave actually emitted by the source, which is made up of the source wave b_{Q1} and a reflected component, should be designated as b_1.

Finally, the power P_4^{abs} is absorbed in the load.

The following section presents ways of calculating the power absorbed in the load and also the maximum power available from the source. Based on these results, important parameters such as efficiency, insertion loss and operating loss will then be presented.

10.1.1 Working with an Equivalent Load

If you want to calculate the maximum power that the general source can deliver, it makes sense to combine the transmission secondary and the load as a so-called equivalent load. This is shown in Fig. 10.2, indicating the progressive wave at the connection port.

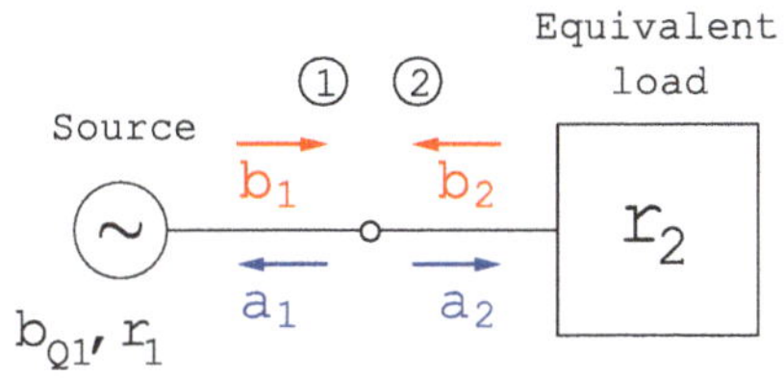

Fig. 10.2: Single-pole representation of a source and an equivalent load (transmission path plus load)

The outgoing waves b_1 and b_2 are calculated via

$$b_1 = b_{Q1} + r_1 a_1 \qquad \text{and} \tag{10.2}$$

$$b_2 = r_2 a_2 \qquad . \tag{10.3}$$

At the connecting port:

$$b_1 = a_2 \qquad \text{and} \qquad b_2 = a_1 \qquad . \tag{10.4}$$

If both equations from (10.4) and Eq. (10.3) are used in Eq. (10.2), the absorbed wave into the equivalent load is obtained

$$a_2 = \frac{b_{Q1}}{1 - r_1 r_2} \qquad . \tag{10.5}$$

Die an der Ersatzlast reflektierte Welle berechnet sich mit Hilfe von (10.3) somit aus

$$b_2 = \frac{b_{Q1} r_2}{1 - r_1 r_2} \qquad . \tag{10.6}$$

In chapter 2 the relationship between waves a and b and the associated power was introduced with the scattering matrix:

$$P_2^{ein} = \frac{1}{2} |a_2|^2 \qquad , \qquad P_2^{aus} = \frac{1}{2} |b_2|^2 \qquad . \tag{10.7}$$

The power absorbed in the equivalent load results from the difference between the incoming and reflected power:

$$P_2^{abs} = P_2^{ein} - P_2^{aus} = \frac{1}{2} |b_{Q1}|^2 \frac{1 - |r_2|^2}{|1 - r_1 r_2|^2} \qquad . \tag{10.8}$$

The value of the equivalent load, at which the power available from the source is maximized, is interesting, as the best signal-to-noise ratio is often achieved under these conditions.

P_2^{abs} becomes maximum when there is a conjugate complex output matching:

$$r_2 = r_1^* \quad .$$

The available power is obtained under this condition:

$$\boxed{P_{1\,max} = P_{2\,max}^{abs} = \frac{1}{2}\,|b_{Q1}|^2\,\frac{1}{1-|r_1|^2}} \quad . \tag{10.9}$$

It can be clearly seen that the maximum available power of a signal source is greater than the source wave power $P_{Q1} = \frac{1}{2}\,|b_{Q1}|^2$, unless the signal source is exactly matched to the system impedance.

Equivalent Load Solution

In a second step, the power P_2^{abs} absorbed by the transmission path as a function of S_{ij}, $(i, j = 2, 3)$ and r_1 and r_4 is to be determined.

If you want to calculate the power P_2^{abs} in this way, you must specify $r_2 = b_2/a_2$ as a function of the parameters S_{ij}, $(i, j = 2, 3)$ and r_4 and insert it into Eq. (10.8). The following calculation steps are very exemplary for system calculations with scattering parameters!

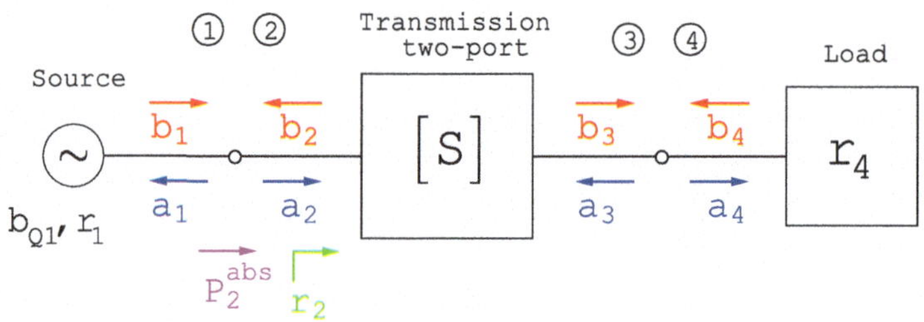

Fig. 10.3: Single-pole representation of a unidirectional transmission path

Fig. 10.3 makes it easy to recognize the relationship between r_4 and the waves a_3 and b_3:

$$b_4 = r_4\,a_4 \qquad \text{and therefore} \qquad a_3 = r_4\,b_3 \qquad \text{resp.} \qquad b_3 = \frac{1}{r_4}\,a_3 \quad . \tag{10.10}$$

If this simple relationship is placed in the second line of the two-port equations

$$\begin{aligned} b_2 &= S_{22}\,a_2 + S_{23}\,a_3 \\ b_3 &= S_{32}\,a_2 + S_{33}\,a_3 \end{aligned} \tag{10.11}$$

of the transmission two-port, a short transformation yields

$$a_3 = \frac{S_{32}\,r_4}{1 - S_{33}\,r_4}\,a_2 \quad . \tag{10.12}$$

Thus, a_3 in the first line of Eq. (10.11) can be eliminated and by dividing a_2 the reflection coefficient is obtained

$$r_2 = S_{22} + \frac{S_{23}\,S_{32}\,r_4}{1 - S_{33}\,r_4} \quad . \tag{10.13}$$

If you insert the result for r_2 into Eq. (10.8), after a short calculation (which is well suited as an exercise) you arrive at the general calculation formula for the power absorbed in the transmission path.

$$\boxed{P_2^{abs} = \frac{1}{2}\,|b_{Q1}|^2\,\frac{|1 - r_4\,S_{33}|^2 - |S_{22} - r_4\,\Delta S|^2}{|1 - r_1\,S_{22} - r_4\,S_{33} + r_1\,r_4\,\Delta S|^2}} \tag{10.14}$$

In the equation (10.14) applies: $\quad \Delta S = S_{22}\,S_{33} - S_{23}\,S_{32} \quad .$

10.1.2 Working with an Equivalent Source

If you want to calculate the power P_4^{abs} absorbed at the load from the network parameters of the transmission path and the source wave b_{Q1}, it is helpful to use an equivalent source according to Fig. 10.4.

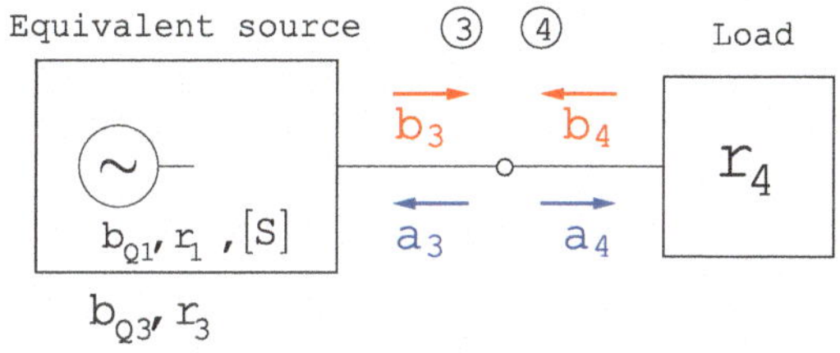

Fig. 10.4: Representation of an equivalent source (from source and transmission path) and a load

The calculation of the source wave of the substitute source b_{Q3} is performed with the termination $r_4 = 0$! Consequently, b_4 and a_3 must also be set to zero.

The source wave can therefore be calculated using the simplified first line of Eq. (10.11).

$$b_{Q3} = b_3 = S_{32}\,a_2 \tag{10.15}$$

The wave a_2 is calculated from the Eq. (10.5) with $r_2 = S_{22}$ (since $r_4 = 0$):

$$a_2 = \frac{b_{Q1}}{1 - r_1\,S_{22}} \quad . \tag{10.16}$$

By inserting Eq. (10.16) into Eq. (10.15), the source wave of the substitute source is obtained

$$b_{Q3} = b_{Q1} \frac{S_{32}}{1 - r_1 S_{22}} \quad . \tag{10.17}$$

The derivation of the input reflection factor r_3 of the equivalent source is exactly the same as the derivation of the Eq. (10.13). Only the indices of the forward and backward directions have to be swapped:

$$r_3 = S_{33} + \frac{S_{23} S_{32} r_1}{1 - S_{22} r_1} \quad . \tag{10.18}$$

The circuit with an equivalent source as shown in Fig. 10.4 is very similar to the circuit with an equivalent load (Fig. 10.2). If you want to calculate the power absorbed in the load P_4^{abs}, you can follow exactly the same procedure as for Eq. (10.8). The indices only need to be swapped from 1 to 3 and from 2 to 4:

$$P_4^{abs} = \frac{1}{2} |b_{Q3}|^2 \frac{1 - |r_4|^2}{|1 - r_3 r_4|^2} \quad . \tag{10.19}$$

The general equation for calculating the power absorbed in the load is again obtained by substituting the expression of the reflection factor of the Eq. (10.18) into the Eq. (10.19).

$$\boxed{P_4^{abs} = \frac{1}{2} |b_{Q1}|^2 \frac{|S_{32}|^2 (1 - |r_4|^2)}{|1 - r_1 S_{22} - r_4 S_{33} + r_1 r_4 \Delta S|^2}} \tag{10.20}$$

The absorbed power can also be easily simulated using a non-linear circuit simulation. However, the result (a numerical value) does not indicate, as Eq. (10.20) does, which measures can be used to increase the power. The practical engineer should be able to recognize that, given a mismatch of r_4, the output power can be increased by maximizing the product $r_4 S_{33}$ on the real number axis.

10.1.3 Characteristics of Transmission Paths

The Efficiency η

The efficiency η is a statement about the quality of an RF transmission path, especially for passive setups with high power.

The efficiency η is the ratio of the RF power delivered to the load to the RF power taken from the source (source power). Expressed mathematically, this results in

$$\eta = \frac{P_4^{abs}}{P_2^{abs}} \quad . \tag{10.21}$$

If we now use the result for P_4^{abs} from Eq. (10.20) and the relationship for the power P_2^{abs} from the Eq. (10.14) to calculate the efficiency, the result is very clear:

$$\boxed{\eta = \frac{|S_{32}|^2 \cdot (1 - |r_4|^2)}{|1 - r_4\, S_{33}|^2 - |S_{22} - r_4\, \Delta S|^2}} \quad . \tag{10.22}$$

The following must apply to a passive RF transmission two-port: $\eta < 1$.

A lossless transmission two-port has the ideal efficiency: $\eta = 1$.

Eq. (10.22) offers several possibilities to optimize the efficiency via the scattering parameters. For example, the parameter $|S_{22} - r_4\, \Delta S|^2$ should be maximized using the optimizer of the circuit simulator.

The Insertion Loss a_E

In practice, the measurement of the transmission (here S_{32}) is often equated with the insertion loss, but this is only correct to a first approximation.

The insertion loss is the ratio between a (reference) power that could be delivered to the load by means of a direct through-connection and the power that is absorbed by the load.

For the calculation or measurement of the reference power, the transmission delay is thus determined by a through-connection with the scattering parameters

$$\begin{aligned} S_{22} &= 0 \quad , \quad & S_{23} &= 1 \\ S_{32} &= 1 \quad , \quad & S_{33} &= 0 \end{aligned} \tag{10.23}$$

and thus the following applies in simplified form $r_2 = r_4$.

$$\frac{a_E}{\text{dB}} = 10 \lg \left(\frac{P_{2\ (r_2 = r_4)}^{abs}}{P_4^{abs}} \right) \tag{10.24}$$

If you use the result for P_4^{abs} from Eq. (10.20) and the relationship for the power $P_{2\ (r_2=r_4)}^{abs}$ from the Eq. (10.14) to calculate the insertion loss, the result is obtained:

$$\boxed{\frac{a_E}{\text{dB}} = 10 \lg \left(\frac{|1 - r_1\, S_{22} - r_4\, S_{33} + r_1\, r_4\, \Delta S|^2}{|S_{32}|^2 \quad |1 - r_1\, r_4|^2} \right)} \quad . \tag{10.25}$$

It can be clearly seen that the reflection coefficients only appear in products and therefore the insertion loss corresponds very well to the forward transmission loss with good matching.

In the practice of RF measurement technology, insertion loss is the most frequently measured quantity. However, it is often the case that the attenuation $a_E < 0$ is also present for a passive network. This is possible if a low-loss transmission path improves the matching between the source and the load.

> Scalar scattering parameter measurements are measurements of the insertion loss and not the (often sought-after) transmission loss!

The Transducer Loss a_T

To calculate the transducer loss a_T, the reference power for the maximum available power of the signal source $P^{abs}_{2\,max} = P^{abs}_{2\ (r_2=r_1^*)}$ is used:

$$\frac{a_B}{\text{dB}} = 10 \text{ lg}\left(\frac{P^{abs}_{2\ (r_2=r_1^*)}}{P^{abs}_4}\right) \tag{10.26}$$

Based on this definition, the transducer loss can be calculated as a function of the network parameters using the Eqs. (10.20), (10.17) and (10.14):

$$\boxed{\frac{a_B}{\text{dB}} = 10 \text{ lg}\left(\frac{|1 - r_1\,S_{22} - r_4\,S_{33} + r_1\,r_4\,\Delta S|^2}{|S_{32}|^2\,(1-|r_1|^2)\,(1-|r_4|^2)}\right)} \quad . \tag{10.27}$$

If a lossless transmission splitter transforms the reflection factor of the load r_4 into $r_2 = r_1^*$, the maximum value of $a_T = 0\,\text{dB}$ occurs.

The following exercise serves as a reference for the exact calculation of any chain circuit.

——— *Calculation Transmission Path* ———

Given: *A transmission path is to be analyzed according to Fig. 10.5.*

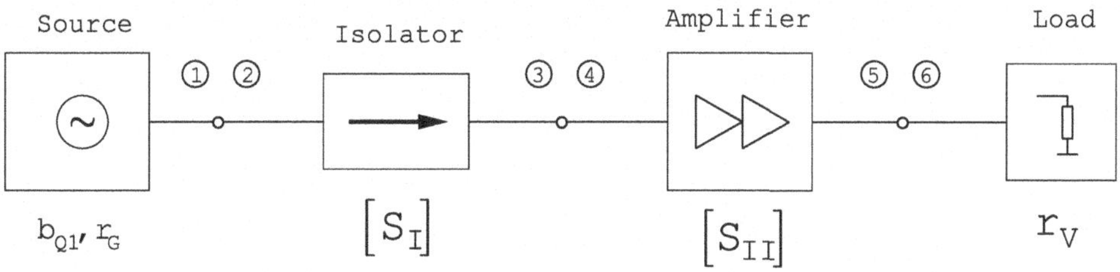

Fig. 10.5: Transmission path to be analyzed

$$b_{Q1} = \sqrt{20\,mW} \quad , \qquad r_G = 0.3 \quad , \qquad r_V = 0.1 \tag{10.28}$$

$$[\mathbf{S_I}] = \begin{pmatrix} 0.1 & 0.1 \\ 0.9 & 0.1 \end{pmatrix} \quad , \qquad [\mathbf{S_{II}}] = \begin{pmatrix} 0.1 & 0.01 \\ 100 & 0.1 \end{pmatrix} \tag{10.29}$$

Find: *Calculate all power values in the units W and dBm as well as the reflection factors dB.*

a) r_3, b_{Q3}, P_3^{Sy} *for port 3 and* P_1^{Sy}

b) r_5, b_{Q5}, P_5^{Sy} *for port 5*

c) *Power absorbed in the load* P_6^{abs} *and the transducer loss* a_T

d) *Calculate the power* P_6^{abs} *in dBm for the case that all components are perfectly matched.*

To a) *Eq. (10.1) can be used to calculate this:*

$$P_1^{Sy} = \frac{1}{2}\,|b_{Q1}|^2 \qquad \Longrightarrow \qquad P_1^{Sy} = 10\,mW \;\equiv\; 10\,dBm \tag{10.30}$$

Eq. (10.17) provides:

$$b_{Q3} = b_{Q1}\,\frac{S_{32}}{1 - r_1\,S_{22}} = 4.15\,\sqrt{mW} \qquad \Longrightarrow \qquad P_3^{Sy} = 8.6\,mW \;\equiv\; 9.3\,dBm \tag{10.31}$$

Eq. (10.18) applies:

$$r_3 = S_{33} + \frac{S_{23}\,S_{32}\,r_1}{1 - S_{22}\,r_1} \qquad \Longrightarrow \qquad r_3 = 0.128 \;\equiv\; -17.9\,dB \tag{10.32}$$

To b) *Using Eq. (10.17) you get:*

$$b_{Q5} = b_{Q3} \frac{S_{54}}{1 - r_3 S_{44}} = 13.3\sqrt{W} \quad \Longrightarrow \quad P_5^{Sy} = 88.3\,W \equiv 49.46\,dBm \quad (10.33)$$

Eq. (10.18) results:

$$r_5 = S_{55} + \frac{S_{45}\, S_{54}\, r_3}{1 - S_{44}\, r_3} \quad \Longrightarrow \quad r_5 = 0.23 \equiv -12.8\,dB \quad (10.34)$$

The output matching is therefore significantly worse than the very good matching of the individual components of $0.1 \equiv -20\,dB$ *in practice.*

To c) *Eq. (10.19) leads to:*

$$P_6^{abs} = \frac{1}{2}\, |b_{Q5}|^2\, \frac{1 - |r_V|^2}{|1 - r_5\, r_V|^2} \quad \Longrightarrow \quad P_6^{abs} = 91.7\,W \equiv 49.62\,dBm \quad (10.35)$$

Eq. (10.26) can be adjusted:

$$\frac{a_B}{\mathrm{dB}} = 10\,\mathrm{lg}\left(\frac{P^{abs}_{2\,(r_2=r_1^*)}}{P_6^{abs}}\right) \quad \Longrightarrow \quad a_B = -39.21\,dB \quad (10.36)$$

with the maximum available power via Eq. (10.9)

$$P^{abs}_{2\,(r_2=r_1^*)} = P^{abs}_{2\,max} = \frac{1}{2}\, |b_{Q1}|^2\, \frac{1}{1 - |r_1|^2} \quad \Longrightarrow \quad P^{abs}_{2\,max} = 11.0\,mW \quad (10.37)$$

The transducer loss is negative because the amplifier raises the level by around $100 \equiv 40\,dB$.

To d) *With ideally matched components, the levels can simply be added up in logarithmic form:*

$$P_6^{abs} = P_1^{Sy} + S_{32}^{dB} + S_{54}^{dB} \Rightarrow P_6^{abs} = 10\,dBm - 0.915\,dB + 40\,dB \Rightarrow P_6^{abs} = 49.09\,dBm \quad (10.38)$$

It can be seen that this simple type of ideal calculation already provides good approximate results.

——— *Calculation Transmission Path* ———

10.2 Signal Flow Method and Diagrams

Under d) in the previous example calculation of a transmission line, it was shown how easily this task can be solved if ideal adjustments are available. In this section, this approach will be extended to more complex problems by using the signal flow diagrams or method based on approximations.

In its original and very complex form, the signal flow method can be applied exactly to all linear problems, [86, 48]. It then replaces the transmission parameters, for example, and is also very attractive for analytical multi-port problems. However, the latter are often solved in practice with circuit simulators, so that the developer is often only required to estimate whether a system approach will work. If a complex system network is simplified by a clever choice of adapted gates, the general signal flow method becomes the **signal tracking method** presented here.

Signal flow diagrams are still very popular today, especially for illustrating more complex problems. Chap. 6 contains an example of the introduction of the M-parameters. Signal flow diagrams consist of nodes (sources and sinks) and directed connecting lines (branches). Three simple rules can be established:

1. Each independent variable a_i (incident wave i) and each dependent variable b_j (outgoing wave j) is represented by a node.

2. Each scattering parameter corresponds to a directed connecting line (branch) between a node a_i (source) and a node b_j (sink). The direction is fixed from a_i to b_j. The indexing of the scattering parameter is given by the indices of the associated nodes.

   Example:

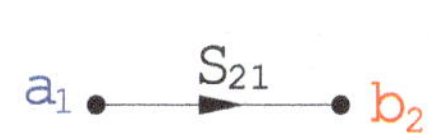

Fig. 10.6: Signal flow graph consisting of source, branch and sink

$$b_2 = S_{21}\,a_1 \tag{10.39}$$

3. Each node corresponds to the sum of the connecting lines entering it, whereby the associated S-parameter for each connecting line must be multiplied by the value of the output node.

   Example:

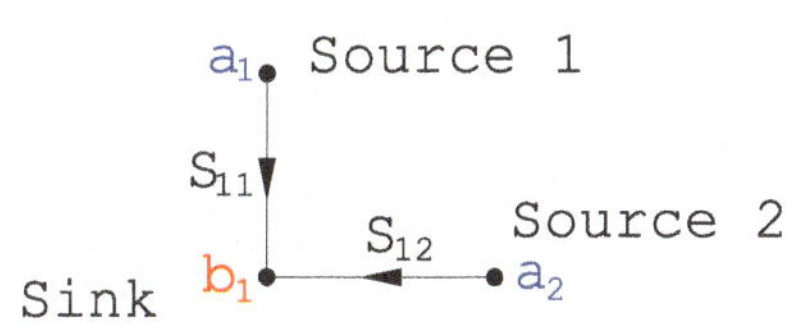

Fig. 10.7: Signal flow graph consisting of two sources and two branches and one sink

$$b_1 = S_{11}\,a_1 + S_{12}\,a_2 \tag{10.40}$$

10.2.1 Components for Signal Flow Diagrams

- Representation of a general two-port (Fig. 10.8):

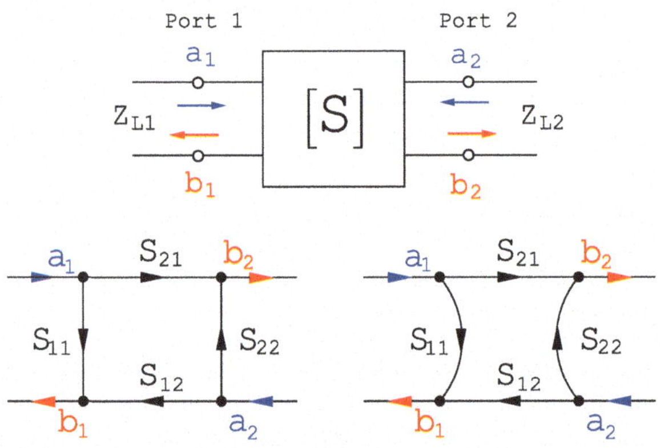

Fig. 10.8: Signal flow diagram of a general two-port in two display modes

Especially in more complex signal flow diagrams, the reflection branches are often represented in arcs. Rule 3) directly results in the familiar S-parameter equations.

$$\begin{aligned} b_1 &= S_{11}\, a_1 + S_{12}\, a_2 \\ b_2 &= S_{21}\, a_1 + S_{22}\, a_2 \end{aligned} \tag{10.41}$$

- Illustration of a two-port matched on both sides (Fig. 10.9):

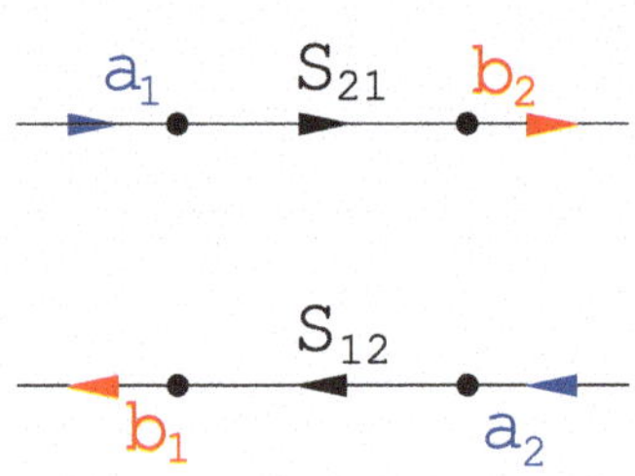

Fig. 10.9: Signal flow diagram of a two-port matched on the input and output side

For practitioners, the use of signal flow diagrams is very interesting when the components are matched. In the usual representation of a two-port, there is no difference between a matched and a mismatched two-port. In the signal flow diagram, however, the vertical branches are omitted. The circuit becomes clearer.

$$b_1 = S_{12}\, a_2 \qquad \text{and} \qquad b_2 = S_{21}\, a_1 \tag{10.42}$$

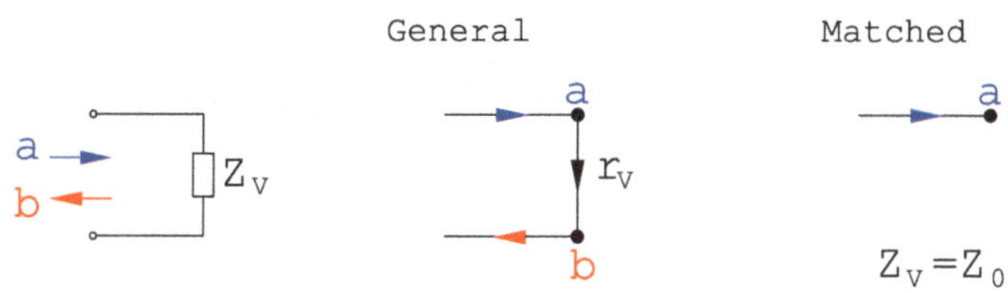

Fig. 10.10: Signal flow diagram of a general single-port load and a matched load

- Representation of a one-port load (Fig. 10.10):

If the load is matched, the vertical connecting line is also omitted here.

The following applies to the general load:

$$b = r_V \, a \qquad \text{with} \qquad r_V = \frac{Z_V - Z_0}{Z_V + Z_0} \quad . \tag{10.43}$$

- Representation of a signal source or a generator (Fig. 10.11):

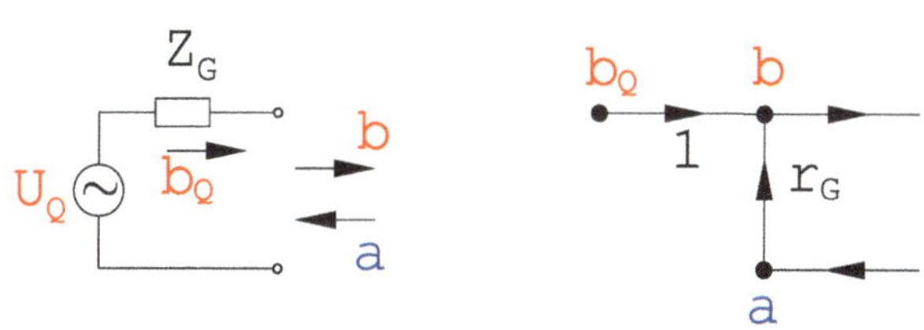

Fig. 10.11: Signal flow diagram of a signal generator

$$b = b_Q + r_g \, a \qquad \text{with} \tag{10.44}$$

$$b_Q = \frac{U_Q \sqrt{Z_0}}{Z_0 + Z_G} \qquad \text{and} \qquad r_G = \frac{Z_G - Z_0}{Z_G + Z_0} \quad .$$

For the design of RF systems, signal generators are usually used that generate a mono-frequency and therefore sinusoidal signal and may only have a finite input matching. In practice, these frequency-stable signal sources can be realized in very good approximation by means of so-called synthesis generators (e.g. [125]).

- Representation of a series impedance (Fig. 10.12):

$$\begin{aligned} b_1 &= r_Z \, a_1 + (1 - r_Z) \, a_2 \\ b_2 &= (1 - r_Z) \, a_1 + r_Z \, a_2 \\ \text{with} \quad r_Z &= \frac{(Z + Z_0) - Z_0}{(Z + Z_0) + Z_0} = \frac{Z}{2 Z_0 + Z} \quad . \end{aligned} \tag{10.45}$$

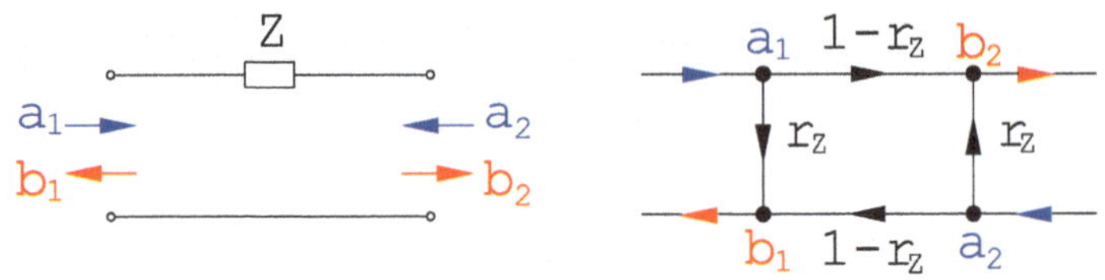

Fig. 10.12: Signal flow diagram of a series impedance

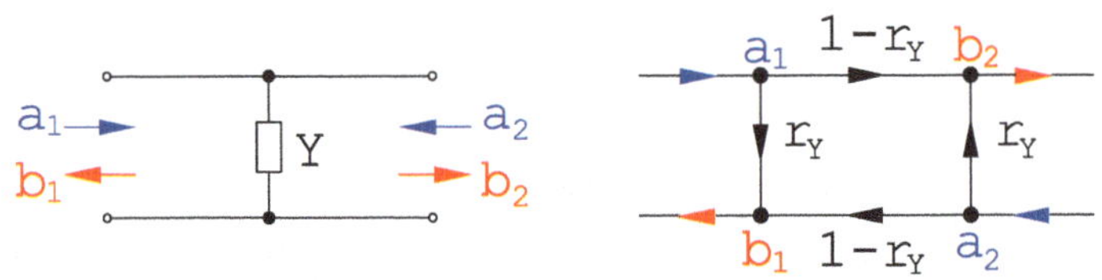

Fig. 10.13: Signal flow diagram of a parallel admittance

- Representation of a parallel admittance (Fig. 10.13):

$$\begin{aligned} b_1 &= r_Y\, a_1 + (1-r_Y)\, a_2 \\ b_2 &= (1-r_Y)\, a_1 + r_Y\, a_2 \\ \text{with} \quad r_Y &= \frac{Y_0-(Y_0+Y)}{Y_0+(Y_0+Y)} = \frac{-Y}{2\,Y_0+Y} \quad . \end{aligned} \tag{10.46}$$

The components presented can be used in the general signal flow method as well as in the signal tracking method described below.

10.2.2 Example Calculation: Signal Tracking Method

Now that a large number of components and their associated signal flow diagrams have been presented, the simple application of the signal tracking method will be illustrated using an example.

With the signal tracking method, a circuit is approximated in such a way that no multiple reflections occur. No closed loops should occur! A small example of calculating the reflection and transmission properties of a transmission chain will serve as an illustration.

The arrows in the signal flow diagram can be regarded as so-called isolators: In the direction of the arrow, the input signal is attenuated or amplified (mathematically multiplied) by the adjacent scattering parameter value and in the opposite direction of the arrow, nothing happens at all.

The example in Fig. 10.14 is intended to illustrate the structure of a transmission chain of a source and an amplifier, which are manufactured on a semiconductor and connected to a line and a load via a bonding wire. In this example, the approximation applies that the source, the amplifier, the line and the consumer are perfectly matched.

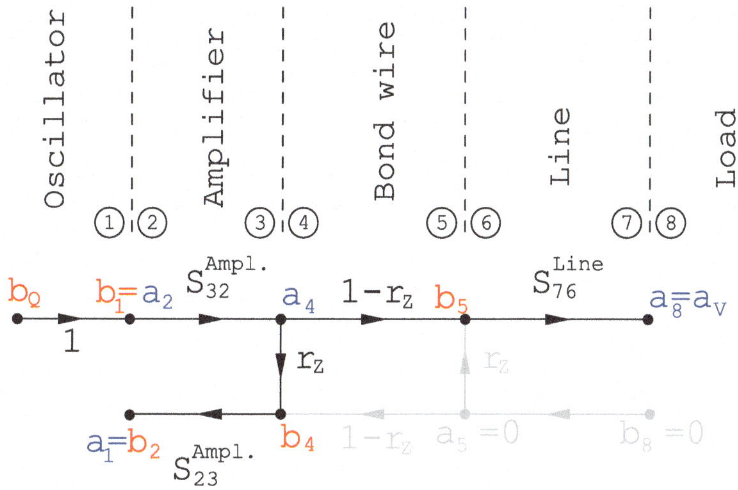

Fig. 10.14: Signal flow diagram of a transmission chain

All scattering parameters shown in the drawing are given. We are looking for the input reflection factor r_{on} as seen from the source and the transmission factor S_{V1} of the entire chain. Both can be specified directly from the Fig. 10.14. The signal flow is followed in the reverse direction:

$$r_{ein} = S_{22} = \frac{a_1}{b_1} = S_{23}^{Ampl}\, r_Z\, S_{32}^{Ampl} \quad , \tag{10.47}$$

$$S_{71} = \frac{a_V}{b_1} = S_{76}^{Lgt}\, (1 - r_Z)\, S_{32}^{Ampl} \quad . \tag{10.48}$$

The power P_i emerging from a network is defined by the relationship

$$P_i = \frac{|b_i|^2}{2} \tag{10.49}$$

and is linked to the wave b_i. The power P_V absorbed at the load as a function of the transmission power P_1 and the scattering parameters is often of interest. This can also be specified directly using the signal tracking method:

$$P_V = \left[\left|S_{76}^{Lgt}\right|\, |(1 - r_Z)|\, \left|S_{32}^{Ampl}\right|\right]^2 P_1 \quad , \tag{10.50}$$

respectively

$$P_V = \left|S_{76}^{Lgt}\right|^2 |(1 - r_Z)|^2 \left|S_{32}^{Ampl}\right|^2 P_1 \tag{10.51}$$

and simplified with logarithmic values:

$$P_V^{dBm} = P_1^{dBm} + S_{76}^{Lgt\,dB} + (1 - r_Z)^{dB} + S_{32}^{Ampl\,dB} \quad . \tag{10.52}$$

A closer look at the Fig. 10.14 shows that the lower branch of the line does not need to be drawn, as the load does not reflect any power and therefore b_V is zero. Further and considerably more complex examples are presented at the end of the sections 10.4 and 10.5 in the form of exercises.

10.3 Important System Components

The aim of this subchapter is to support RF engineers in developing systems using state-of-the-art RF components in a wide variety of environments, especially planar ones.

However, a large number of RF components can only be developed and produced by specialist companies (e.g. a semiconductor company). These components, which are often required in systems but are "only" purchased by most developers, will be briefly presented below. The important aspects for system integration will be highlighted. A pure S-parameter description is sufficient for many components.

The techniques for implementing these components optimally at their RF ports by adapting them in a system were described in detail using the Γ transformer and in particular the use of the Smith Chart tool.

10.3.1 Non-Reciprocal Passive Components

Non-reciprocal media are required to realize passive non-reciprocal RF components. These media are marked with a direction-dependent material tensor:

$$\overset{\leftrightarrow}{\varepsilon} \neq \overset{\leftrightarrow}{\varepsilon}^T \quad , \quad \overset{\leftrightarrow}{\mu} \neq \overset{\leftrightarrow}{\mu}^T \quad , \quad \overset{\leftrightarrow}{\kappa} \neq \overset{\leftrightarrow}{\kappa}^T \quad . \tag{10.53}$$

These materials are called: *Gyrotope media* .

Examples of gyrotropic media are:

- Pre-magnetized ferrites $\overset{\leftrightarrow}{\mu}$
- Pre-magnetized plasmas $\overset{\leftrightarrow}{\varepsilon}$
- Pre-magnetized semiconductors $\overset{\leftrightarrow}{\kappa}$

In RF technology, components with pre-magnetized ferrites are often used in measurement technology and in front-end circuits. These are the circulator and the isolator.

<u>The Circulator</u>

The symbol of a circulator is shown in Fig. 10.15.

The scattering parameters of an ideal circulator

$$[\mathbf{S_{Circ}}] = \begin{pmatrix} 0 & 0 & |1| \\ |1| & 0 & 0 \\ 0 & |1| & 0 \end{pmatrix} \tag{10.54}$$

only have the absolute values zero or one. In practice, circulators with losses between 0.5 and 3 dB and isolations between 10 and 35 dB are commercially available as housed components or as SMD components.

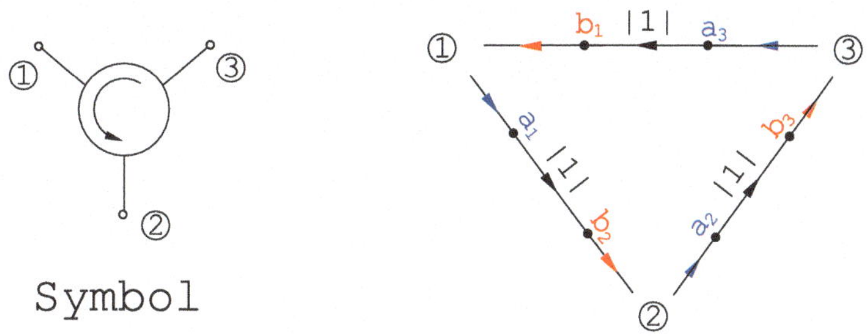

Fig. 10.15: Left: Symbol of a circulator; right: signal flow diagram of a lossless circulator with perfect matching and isolation

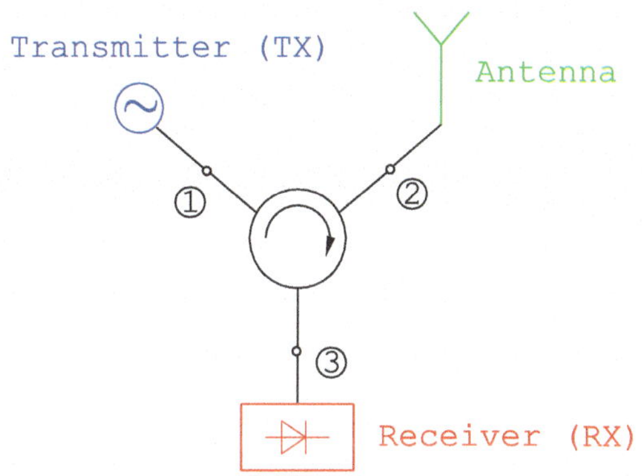

Fig. 10.16: Transceiver system with only one antenna and continuous transmit/receive operation in the same frequency range

The example of a transceiver application in Fig. 10.16 shows why the phase values do not play a major role in a circulator. It also shows that simultaneous transmit and receive operation is possible using only one antenna with a circulator.

The large signal from the transmitter is coupled to the antenna without being heavily attenuated and without causing major interference to the receiver. At the same time, the small received signals detected by the antenna arrive at the receiver with low loss. In practice, however, the cross-talk of the transmit signal is so great that the receive band is in a different frequency band. At least the circulator ensures that the receiver is not driven into compression.

Fig. 10.17 shows a top view of a microstrip line to explain how a circulator works. Between the metallization layer shown above and the ground layer, a pre-magnetized ferrite is located centrally under the metal surface, which generates the continuous field $\vec{H}_0$.

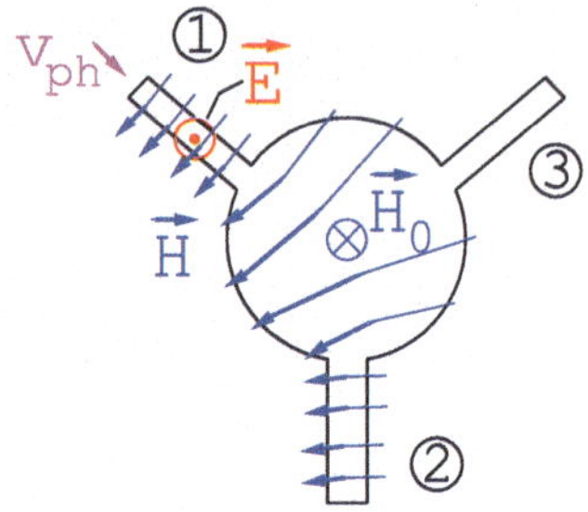

Fig. 10.17: Representation of the orientation of the vectorial $\vec{E}$ and $\vec{H}$ fields in a microstrip line circulator

An electromagnetic wave incident at port 1 with the phase velocity v_{ph} and an electric field (E-field), which points out as an instantaneous absorption, as well as an H-field that propagates parallel to the surface, encounters the permanent field pre-magnetized by $\vec{H}_0$, which is permanently directed into the plane. $\vec{H}_0$ changes the direction of the alternating field $\vec{H}$, which is now perpendicular to port 3 in the center of the circulator. As the line of port 3 is no longer enclosed by the magnetic field lines, no energy is coupled into port 3. All the energy is transmitted to port 2.

The Isolator

Fig. 10.18 shows the symbol of an isolator.

Fig. 10.18: Left: Symbol of an isolator; right: signal flow diagram of a lossless isolator with perfect matching and backward isolation

An ideal isolator is matched at each port and transmits a signal only in the forward direction, as shown by the S-parameters of the matrix:

$$[\mathbf{S_{Iso}}] = \begin{pmatrix} 0 & 0 \\ |1| & 0 \end{pmatrix} \quad . \tag{10.55}$$

Isolators are used between the transmitting amplifier and the antenna, particularly in communication systems with relatively high linearity requirements such as LTE and WLAN, so that antenna mismatches that occur due to changing environmental conditions do not distort the operating point of the amplifier.

Isolators are widely used in RF measurement technology. In most applications, the aim is to improve the matching of the individual measuring branches so that they can be approximated as ideally matched. The reflected waves should have no influence on the power output. This is ensured by the use of an isolator.

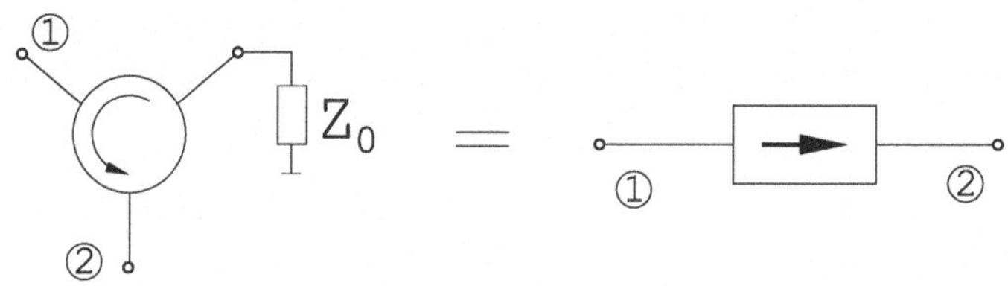

Fig. 10.19: Realization of an isolator

An isolator consists of a circulator that is terminated at a port with a broadband wave sump (possibly 50 Ω-SMD resistor as shown in Fig. 10.19).

In modern highly integrated circuits, it is not desirable to use ferrites due to the different technology, costs, weight, losses and size. In the simplest case, an isolator can also be replaced by an amplifier and possibly an attenuator. However, this only works for simple systems.

However, when considering a communication system such as 6G, the power amplifier must meet the highest requirements in terms of linearity. An isolator is used to ensure that the linear transmission behavior of the amplifier is not disturbed by mismatching of the antenna. A simpler solution is to use the ±90° coupler, which is described on page 184 and also in detail in [61]. Only 3 SMD components and 2 symmetrical antennas are required here.

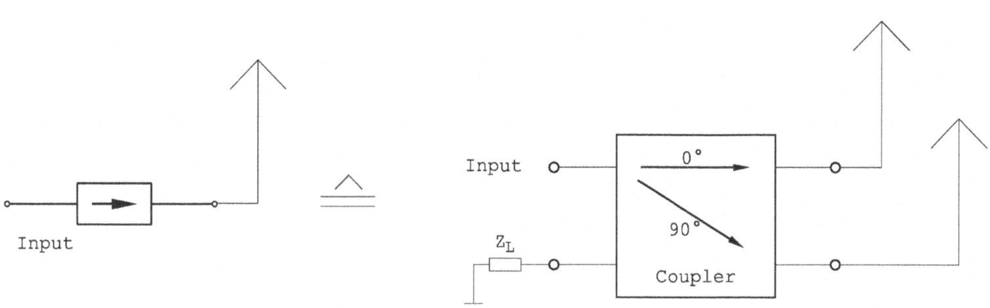

Fig. 10.20: Front-end circuit with 90° coupler, which in the simplest case can be realized by a coil and an inductor

The mode of operation is similar to that of a so-called *balanced amplifier*: If there is a reflector near the two identical antennas, the reflected energy is reflected once with 0° and once with 180° phase rotation to the input (output of the power amplifier).

Both reflected waves cancel each other out and the power amplifier is very well matched at the output. On the other hand, the reflected energy reaches the resistor with a 90° phase rotation, overlaps there constructively and is absorbed there. The good quality of such a circuit is published in detail in [104].

10.3.2 Detectors

A detector converts the current power of a RF signal into a low-frequency (often DC) voltage signal U_{det}. It is therefore to be regarded as a termination for the RF input signal and as a voltage source for the low-frequency output signal. This is symbolically represented with the possibly non-linear conversion factor α in Fig. 10.21.

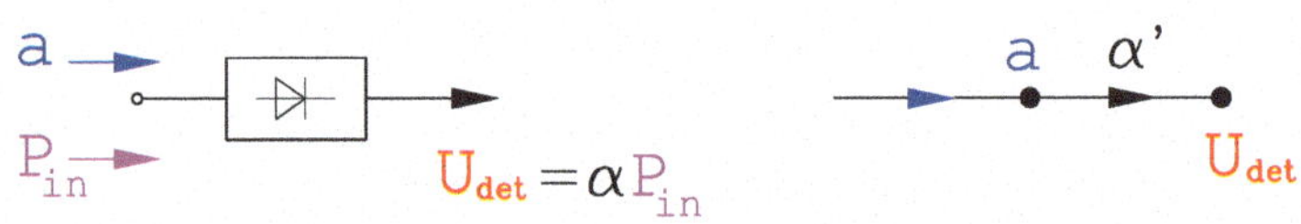

Fig. 10.21: Symbol and signal flow diagram of a detector

At the beginning of the pioneering days of radio, detectors were connected directly to the radio receiving antenna as simple rectifiers. With the simple design: Antenna, tunable LC circuit, detector and headphones made it possible to realize the first radio receivers that did not require a power supply and merely rectified the amplitude (voice) modulated RF signal.

Detectors from the low-frequency range through to optics are now used as receivers. They are used to detect signals with analog and discrete amplitude modulation (AM) in transmission technology as well as to measure unmodulated RF signals in measurement technology.

In optical signal transmission for fiber optic technology, the laser light is subjected to discrete on/off modulation on/off modulation, but this is carried out at switching frequencies well into the GHz range. In this case, the output of the optical detectors is a microwave signal.

Otherwise, a low-frequency signal in the kHz or lower MHz range is usually present at the output, so that one speaks of a video signal. In measurement technology, the output signal is often just a DC voltage.

In the RF case, a component with a non-linear characteristic curve is required for power detection. The most sensitive components in semiconductor electronics are Schottky diodes. However, transistors are also used in semiconductor electronics for integrated solutions.

If you need highly sensitive detectors for narrow frequency ranges, a discrete solution with Schottky diodes is interesting, [2]. These discrete circuits generally work in the area of quadratic dependence between input power and output voltage.

However, in many cases the sensitivity of integrated broadband detector circuits, which may amplify the video signal logarithmically, is sufficient so that there is a linear relationship between the RF power and the video signal over a wide power range, [99].

In addition to the sensor, modern sensor electronics also have a processor and a small memory. A sensor for more precise measuring tasks should be calibratable. Only a good calibration measurement gives the sensor a good measurement quality.

The following criteria are used to assess a detector circuit:

- Sensitivity in the form of the minimum detectable energy at a given signal bandwidth,
- Sensitivity in the form of the slope or the correlation between input power and detector voltage,
- Linearity range and linearity error,
- Compression and destruction performance,
- Input matching,
- For precise measurement tasks, the sensor must be calibratable.

For the further considerations of systems, we assume that the detectors are matched on the input side (as shown in Fig. (10.21)). Furthermore, the detectors should be time-invariant, which is also the case in practice, especially with integrated products and low temperature variation.

10.3.3 Mixer

An RF mixer converts either a low-frequency signal into an RF signal or vice versa in a highly linear manner. The former is called upward mixing and the latter downward mixing.

For this purpose, a relatively large sinusoidal RF signal, known as a local oscillator signal or pump signal, is used to change the operating point of a non-linear component significantly on a non-linear characteristic curve. The representation of a downward mixer with indication of the three signal powers

P_s: RF signal,

P_{LO}: local oscillator signal and

P_{IF}: Intermediate frequency or low frequency signal

can be seen in Fig. 10.22.

Due to the high linearity, low conversion losses, the preservation of complex information and the very high sensitivity, mixers have prevailed over detectors in many applications.

For example, the technically much more complex mixer concepts are used for radio transmission in all consumer products such as radios, televisions and cell phones.

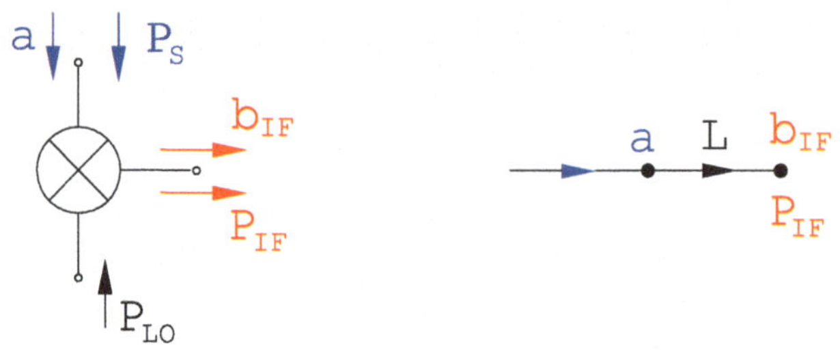

Fig. 10.22: Symbol and signal flow diagram of an RF mixer with the conversion loss L and the power directions for downward mixing

However, the highly linear frequency conversion of a mixer is not unique. There are two signals around the LO signal that can be converted, Fig. 10.23. A distinction is made here between the two operating modes DC and reverse.

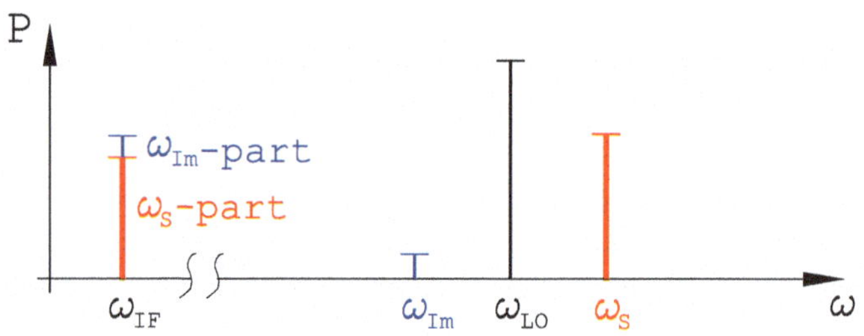

Fig. 10.23: Frequency diagram for a downward mixer in equal site

In the equal site applies to the signal to be implemented:

$$\omega_{IF_{part1}} = \omega_S - \omega_{LO} \quad . \tag{10.56}$$

However, in addition to the equal site signals, signals from the so-called image site ω_{Im} are also converted. The following applies to the conversion of this so-called image frequency:

$$\omega_{IF_{part2}} = \omega_{LO} - \omega_{Im} \quad . \tag{10.57}$$

The signals from IF_{part2} distort the actual useful signal from IF_{part1}.

Since the receiver cannot separate the two signal components in the intermediate frequency range (e.g. an A/D converter band), it must be ensured in the RF range around the local oscillator frequency that this image frequency is suppressed.

This can be achieved using filters or so-called single-sideband converters, which are discussed in the following section.

If the frequency range below the LO frequency is to be converted, this is referred to as the image site.

For the linear description of a mixer with perfect image frequency suppression, the Y-matrix is used for historical reasons with the conductance values G_0 and G_{LO}.

$$\begin{aligned} I_S &= G_0\, U_s + G_{L0}\, U_{IF} \\ I_{ZF} &= G_{LO}^*\, U_s + G_0\, U_{IF} \end{aligned} \tag{10.58}$$

Since the conductance G_{LO} is almost real, a mixer is reciprocal, which applies without restriction, especially for power considerations.

Although the currents and voltages have three different frequencies, a mixer can be treated like any other linear circuit.

In practice, the LO signal does not appear for the system design of the signal branch, so that ultimately the RF transmission is taken into account as simply as the signal flow diagram in Fig. 10.22.

The mixer behaves like an attenuator with the conversion losses L.

However, the simplest mixers, based on a Schottky diode, are only used in the microwave range (MW range). Commercially available RF semiconductor mixers are in practice highly integrated circuits based on a variety of concepts, [50, 92], and sometimes include LO amplifiers.

The simplest mixers generate a large number of additional frequencies due to the nonlinear behavior of the components. With more complex mixer architectures, these additional mixing products only occur in negligible quantities. Intelligent phasing suppresses the unwanted mixing products. These complex architectures also help to suppress amplitude fluctuations in the LO signal.

Such mixers are often referred to as single, dual or quad balanced mixers. The most popular transistor mixer is the Gilbert mixer. This mixer, like the double-balanced ring-diode mixers, suppresses the local oscillator signal at the signal port very effectively.

The noise figure of a mixer F_{mixer} is approximately the same as the conversion loss L:

$$F_{Mischer} \approx L \quad . \tag{10.59}$$

A mixer is described by the criteria:

- Conversion loss and noise figure,
- Required LO level and frequency range,
- Isolations between the three ports,
- Linearity through compression point and/or IM3 specifications,
- Matching (especially the LO and signal port).

Each mixer can be considered as a multiplier in the time domain for the system analysis. For upward and downward mixing, the simple addition theorem of a multiplication applies to the mixer:

$$\cos(\omega_{LO}t)\,\cos(\omega_{IF}t) = \frac{1}{2}\,[\cos(\omega_{LO}t - \omega_{IF}t) - \cos(\omega_{LO}t + \omega_{IF}t)] \quad . \tag{10.60}$$

10.3.4 Single Sideband Converters and IQ Modulators

In the mixer section, the problem of the image frequency was discussed in detail and it was pointed out that a single-sideband converter or offset converter does not have this problem.

Single sideband converters and the closely related IQ modulators (also known as vector modulators) are based on interconnected mixers, which are regarded as multipliers. The structure of a single sideband shifter with two mixers and passive components, all of which have already been presented in detail, is shown in the following Fig. 10.24.

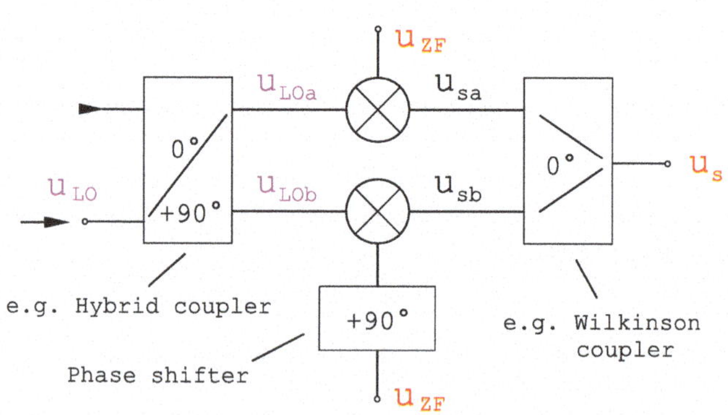

Fig. 10.24: Construction of a single-sided band converter in the image position

The local oscillator signal

$$u_{LO} = \hat{u} \cos(\omega_{LO} t) \tag{10.61}$$

is divided into u_{LOa} and u_{LOb} by means of a 0°/90° coupler and shifted in phase by 90°s in the lower branch. This applies to the two signals reduced by 3 dB:

$$u_{LOa} = \frac{1}{\sqrt{2}} \hat{u} \cos(\omega_{LO} t) \quad \text{and} \quad u_{LOb} = -\frac{1}{\sqrt{2}} \hat{u} \sin(\omega_{LO} t) \quad . \tag{10.62}$$

Assuming the transmission case with an IF transmission signal of the simple form

$$u_{IF} = \hat{u}_{IF} \cos(\omega_{IF} t) \tag{10.63}$$

and the conversion losses L, the multiplication results in the two partially phase-shifted signals with the voltages:

$$u_{sa} = \frac{\hat{u}_{IF}\,\hat{u}}{\sqrt{2}} L \cos(\omega_{LO} t) \cos(\omega_{IF} t) \quad \text{and} \quad u_{sb} = \frac{\hat{u}_{IF}\,\hat{u}}{\sqrt{2}} L \sin(\omega_{LO} t) \sin(\omega_{IF} t) \quad . \tag{10.64}$$

Addition by means of the 0° coupler using the addition theorems results in

$$\cos(\alpha - \beta) = \cos(\alpha) \cos(\beta) + \sin(\alpha) \sin(\beta) \tag{10.65}$$

for transmit signal holds

$$u_s = \sqrt{2}\, \hat{u}_{IF}\, \hat{u}\, L \cos(\omega_{LO} t - \omega_{ZF} t) \quad . \tag{10.66}$$

The amplitude and phase of the transmitted signal is directly proportional to the IF signal. The upper side-band was extinguished by the two 90° phase rotations. In practice, the suppression of the upper side-band is typically 40 dBc.

A similar calculation can also be performed for the single side-band converter in the case of receiving.

The only difference between an IQ modulator and a single-side-band converter is the lack of a phase shifter in the IF branch, Fig. 10.25.

However, the fact that the appearance of this IQ modulator looks very different is partly due to the fact that IQ modulators are manufactured entirely using semiconductor technology. This is why there are no large distributed reciprocal couplers in these circuits and why the IQ transmitter modulator must be distinguished from the IQ receiver when handling them.

The transmitter and receiver units, as well as the oscillator technology described in the next section, are now integrated on a single so-called transceiver IC (consisting of transmitter and receiver).

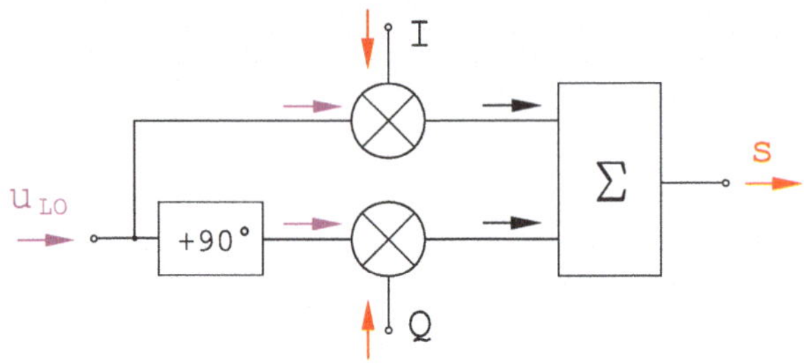

Fig. 10.25: Construction of an IQ modulator for the transmission case in the equal site

The mathematical description of the IQ modulator is similar to that of the single sideband transmitter. Modern data transmission and communication systems such as WLAN and 6G work with simultaneous amplitude ($A_{(t)}$) and phase modulation ($\phi_{(t)}$), which, however, can only assume discrete values. The modulated and up-mixed signal is to be transmitted as the transmission signal ($s_{(t)}$):

$$s_{(t)} = A_{(t)} \cos\left(\omega_{LO} t - \phi_{(t)}\right) \quad . \tag{10.67}$$

If the addition theorems are again applied to the equation (10.67), the transmitted signal can be divided into a component $I_{(t)}$, which is in phase with the LO signal, and a component $Q_{(t)}$, which is rotated in phase with the LO signal by exactly 90°:

$$s_{(t)} = I_{(t)} \cos\left(\omega_{LO} t\right) - Q_{(t)} \sin\left(\omega_{LO} t\right) \qquad \text{with} \tag{10.68}$$

$$I_{(t)} = A_{(t)} \cos\left(\phi_{(t)}\right) \qquad \text{and} \qquad Q_{(t)} = A_{(t)} \sin\left(\phi_{(t)}\right) \quad . \tag{10.69}$$

This generates the missing phase shifter in the IF branch in the computer. The fact that the Q signal is shifted by -90° to the I signal results in a conversion in the equal site.

10.3.5 Oszillators and Synthesis Generators

Anyone who has tried to build an RF amplifier without a stability calculation (k-factor) knows how easy it is to make an oscillator. Unfortunately, such oscillators do not oscillate at the desired frequency and the other properties are also inadequate.

A good oscillator consists of an amplifying unit and a resonator with the highest possible quality. If the frequency of this resonator can be varied via at least one varactor diode, it is referred to as a voltage-controlled oscillator, for short VCO.

An oscillator exhibits fluctuations in the output amplitude over time (AM noise) on the one hand and (in practice even more serious) fluctuations in the phase around the target frequency on the other. This phase noise is usually displayed for the upper sideband with the difference between the noise value and the maximum oscillator signal (which is not shown), as can be seen in Fig. 10.26.

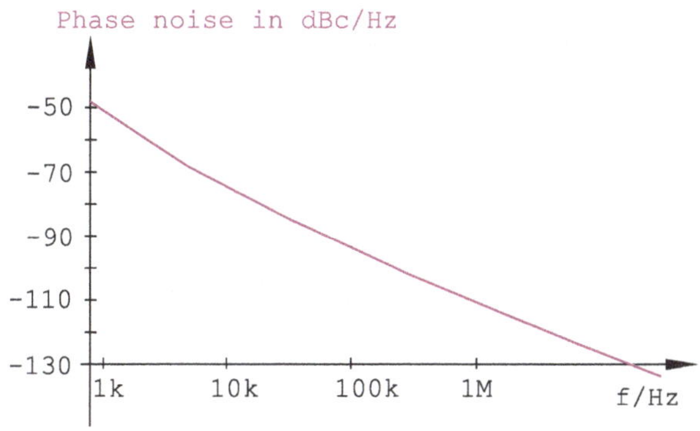

Fig. 10.26: Typical phase noise of an unstabilized oscillator with LC resonator

Of the large number of parameters to be considered when using an oscillator, this phase noise often has the greatest influence on the system characteristics. For example, if you want to mix down a signal at a distance of 1 MHz from the LO signal and consider the mixer as a multiplier, the mixer also mixes down the phase noise signal of the LO oscillator. Consequently, the phase noise should be below the expected signal level at the receiver (see section 8.6.4).

When assessing and using an oscillator, there are a number of points to consider in addition to phase noise:

- Output power and its frequency response at the VCO,
- Power consumption,
- Generation of harmonics with a typical distance of 20 dBc from the fundamental and, if necessary the subharmonics (half the frequency of the fundamental),
- *Frequency Pushing*: A variable that characterizes how much the frequency changes with fluctuations in the supply voltage,
- *Frequency Pulling*: Indicates how much the frequency changes with a mismatch occurs at the output,
- Temperature response or drift of the output power and frequency,
- Tuning characteristics in MHz/V of a VCO and their linearity and maximum changeability (speed).

A detailed, modern and very practical description of oscillators can be found in [50, 113].

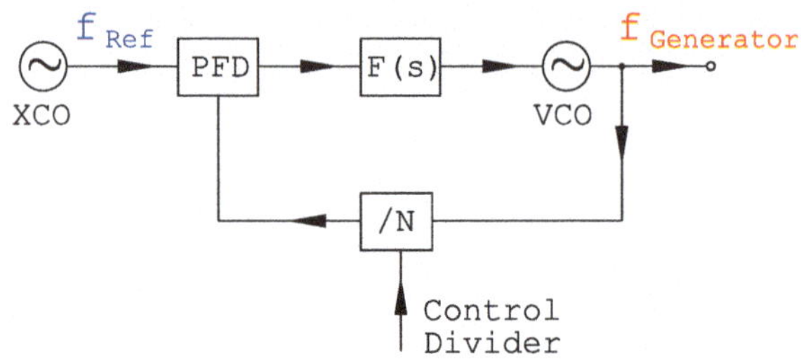

Fig. 10.27: Block diagram of a single-loop synthesis generator

Many of these properties are significantly improved if an RF VCO is stabilized to the very good properties of a crystal oscillator (XCO) using a phase-locked loop (PLL). The entire circuit is called a step or synthesis generator or synthesizer for short. The simplest implementation of a synthesis generator is shown in Fig. 10.27.

The output signal of the RF VCO is divided to a lower frequency by means of an RF divider (/N) with switchable divider values and is then applied to the phase frequency discriminator (PFD).

The signal from the quartz reference oscillator (XCO), which is very low-phase noise and temperature-stable, is fed to a second input of the phase-frequency discriminator. This PFD provides an output signal if the two input signals are not in phase. This output signal is smoothed using the filter F(s) and fed to the control input of the RF VCO. The frequency $f_{Generator}$ of the voltage-controlled oscillator is determined via the frequency of the reference oscillation f_{Ref} (reference frequency) and the value N of the frequency divider as follows:

$$f_{Generator} = f_{Ref} \cdot N \quad . \tag{10.70}$$

Changes in the integer divider value N cause a change in the output frequency in the increment f_{Ref}. In order to achieve a small step width, the reference or comparison frequency must be kept low.

This leads to a poorer phase noise of the PLL, because the noise voltage of the phase comparator and the comparison frequency is amplified by the factor N and becomes visible at the VCO output.

Although residual components of the reference frequency and their harmonics that occur at the phase comparator output are attenuated by the subsequent loop filter, they still influence the VCO and appear as side lines of the useful signal at a distance of f_{Ref} and their multiples.

A larger ratio of comparison frequency to loop bandwidth improves the suppression of these side lines. Since the loop bandwidth is related to the settling time, this classic PLL forces some compromises in terms of sideline rejection, frequency step size, phase noise and settling time.

The fractional-N synthesizer offers a way out of these compromises.

Functional Principle of a Fractional-N Synthesizers

With a divider that is not limited to integer divider ratios, but also allows fractional values, the step size of the synthesizer can be made smaller than the reference frequency [36, 114]. The calculation of f_{VCO} according to Eq. (10.70) remains valid. However, N is then no longer restricted to integer values.

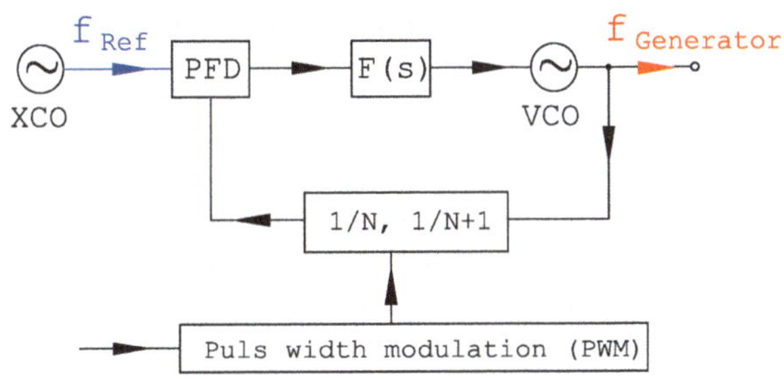

Fig. 10.28: Block diagram of a fractional-N synthesizer

A fractional divider ratio is achieved by switching between two or more integer values (a pulse width modulation of the divider ratio, so to speak). The fractional-N synthesizer is illustrated in Fig. 10.28.

When switching between two divider values N_1 and N_2 with the pulse durations T_1 and T_2 the mean value of the divider ratio is then

$$\bar{N} = \frac{N_1 \cdot T_1 + N_2 \cdot T_2}{T_1 + T_2} \quad . \tag{10.71}$$

The frequency of the VCO should then be at the reference frequency multiplied by the average value of the divider ratio. Switching the divider ratio causes interference phase modulation on the phase comparator. The smallest time unit in which T_1 and T_2 can vary is T_{pulse}. The minimum frequency of the interfering phase modulation (IPM) is therefore dependent on f_{pulse} and the required smallest adjustment step size N of:

$$f_{IPM} = f_{pulse} \cdot \Delta N \quad . \tag{10.72}$$

The minimum frequency of the phase disturbance caused by the divider switching thus corresponds to the smallest step size f_s of the frequency synthesizer. In terms of interference, the fractional-N method differs only slightly from the classic synthesizer with a fixed divider.

If you try to suppress the interference using a narrow loop filter, you have to make the same compromises as in the previous section. The only advantage would then be the lower multiplication factor of the phase comparator and reference noise.

Since the course of the disturbance can be calculated in advance, it can be compensated for by adding an antiphase voltage. The loop filter bandwidth can then be selected to be greater than the smallest frequency step size f_s.

This enables direct digital angle modulation by digitally reproducing the modulation signal. The small step width with a short settling time also allows, for example, predictable Doppler shifts to be precisely readjusted.

Compensation of the phase interference by the divider switching requires a good DA converter to generate the compensation voltage in addition to the calculation logic.

An alternative recently used in the RF sector is the $\Sigma\Delta$ process, which has already become established in CD audio technology for digital-to-analog and analog-to-digital conversion. Due to its ease of integration, it already plays an important role in RF technology as a digital frequency converter.

$\Sigma\Delta$-Fractional-N Synthesis Generator

As explained above, filtering out the interference has few advantages, as the loop filter bandwidth must be made smaller than the frequency step width, [111].

The $\Sigma\Delta$ fractional N method, like the conventional fractional N method, causes side lines at the distance of the frequency step width. However, instead of the usual pulse width modulator, a $\Sigma\Delta$ modulator is used, whose spectral energy maximum is at RF frequencies.

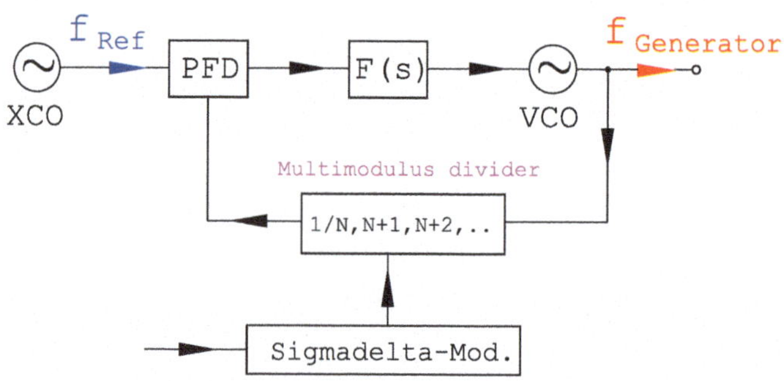

Fig. 10.29: Block diagram of a $\Sigma\Delta$ fractional-N synthesizer

The changes in the divider value of the PLL are controlled by the $\Sigma\Delta$ modulator. This provides a quasi-random data stream whose arithmetic mean value can be set digitally.

Because the energy is not evenly distributed spectrally, but lies for the most part in the higher frequency range, so little interference energy is obtained near the carrier that the analog noise of the phase comparator and the reference predominates. The loop filter can therefore be selected wider than the frequency step width without compensation measures.

$\Sigma\Delta$-Modulators can be realized in different structures. An interesting variant is the MASH modulator, as described in [94], because it does not require a feedback network and therefore has no stability problems.

Direct Digital Modulation

The small step size of a $\Sigma\Delta$ fractional-N synthesizer enables direct digital frequency and phase modulation [36].

This is an exact D/F converter. The advantages of a direct digital modulator are obvious:

- No D/A conversion of digital modulation data is necessary.
- It is possible to change the type of modulation and the modulation range via software.
- Large modulation boosts without distortion can be realized (non-linearities of the VCO characteristic curve have no influence).
- There are no longer any analog adjustment elements for setting the stroke.
- The modulation deviation is independent of component tolerances.

The $\Sigma\Delta$ synthesizer has long been used for GMSK modulation in GSM handsets on a large industrial scale.

Synthesis generators can be found in almost all RF circuits that are manufactured in large quantities.

The costs of integrated circuits are now very low. However, the development time required to implement a computer-controlled synthesizer must be taken into account.

German and English-language literature for a more detailed introduction includes [50] is recommended.

10.3.6 Antennas

Antennas are essential components in RF technology, serving as the interface between guided electromagnetic waves in a transmission line and radiated waves in free space. They play a critical role in determining the efficiency, range, and reliability of wireless communication systems.

The performance of an antenna directly affects signal strength, bandwidth, and the overall quality of data transmission.

Different applications require specifically designed antennas, from compact microstrip antennas in mobile devices to large parabolic reflectors in satellite communication. In modern RF systems, antennas must be optimized not only for frequency and gain but also for size, weight, and environmental robustness.

Emerging technologies such as 5G, IoT, and autonomous vehicles further increase the demand for advanced antenna designs with beamforming and multi-band capabilities. Ultimately, the antenna is a key factor in the success of any RF system, translating electronic signals into efficient and reliable wireless connections.

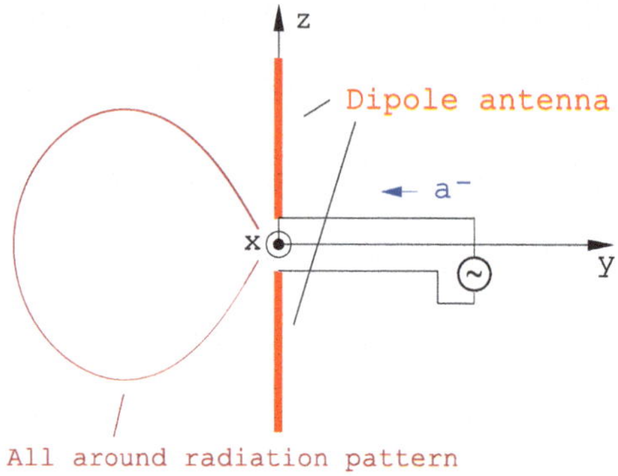

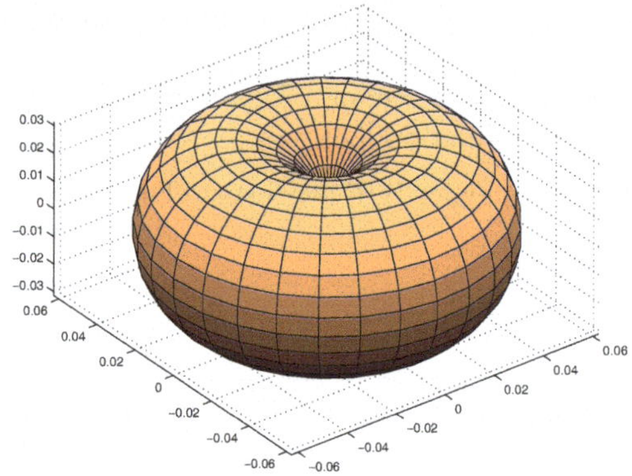

Fig. 10.30: Representation of a (symmetrical) dipole antenna along the z-axis and the associated radiation pattern (shape of a toroid) of the main lobe driven by a symmetrical source using a push-pull signal (left: 2D, right: 3D)

In practice, antennas can often be manufactured inexpensively using simple means. A very simple antenna design is the dipole antenna shown in Fig. 10.30, which is also known as a Hertzian dipole, elementary radiator and short linear antenna.

If knowledge of the radiation characteristics is not very important, antennas can be developed using a very simple "practical approach".

——— Matching of Antennas ———

1. Select the design. The dipole antenna shown or the planar patch antenna is very popular, [17, 39, 50, 80].

2. Then realize this antenna with the given approximation data from the book specifications. With a dipole, for example, it is sufficient to select the length of the antenna wires.

Maximum radiation in the main direction (perpendicular to the antenna alignment) is achieved by selecting $\lambda/2$ for the total length of the two wires. As a minimum, the wire lengths should be around $\lambda/10$.

3. The next step is to measure the input reflection coefficient of the antenna. Many antennas, including the dipole antenna, are symmetrical.

For these antennas, the four unbalanced S-parameters must be measured using a classic unbalanced two-port network analyzer and calculated from these M_{11}^{-}. M_{11}^{-} is related to $100\,\Omega$.

4. Now a matching circuit is developed, which often includes an impedance transformation in addition to the balun functionality. This was explained in detail in the previous Chap. 4.

——— Matching of Antennas ———

Fig. 10.31: Head section of an 868 MHz transceiver circuit with integrated dipole antenna

An example of this procedure is shown in the following pictures. Fig. 10.31 shows a transceiver circuit with integrated dipole antenna.

The associated simulation based on the two-port measurement using M-parameters shows the following Fig. 10.32.

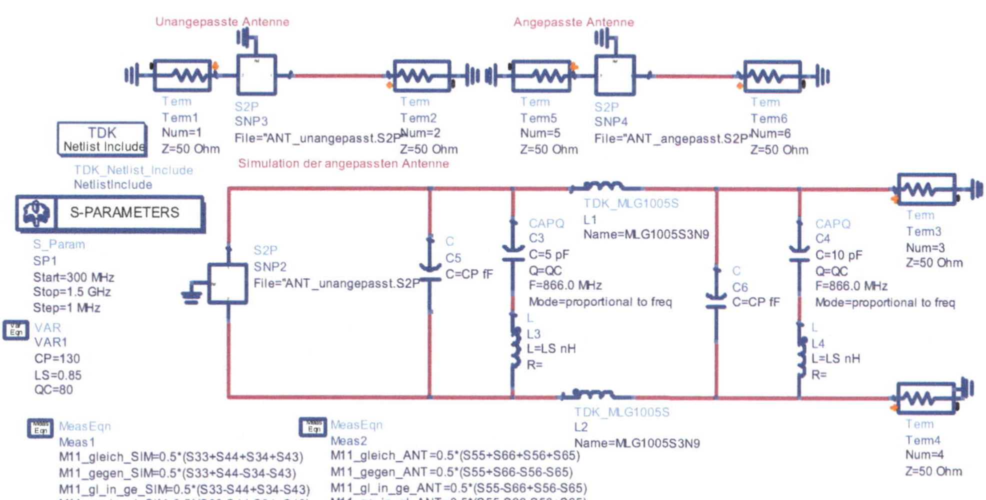

Fig. 10.32: Illustration of the matched antenna circuit and the ADS-simulations of the measured dipole antennas

The result in Fig. 10.33 shows that a very good agreement between simulation, model and measurement results can be achieved in this way.

If the antenna with its low-loss metallization surfaces is well matched, then it also radiates most of the energy. An antenna is passive and reciprocal. Consequently, an antenna that radiates well also receives well.

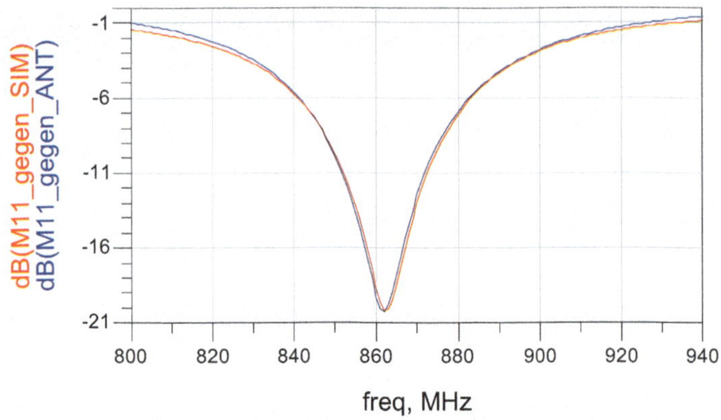

Fig. 10.33: Reflection behavior calculated in the simulation compared to the measured values of an integrated dipole antenna

However, if you want to predict and optimize the radiation characteristics of an antenna or even swivel the antenna beam electronically, it is necessary to study the very extensive fundamentals, [17, 39, 48, 50, 80, 131].

For the use of antennas in measurement technology and in RF-ID systems, it is important to know the so-called circular antennas. A distinction is made between right- and left-polarized or right- and left-circular antennas.

If both are aligned in the same direction, there is a large degree of isolation between the antennas. If both antennas are aligned in opposite directions and in the respective main beam of the other antennas, there is maximum transmission.

These circular antennas help to identify objects, since a metallic object rotates the reflected waves in polarization and dielectric objects have reflections with the same polarization.

With highly linear mobile radio systems such as 6G, there are problems with mismatching antennas due to incorrect use caused by shielding with one hand. A very interesting workaround for this was presented in the insulators subsection (page 342).

10.4 Scalar S-Parameter Measuring Systems

Scalar network analysis emerged in the mid-20th century as a key measurement technique for characterizing microwave components in the GHz range. Unlike vector network analysis, which measures both magnitude and phase, scalar network analysis focuses solely on measuring the amplitude response of a device under test.

Early scalar network analyzers were based on swept-frequency generators and crystal detectors, allowing engineers to measure insertion loss, return loss, and standing wave ratio (SWR).

In the 1960s and 1970s, the rapid development of solid-state microwave sources and detectors significantly improved the accuracy and frequency range of scalar measurements. The technique became especially popular in manufacturing environments due to its lower cost and simpler operation compared to vector analyzers.

Although it cannot capture phase information, scalar network analysis provided valuable insights for filter design, antenna matching, and waveguide component evaluation in the GHz range.

By the 1980s, digital processing began to enhance measurement speed and dynamic range, making scalar network analyzers more versatile.

With the rise of affordable vector network analyzers in the 1990s, scalar network analysis gradually shifted towards niche applications where only magnitude data was needed. Today, scalar network analyzers remain relevant in cost-sensitive or field-based GHz measurements, valued for their simplicity and robustness.

Even with small quantities, the cost of manufacturing a narrow-band scalar S-parameter measuring circuit can be kept below 5 €. This trend, due to the very low-cost components resulting from the mobile phone boom, makes it possible to use RF circuits in areas that were previously covered by other sensors.

The two lower free frequency bands for industrial applications with the following cut-off frequencies are particularly suitable for the realization of such circuits:

$$2.4000 - 2.4835\,\text{GHz} \quad \text{and} \quad 5.725 - 5.875\,\text{GHz} \quad .$$

The minimum effort required to set up a scalar S-parameter measurement system consists of an oscillator and a detector.

Scalar measurement systems have a wide range of applications:

- Industrial measurement technology
  - Microwave barriers as a replacement for light barriers (transmission measurement) and magnetic barriers (reflection measurement)
  - Moisture measurement (non-contact, during production)
  - Dust measurement
  - Room monitoring
  - Object recognition
- Microwave electronics

- Control of the amplifier power
- Monitoring the antenna matching

- Scalar network analyzers for S-parameter measurement

RF systems always offer the advantage over conventional systems that many materials, such as walls, are more or less transparent to the radio signal. This means that a microwave barrier can be installed behind two walls and is therefore invisible to people in the area to be monitored, Fig. 10.34.

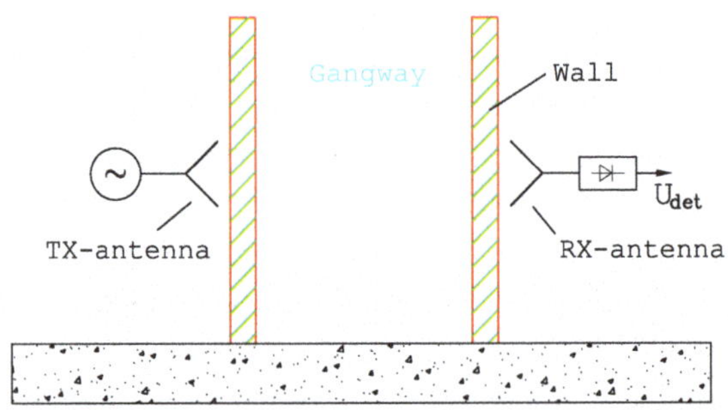

Fig. 10.34: Top view of a microwave barrier mounted behind two walls for passage control

RF sensors also have the advantage that they can be contactless. As a result, RF measurement systems have already become established for measuring moisture in food and cosmetic products during ongoing production.

Despite the advantages that these systems can be developed very quickly and are extremely inexpensive to manufacture, it should not be forgotten that scalar systems have the following disadvantages:

- the measurement errors are relatively large,
- the measurement dynamic range is relatively low and
- the measurements are sensitive to interference when using broadband detectors.

10.4.1 Transmission Measurements

The simplest scalar transmission measurement is the measurement that every RF developer performs if he only has a spectrum analyzer or power meter and a signal source available to determine the transmission scattering parameters of a two-port. This procedure is referred to as **substitution method**. This method is based on the approximation that both the source and the detector are perfectly matched[1].

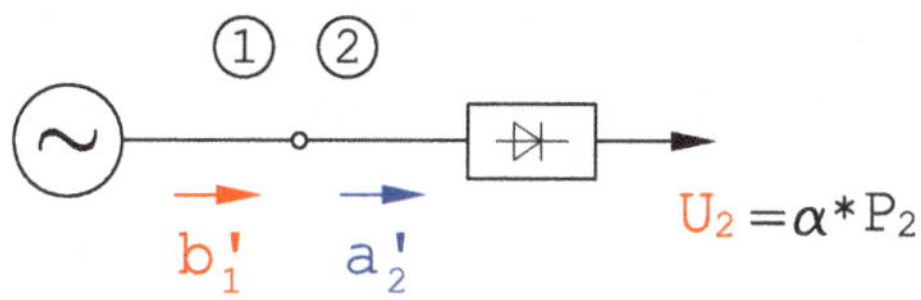

Fig. 10.35: Diagram of the reference measurement

In the first step, a REFERENCE MEASUREMENT is carried out, as shown in Fig. 10.35.
For the power P_2 emitted by the source and measured by the detector, applies:

$$P_2 = \frac{|a_2'|^2}{2} = \frac{|b_1'|^2}{2} \quad . \tag{10.73}$$

In the second step of the substitution method, the OBJECT MEASUREMENT is carried out according to Fig. 10.36.

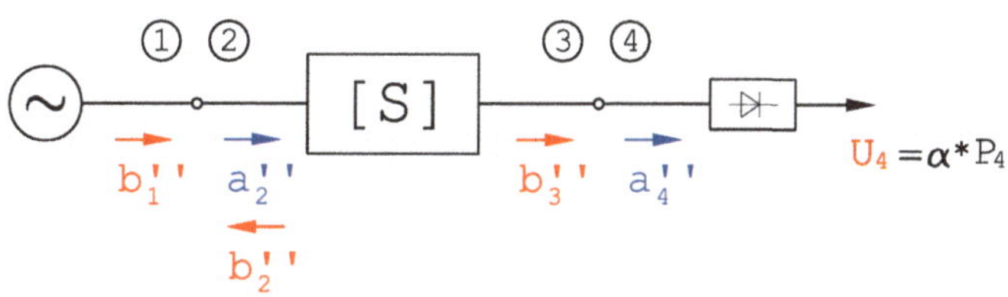

Fig. 10.36: Diagram of the object measurement

The following applies to the power P_4 that is transmitted through the measured object:

$$P_4 = \frac{|a_4''|^2}{2} \quad . \tag{10.74}$$

The amount of forward transmission can be calculated directly from the two detected values using the approximation $|b_1''| = |b_1'|$ (applies, for example, if $b_2'' = 0$ and the detector is also perfectly matched):

$$|S_{21}| = \frac{|b_3''|}{|a_2''|} = \frac{|a_4''|}{|b_1'|} \quad \Rightarrow \quad |S_{21}|^2 = \frac{|a_4''|^2}{|b_1'|^2} = \frac{P_4}{P_2} = \frac{U_4}{U_2} \quad . \tag{10.75}$$

In practice, logarithmic values are preferred:

$$S_{21}^{dB} = 10 \lg \left(\left|S_{21}^2\right|\right) = 10 \lg \left(\frac{P_4}{P_2}\right) = 10 \lg \left(\frac{U_4}{U_2}\right) \quad , \tag{10.76}$$

$$S_{21}^{dB} = P_4^{dBm} - P_2^{dBm} = U_4^{dBV} - U_2^{dBV} \quad . \tag{10.77}$$

The substitution method has two problems:

- The output power of the source must be time-invariant.
- Two measurements and two contacts are required.

The following methods do not have these problems. If the measuring accuracy plays a subordinate role due to large level differences, such as with microwave barriers, this substitution method is very interesting as it requires the least amount of components.

[1]This has already been discussed in detail in the section 10.1.

For transmission measurements with higher accuracy, it is necessary to detect a measure of the power emitted by the source. A second detector is used for this purpose and a directional coupler is used for higher frequencies or a bridge for lower frequencies. The structure of the directional coupler method is shown in Fig. 10.37.

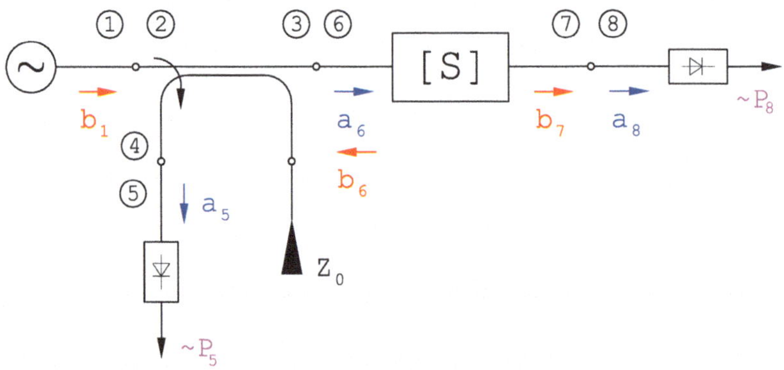

Fig. 10.37: Illustration of the implementation of a transmission measurement using the directional coupler method

The relationship between the two detected powers P_5 and P_8 with the source power P_1, the coupling transfer factor κ and the passband factor τ is determined using the signal tracking method for two identical detectors:

$$P_5 = |\kappa|^2 \, P_1 \qquad \text{and} \qquad P_8 = |S_{21}|^2 \, |\tau|^2 \, P_1 \quad , \tag{10.78}$$

$$\Rightarrow \quad |S_{21}|^2 = \frac{|\kappa|^2}{|\tau|^2} \frac{P_8}{P_5} \quad , \tag{10.79}$$

$$\Rightarrow \quad S_{21}^{dB} = P_8^{dBm} - P_5^{dBm} + \kappa^{dB} - \tau^{dB} \quad . \tag{10.80}$$

——— *Exercise: Application of the directional coupler Method* ———

Given: *A low-loss coupler is measured with:* $\kappa^{dB} = -10\,dB$ *and* $\tau^{dB} = -0.5\,dB$. *The following two measured values are displayed:* $P_8^{dBm} = -25.5\,dBm$ *and* $P_5^{dBm} = -20\,dBm$.

Find: *a) Calculate the transmission power* P_1^{dB}.

b) Enter the scalar transmission value $\left|S_{21}^{dB}\right|$ *of the measurement object.*

To a) *From the Eq. (10.78) follows:*

$$P_1^{dBm} = P_5^{dBm} - \kappa^{dB} = -20\,dBm - (-10\,dB) = -10\,dBm \quad . \tag{10.81}$$

To b) *This results directly from the Eq. (10.80):*

$$S_{21}^{dB} = -25.5\,dBm - (-20\,dBm) - 10\,dB + 0.5\,dB = -15\,dB \quad . \tag{10.82}$$

——— *Exercise: Application of the Directional Coupler Method* ———

The layout of directional couplers is very large, especially at lower frequencies. The costs are proportional to the board area. Fewer costs are associated with the implementation of the signal divider method. Fig. 10.38 shows the simple setup, which only requires two resistors instead of the directional coupler.

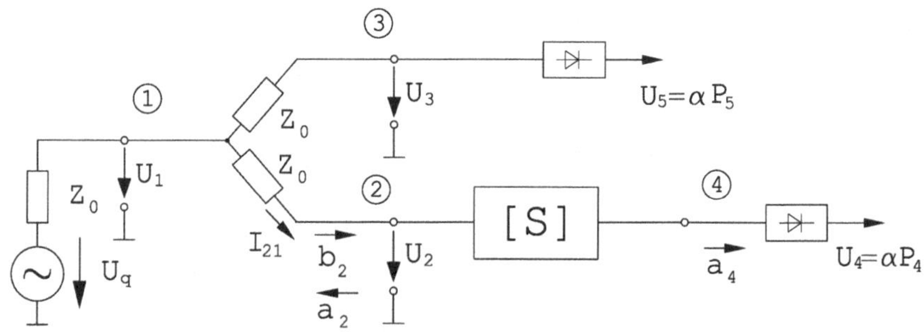

Fig. 10.38: Illustration of how to perform a transmission measurement using the signal divider method

This circuit can only be implemented if the resistors are almost ideal. This applies to 0402 SMD resistors in a good approximation up to around 2.5 GHz. The principle of operation, that this simple arrangement is a transmission measurement, can only be shown via the mathematical description.

The following applies between the voltages at ports 1 and 2:

$$U_1 = Z_0 \, I_{21} + U_2 \quad . \tag{10.83}$$

If Eq. (10.83) is expanded with $\frac{1}{\sqrt{Z_0}}$, normalized voltages and currents are obtained, which are indicated by lower case.

$$\frac{U_1}{\sqrt{Z_0}} = \sqrt{Z_0}\, I_{21} + \frac{U_2}{\sqrt{Z_0}} \quad \Rightarrow \quad u_1 = i_{21} + u_2 \quad . \tag{10.84}$$

The relationship between the normalized voltages, currents and the wave sizes is as follows:

$$i_{21} = b_2 - a_2 \quad , \quad u_2 = b_2 + a_2 \quad . \tag{10.85}$$

Thus, Eq. (10.84):

$$u_1 = b_2 - a_2 + b_2 + a_2 = 2\, b_2 \quad . \tag{10.86}$$

Since the detectors are matched to Z_0, U_3 applies:

$$U_3 = \frac{U_1}{2} = b_2 \sqrt{Z_0} \quad . \tag{10.87}$$

In this most important step, it can be shown that the power detected in the upper branch is directly proportional to the wave entering the measurement object:

$$P_5 = \frac{|U_3|^2}{2\,Z_0} = \frac{|b_2|^2}{2} \quad . \tag{10.88}$$

The same applies to the power applied to the lower detector:

$$P_4 = \frac{|a_4|^2}{2} \quad . \tag{10.89}$$

Finally, you reach the transmission via:

$$|S_{21}|^2 = \frac{|a_4|^2}{|b_2|^2} = \frac{P_4}{P_5} = \frac{U_4}{U_5} \quad , \tag{10.90}$$

$$S_{21}^{dB} = P_4^{dBm} - P_5^{dBm} \quad . \tag{10.91}$$

A slight disadvantage of the signal splitter method is that the signal splitter has 3 dB additional losses.

10.4.2 Reflection and Unidirectional Measurements

Transmission measurements alone are not always sufficient and in some cases it is not possible to carry out transmission measurements due to local conditions.

For example, the moisture content of pastes on metallic conveyor belts can only be determined by reflection measurements, as there is no transmission.

The measured variables that can be traced back to metallic, magnetic and/or dielectric objects can usually be identified by transmission and reflection measurements. Transmission measurements generally have a larger dynamic range and therefore lower measurement errors than reflection measurements.

If both measurements are carried out simultaneously, the information content increases. The measurement of reflection and transmission parameters in only one direction is called unidirectional measurement.

Measurements in the second direction are generally not carried out with scalar arrangements, as the measurement objects are often passive and symmetrical (and therefore reciprocal and reflection-symmetrical) and therefore only two unknown S-parameters can be detected.

The double directional coupler method according to Fig. 10.39 is recommended for pure reflection measurements as well as for unidirectional measurements.

This is due to the fact that the detectors are often less well matched than wave sumps, which in the lower GHz range are simply realized by a 51 Ω-0402 SMD resistor with good matching. Good matching is required for the couplers to function optimally.

This is particularly necessary for the reflection measuring point (Fig. 10.39, port 5), as the size of the crosstalk from the transmitting side determines the reflection measurement dynamics.

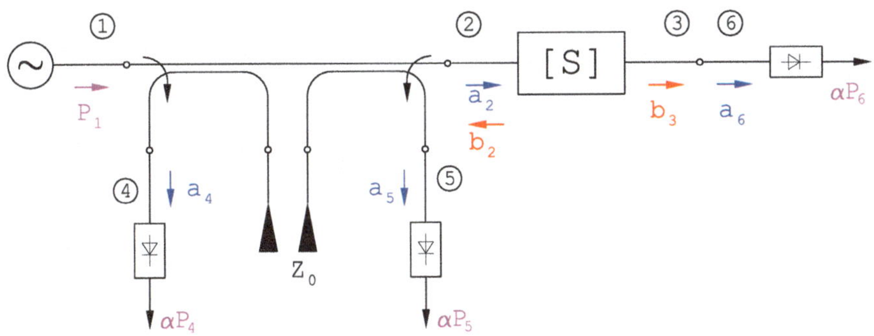

Fig. 10.39: Unidirectional reflection and transmission measurement using the double directional coupler method

In practice, the influences of this crosstalk affect the measurement dynamics more than the detector dynamics.

The double directional coupler method can be described very simply mathematically in analogy to the directional coupler method using two identical couplers and identical detectors:

$$P_4 = \frac{1}{\alpha}\,|\kappa|^2\,P_1 \quad , \tag{10.92}$$

$$P_5 = \frac{1}{\alpha}\,|\kappa|^2\,|S_{11}|^2\,|\tau|^4\,P_1 \quad , \tag{10.93}$$

$$P_6 = \frac{1}{\alpha}\,|S_{21}|^2\,|\tau|^4\,P_1 \quad , \tag{10.94}$$

$$\Rightarrow \quad S_{21}^{dB} = P_6^{dBm} - P_4^{dBm} + \kappa^{dB} - 2\,\tau^{dB} \quad , \tag{10.95}$$

$$\Rightarrow \quad S_{11}^{dB} = P_5^{dBm} - P_4^{dBm} - 2\,\tau^{dB} \quad . \tag{10.96}$$

The first coupler could again be replaced by a signal splitter, but this would slightly worsen the properties of the second coupler due to the poorer matching of the transmission port.

10.5 Vectorial S-Parameter Measuring Systems

In contrast to vectorial measurements, scalar measurements have very large measurement errors that cannot be corrected by intelligent post-processing in the computer. The background to the further processing of the measured values in the computer, which is standard in vector network analysis, is briefly explained in the last section.

This section deals with the hardware of vector systems. Three objectives were pursued in the presentation:

1. The aim is to develop a basic understanding of the concepts of modern transmission technology based on homodyne or heterodyne principles.
2. The realization and design of simple vectorial measuring devices will be explained.
3. An understanding of the commercially available vector network analyzers for precise scattering parameter measurement and the different basic concepts should be developed.

In contrast to modulated transmission systems, where the adjacent channels have to be taken into account in practice, vector network analysis provides a very clear scenario for frequency conversion. There is only the sinusoidal mono-frequency measurement signal (frequency ω_S) at the measurement object.

As introduced in the section 10.3.3, a local oscillator signal (LO signal) is required so that the measurement signal can be converted to a lower frequency range for display or further processing. The signals applied to the frequency-converting components (mixer or single-sideband receiver) with the simple relationship $\omega_{ZF} = \omega_S - \omega_{LO}$ are illustrated in Fig. 10.40.

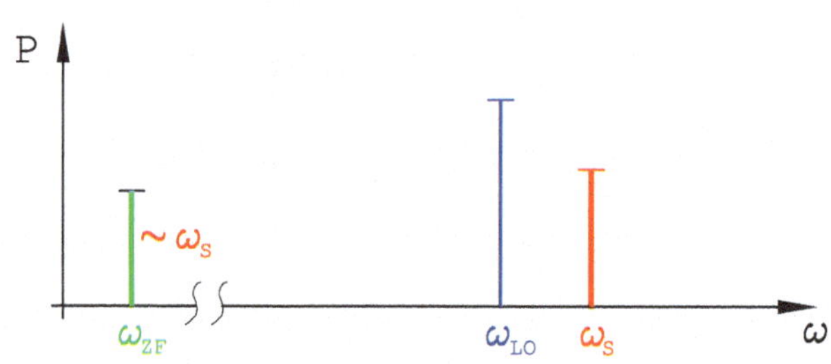

Fig. 10.40: Frequency scheme in the vector RF network analysis for downward mixing

Vector network analyzers as well as the transmission paths for amplitude and phase modulated signals are divided into two concepts:

1. Homodyne systems consisting of an RF transmitter and, for example, a single sideband transmitter and several single sideband receivers.
2. Heterodyne systems consisting of two RF transmitters and mixers for downward mixing.

Both concepts have *vector receivers*. These vector receivers should be symbolized for system consideration, as shown in Fig. 10.41.

In the following, we first explain how to arrive at the complex signal U_i and then describe some homodyne concepts and the basic concept for heterodyne systems.

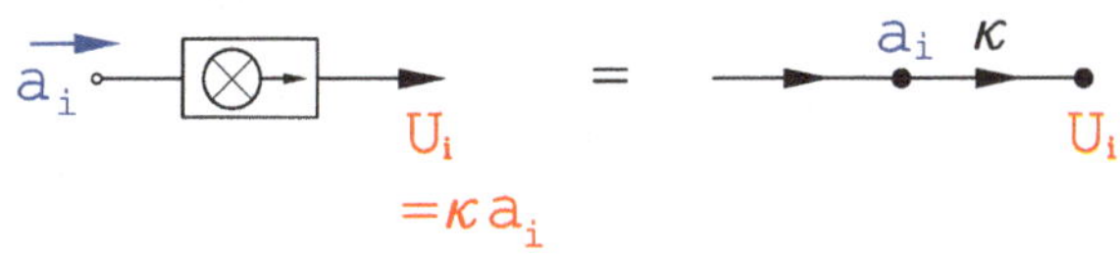

Fig. 10.41: Symbol and signal flow graph of a vector receiver that converts the complex RF wave a_i into the complex voltage U_i

The IF Signal Processing

In vector network analysis, synthesis generators with their excellent frequency stability are used as transmitters without exception. As a result, the measurement signals can be filtered very narrowly, allowing network analyzers to achieve high measurement dynamics of well over 100 dB.

Narrow-band bandpass filters, for which crystal filters can be used in the IF frequency range, form the beginning of the signal processing chain, which is implemented at the IF output of the mixer or single-sideband receiver (frequency converter for short), as shown in Fig. 10.42.

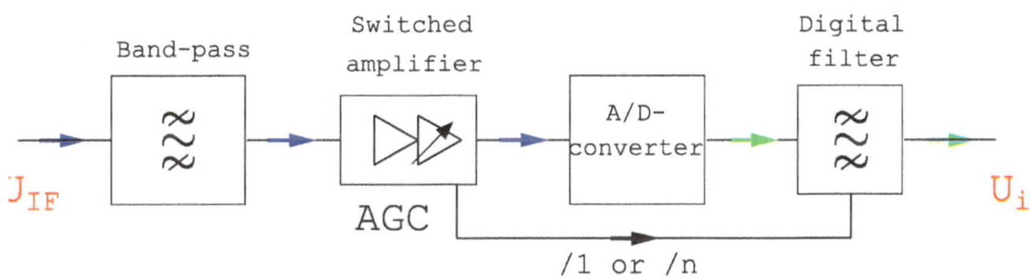

Fig. 10.42: Signal processing chain of a vectorial receiver between frequency converter and digital processing as complex voltage U_i

As the dynamic range of the A/D converter is often insufficient, many receiver chains contain switchable amplifiers (AGC: Automatic Gain Control), which are activated at very low input levels. It should be noted during development that these operational amplifiers only precisely change the amount of voltage amplification by a factor of 1 or n during switching and do not perform any additional phase rotations. Changes in the input and output impedances around the wired operational amplifiers cause such phase shifts.

Digital filtering is used in modern measuring devices to reduce the bandwidths to 1 Hz and less. If amplification by the factor n has been carried out, this must be reversed by dividing the voltage signal in the computer.

The IF frequency ($f_{IF} = 1/T_p$) is typically selected between 10 and 100 kHz. The maximum sampling frequency of the high-resolution A/D converter is the limitation here.

The sinusoidal received signal is sampled four times, as can also be seen in Fig. 10.43. The sampling time is firmly linked to the reference crystal oscillator signal, but is in principle arbitrarily selected. All receivers are scanned simultaneously so that a time-synchronized reference measurement is always carried out. The reference measuring point (Fig. 10.39, port 4) is always selected as the reference.

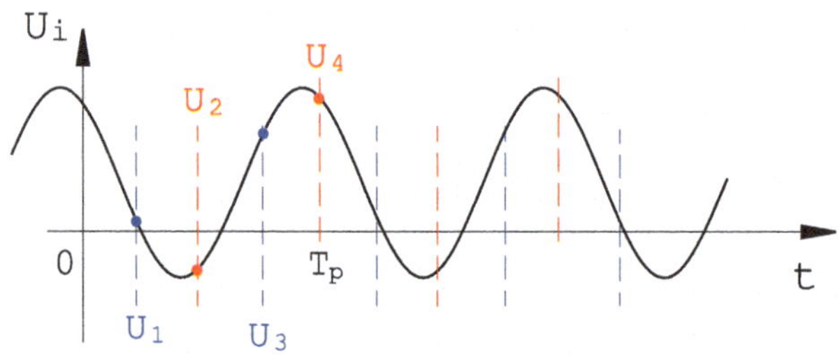

Fig. 10.43: Sampling of the sinusoidal signal of a vectorial receiver to determine the complex voltage U_i

The first sampled value U_1 is defined for the real part and the second sampled value U_2, shifted by 90°, for the imaginary part. Due to the quadruple sampling, a possible DC voltage component can also be calculated out, as the following must apply:

$$U_1 = -U_3 \qquad \text{and} \qquad U_2 = -U_4 \quad . \tag{10.97}$$

This means that the peak values of the amount $\hat{U}_i$ and the phase $\angle U_i$ of the complex signal U_i can be calculated directly from the sampled values:

$$\left|\hat{U}_i\right| = \frac{1}{2}\sqrt{(U_1 - U_3)^2 + (U_2 - U_4)^2} \qquad \text{and} \qquad \angle U_i = \arctan\left(\frac{U_1 - U_3}{U_2 - U_4}\right) \quad . \tag{10.98}$$

10.5.1 Homodyne Concepts

Homodyne systems are transmission systems that use only one RF transmitter. Heterodyne systems use two RF transmitters. In recent decades, the production of synthesis generators has been very expensive, so intensive research has been carried out into new homodyne processes.

This means that there is now a very wide range of measurement concepts with just one RF transmitter:

1. Direct frequency synthesis and sampling (software radio),
2. Single-sideband transmit-receive method,
3. Single sideband offset by phase modulators,
4. Phase switch method,
5. Six-port procedure.

Only the first two procedures will be discussed here. For special applications and in the very high frequency range, the other methods are very useful and can be found in full in [125]. Intensive research is being carried out in the field of six-port methods, particularly for multi-standard receivers, [125, 127, 152].

Direct Frequency Synthesis and Sampling (Software Radio)

It is possible to generate a sinusoidal signal directly via D/A conversion and to sample it directly without frequency conversion at very low cost well into the MHz range. The associated generators are called DDS (Direct Digital Synthesizers) for short. The disadvantage is that you can only use digital filters, as there is no longer a fixed IF frequency.

Measuring devices based on this concept have become firmly established on the market as so-called LCR meters. A very simple structure of an LCR meter with a broadband resistive coupler is shown in Fig. 10.44[2].

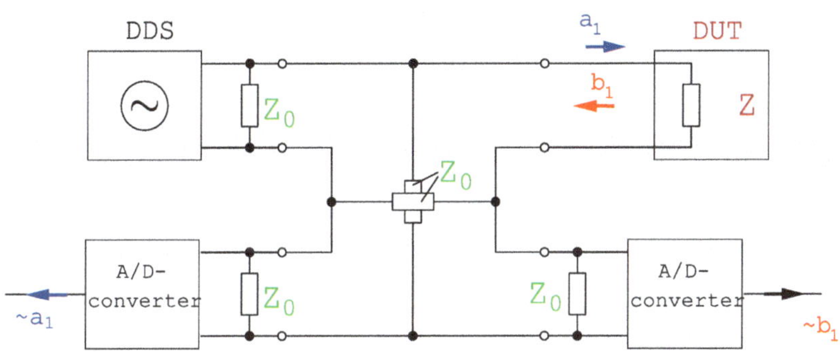

Fig. 10.44: Possible design of an LCR meter as a homodyne reflectometer

This is a simple reflectometer for S_{11} measurement. However, only the converted Y and Z parameters are displayed to the user.

The advantage is that Z_0 can be switched and the device is therefore maximally selective for different impedances.

Many system specialists hope that the so-called software radio, whose hardware will consist solely of D/A converters and A/D converters, will be the solution for the future of RF communication technology.

However, if you examine the system requirements in detail, such as spectral purity, transmission power, reception sensitivity, etc., you will quickly realize that software radios will not meet these requirements for frequencies above 2 GHz in the near future, [132].

Single-Sideband Transmit-Receive Method

The circuit for setting up a single-sideband transmitter or receiver is shown in the diagram 10.24. For the single sideband transmitter, the signal U_{ZF} is derived from the reference crystal signal of the RF transmitter (typically 10 MHz) by frequency division.

[2]The measurement object is identified by the common abbreviation DUT (Device Under Test).

If integrated circuits are available, for example in the form of a vector or IQ modulator (Fig. 10.25), which have the functionality of single sideband suppression, extremely inexpensive homodyne analyzers can be produced, [35, 93, 125, 130].

Fig. 10.45 shows a homodyne network analyzer, where all vector receivers and the modulator are to be realized by a vector modulator according to Fig. 10.25. As an exception, the LO oscillator and the associated connections have been drawn in this picture.

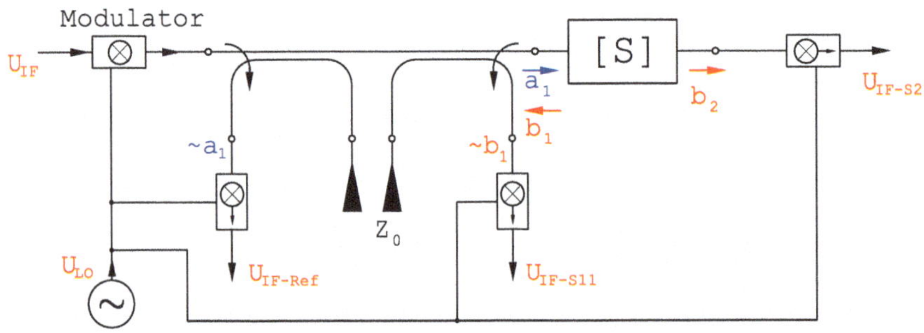

Fig. 10.45: Block diagram of a unidirectional vectorial homodyne network analyzer

Most mobile radio concepts are based on this principle. For this reason, a large number of vector modulators with excellent image frequency suppression properties can be found in the lower GHz range. If both the transmitter and the receiver suppress the image frequency by 50 dB, you still have good conditions with a measurement dynamic range of 80 dB (= 2·50 dB -20 dB).

Homodyne concepts are reluctant to be used for precision measuring devices in RF measurement technology, as they exhibit too many non-eliminable measurement errors due to the image frequency problem and other finitenesses. In this case, a heterodyne concept is selected.

10.5.2 Heterodyne Concepts

The Classic Heterodyne Concept

In the classic heterodyne concept, two synthesis generators are operated directly via just one reference oscillator (quartz oscillator) to generate the two RF frequencies ω_s and ω_{LO} according to Fig. 10.46.

This ensures that the IF frequency is maintained with high precision as a frequency spacing in the RF range over the full range of the synthesis generators.

In the past, several frequency-converting stages were used, for example to achieve shorter settling times for RF transmitters. This is rarely done for cost reasons.

In mobile radio technology, the heterodyne concept was used in the early days of GSM. The modulation signal was applied to a transmitter. Nowadays, heterodyne systems are only used in the research stage for multi-standard systems in mobile communications.

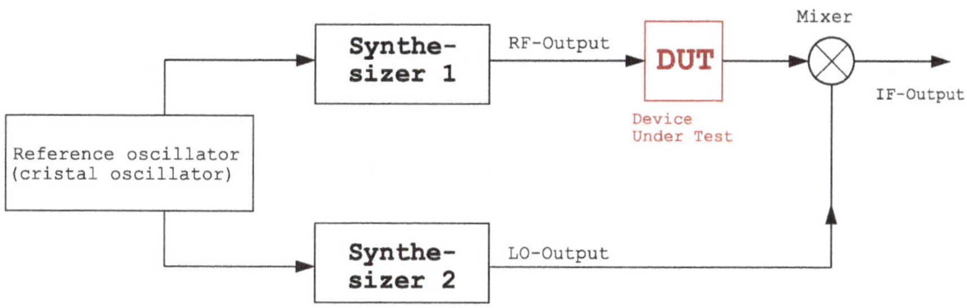

Fig. 10.46: Block diagram of a heterodyne measuring system for the substitution method

In RF measurement technology, this classic heterodyne concept is the standard for the manufacture of both spectrum analyzers and network analyzers.

The $\Sigma\Delta$-Transmitter

The novel concept of the $\Sigma\Delta$ transmitter is a special form of heterodyne transmission system, [60]. This concept is based on two $\Sigma\Delta$ synthesizers. The basic concept of such a highly flexible $\Sigma\Delta$ synthesizer is shown in Fig. 10.47.

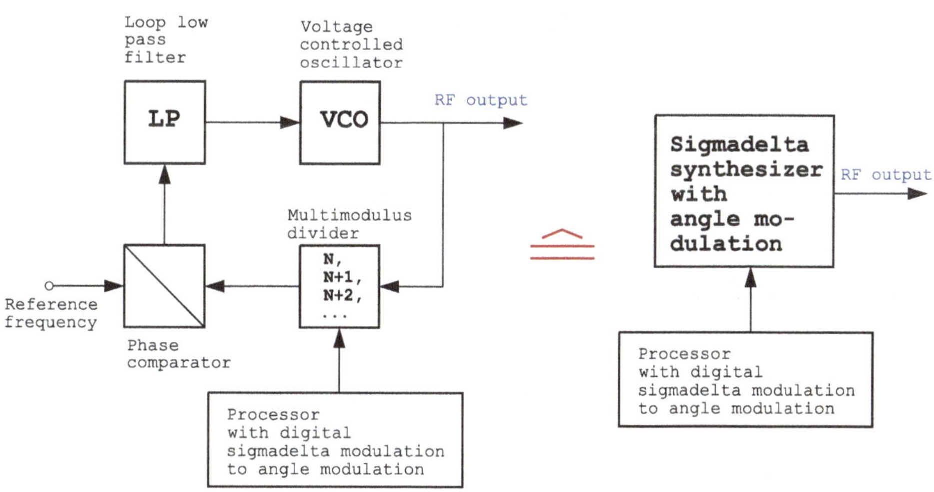

Fig. 10.47: Block diagram of a $\Sigma\Delta$ synthesizer controlled by the processor

As already explained in the synthesis generator section, such a synthesizer can perform any angle modulation via the purely digital control and thus generate an RF signal of the form

$$y_{1(t)} = A_1 \sin\left(\omega_t t + \varphi_{1(t)}\right) \quad , \tag{10.99}$$

where A_1 is a fixed amplitude, ω_t contains the RF carrier angular frequency, t is the time variable and $\varphi_{1(t)}$ contains the angle-modulated information.

If you want to design a signal source (transmitter) that can in principle be used for any communication standard, its output signal must have both angle or phase modulation (PM for short), which is related to frequency modulation (FM), and amplitude modulation (AM). Mathematically, the signal must be designed as follows:

$$s_{(t)} = A_{(t)} \sin\left(\omega_t t + \varphi_{(t)}\right) \quad , \tag{10.100}$$

where $A_{(t)}$ describes a modulated amplitude.

The $\Sigma\Delta$ transmitter concept now provides for two $\Sigma\Delta$ synthesizers to be connected directly together and the addition of both output signals corresponds to the signal $s_{(t)}$. This $\Sigma\Delta$ transmitter concept is illustrated in Fig. 10.48.

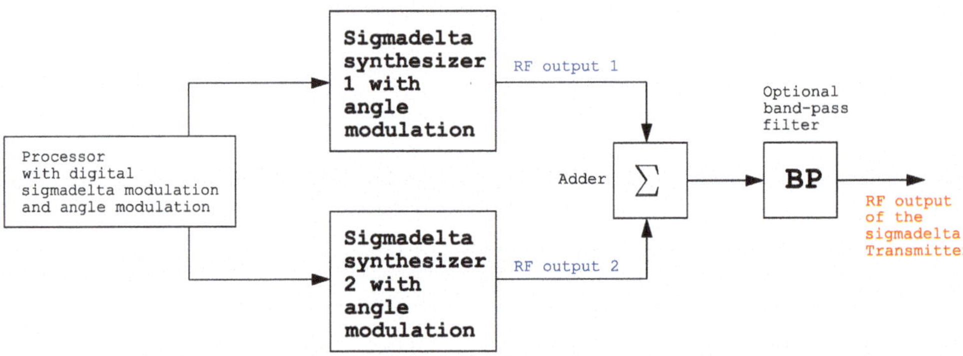

Fig. 10.48: Block diagram of the $\Sigma\Delta$ transmitter controlled by the processor

Mathematically, it should now be deduced that the addition of two angle-modulated signals results in an amplitude- and angle-modulated signal. The two $\Sigma\Delta$ synthesizers 1 and 2 should provide the maximum output amplitudes $A_1 = A_2 = 1$ for this case to simplify the mathematics.

In this case, the maximum amplitude of the desired AM amplitude must not be greater than 2. The initial equation is therefore:

$$A_{(t)} \sin\left(\omega_t t + \varphi_{(t)}\right) = \sin\left(\omega_t t + \varphi_{1(t)}\right) + \sin\left(\omega_t t + \varphi_{2(t)}\right) \quad . \tag{10.101}$$

The $\omega_t t$ dependency can be eliminated using the complex calculation. A short calculation leads to the intermediate result:

$$A_{(t)} \cos\left(\varphi_{(t)}\right) = \cos\left(\varphi_{1(t)}\right) + \cos\left(\varphi_{2(t)}\right) \quad , \tag{10.102}$$

$$A_{(t)} \sin\left(\varphi_{(t)}\right) = \sin\left(\varphi_{1(t)}\right) + \sin\left(\varphi_{2(t)}\right) \quad . \tag{10.103}$$

A further short calculation then leads to the closed analytical equations, which can be used to calculate the phase modulations of both $\Sigma\Delta$ synthesizers from the given AM and PM values:

$$\varphi_{1(t)} = \varphi_{(t)} + 1/2 \arccos(A^2_{(t)}/2 - 1) \quad , \tag{10.104}$$

$$\varphi_{2(t)} = \varphi_{(t)} - 1/2 \arccos(A^2_{(t)}/2 - 1) \quad . \tag{10.105}$$

In contrast to conventional transmitter concepts, this type of transmitter eliminates all mixers and D/A converters as well as the necessary phase shifters for single-sideband suppression.

Furthermore, significantly fewer switches are required. The entire transmitter RF block consists only of two identical VCOs, an adder network, any power amplifiers connected behind it and, if necessary, a filter.

Both synthesis generators are connected to a reference oscillator, as is also done in the classic heterodyne concept. This ensures that the two signals can be adjusted with high precision, particularly in terms of their phase position in relation to each other.

The special $\Sigma\Delta$ algorithms for controlling the multi-modulus dividers also ensure a best quality transmission signal including low noise in and around the transmission band.

The hardware costs are significantly lower than the costs of known concepts, but involve new optimization criteria. For example, the outputs of the (power) VCOs must have the same amplitude. This criterion, which is certainly easy to fulfill, has not yet been set for any VCO pair.

Although this system allows AM, the power amplifiers can be operated in compression and thus at maximum efficiency. With the latest standards with linear digital modulation such as 6G and WLAN, this allows efficiency levels to be increased from currently around 40 % to over 60 %.

If the power amplifier is optimized in such a way that the transistor output stages can be connected directly together, this also results in an extremely broadband solution with high efficiency. Research work on this topic can be found under the heading of so-called LINC technology.

Transceiver circuits are often required. If the $\Sigma\Delta$ concept is selected for the transmitter path, the receiver path can be realized with a simplified single sideband receiver. The two oscillators can be used to provide the two local oscillator signals offset by 90° when receiving.

10.5.3 Network Analyzer Concepts

The qualified user of a network analyzer must be familiar with its conceptual design, which is explained below, and the associated advantages and disadvantages.

GSOLT Concepts

While a scalar network analyzer (SNA) is at best a unidirectional device with reflection and transmission measurement properties, this design is the minimum effort for a purchasable vector network analyzer (VNA). Fig. 10.49 illustrates the concept of a VNA that can be used to measure S_{11} and S_{21}.

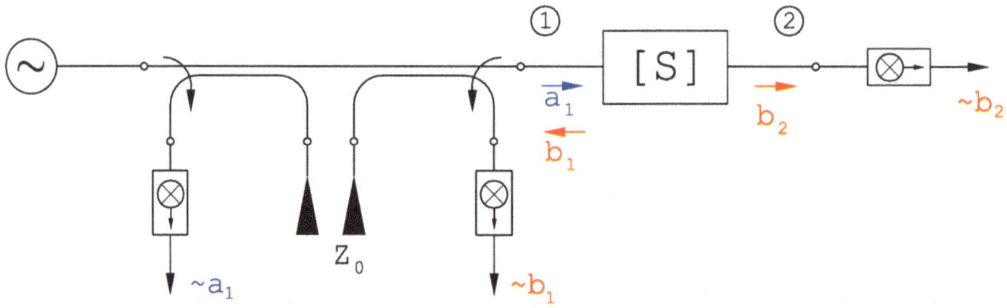

Fig. 10.49: Block diagram of a unidirectional vector network analyzer based on the GSOLT concept

Such simple and therefore inexpensive analyzers may be used in production, where only one hundred percent measurement of the forward variables is required.

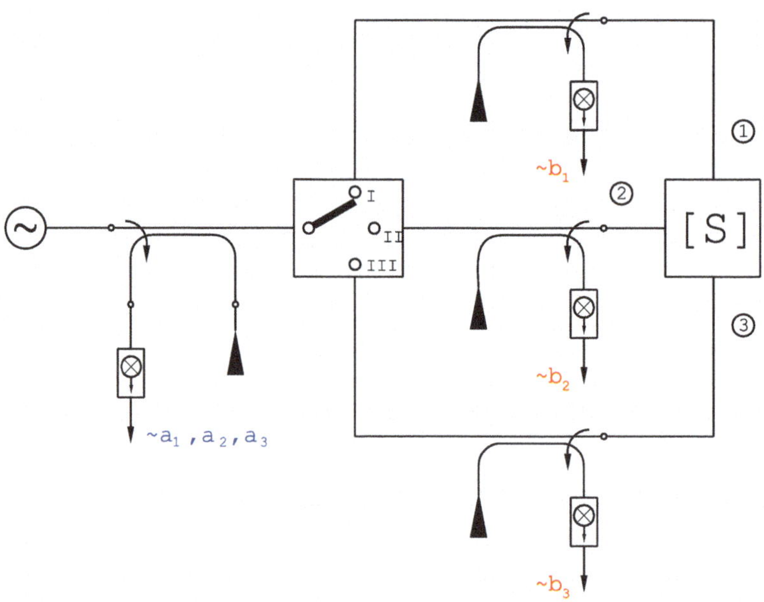

Fig. 10.50: Block diagram of a vectorial multi-port network analyzer according to the GSOLT concept

This VNA can measure all 9 S-parameters of the measurement object in all three switch positions after the switch has been switched.

For the developer, it is particularly important for the design of active circuits to also determine the reverse variables. In addition to two-port devices, a modern measuring device should also be able to measure three-port devices such as signal splitters and so-called duplexers or even four-port devices for mixed-mode applications.

There are two concepts to meet these requirements. The more cost-effective concept is shown in Fig. 10.50 as a three-port VNA.

A single reference measuring point detects a measurement for the outgoing shaft regardless of the switch position. The subsequent RF changeover switch must pass through the three switch positions with high repeatability and must not have any crosstalk.

In practice, this means that the switch must have an isolation of far more than 120 dB. The reflected and transmitted waves are measured at the measuring points behind the switch.

The advantage of this concept lies in the small number of measuring points. Only n+1 measuring points are required for an n-port VNA. The name GSOLT concept comes from the only error correction method that can be used for this concept, [55].

Most two-, three- and four-port network analyzers are based on this GSOLT concept.

Reflectometer Concepts

The reflectometer concept, as shown in Fig. 10.51, is somewhat more attractive in application, but also more expensive.

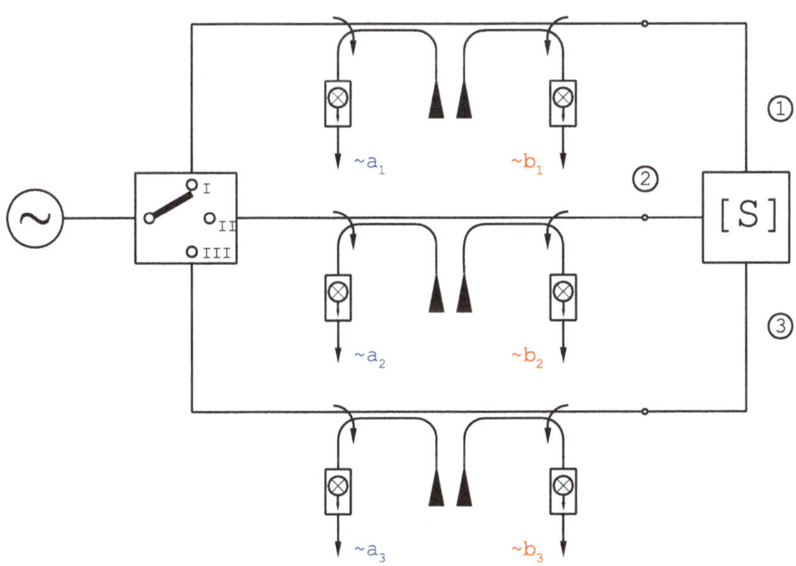

Fig. 10.51: Block diagram of a vector network analyzer based on the reflectometer concept for three-port measurements

In the past and with many current devices, a changeover switch is still used. This switch can exhibit crosstalk, which could theoretically be eliminated using mathematical correction methods, but this is avoided because the critical measured values are in the noise range and the mathematical models are no longer fulfilled in this range.

A great advantage of this reflectometer concept, however, is that the switch can also drift for precision measurements over time and even during a measurement. In this concept, the reference measuring points are located behind the switch and therefore the switch can be regarded as part of the signal source.

It has been noted several times in this chapter that the manufacturing costs for signal sources are increasingly falling. This trend now also makes it possible to eliminate the very expensive changeover switch.

Fig. 10.52 shows how simple a three-port VNA with a total of four signal sources[3].

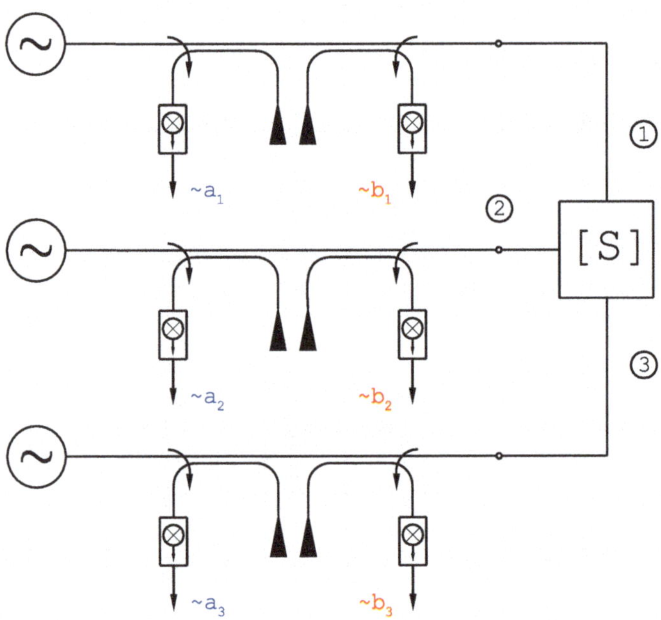

Fig. 10.52: Block diagram of a vector network analyzer based on the reflectometer concept with four signal sources

As we will see in the following section, a large number of so-called calibration methods can be used to calculate the system errors for this hardware concept. The low demands on the manufacturability of the necessary calibration standards make this concept attractive.

The only disadvantage of this concept is the high cost of measuring points. For an n-port VNA you now need 2·n measuring points.

[3]The signal source for the LO signal, which is necessary to realize the vectorial measuring points, is not shown.

sr-Concepts

To reduce costs, VNA solutions that use a so-called "*switched-receiver* concept" (sr-concept for short) are increasingly being used. With this concept, the measuring points are switched to a single receiver and a number of receivers. With the switches, care is taken to ensure that the switch inputs are always terminated (even with isolation).

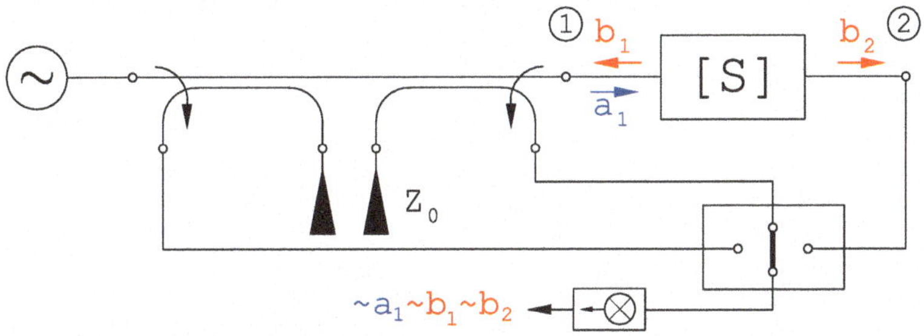

Fig. 10.53: Block diagram of a unidirectional vectorial network analyzer based on the sr-concept with only one switched receiver

These VNAs measure more slowly, but are good enough for many applications. If you reduce the VNA to just **one** receiver, there are two concepts:

1. You must always switch between the measuring points at each frequency point.
2. The transmitter must have very high reproducibility in terms of power and phase value (compared to the 10 MHz reference signal).

The 1st concept is extremely slow and is therefore not used in practice.

The 2nd concept is currently offered by [72] and is shown in Fig. 10.54.

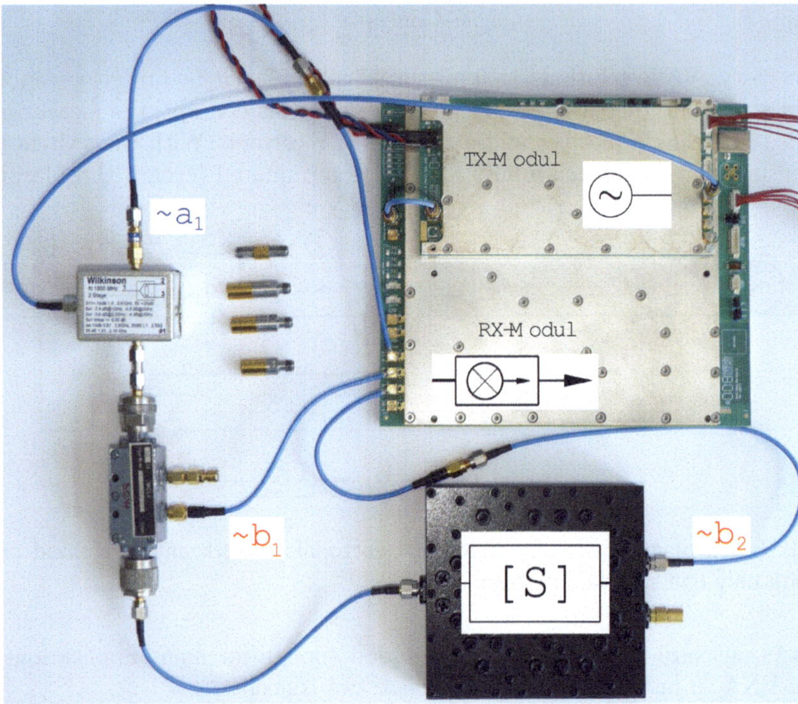

Fig. 10.54: Photo of a unidirectional vectorial network analyzer based on the sr-concept with only one switched receiver and TX and RX module from [72]

——— *Exercise: Design of a Reflectometer* ———

Given: *A simple reflectometer circuit according to Fig. 10.55 with ideal couplers (matched, perfectly isolating and lossless) with the coupling transfer factors* κ_1 *(left coupler) and* κ_2 *(right coupler) as well as the corresponding transmission factors* τ_i*, ideal lines (matched and lossless), ideal wave sumps (matched), ideal receivers with conversion factor 1 (matched,* b_i= *output signal) and an ideal signal source (matched).*

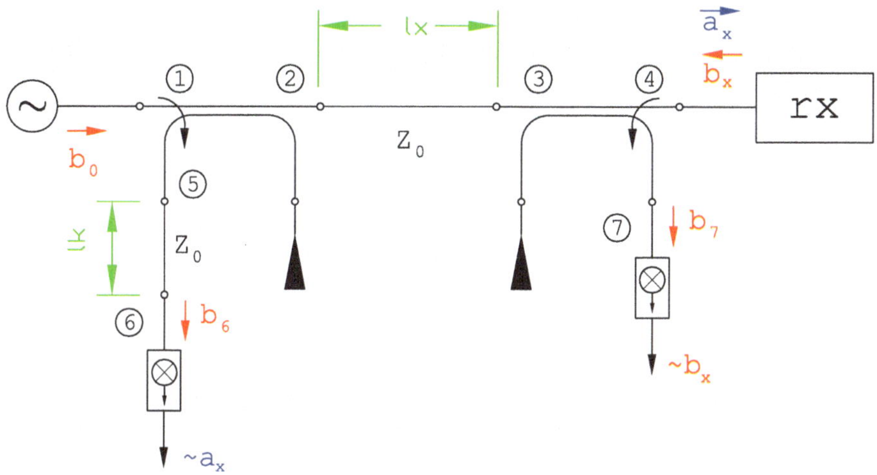

Fig. 10.55: Simple reflectometer for vector measurements

The correlation applies to the coupling attenuations: $C_i = 20\lg\left(\frac{1}{|\kappa_i|}\right)$ dB. *The following applies to the electrical lengths of the couplers:* $\kappa_i = |\kappa_i|\,e^{-j\beta l_\kappa}$ and $\tau_i = |\tau_i|\,e^{-j\beta l_\tau}$. *The correlation applies to the insolation attenuations:* $D_i = 20\lg\left(\frac{1}{|\sigma_i|}\right)$ dB.

Find:

a) *Use the signal tracking method to calculate the ratio* b_7/b_6 *as a function of* κ_i, τ_i, l_k, l_x *and* r_x.

b) *The following should apply:* $C_2 = 20\,\text{dB}$. *Calculate* C_1 *for* $|r_x| = \left|\frac{b_7}{b_6}\right|$.

c) *Based on a) and b),* l_k *is to be calculated as a function of the variables* l_x, l_κ *and* l_τ *for* $r_x = \frac{b_7}{b_6}$ *is to be calculated.*

d) *The influence of a finite isolation attenuation of* $D_2 = 40\,\text{dB}$ *at the 2nd coupler is to be investigated on the measurements* $r_{x1} = 0.3\,e^{j\varphi_x}$ *and* $r_{x2} = 0.1\,e^{j\varphi_x}$.

To a) *The following applies to* b_7:

$$b_7 = \kappa_2\, r_x\, \tau_2\, e^{-j\beta l_x}\, \tau_1\, b_0 \quad . \tag{10.106}$$

For b_6 *applies:*

$$b_6 = \kappa_1\, e^{-j\beta l_k}\, b_0 \quad . \tag{10.107}$$

Consequently, the following applies to the relationship between the two waves:

$$\frac{b_7}{b_6} = \frac{\kappa_2 \, \tau_1 \, \tau_2}{\kappa_1} \, e^{-j\beta(l_x - l_k)} \, r_x \qquad . \tag{10.108}$$

To b) *The following applies to an ideal coupler 2:*

$$|\kappa_i|^2 + |\tau_i|^2 = 1 \qquad . \tag{10.109}$$

Rearranged and inserted results:

$$|\tau_2| = \sqrt{1 - |\kappa_2|^2} = \sqrt{0.99} \qquad . \tag{10.110}$$

From the equation (10.108) follows the fulfillment of the reflection condition:

$$1 = \left| \frac{\kappa_2 \, \tau_1 \, \tau_2}{\kappa_1} \right| \qquad . \tag{10.111}$$

The required coupling factor κ_1 *or the corresponding attenuation* C_1 *can now be calculated from the Eq. (10.111) using the Eq. (10.109):*

$$\underbrace{|\kappa_2|^2 \, |\tau_2|^2}_{=0.0099} (1 - |\kappa_1|^2) = |\kappa_1|^2 \quad \Rightarrow \quad C_1 = 20.09 \, dB \qquad . \tag{10.112}$$

To c) *From Eq. (10.108) one can directly calculate the phase condition can be set up:*

$$\angle \frac{b_7}{b_6} = -\beta l_{\kappa_2} + \varphi_x - \beta l_{\tau_2} - \beta l_x - \beta l_{\tau_1} + \beta l_{\kappa_1} + \beta l_k = \varphi_x \qquad . \tag{10.113}$$

It follows directly from this:

$$l_k = l_x + 2 \, l_\tau \qquad . \tag{10.114}$$

To d) *Using the signal tracking method, you can calculate the new signal* b_7' *at port 7:*

$$b_7' = b_7 + \sigma_2 \, \tau_1 \, e^{-j\beta l_x} \, b_0 \qquad . \tag{10.115}$$

Then divide by b_6*, insert* $r_x = \frac{b_7}{b_6}$ *and eliminate* b_0 *using Eq. (10.107):*

$$\frac{b_7'}{b_6} = r_x + \underbrace{\frac{\sigma_2 \, \tau_1}{\kappa_1} \, e^{-j\beta(l_x - l_k)}}_{\text{Measurement error} \quad m_f} \qquad . \tag{10.116}$$

With the values $|\kappa_1| = 0.099$*,* $|\tau_1| = 0.949$ *and* $|\sigma_2| = 0.01$ *and the most unfavorable phase position of 0° , this results in a measurement error of* $|m_f| = 0.0959$*. For the reflection coefficient* r_{x1}*, this means that the error is around 32 %, while the error for* r_{x2} *is around 96 %. This means that, despite very good couplers, it is not possible to measure objects with more than 10 dB matching with such a measuring system.*

——— *Exercise: Design of a Reflectometer* ———

10.5.4 Calibration of Vector Network Analyzers

The process of training an artificial intelligence model closely parallels the calibration of a vector network analyzer (VNA). In both cases, the system begins in an unoptimized state, where raw outputs are distorted by internal biases, noise, or systematic errors.

Calibration in a VNA involves applying well-characterized reference standards to mathematically remove measurement errors, just as AI training uses labeled datasets to adjust model parameters and minimize prediction errors.

Both require iterative refinement? A VNA performs multiple calibration steps to converge on accurate measurements, while AI repeatedly updates weights through optimization algorithms. The quality of the references or training data directly determines the final performance: Inaccurate standards in calibration or poor datasets in AI lead to unreliable results.

Ultimately, both processes transform an uncalibrated or untrained system into a precise one, dependable tool for analyzing complex, real-world signals.

In practice, the measuring accuracy of modern network analyzers depends solely on calibration and its correct implementation. Calibration is carried out as follows:

1. A calibration procedure is selected. Depending on the conceptual design and software equipment of the VNA, more or fewer procedures are available.
2. In accordance with the regulations of the calibration procedure, partial or known calibration standards are connected and measured.
3. The network analyzer calculates the so-called error coefficients for the valid error model of the calibration method used.
4. Finally, the measurements of unknown measurement objects are carried out and corrected using the error coefficients and the algorithm associated with the error model.

A general German-language introduction to the aspects relevant in practice can be found in [118][4]. For users working on planar circuits, the more detailed application note [119] is suitable.

For those who want to gain a basic understanding of the calibration procedures, the German-language books [29, 38, 39, 49, 125] and the publications [25, 30, 31, 51, 52, 55] should be mentioned as the most important coaxial calibration procedures.

Interesting methods for on-wafer measurements are described in the publication [53, 54, 56, 59]. On the other hand, planar GHz circuits are best measured using a test socket, as shown in Fig. 10.56.

The multi-port measuring device with an almost unlimited number of measuring ports that has been on the market for the longest time is shown in [7].

If you want to control a setup with a switching matrix, this is done using the so-called ZVR-K11 software [116]. Multi-port measurements are carried out using the methods presented in [33, 55, 56, 59].

[4] This application note can be obtained from Rosenberger Hochfrequenztechnik.

Fig. 10.56: Test fixture for three-port and multi-port measurement manufactured by Rosenberger Hochfrequenztechnik

10.6 Description of the Dual-Mode Radio

Finally, a new type of radio standard will be presented, which was introduced as dual-mode radio and is based on the fundamentals of dual-mode transmission technology.

This new concept for radio transmission was developed at the IMP and used for several applications. It has many advantages over the known concepts. In addition to the low component costs, it is also immune to Doppler effects.

Dual-mode radio is generally characterized by the fact that two free-space modes are used and the differences between the signals transmitted in parallel are evaluated. Instead of free-space modes, conduction modes such as common mode and push-pull mode can also be used.

This requires two conductors and a ground (three-wire system). Dual-mode radio (DM radio for short) allows all known basic modulation methods such as amplitude modulation (AM), phase modulation (PM) and frequency modulation (FM).

These are presented below using examples of individual and combined modulation. In the most complex expansion stage, all three basic modulation types can even be combined for the first time.

As the difference between two signals is evaluated in DM radio, all converted modulations are referred to as difference modulations (DM) in the following applications.

Most of the applications presented below have been reduced to the basic concept, which is also sufficient for simple applications. In practice, however, amplifiers and (bandpass) filters are added to the DM radios for various reasons.

Description of Amplitude Difference Modulation (ADM)

A realization form for setting up a unidirectional DM radio link for amplitude-modulated signals is shown in Fig. 10.57. Components that are often necessary in practice, such as amplifiers and bandpass filters, were not included.

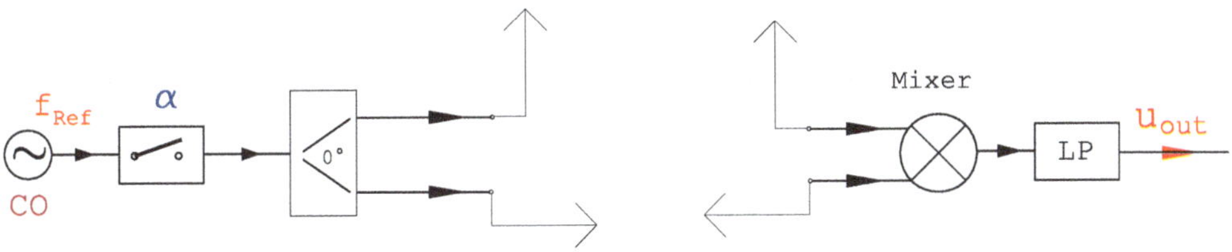

Fig. 10.57: Block diagram of a dual-mode radio link for amplitude-modulated signals

The mono-frequency RF oscillator (CO: continuous oscillator) sends a signal with the reference frequency f_{Ref}. The amplitude of this RF signal is then modulated by the digital on/off circuit. This is done by an RF switch.

The AM-modulated RF signal is split into two paths using the signal splitter (0° coupler). The two amplitude-modulated signals are emitted via the antenna pair with vertical and horizontal polarization.

The second pair of antennas receives both signals and feeds them to the mixer, which is designed as a multiplier. In practice, the Gilbert cell and ring diode mixer are possible ways of implementing the mixer. The mixer has output signals in the double frequency range and in the baseband range. The latter is only determined by the modulation bandwidth. A simple low-pass filter (LP) only allows the lower frequency range to pass.

The signal u_{out} (Fig. 10.58) can be used directly as a digital receive signal in the subsequent electronics.

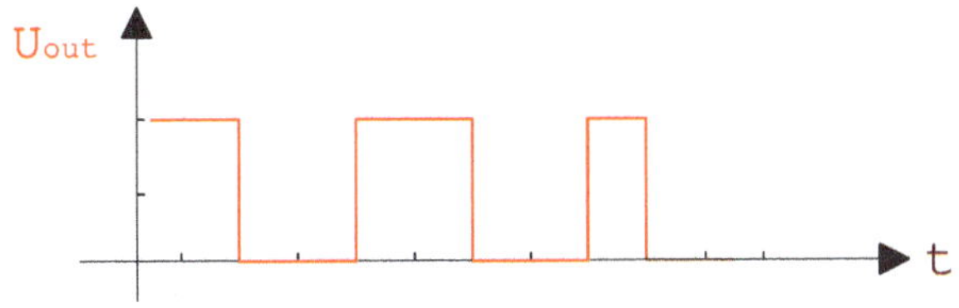

Fig. 10.58: Receive voltage at the output of the DM radio link for amplitude-modulated signals

In contrast to classic AM, the transmission signal is split and emitted twice. This means that ADM requires more components. This ADM has the advantage over a classic AM that the multiplier is significantly more sensitive than a power detector. The minimum detectable signals for multipliers are typically -120 dBm, while for power detectors they are typically only -60 dBm.

Description of Phase Difference Modulation (PDM)

The simplest PDM is classified as more robust than the ADM (based on the BPSK). This is why the first demonstrators were also equipped with this form of modulation. The simplified structure of a dual-mode radio link for phase-modulated signals is shown in Fig. 10.59.

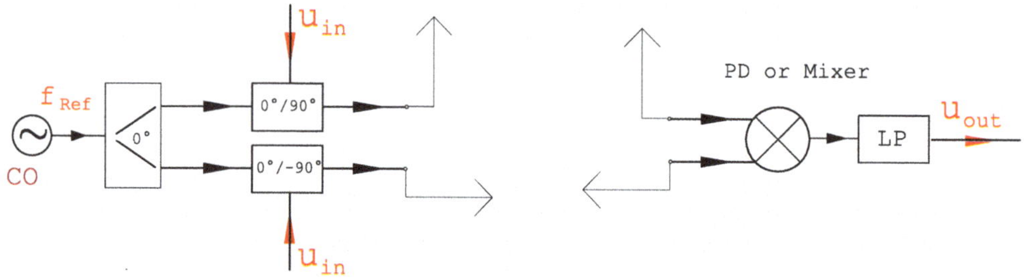

Fig. 10.59: Block diagram of a dual-mode radio link for phase-modulated signals

The mono-frequency RF oscillator (CO) transmits the signal

$$s_{osc(t)} = A_{osc(t)} \cos\left(\omega t + \phi_{osc(t)}\right) \tag{10.117}$$

with the angular frequency $\omega = 2\pi\, f_{Ref}$. $A_{osc(t)}$ contains the amplitude noise and $\phi_{osc(t)}$ the phase noise of the fixed oscillator signal. This signal is split into two paths using the signal splitter, thereby attenuated by at least 3 dB and rotated in phase.

$A_{osc(t)}$ can be replaced by $A_{div(t)}$ and $\phi_{osc(t)}$ by $\phi_{div(t)}$. A 0°/90° and a 0°/-90° phase modulator are used to change the phase values of the two mono-frequency RF signals. The 0°/90° modulator can, for example, be realized with two changeover switches, a through-connection and a high-pass filter with 90° phase rotation. The 0°/-90° modulator, on the other hand, can consist of two changeover switches, a through-connection and a line with -90°=270° phase rotation.

To implement BPSK modulation, the control logic of the two phase shifters is such that either both phase shifters have the value of 0° or the magnitude value of 90° or -90°. Equal losses and the same basic phase shift of the phase shifters allow $A_{div(t)}$ to be replaced by $A_{mod(t)}$ and $\phi_{div(t)}$ by $\phi_{mod(t)}$.

For modulation with a signal whose frequency is significantly lower than the carrier signal, the result is for a logical "0" results in the common mode signal

$$s_{mod(t)} = A_{mod(t)} \cos\left(\omega t + \phi_{mod(t)}\right) \tag{10.118}$$

on both modulation channels. On the other hand, the push-pull signal results for a logical "1"

$$s_{mod(t)} = A_{mod(t)} \cos\left(\omega t + \phi_{mod(t)} \pm \pi/2\right) \tag{10.119}$$

with the different phase rotations for the two channels.

The two phase-modulated signals are emitted via the vertically and horizontally polarized antenna. If the antenna pairs are orthogonally aligned, $A_{mod(t)}$ can be replaced by $A_{ant(t)}$ and $\phi_{mod(t)}$ by $\phi_{ant(t)}$.

The two signals of the receiving antenna pair are fed to the mixer, which is designed as a multiplier. If the same transmission channel can be assumed for both signals, they in turn experience the same attenuation and phase shift.

You can replace $A_{ant(t)}$ with $A_{mix(t)}$ and $\phi_{ant(t)}$ with $\phi_{mix(t)}$. The mixer has output signals in the double frequency range and in the IF frequency range. The latter is only determined by the modulation bandwidth. The low-pass filter only allows the lower frequency range to pass.

For a logical "0", this results in the signal

$$s_{mix(t)} = \frac{A^2_{mix(t)}}{2} \left[1 + \cos\left(2\,\omega t + 2\,\phi_{mix(t)}\right)\right] \tag{10.120}$$

and after low-pass filtering the positive DC voltage $u_{out(t)} = \frac{A^2_{mix(t)}}{2}$.

For a logical "1", this results in the signal

$$s_{mix(t)} = \frac{A^2_{mix(t)}}{2} \left[-1 + \cos\left(2\,\omega t + 2\,\phi_{mix(t)}\right)\right] \tag{10.121}$$

and after low-pass filtering the negative DC voltage $u_{out(t)} = -\frac{A^2_{mix(t)}}{2}$.

The output signal u_{out} (e.g. as shown in Fig. 10.60) can be used as a differential digital receive signal directly in the subsequent electronics.

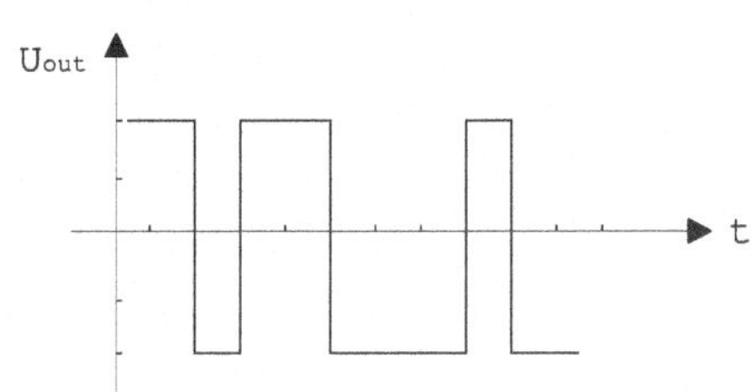

Fig. 10.60: Receive voltage U_{out} at the output of the dual-mode radio link for phase-modulated signals

Although 0°/180° phase modulation is used here, the two channels only need to be modulated at 90° in the phase, which in practice almost halves the necessary bandwidth.

It is also interesting to note that the two signals between the modulators and the mixer correspond to a common mode and a push-pull mode. There are a lot of circuit developments for these two modes, which simplifies the implementation of any further hardware components that may be required. For this reason, the PDM-BPSK method is also very interesting for wired signal transmission.

The common mode and push-pull mode resistances of the cables are easy to calculate. Furthermore, with this transmission it is also immediately apparent that dispersion is not a problem in the transmission channel.

Even two common-mode signals that have been equally modified by non-linear components result in another common-mode signal.

The same applies to the push-pull signal. In the end, the receiving mixer only evaluates whether the signal is a direct or a push-pull signal. The signal shape (sine, rectangle, triangle or other) plays no role at all.

Therefore, non-linearities in the system do not have any particular effect. Signal spreading with a noise signal is easy to implement.

If both phase modulators are used independently of each other, the phase differences have the values values 0°, 90°, -90° and 180°. This is a QPSK modulation. Uout now corresponds to the signal as illustrated in Fig. 10.61.

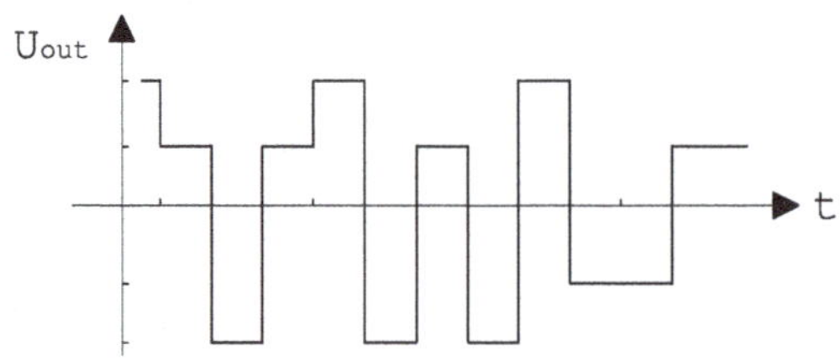

Fig. 10.61: Receive voltage U_{out} at the output of the dual-mode radio link for QPSK and APDM signals

As a result, twice the amount of information can be transmitted with the same bandwidth. However, the output signals cannot be processed as easily as the output signal of the BPSK system. The latter can be processed digitally directly.

The simple receiver shown so far is similar in sensitivity to a detector receiver. If you want to realize sensitivities that are in the range of classic heterodyne systems, an additional conversion stage must be added, Fig. 10.62. In both receivers shown, the oscillator does

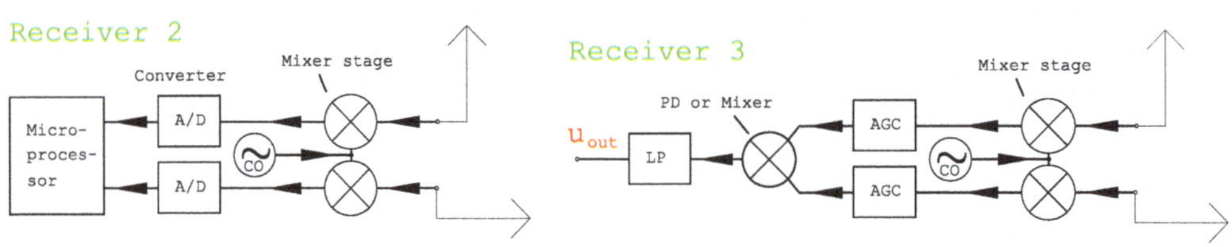

Fig. 10.62: Illustration of second receiver with additional conversion stage

not have to be stabilized. Even its noise is not included in the reception quality. At the IMP, the receiver 3 was installed next to the detector receiver successfully built for a 2.45 GHz system using SMD technology.

Brief Description of Amplitude Phase Difference Modulation (APDM)

In [63] it is shown how the combination of AM and PM modulators to create a dual-mode radio link for amplitude and phase-modulated signals.

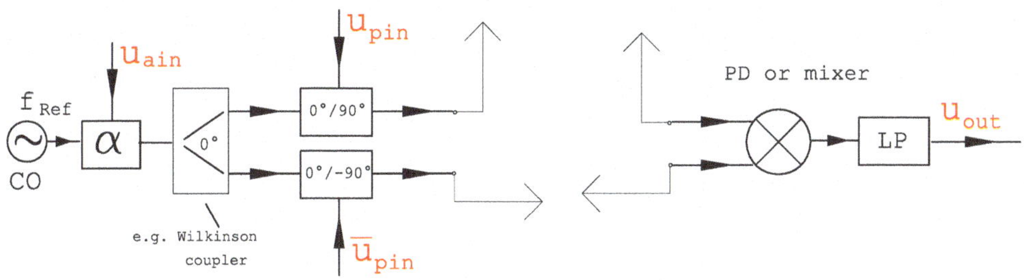

Fig. 10.63: Block diagram of a dual-mode radio link for AM and BPSK modulation

A possible setup of a dual-mode radio link for amplitude and phase-modulated signals is shown in Fig. 10.63.

The RF oscillator 10 transmits a signal with the frequency f_{Ref}. The amplitude of this signal is then modulated by Uain, e.g. using a switchable attenuator (e.g. with 0 dB and 3 dB insertion loss).

This modulated signal is split into two paths using the signal splitter. By means of a 0°/90° and a 0°/-90° phase modulator (12 and 13), the phase values of the two monofrequency amplitude-modulated RF signals are changed by Upin.

To implement a BPSK modulation, the control logic of the two phase shifters is such that either both phase shifters have the value of 0° or the magnitude value of 90°.

The two amplitude- and phase-modulated signals are emitted via the TX antennas. The RX antennas receive both signals and feed them to the mixer. The low-pass filter allows the lower frequency range of the mixer signal to pass. The Uout signal of an APDM system contains four states and therefore two bits. This signal can be evaluated and used by a computer or subsequent electronics.

Further modulation options in conjunction with frequency modulation are shown in [63].

Although 0°/180° phase modulation is used here, the two channels only need to be phase-shifted by 90°, which reduces the bandwidth in practice. Significantly more bits per signal value can again be achieved by uncorrelated control of the phase shifters and/or additional amplitude values (e.g. U = 0 V, then without greater hardware effort).

Description of Frequency-Difference Modulation (FDM)

Since this invention is suitable for implementing the simplest radio systems, so-called static frequency modulation was introduced in [63] in addition to classic frequency modulation (FM). Fig. 10.64 shows an initial block circuit for a possible implementation of both frequency modulations.

With static FM, f1 and f2 are never changed above the frequency. Consequently, only one piece of information (not one bit) can be transmitted. The information lies in the difference between the frequencies. This very simple process is very interesting for RFID applications, for example. The advantage is that the dividers (24 or 25) cannot be set via a computer, but only via a digital word (e.g. dip switches or bonding wires).

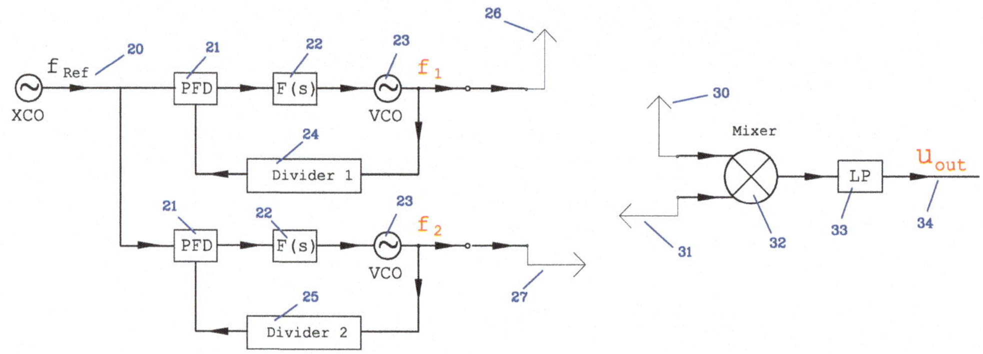

Fig. 10.64: Block diagram of a dual-mode radio link for static or classic FSK modulation (particularly suitable for the lower GHz range)

The realization form in the Fig. 10.64 is interesting for applications in the one and lower two-digit GHz range, as integrated phase-locked loops (PLLs) and even integrated low-cost synthesizers are available there.

The illustrated setup with two synthesizers connected to the same crystal oscillator (XCO: crystal controlled oscillator) 20 shows a unidirectional dual-mode radio link for frequency-modulated signals. The two voltage controlled oscillators (VCO: voltage controlled oscillator) 23 provide the output signals with the frequencies f1 and f2. To ensure that these frequencies are quartz-stable, the RF signals are converted into signals with a frequency of the order of f_{Ref} via the frequency dividers 24 and 25 and the associated divider ratios (e.g. by 1/N and 1/(N+1)).

If there are frequency or phase differences between the signals involved and the reference frequency, the phase frequency discriminators (PFD) 21 output a control signal that passes through the filter 22 and adjusts the VCO if necessary.

The two mono-frequency signals are emitted via the antennas 26 and 27. Antennas 30 and 31 receive both signals and feed them to mixer 32. The mixer has the two output signals with the frequencies f1+f2 and df=f1-f2. The maximum frequency of df is only determined by the modulation bandwidth. In the case of static frequency modulation, this is the maximum permitted value for df. A low-pass filter 33 only allows the difference frequencies df to pass.

The signal Uout is now a low-frequency sinusoidal signal with the frequency df. This alternating signal can be easily evaluated by an electronic frequency counter. If a frequency detector (or PFD) is used instead of a mixer, the output signal corresponds directly to that of a frequency counter with an analog output signal.

In classic frequency modulation, the frequency divider, which may also be controlled as a fractional divider with or without sigma-delta modulation, of one or both synthesizers is modulated quickly. Here you have double the effort for the sender, but significantly less effort for the receiver. If there are significantly more receivers than transmitters, the DM radio is the more cost-effective solution.

Another advantage is that both synthesizers can now be modulated, meaning that each synthesizer only has to synthesizer only has to manage half the frequency deviation. Fast modulation can only be performed within the bandwidth of a phase control loop. With dual-mode radio, twice the bandwidth is available.

The second form of implementation, shown in Fig. 10.65, is a very interesting concept, especially for the two- and three-digit GHz range.

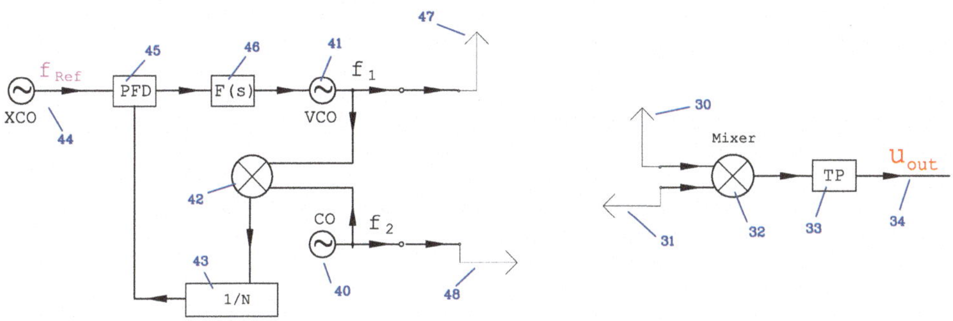

Fig. 10.65: Block diagram of a dual-mode radio link for static or classic FSK modulation (particularly suitable for the upper GHz range)

A fixed oscillator 40 transmits the mono-frequency RF signal with the frequency f2. The mixer 42, the divider 43 and the PFD 45 ensure that the frequency spacing of the output signal of the VCO 41 with the absolute frequency f1 to the frequency f2 is realized in a crystal-stable manner.

The divider 43 must only be able to reliably process the distance df as the maximum frequency. In practice, the divider values of 43 are significantly lower than the divider values of 24 and 25.

As a result, the PLL regulates the phase noise of the VCO 41 much better. This circuit will therefore offer better transmission characteristics than the circuit shown in Fig. 10.64. The receiver is constructed in exactly the same way as before.

Brief Description of Amplitude-Phase-Frequency Modulated DM Radio (APFDM))

In contrast to the frequently used IQ modulators, where there is no real difference between PM and FM, the identical receivers presented here provide different output signals for phase and frequency modulation.

A DM receiver can therefore also demodulate all three pieces of information. However, a so-called clock recovery circuit must be implemented to evaluate such an output signal. This circuit is standard in current radio transmission links. The advantage of this is that the amount of information increases significantly with the same bandwidth.

Brief Description of 4-Mode Differential Modulation (X4DM)

All the methods described above can be implemented directly for an application with right- and left-circular free-space modes. This will be illustrated using the example of phase modulation of all four free-space modes.

Example: P4DM

On the receiver side, the familiar receiver (e.g. Fig. 10.59) is set up twice and implemented with one linear and one circularly polarized antenna pair. On the transmitter side, a 1 to 4 signal divider can be used instead of the signal divider 11.

By using the phase shifters 12 and 13 twice, linear and circularly polarized waves are now fed using a PDM. If, for example, a double BPSK is performed, the output signal shown in Fig. 10.60 is lowered or raised in both channels by the second modulation to a minimum of half the level or a maximum of one and a half times the level. In any case, the voltage is still significantly above or below zero volts.

Consequently, twice the amount of information can be transmitted in this way, even with the same bandwidth. However, the signal quality and therefore the bit error rate is also reduced in this example.

If you use a cross dipole antenna with 90° phase shifters, you can use this one antenna for all transmit or receive all four signals.

10.6.1 Advantageous Design of the Dual-Mode Radio

The purpose of DM radio is to set up a transmission path for two free-space modes by utilizing the correlation of the signals. With this DM radio, the full transmission capacity of the radio channel is always used. If all four modes of the free-space transmission path are used, the transmission capacity can be doubled again.

For the simpler modulation forms of DM radio, no additional signal processing is required for the transmission of purely digital signals. Both modulation with digital signals and demodulation in digital signals are achieved using simple hardware solutions. A unidirectional DM radio transmission system requires a minimum of only one oscillator for the transmitter and no oscillator in the receiver. The phase noise of the oscillator has no influence on the sensitivity and therefore the range of the transmission system.

By evaluating a correlated signal, it can be assumed that even the slight non-linearities in the form of adjacent channel crosstalk have a much smaller effect. This means that a synthesizer is only required for the transmitter if it is necessary to ensure that a radio channel is maintained.

All in all, the components required for the transmitter and receiver are reduced. Depending on the selected modulation, the required components are reduced to an RF detector designed as a multiplier.

The immense reduction in hardware costs is therefore also reflected positively in the pricing. The fundamental elimination of a low-noise synthesizer also allows the implementation of broadband radio transmissions in the high two-digit and three-digit GHz range.

In addition to the price reduction for current RF transmission systems, the cheaper manufacturability of RF transmission systems offers applications, there is also scope for new applications that are still unattractive due to current costs.

All known modulation methods with the known characteristics of bandwidth requirements, bit error rates and the like can be used. However, the associated modulators are significantly simplified. For example, a multiplier (Gilbert mixer, ring mixer, etc.) can be used to produce very good receivers for the three basic modulations AM, PM and FM with different output signals. Therefore, these three modulation types can even be combined at the same time, which saves even more bandwidth.

Doppler Effect

The fact that only the correlation between two signals is evaluated means that the Doppler effect no longer has any influence on signal transmission. If both modulated and correlated transmission signals originate from the same oscillator, its phase noise no longer has any influence. Broadband noise in the oscillator signal is used to spread the signal (similar to OFDM), which reduces fading effects in the radio channel.

In the receiver of dual-mode radio systems, the two signals from both antennas are usually converted into the intermediate frequency range (IF range) using a multiplier or mixer. This ensures that the logarithmic measurement has twice the sensitivity of an (amplitude) detector receiver.

As the receivers can be extremely simple in design, they generally require very little energy for the power supply. For simple systems, such as those used in RFID applications, you can get by without any additional power supply. For example, two mono-frequency signals of the same frequency can be used in parallel with a so-called FDM signal to supply power via dual-mode radio. The two signals are rectified using a passive ring mixer and made available to the system via voltage multiplication.

Various transceiver units designed as phase-difference modulation have been set up as part of Bachelor's and Master's theses and in a first spin-off company (www.hhft.de).

An eye diagram and the corresponding spectrum are shown in Fig. 10.66 to illustrate the reception quality.

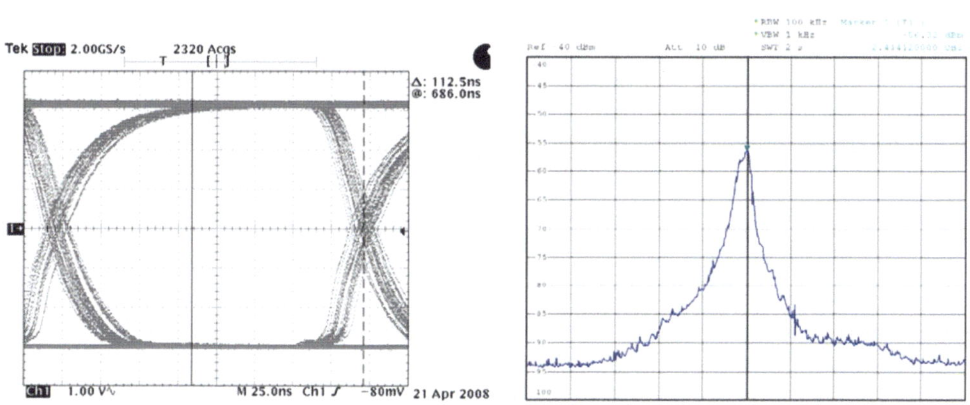

Fig. 10.66: Eye diagram and associated spectrum for the first demonstrator with a signal modulation bandwidth of 5 MHz and an additional noise bandwidth of 10 MHz

The first results already showed that the DM radio works and were summarized in [64].

The noise signal at the input of the oscillator (here VCO) can be used to generate a signal spread including level reduction. As you can see, the classic modulation spectrum is no longer recognizable. The signal can only be demodulated with a DM receiver and is therefore very tap-proof.

The structure was even robust against antenna rotations of ±25°.

It has been shown that symmetry errors in the structure of the DM system of up to 60° do not cause any measurable deterioration in the received signal. The attenuation difference may also be up to 5 dB.

A large number of DM systems have already been set up and successfully tested, including direct connection to the CAN bus. Fig. 10.67 shows a 2.45 GHz DM transceiver. The surrounding ICs are only used for further signal processing. The actual transceiver is constructed internally with individual transistors and discrete SMD components.

Fig. 10.67: DM transceiver from [72] in use as a tap-proof radio link

The results of investigations for radio relay applications are presented below. Two linearly polarized Yagi antennas were used as antennas for the study. In other words, one was aligned vertically and one horizontally. These were used to show that rotating an antenna by up to 35° also did not result in a deterioration of the reception signals.

Based on these results, it is ensured that the new DM radio for radio relay applications works with the PDM.

A master's thesis on PDM signal transmission via cables was also completed. With improved modulators, an error-free transmission rate of 60 Mbit/s was achieved for the frequency range 30-130 MHz (above DSL standards) for a simple telephone cable.

DM technology can therefore also be used advantageously for the use of power lines for data communication.

Chapter 11

Radio Frequency Identification (RFID)

In `Radio Frequency IDentification` RFID technology, a German author (Klaus Finkenzeller, [34]) is the leading international author for a GHz application for the first time. In Germany, RFID technology is now the top-selling technology in the RF sector.

This chapter provides a brief overview of RFID technology as a whole, [5][1]. The UHF system is described in more detail below. Many basic techniques for system design from a hardware development perspective are explained using a UHF example.

11.1 Brief Overview of all RFID Systems

The first major innovation in logistics was the introduction of the barcode. This innovation triggered a veritable revolution in many areas of product recognition, as this barcode made automatic detection possible for the first time. Nowadays, this code can be found on almost all everyday products, from yogurt pots in the supermarket to luggage at the airport.

Optical reading of the code is possible without contact at distances in the centimeter range. The disadvantage of the barcode is its limited data capacity and the fact that it is not possible to reprogram a barcode, e.g. to add information.

Large data capacity and reprogrammability characterize the storage of data on silicon chips, in short chip cards. The best-known example is the SIM card (`Subscriber Identity Module`) in smartphones. This makes it possible to store and change large amounts of data. The only limitation with chip cards is the need for electrically conductive, mechanical contact between the card and the reader for reading and writing the information.

[1] Here I would like to express my sincere thanks to Mr. Andreas Backes for the release of his master's thesis. Many passages in this chapter are based on the work of Mr. Backes.

H. Heuermann, *Radio Frequency Technologies*,
https://doi.org/10.1007/978-3-658-50170-9_11

The advantages of barcodes and chip cards are combined in contactless RFID radio systems with an integrated silicon chip as data memory. Ideally, the energy required to read and write the data is also provided via the radio field in addition to the data.

The transceiver unit is known as a `reader` and the wireless memory chips in combination with an antenna are known as transponders or `tags` for short. Use in the mass market always goes hand in hand with the desire for more cost-effective devices. While tags are inexpensive, readers are relatively expensive.

Advantages of RFID Systems

RFID systems are used in a wide variety of automatic identification applications. The use of RFID systems is particularly advantageous in the field of industrial automation. The growing requirements of quality management and quality assurance in automated production can be easily realized with RFID systems.

Production and measurement data can be easily stored and stored on the object itself during production. They are thus available and documentable at any object at any time. This also makes production more flexible. Production parameters can be saved on the object and read out directly in the respective production line.

Objects can be removed from the series at any time and added again later without confusion. A failure of the central computer also poses no problem with regard to the unique identification and localization of objects due to the decentralized storage of data on RFID tags [34].

Fig. 11.1: Illustration of a typical reader application, photo by Pepperl + Fuchs, product overview, optical identification, RFID

Since the information is stored directly on the object, the distance between the so-called information level and the object level is very small, [83]. With barcode systems, considerably more steps are required between object detection and information retrieval.

For example, the system must request the information from the stored database after manual scanning. With RFID systems, it is directly available.

Objects often have to be specially aligned to capture barcodes, as the scanner needs direct optical access to the barcode. In many areas, scanners are guided by hand for this purpose. Compared to other Auto-ID processes, RFID technology requires less manpower (see Fig. 11.1), which brings with it great potential for rationalization [34].

Another advantage over barcodes in particular is the fact that RFID tags are not impaired in their function by soiling, as occurs during production. High temperatures, humidity, dust and chips do not affect RFID systems. In contrast, optical scanner systems such as barcode readers are severely restricted in their functionality by dirt [34].

11.1.1 RFID Standards

Standardization is of fundamental importance for all communication systems, including connectors. By defining certain frequency ranges, communication protocols (also referred to as `coupling` in RFID), modulation methods and transmission power, smooth operation between readers and tags from different manufacturers is guaranteed. The two most important standardization bodies are the

- European Telecommunications Standards Institute (ETSI)

for Europe and the

- International Organization for Standardization (ISO)

at international level. UHF RFID systems, for example, regulate the standards

- ETSI EN 302 208-1 and
- ISO-IEC_CD 18000-6C .

The ISO standard is the basis of the Electronic Product Code Standard EPC Class 1 Gen 2 for RFID tags and readers of the EPCglobal™[2].

[2] Read more at: www.epcglobalinc.org/standards/uhfc1g2.

The aim of EPC is to increase the transparency of the flow of goods within the supply chain and to improve the exchange of quality information on products within the chain. RFID systems offer the best prerequisites for this.

11.1.2 RFID Frequency Ranges

The following frequency ranges are approved for commercial RFID systems:

- 135 kHz, low frequency, LF
- 13.56 MHz, high frequency, HF
- 868 MHz in Europe, 915 MHz in den USA, ultra high frequency, UHF
- 2.45 GHz, radio frequency, RF; microwave, MW
- 5.8 GHz, RF, MW

The type of physical coupling of the system is directly linked to the operating frequency. Coupling refers to the physical effect that enables communication between reader and tag basically. In the case of contactless identification systems, these are either

electrical, magnetic, or electromagnetic

fields. Electrical and magnetic coupling is often referred to as capacitive or inductive coupled RFID systems, which are not discussed further here.

Systems with electromagnetic coupling are generally referred to as microwave systems. In accordance with their functional principle, microwave systems are often referred to as `backscatter systems`.

The function of backscatter systems is described in detail in section 11.4 of this chapter.

Capacitively and inductively coupled systems in the LF and HF frequency ranges operate in the near field of the reader antenna. The UHF system and the microwave systems use the far field of the antenna for electromagnetic coupling. The transition from the near field to the far field at a distance r from the antenna is approximated via

$$r \approx \frac{\lambda_0}{2 \cdot \pi} \tag{11.1}$$

and linked to the wavelength λ_0 of the free space. This approximation formula applies to antennas whose dimensions are small compared to the distance r. Frequency f and wavelength are given by

$$f = \frac{c_0}{\lambda_0} \quad , \tag{11.2}$$

where c_0 is the speed of light.

At the start of the far field, the electromagnetic wave `non-reactive` detaches from the antenna. This point therefore defines the maximum theoretical read range for capacitively and inductively coupled systems. At 13.56 MHz, for example, the near field ends at a distance of approximately 3.5 m.

Close coupling systems with ranges of a few centimeters and remote coupling systems with ranges of up to one meter are commercially available. Initially, the majority of RFID systems used were inductively coupled remote coupling systems.

In quantitative terms, these make up the majority of systems and have the largest market share in the automotive immobilizer sector. also have a qualitatively large field of application. UHF systems are growing from year to year and are now well above LF systems and slightly above HF systems in terms of sales.

In electromagnetically coupled systems, e.g. in the 5.8 GHz range, the far field already begins after a distance of 5.2 cm from the antenna. Electromagnetically coupled `long-range systems` in the UHF and MW frequency range achieve ranges of 10 m with passive tags and up to 40 m with active tags using the backscatter method. In [50] an RFID system is described in which you can locate the tag up to 5900 m and also receive the distance information.

The propagation of the electromagnetic wave in the far field or free space shows additional dependencies on the frequency. Depending on the frequency, the antennas generate complex transmission and receive characteristics of varying intensity.

These characteristics also vary across all solid angles. The objects in the open space can also have different reflection properties over the frequency. These interactions depend on the ratio of the wavelength to the size of the object that the radiation hits.

At frequencies well below 1 GHz, there are many coupling and propagation effects based on refraction and diffraction of the electromagnetic wave. At higher frequencies, the propagation properties of electromagnetic waves become increasingly similar to those of visible light.

Since the propagation behavior of light has the character of rays, there are only reflections at high frequencies and increased losses when penetrating media, as known from optics.

In free space, the energy density decreases with increasing distance due to the increasing spatial distribution of the electromagnetic wave in the far field. The resulting `free space attenuation` α_F increases with the distance r and the frequency f according to

$$\alpha_F^{dB} \approx -147.6\,dB + 10 \cdot \lg(r) + 20 \cdot \lg(f) - G_T^{dB} - G_R^{dB} \tag{11.3}$$

The so-called antenna gains of the tag G_T and the reader G_R, which can be used to increase the range, counteract the free-space attenuation. However, the illuminated area becomes smaller with higher profits, which is often undesirable.

Especially during the day, you often don't know where the reader is. This is why antenna structures are used here that have an almost spherical radiation pattern and therefore a gain of 1.

The attenuation in the far field is considerably lower than the attenuation in the near field. In the case of electromagnetically coupled systems, only the far-field attenuation applies to the much larger part of the radio link, whereas inductive systems are subject to the attenuation of the near field over the entire radio channel.

The operating frequency and the associated physical coupling of an RFID system therefore have a major influence on its properties. In general, UHF and MW systems allow a greater range than inductively coupled methods due to their higher frequency. If only this group is considered, the range naturally decreases with increasing frequency according to the formula (11.3).

In the following, readers, tags and their communication are presented with the main focus on the UHF band at 868 MHz, as these have become the most popular.

This is partly due to the very low costs for the UHF tags, which are only around 0.10 € . Tags at 13 MHz cost at least 0.50 € and the other LF tags at least 1 € to 25 € .

11.2 Reader Concepts for the UHF Band

Readers are read-write devices and enable access to the information stored on the tag. They have the function of both the software connection to the peripherals and the radio connection to the tags. They therefore consist of a digital and an analog part. The digital part processes the messages and the analog part transmits them via the radio channel.

The following section describes the separation of the analog front end and message processing using a digital signal processor.

11.2.1 Analog Front End with Digital Signal Processor

The classic approach to implementing an RFID reader is shown in Fig. 11.2 as a block diagram. This approach is therefore suitable for implementing almost any wireless communication unit. The core element is a digital processor, possibly even a signal processor, or DSP for short. The DSP is used for signal generation, digital filtering, control of the synthesizer and demodulation of the received signal. It also provides the information for the readers environment.

The remaining components form the analog send/receive path, the front end. The front end is used for frequency conversion, amplification and filtering. The components of the front end limit the flexibility of the reader in terms of the possible frequencies.

The filters in particular can generally only be used for a single RFID band. The requirement for high power generation with simultaneous linearity conflicts with broadband use over large frequency ranges, such as 868 MHz and 2.45 GHz.

When transmitting, the DSP generates a modulated baseband signal. The modulation of the baseband signal affects the spectrum of the upward mixed signal in the 868 MHz band. The aim is to increase spectral efficiency through baseband modulation. This minimizes crosstalk between the reader and adjacent channels.

The generated baseband transmit signal is mixed with the pump signal of the synthesizer, which is also referred to as local oscillator signal (LO), in the RF band around 868 MHz.

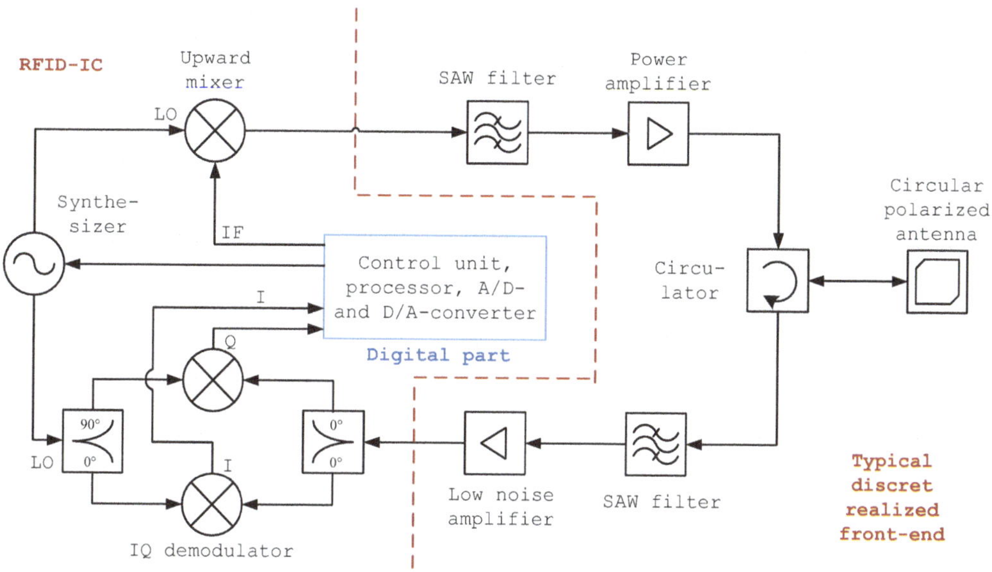

Fig. 11.2: Illustration of a typical reader concept and its subdivision

The harmonics generated during mixing are suppressed with the surface acoustic wave (SAW) filter. In principle, it would also be desirable to suppress the harmonics generated by the power amplifier (PA) during amplification.

However, this often fails due to the low power compatibility of current SAW filters. A maximum of 33 dBm (corresponds to 2 W) may be emitted in the UHF band. However, SAW filters are typically only specified with a maximum input power of 10 dBm.

Either particularly high IP3 requirements are placed on the PA due to the harmonic problem or a dual-circuit resonator filter is realized using SMD components.

The transmitted signal reaches the antenna via the circulator, which usually also has a band-pass characteristic. The received signal is received via the same antenna and fed into the receive path via the circulator. Alternatively, a directional coupler or two individual antennas can be used instead of a circulator.

In the receive path, the SAW filter filters the spectral components of applications in directly adjacent frequency ranges from the received signal. This also prevents power-intensive signals, such as GSM signals at 900 MHz, from driving the low-noise amplifier (LNA) into compression.

In the two paths of the IQ demodulator, the signal is downmixed with the DSP after division into the intermediate frequency range for digital processing. Because the LO signal is mixed directly once - i.e. in phase - and once with a 90° phase shift, two signals are present at the DSP.

A signal is in phase (In phase, I) and a signal is shifted by 90° (Quadrature phase, Q). By a further digital phase shift of -90° of the Q signal in the DSP, a mirror frequency suppression of approx. 50 dB is realized [50].

As the receive signal is downmixed with the same signal that is used to upmix the transmit signal, the carrier frequency after downmixing is in the IF band[3].

After this down-conversion, the data signal modulated onto the carrier is in the range from a few kHz to the MHz range, depending on the data rate.

This places high demands on the phase noise of the LO synthesizer. The phase noise should be below the receiver noise, otherwise it could overshadow signal components. This mixing concept allows the carrier frequency to be suppressed by a high-pass filter in the intermediate frequency range.

The signal is filtered in the DSP by a digital bandpass-filter and then demodulated. The center frequency, bandwidth and slope of digital filters can be varied quickly using software by varying their coefficients.

11.2.2 RFID Transceiver

As the use of digital signal processors involves a great deal of programming, special RFID transceiver chips are an interesting alternative. These integrate the majority of the hardware described in section 11.2.1 and supplement it with logic specially tailored to RFID requirements.

In addition to signal generation in the baseband, these chips also demodulate the received signal and provide pre-implemented digital filters.

An example of this type of chip is the R1000 UHF RFID transceiver originally developed by Intel® and now acquired by IMPINJ[4], which combines this functionality in an 8 x 8 mm housing.

In the block diagram of the R1000 in Fig. 11.3, many striking elements of the approach presented in section 11.2.1 can be found. In addition, an IQ modulator can be used for upward mixing, which allows single sideband and phase-reversal modulation.

After the reader, RFID tags for the UHF band are presented below.

11.3 Transponder for the UHF Band

Modern RFID tags only consist of a chip and an antenna. The antennas are often metallic structures that resemble a symmetrical dipole antenna in the broadest sense.

These are often printed on self-adhesive films. One example is from the company Avery Dennison, which is shown in Fig. 11.4.

In addition to the antenna function, the printed structure (black) also adapts to the base impedance of the chip. The logic for communication with the reader and a data memory are integrated in the chip (gray).

[3] If stands for *intermediate frequency* and is so low frequency that it can be sampled with an A/D converter.

[4] This IC described here already has the successor R2000.

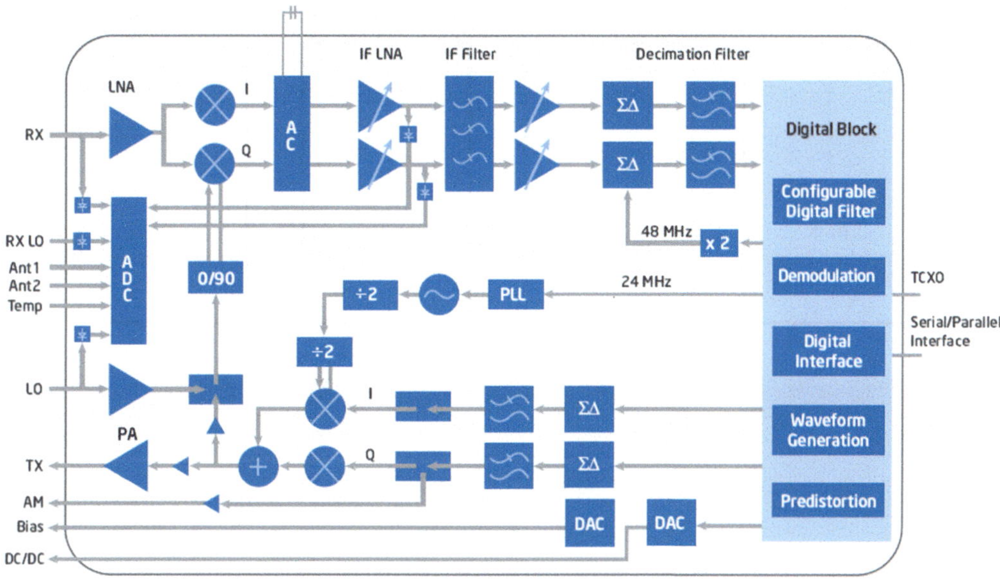

Fig. 11.3: Functional block diagram of the R1000 reader IC from IMPINJ

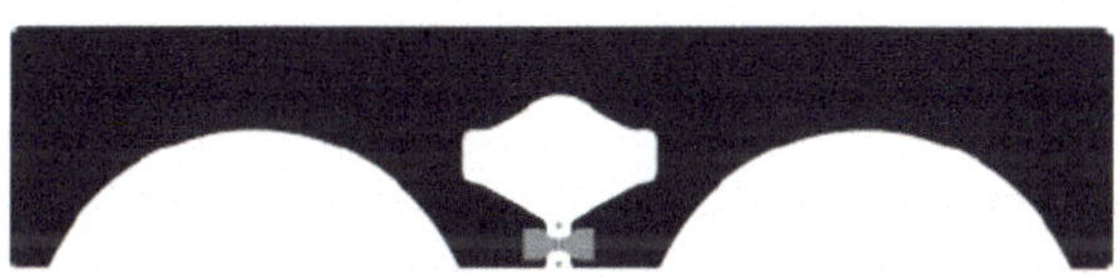

Fig. 11.4: Sketch of the UHF tag from Avery Dennison (type AD-662uDNA) consisting of an omnidirectional symmetrical antenna and a semiconductor circuit with the dimensions of 90 mm·19 mm

The tag ICs are also offered individually and their data sheet also includes designs for antennas, including their radiation characteristics, as with the LXMS21ACMF-183 model from the Murata company.

The basic internal structure of these tags is illustrated in the Fig. 11.5. It is important to note that the front end contains a sensitive rectification circuit that generates the energy for the tag from a UHF supply signal, which is explained in more detail below.

Tags are divided into several categories according to their performance. The categorization of the tags is based on the class definition of the EPCglobal™standard [24] or also in the ISO/IEC 18000-6:2013.

The EPC describes the interface protocols, which were first published in 2004 and are freely available in the latest version Gen2 v2.1 in 2018.

This establishes four classes of tags.

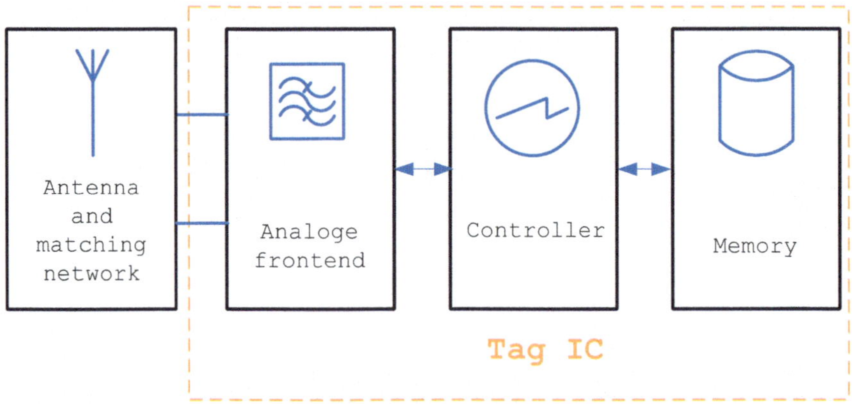

Fig. 11.5: Basic structure of a UHF tag

Class 1: This simplest form involves passive tags without their own energy source, which are referred to as `Identity Tags`. As a minimum, they must provide memory for an electronic product code (EPC) and a tag identifier (TID).

In addition to the command set required for communication, a function for permanently deactivating the tag is implemented (`"kill" command`). An extended user memory and password protection are also provided as options in the standard.

All higher tag classes must also support this basic functionality and must not affect the function of class 1 tags or their communication. Unlike class 1, classes 2 to 4 have so far only been of an informative nature. They are not yet standardized.

Class 2: These so-called `Higher-Functionality-Tags` should have an extended memory and a longer TID than class 1 tags.

Class 3: Because of their integrated energy storage, class 3 tags are referred to as semi-passive tags. The energy from the storage device, e.g. a battery, is used to power the tag IC.

With class 1 and 2 tags, this energy must be taken from the radio field, which reduces the range compared to semi-passive tags. Semi-passive tags cannot transmit independently; communication is still based on the backscatter principle. They have all the features of class 2 tags.

Class 4: In addition to reader-to-tag communication, fully active class 4 tags also allow tag-to-tag communication and support the functionality of all subordinate classes. In the case of tag-to-tag communication, they use energy from their memory to actively send information, which distinguishes them from all other classes.

11.4 Basic Conditions for the Protocols and Regulations of UHF Band RFID Devices

Communication Between Reader and Tag in the UHF Band

As the class definitions in the previous section show, communication - with the exception of the special case of class 4 tags - only takes place between reader and tag. Here, all communication starts with the reader. The reader is the "interrogator" and the tag answers. Such systems are referred to as `interrogator talks first (ITF)` systems.

In addition to the question itself, the reader also specifies the exact form in which the tag should respond during the request and which tag should respond at all. There are usually several tags in the radio field of the reader. The communication process described below is shown schematically in Fig. 11.6.

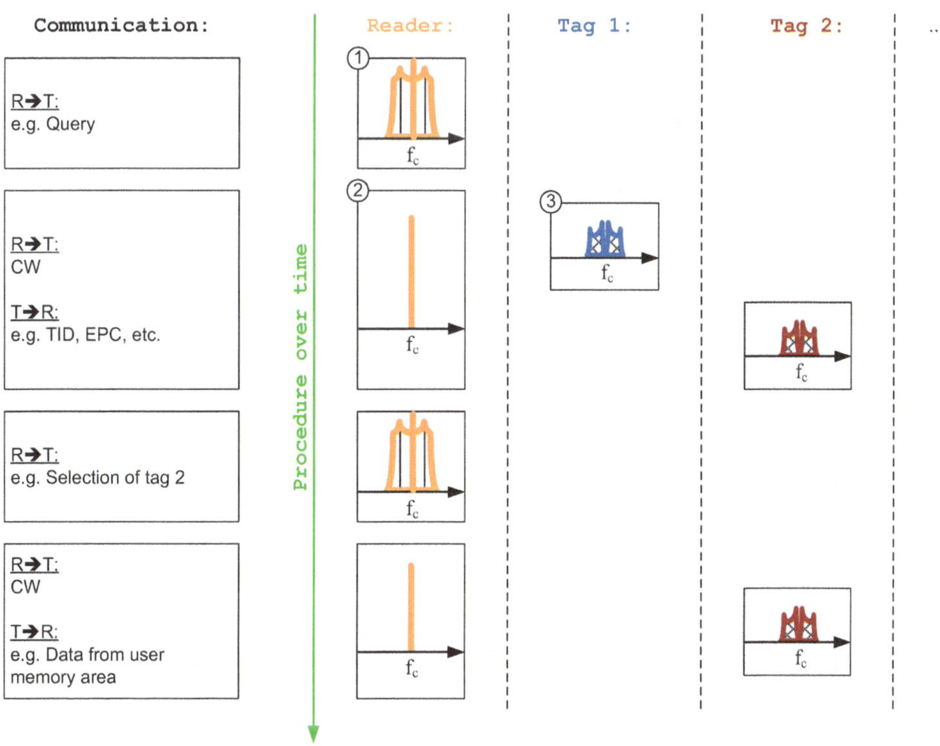

Fig. 11.6: Display of spectra over time during communication between reader and tags

At the start of a communication, the reader starts a query in which it requests all tags in its radio field to transmit their TID or EPC number at a defined data rate and data coding.

The reader transmits a double-sideband, single-sideband or phase-reversal amplitude modulated signal (DSB, SSB or PR-ASK) with pulse interval data encoding and a data rate in the range of 40 to 160 kHz.

This signal induces a voltage at the tag's antenna which, in addition to transmitting data after rectification with Schottky diodes, also serves to supply the tag with power [34]. In Fig. 11.6, this signal is marked with a (1).

After the request, the reader emits a constant sinusoidal signal (or continous wave: CW), see Fig. 11.6, (2). This signal is still partly used to supply the tag with power.

Another part of the signal is reflected by the backscatter cross-section σ of the tag antenna and thus returns to the reader. The tag transmits the requested information about the change in this backscatter cross-section (Fig. 11.6, (3)).

This modulation causes the tag to change the amplitude or phase of the reflected CW signal. For this purpose, an additional real or complex impedance is switched on or off in addition to the input impedance of the tag. The change in the backscatter cross-section gives this method the name σ `modulation` or `backscatter modulation`.

The tag executes this modulation with the data rate specified by the reader and FM0 baseband or Miller coding. This is also specified by the reader within the request. The modulation of the tag is either an amplitude modulation `Amplitude Shift Keying` or a phase modulation `Phase Shift Keying`; this is determined by the manufacturer of the respective transponder.

The responses of the passive tags in the radio field are thus modulated onto the CW signal transmitted by the reader. If two or more tag responses overlap when they are received in the reader, the implemented coding results in invalid symbols.

In this case, the reader sends a repeat request. If the reader has all the TIDs or EPCs of the tags in its radio field, it can put specific tags into sleep mode, read and write individual tags or deactivate them permanently.

This is done via several precisely defined sequences of reader command blocks and CW blocks, within which the tag or tags respond. The exact time sequence and all commands can be found in the standard [24].

Regulations in the UHF RFID Sector and the Resulting Filter Requirements

The regulations of ISO-IEC_CD 18000-6C are relevant for the design of a UHF RFID system. For single-channel operation, this standard stipulates that the reader and tag share one of eight possible channels between 865.6 MHz and 867.6 MHz during communication using the `Time Domain Multiplex (TDM)` method.

This corresponds to a channel bandwidth of 250 kHz. The request from the reader and the response from the tag are separated in time.

During the tag response, a type of `Frequency Domain Multiplex` procedure is used, as the reader transmits a CW signal in the center of the channel and the tag responds simultaneously offset to the right and left in the band via backscatter modulation of the CW signal, as presented in section 11.4.

The distance of the tag response from the center of the channel depends on the data rate used. This spectral distribution is shown schematically in Fig. 11.7.

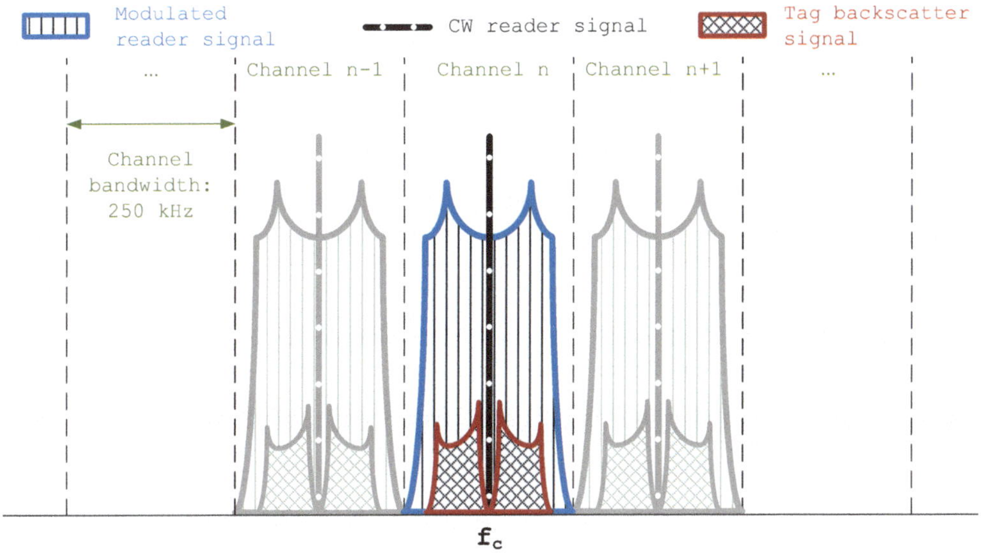

Fig. 11.7: Spectrum distribution over the 8 channels in the RFID UHF band

A channel filter is required to protect the tag response of the channel used against adjacent channel interferers - see Fig. 11.7 in gray. Ideally, this filter would have a passband width corresponding to the channel width and strongly suppress the adjacent channels outside the band.

This would require an infinitely steep filter slope, which is not technically feasible. In order to sufficiently suppress the immediately adjacent channels, the passband width of the channel filter is therefore selected to be smaller than the channel width.

Since communication between reader and tag should be possible in each of the eight channels, the center frequency of the filter must be variable so that it can be tuned to each channel.

For the European area, the ETSI provides a special regulation for communication. Before sending a request, readers must check whether the channel they are to use is free. This `Listen-Before-Talk (LBT)` check must be carried out before each communication sequence.

The channel is considered free if no signal stronger than -96 dBm is measured in the channel. During a communication sequence, the reader permanently blocks the channel for other readers by alternating between a strong, modulated data signal and an equally strong CW signal.

11.5 Development of Front-End Circuits for UHF Band RFID Devices

Design of Standard UHF Band RFID Devices

The development of a reader for shorter range requirements is relatively short. This starts with a current market analysis of the existing reader ICs if you want to produce large quantities yourself, or directly with the module, which is already available for prices of around EUR 20 €.

This is where the difficulty of comparing these ICs or modules arises, as the manufacturers generally do not provide simple range specifications in use with associated tags.

This means that you have to compare the properties generally to be observed for transceivers from the individual data sheets in a table you have created yourself:

- Frequency range, modulation forms, sensitivity, RX amplification
- Data rates, digital interface (e.g. PCI interface)
- TX output power, input and output reflection coefficients
- RFID standard conformity
- Baseband properties (digital filter channel filtering if necessary)
- Blocking immunity, noise figure
- If necessary: Operating voltage and power consumption

As an example, a reader IC or module from Murata is shown in Fig. 11.8 as a block diagram.

The corresponding module is shown in Fig. 11.9.

These modules have a single band-pass filter, which is usually realized as a SAW filter and is permeable for the entire RFID band. This filter, including the necessary matching circuits and the antenna, forms the entire front end.

The CW block problem: The transmission power (for the CW signal) is typically 10 dBm and the matching values of the TX chain are at best -20 dB. This results in a back reflection of the TX signal of -10 dBm.

With a good blocking immunity of 50 dB, the receiver can still detect -60 dBm. However, the IC itself has significantly better sensitivities, which are well below -100 dBm.

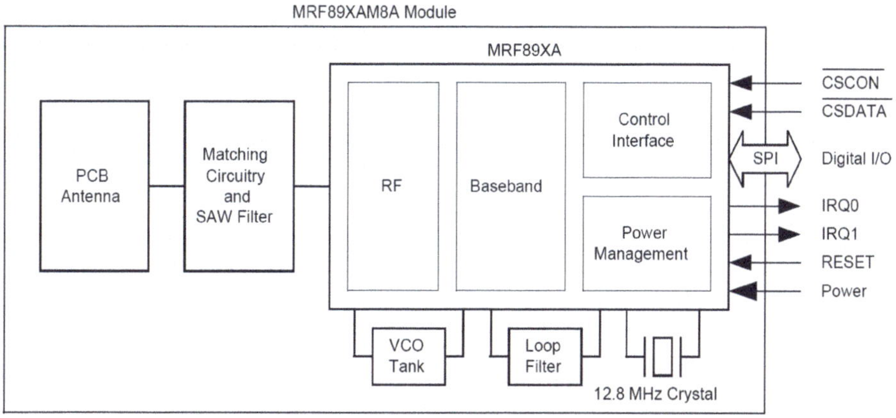

Fig. 11.8: Block diagram of a reader IC/module from Murata for the RFID UHF band

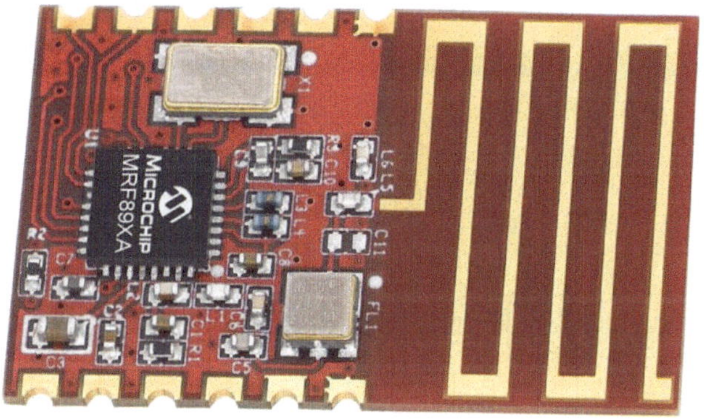

Fig. 11.9: Photo of a module from Murata for the RFID UHF band with a board length of just 2.7 cm

The possibilities of the IC are far from fully utilized with this very simple front end. If you want to develop systems that have a significantly greater range and perform applications reliably, as shown in Fig. 11.1, then you need to develop a significantly improved front end.

Use of Power Amplifiers and Extended CW Block Problems

In order to supply the tags at greater distances with sufficient energy via the CW signal, the use of a power amplifier (PA), as illustrated in Fig. 11.2, is urgently required. This PA is set to the maximum permissible output power of 33 dBm.

In this case, the TX and RX chains must be set up separately. If you can use two antennas, the reader will be larger, but the separation will be slightly better than if you work with a circulator.

With this system, the noise figure can also be improved by an additional LNA. However, the already introduced CW block problem remains. In practice, this block problem is much greater than already imagined due to propagation effects.

The 2 W-CW signal from the reader enters the receive chain as a TX-RX cross-talker with typically 25 dB attenuation via the circulator (path: path 3 shown in Fig. 11.10). Furthermore, as already explained, this signal reaches the RX path via the antenna mismatch with at best 20 dB attenuation.

Ultimately, the 2 W-CW-TX signal reaches the transceiver input via reflections on metallic surfaces or direct transmit path-receive path crosstalk (Fig. 11.10). With large metal surfaces, it can even be the largest interferer.

This crosstalk is significantly reduced by choosing a circular antenna, as the change in the direction of rotation of the fields through the wall means that the polarization no longer matches that of the reader antenna.

The received signal is suppressed quite strongly (typically 20 dB). However, in the worst case, these three crosstalkers can add up constructively.

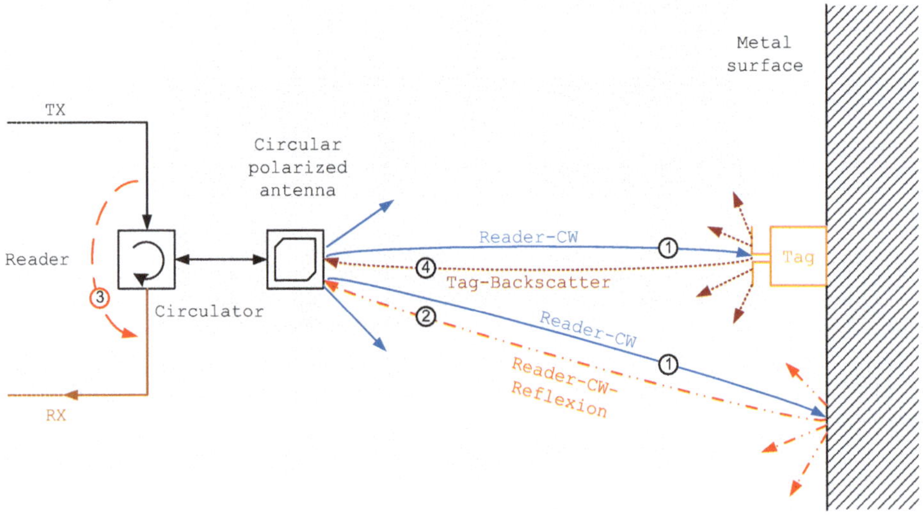

Fig. 11.10: Possible crosstalk and reflections of the 2 W-TX signal in the receive path

The crosstalk components of the CW signal that enter the receive path must be suppressed there to such an extent that they satisfy the blocking immunity of the receive IC. To achieve a sensitivity of -110 dBm with a blocking immunity of 50 dB, no higher power than -60 dBm may be applied to the receiver.

The 33 dBm Tx signal arrives at the RX branch with around 18 dB attenuation and a power of 15 dBm. There it would drive the LNA in compression with this high input power.

It must therefore be suppressed at this point.

The Electronic Circulator

The crosstalk energy of the CW signal can be suppressed by an antiphase signal with the same amplitude. The corresponding idea was published in [67] and implemented as part of a final thesis at the IMP, [46]. The rather complex solution for this is shown schematically in Fig. 11.11.

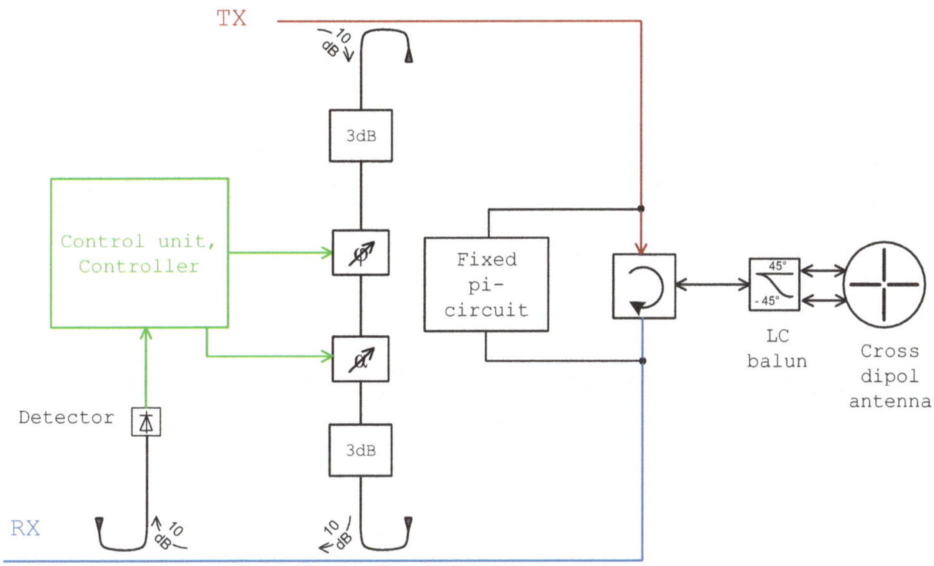

Fig. 11.11: Circuit solution for crosstalk cancellation using the electronic circulator

In general, the first thing to do is to minimize all possible interference signals. Therefore, a very good adjustment of the antenna is necessary here. In the same way, the over-speaker on the circulator can be reduced in advance by means of a permanently installed compensation circuit.

This circuit of a phase shifter with high-impedance input impedance is shown as a fixed π-LC circuit in Fig. 11.11. With such a simple hardware solution consisting of 3 SMD components, an interference reduction of around 20 dB is achieved by phasing out the signal.

The precise measurement of the circulator (with 50 Ohm termination instead of the antenna) carried out beforehand is the essential first step in this hardware crosstalk reduction step.

The further electronically performed reduction step of the electronic circulator, which is shown in the left part of the Fig. 11.11, is carried out as follows:

On the TX coupler, a small part of the TX signal is decoupled for the purely electronic part of the circulator, then rotated in phase and adjusted in amplitude so that this signal cancels out with the crosstalk signal after coupling into the RX coupler.

The two 3 dB attenuators are necessary because the adjustable phase shifter and the adjustable attenuator are not well adapted for all setting values. This mismatch would prevent the couplers from working properly.

The associated control is performed by a microprocessor, which also provides information about the depth of the depth of the erasure by means of a detector measurement.

The setting values for the phase shifter and the controllable attenuator for different frequency values are determined after the reader is switched on. To do this, only the 2 W-TX signal is applied and the phase and magnitude are changed by a more or less intelligent algorithm until the signal at the detector output can no longer be minimized.

This circuit then runs stably with suppression values of typically 50 dB. Fig. 11.12 shows measured values of a circuit set up in this way.

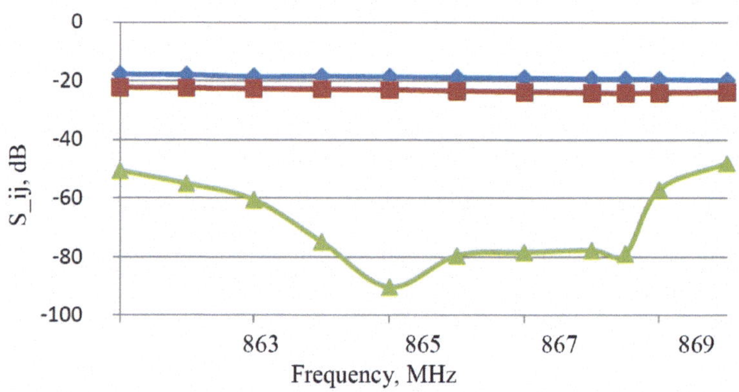

Fig. 11.12: Isolation (in green) and matching values (red and blue) of the electronic circulator after crosstalk cancellation

This measure now reduces the crosstalk signal from 15 dBm to around -45 dBm. This means that the crosstalk signal is minimized by around 60 dB. The first detailed description is published in [67].

The amplification required for a good noise figure using the LNA can now be carried out. After that, however, the crosstalk at around -25 dBm is far too high to achieve any significant system dynamics of the reader, as the blocking immunity is at best 50 dB and the -75 dBm is not within the sensitivity range of the receiver.

One possible measure for fully utilizing IC performance is presented in more detail in the following section. The bypass circuit of the electronic circulator is a general electronic circuit for the GHz range.

If an IC manufacturer implements this circuit in a single IC, then it will not only be used in front-end circuits for antenna cross-talkers or circulators.

Another very important application is couplers for measurement and control purposes. Their isolation can also be significantly increased.

This circuit has a similar range of applications to noise-canceling circuits in the audio sector. In the most modern *full-duplex* research work, this approach is adapted to broadband systems, [75].

11.5.1 The Tunable Channel Filter with Strong CW Signal Suppression

Many SAW filters for use in UHF readers have typical bandwidths of 7 MHz, which far exceed the 2 MHz of the total bandwidth for European solutions.

The channel bandwidth is only 0.25 MHz. This means that the noise can be reduced by around 15 dB and the range significantly increased if a channel filter with a bandwidth of just 0.25 MHz is used instead of 7 MHz.

In addition, a receive filter should provide a CW signal suppression of significantly more than 40 dB. This extends the suppression circuit described above. A very narrow-band bandstop at the center frequency of the channel is necessary here.

However, this CW signal suppression is not desired if the reader performs the Listen-Before-Talk test (LBT test for short) before communication. Here it makes sense to measure the signal energy at the center frequency of the channel.

In this LBT analysis, the best filter is a band-pass filter with the narrowest possible bandwidth at the center frequency of the channel.

A switchable filter is required to ensure the best communication with CW signal suppression and the best LBT examination. The symbols of these two filters are shown in Fig. 11.13.

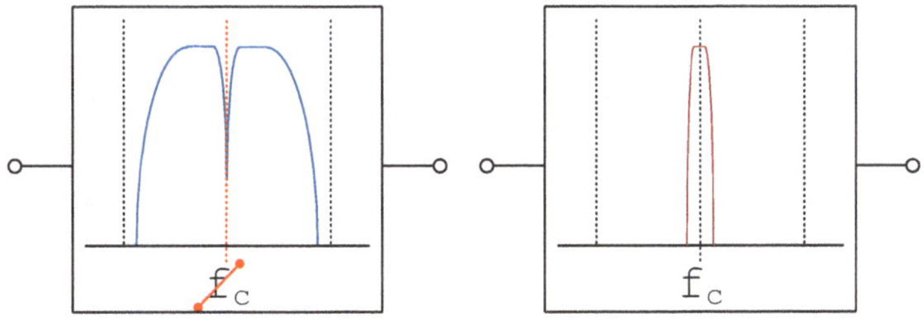

Fig. 11.13: Symbols for the channel filter for communication with CW signal suppression (left) and the LBT filter (right)

Such narrow-band filters can only be realized in the UHF range with SAW resonators and are then not tunable. To realize these desired filter characteristics, the detour via a down-up mixer stage to an intermediate frequency is a sensible approach.

The basic approach is shown in Fig. 11.14 as a block diagram.

During the mixing process, the RF signal is converted into a low-frequency signal (IF) using a local oscillator signal (LO).

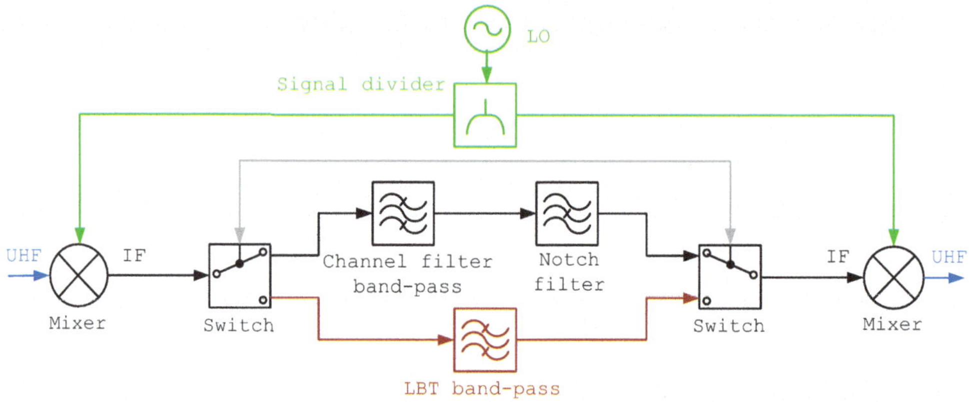

Fig. 11.14: Basic structure of a UHF RFID channel filter with CW signal suppression and an LBT filter

By varying the LO frequency, RF signals at different frequencies can always be mixed exactly in the middle of the channel filter. The IF frequency of the channel filter can be selected to be constant.

After filtering at the intermediate frequency, the signal is mixed back to the original UHF frequency using the same LO signal. This ensures the tunability of the filter.

Mixing in the IF range considerably reduces the requirements with regard to the quality (Q) of the band-pass filter to be realized. If you want to realize a 250 kHz band-pass filter in the RF range at 1 GHz, then according to

$$Q = \frac{f_c}{\Delta f} \tag{11.4}$$

a Q-factor of 4000 is necessary. This high quality requirement is reduced to only 40 when mixed to an intermediate frequency of 10 MHz. Expressed in relative bandwidths, this corresponds to a reduction from 0.025 % to 2.5 %.

The carrier suppression (CW signal suppression) is realized by a very narrow notch filter (`notch filter`) with the same center frequency as the band-pass, i.e. also in the intermediate frequency range.

This notch filter, which is intended to achieve a high TX signal suppression, can be realized advantageously by a dual-circuit filter with quartz resonators.

For the list-before-talk process, the IF range is switched to an even steeper, but also narrower band-pass filter. The channel band-pass filter cannot be used for this because of the notch in the middle of the channel.

To complete the RX path of the front end, a low-noise amplifier and, if necessary, SAW filters are connected upstream or downstream of the filters.

Using this approach, an electronically tunable high-Q channel filter with carrier suppression for UHF RFID front-ends was realized, published in more detail, [68], [123].

Fig. 11.15 shows the transfer function for the various channels of this tunable filter impressively, whereby the realized suppression of around 50 dB for the 2 W CW signals cannot be seen in the triggering.

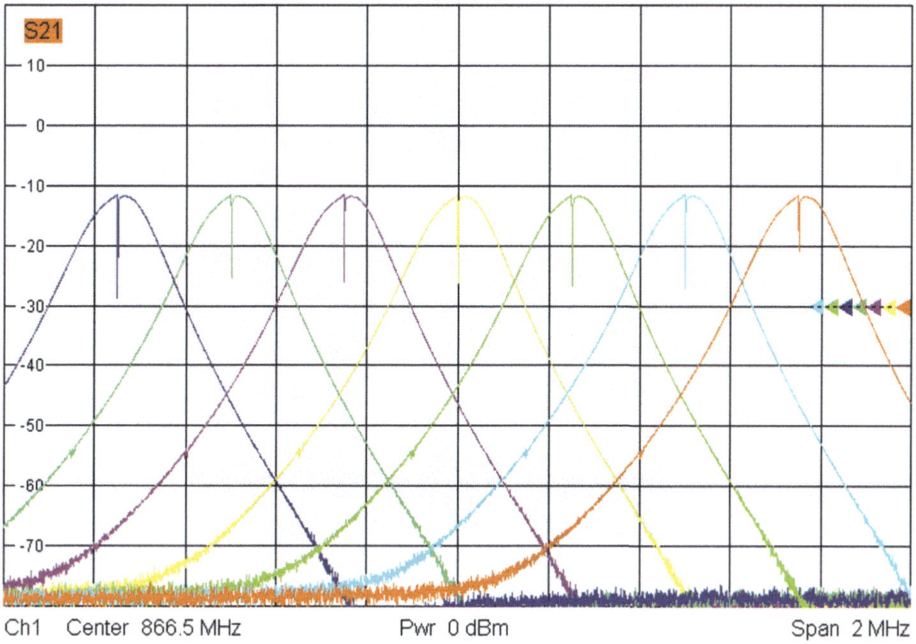

Fig. 11.15: Transmission measurement of the realized UHF RFID channel filter for the different channels

The realized reader circuit in Fig. 11.16 shows that the two band-pass filters and the notch filter take up almost 50 % of the board area.

Fig. 11.16: Photo of the realized UHF RFID reader with the multifunctional channel filter

11.6 Level Plans for the Development of Front-End Circuits Using the Example of UHF Band RFID Devices

Throughout the previous section, the power of the 2W CW signal was traced through the signal chain from TX to the input (RX) of the IC. The aim is to maintain a maximum value that is above the sensitivity by the value of the block immunity.

When developing an RFID device, there are still a large number of `boundary conditions` that result from the product objectives such as:

- Selling price, production costs,
- Range, special application fields,
- Size, weight, operating voltage and power consumption.

All level plans must be carried out for the two different operating modes `Communication` and `LBT query`.

The level plans are derived from the following criteria, taking into account the boundary conditions listed above.

- Compliance with RFID UHF specifications for eligibility for approval
- Sufficient power supply for the passive tag via the CW signal
- Reduction of own and external interference signals (e.g. GSM900)
- Best possible signal-to-noise ratio at the receiver
- Stability of the circuit (in practice no more than 40 dB gain in one frequency range)

For the following signals, the level values, transmission values and in some cases also adjustments between each component and in some cases also from the component are specified.

- Power level of the 2 W CW signal, TX transmit signal and RX receive signal with the Specification of the transmission characteristics of the components between the points (also applies to the following points)
- Noise power of the RX receive signal, noise power in the 1/f frequency range
- Noise figure along the receive chain
- Power level of own and external interference signals
- Power level of the harmonics in the TX chain

These plans must be drawn up for the design of the initial solution as well as for measurements. It is only when you compare these two plans that you recognize shortcomings that are often not serious as a single factor, but which have such a strong impact when they occur repeatedly that, for example, they are not eligible for approval.

Successful device development is not possible without such level plans. In this area of RF development, people often work with Excel spreadsheets they have created themselves.

Chapter 12

Localization and Positioning

This chapter presents modern RF systems from a hardware development perspective that can be used to localize your own location and/or locate other objects. The first is generally based on a navigation system and the measurement of data signals from multiple sources (e.g. satellites) from a known location. The latter is usually based on a radar system and distance measurement.

The localization or positioning of the user's own location is usually carried out either on a global coordinate system or on a reference point. The former is usually carried out for outdoor localization and the latter for indoor localization.

In robotics, the aim is to determine position and orientation. These two variables are summarized as `pose`. The basis of navigation is the determination of the pose.

The fields of application of these associated RF systems are now immense, which means that a normal 4-person household typically has at least 6 systems in smartphones, bicycle speedometers and driver assistance systems in cars. The majority of these are GPS (Global Positioning System) positioning systems, which are explained below, as is the new European Galileo system.

In industrial technology, especially robotics, as well as for autonomous driving, the GPS system is not sufficient in terms of precision and often also due to the indoor application. The same applies to locating objects around the car.

There will be an increasing number of further and new developments in radar systems, which are described in more detail in the second part of this chapter.

H. Heuermann, *Radio Frequency Technologies*,
https://doi.org/10.1007/978-3-658-50170-9_12

A large number of RF navigation systems are also used in aviation and military applications. Nowadays, the ILS system introduced in the 1930s, whose simple technology has been published many times, is used for the additional control of a landing at night. All of these military systems will not be discussed here.

12.1 Global Navigation Satellite Systems (GNSS)

The Global Positioning System (GPS)

GPS was developed by the American military and released for civilian applications in the 1980s. It is one of several satellite systems for self-positioning, which now number over 30.

There is no doubt that GPS is based on a sophisticated basic communication technology that uses CDMA modulation to ensure simultaneous communication between all available satellites and the receiver.

Many iteration methods have also been developed to calculate the system time t_0 as well as the three location coordinates x_0, y_0 and z_0. At "$https://de.wikipedia.org/wiki/GPS-Technik$" you can even find a closed-form solution for calculating these four variables from the signals of four satellites.

If the receiver's clock is synchronized with the GPS system, the signals from three satellites are sufficient to determine the location.

The carrier frequencies for both the GPS and Galileo systems are

- 1575.42 MHz for the civilian unencrypted signals and for
- 1227.60 MHz (GPS only) for the military encrypted signals and
- 1176.45 MHz (currently under construction) for civil signals and aviation as well as rescue services.

GPS-Satelliten senden mit nur 27 W in einer Entfernung von 20200 km. At typically -125 dBm, the reception power is already in the Atto-Watt range.

Modern receiver ICs have a sensitivity of up to -160 dBm after a good LNA and a narrowband filter stage, whereby these then average over 20000 values.

At the start (without GSM support), these modern receiver ICs still need around 30 seconds to determine the first position and then another 1 second for each additional position determination.

In addition to the Russian GNSS GLONASS, there is also the Chinese BeiDou system and the European Galileo system. India and Japan also have their own systems.

The Galileo System

The Galileo system uses a much more complex modulation technique than the older GPS and is therefore not compatible. The previous GPS signal processing ICs cannot process Galileo signals, but the new Galileo signal processing ICs can evaluate GPS signals.

Galileo also offers the High Accuracy Service (HAS), which provides a much more precise position determination than the usual 4 m. This HAS service is offered in the frequency ranges 1164-1214 MHz, 1260-1300 MHz and 1563-1591 MHz.

For the armed forces, police, intelligence services and rescue services, the two bands in the ranges 1260-1300 MHz and 1563-1591 MHz are also used in encrypted form if necessary.

For search and rescue (SAR) services, a return channel from the emergency call transmitters of ships or aircraft to the satellite system in the frequency range from 406.0 to 406.1 MHz will also be made available for the first time.

The RF Front End

The front-end circuits for GPS and Galileo are often the same. In general, there is a very large number of components for creating these GHz receivers available on the market.

Antenna: In general, the antenna efficiency is rarely specified in the data sheets. High efficiency is very important, as the losses of the antennas have a direct effect on the noise figure and therefore the sensitivity of the receiver.

The efficiency values of the antennas are often only between 45 and 75 %. But with many products you have to estimate the efficiency yourself. This efficiency is high if the antenna fulfills the following criteria:

- Structure: As much air as possible, as large as possible, as much metal as possible
- Position: Metal element preferably lambda/4 above ground
- Radiation diagram: Hemisphere as uniform as possible

Unfortunately, there are no balanced GPS antennas for sale. In a balanced receive path, interference signals occur as common mode signals and are therefore optimally suppressed. This is explained in detail at the beginning of Chap. 5.

Among other things, a loop antenna fulfills the desired requirements well. Fig. 12.1 shows such a basically balanced antenna, which still contains a simple balun in the coupling plate for connection.

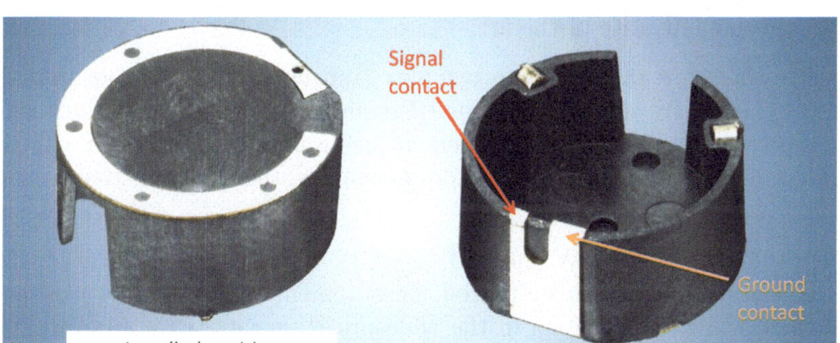

Fig. 12.1: Illustration of a GPS loop antenna from the company TE-Connectivity

The development process for a symmetrical or balanced 868 MHz antenna was described in section 10.3.6. The directional pattern of an antenna can be reliably calculated in advance using a field simulation and optimized if necessary. The necessary basics for field simulation and details on antenna technology are presented in [50].

After the antenna, a standard front end has an LNA (or two LNAs) specially designed for GPS and a narrowband bandpass filter. The latter is a SAW filter with a typical bandwidth of 2 MHz and lowest losses of 1 dB. Balanced filters are available on the market.

The most sensitive front end is obtained if it is continuously balanced from the antenna to the receiver IC. The background is explained in detail in Chap. 5.

If possible, the LNA stage should be balanced and designed with the lowest possible noise figure. A discrete solution is often better, as the passive SMD components have a better quality than coils and capacitors realized in semiconductors.

As the satellites are always far away, several LNA stages can always be used. These improve the signal-to-noise ratio. This is at the expense of the number of components and power consumption.

State-of-the-art receiver ICs have sensitivities of less than -140 dBm, which can be increased to -155 dBm and less by coupling the time base of communication systems (from GSM to CDMA systems) as well as by tracking algorithms.

12.2 The FMCW Radar System

The classic radar system is pulse radar, which is based on a time-of-flight measurement of an RF or microwave frequency pulse and can be classified as a time-domain measurement of the impulse response. Modern systems measure the transmitted signal and correlate it with the received signal, which provides a great improvement in distance measurement and much more additional information.

However, these time-of-flight radar systems can only be used for longer ranges. These systems can be found in aircraft and ship radar systems. For measurements over shorter ranges (a few kilometers), measuring the transfer function in the frequency domain is more advantageous and state of the art.

If you want to measure the complete complex transfer function (band-limited), this is referred to as the SFCW method (**S**tepped **F**requency **C**ontinuous **W**ave), which corresponds to the vectorial reflection measurement using a network analyzer. As the associated hardware is still complex and therefore expensive, these systems have only recently found their way into products.

For many decades, the FMCW method (FMCW **F**requency **M**odulated **C**ontinuous **W**ave) has been established in industrial measurement technology as well as automotive technology, which is described in the widespread literature via an evaluation of the difference frequency between the transmitted and received signal.

This radar differs from simple Doppler radar only in that the GHz oscillator it contains can be tuned over a certain bandwidth.

Fig. 12.2 shows such an FMCW radar, which contains only one mixer and is therefore simpler than the previously shown RFID concept with an IQ receiver, which also consists of two mixing stages and other components.

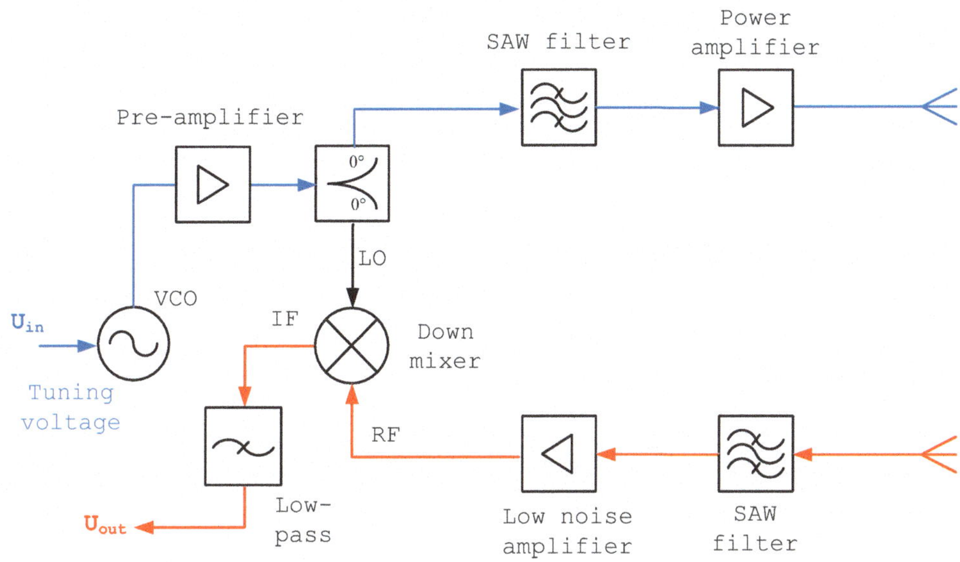

Fig. 12.2: Representation of a typical FMCW radar

The pure VCO structure shown here is used in inexpensive systems. The VCO-only solution is particularly advantageous for short-range applications, as it can be swept quickly and the noise of the LO and RF signals is well correlated and therefore does not generate any major measurement errors.

Since synthesizer ICs are already very inexpensive, their use is not unusual. The advantage is the reduced phase noise, which improves the sensitivity of the radar. The reduced frequency sweep time can be a disadvantage.

The next chapter describes this FMCW process in more detail, particularly in terms of the theoretical description.

12.2.1 Discrete Sampling in the Frequency Range

In communications engineering, the sampling of an electrical signal in the time domain (at different time pionts) is very common.

In electrical measurement technology, on the other hand, the sampling of electrical signals over different frequency points is of interest, especially when measuring components and transmission paths in the higher frequency range[1].

[1]LCR-meters are also using a sampling in the frequency domain.

Similar to communications engineering, a frequency-limited signal can be completely reconstructed by the sampled values in the time domain. This is utilized in radar technology: A time-limited impulse response can be described completely from samples of the transfer function. Alternatively, the distance can also be calculated from the transfer function. For the very modern harmonic radar, this is also presented in detail in [50].

If the complex transfer function of a system is measured in the frequency domain at discrete values at intervals of $\Delta\omega$, here denoted by $H_d(\omega)$, the measurement results can be represented as follows using the continuous transfer function $H(\omega)$ and the delta function $\delta(\omega)$ as well as the sampling interval $\Delta\omega$:

$$H_d(\omega) = H(\omega) \sum_{n=-\infty}^{\infty} \delta(\omega - n\Delta\omega) = \sum_{n=-\infty}^{\infty} H(n\Delta\omega)\,\delta(\omega - n\Delta\omega) \tag{12.1}$$

with the sampling interval

$$\Omega = \frac{2\pi}{T} \quad . \tag{12.2}$$

If the delta function is under an integral, the so-called fade-out property of the delta function is sometimes used:

$$\int_{-\infty}^{\infty} F(\omega)\,\delta(\omega - \omega_0)\,d\omega \;=\; F(\omega_0) \quad . \tag{12.3}$$

The sampling function $P(\omega)$ is an infinite periodic sequence of delta functions in the frequency range:

$$P(\omega) = \sum_{n=-\infty}^{\infty} \delta(\omega - n\Delta\omega) \quad . \tag{12.4}$$

Eq. (12.1) can therefore also be interpreted as

$$H_d(\omega) = H(\omega)\;P(\omega) \quad . \tag{12.5}$$

As is well known, a periodic time signal has an infinite discrete frequency spectrum, which consists of equidistant spectral lines whose amplitude can be calculated using a Fourier series expansion. This periodic signal can be developed into a series with the coefficients of the discrete spectrum.

Similarly, you can also perform a series expansion of a periodic function in the frequency domain. For example, for a periodic function $\gamma(\omega)$ with the period $T = \frac{2\pi}{\Delta\omega} = \frac{1}{\Delta f}$ we obtain the series

$$\gamma(\omega) = \sum_{n=-\infty}^{\infty} \Gamma_n \; e^{jnT\omega} \qquad . \tag{12.6}$$

The coefficients Γ_n are obtained from the relationship

$$\Gamma_n = \frac{1}{\Delta\omega} \int_{-\Delta\omega/2}^{\Delta\omega/2} \gamma(\omega) \; e^{-jnT\omega} d\omega \qquad . \tag{12.7}$$

Similarly, the periodic sampling function $P(\omega)$ can be converted into a series

$$P(\omega) = \sum_{n=-\infty}^{\infty} V_n \; e^{jnT\omega} \tag{12.8}$$

with the coefficients V_n from

$$V_n = \frac{1}{\Delta\omega} \int_{-\Delta\omega/2}^{\Delta\omega/2} \left[\sum_{n=-\infty}^{\infty} \delta(\omega - n\Delta\omega) \; e^{-jnT\omega} \right] d\omega \qquad . \tag{12.9}$$

Because most of the sum terms lie outside the integration interval, the following applies:

$$V_n = \frac{1}{\Delta\omega} \int_{-\Delta\omega/2}^{\Delta\omega/2} \delta(\omega) \; e^{-jnT\omega} d\omega = \frac{1}{\Delta\omega} \qquad . \tag{12.10}$$

This results in the scanning function:

$$P(\omega) = \sum_{n=-\infty}^{\infty} \frac{1}{\Delta\omega} \; e^{jnT\omega} \qquad . \tag{12.11}$$

Let $h(\theta)$ be the inverse Fourier transform of $H(\omega)$. With the frequency $f = \frac{1}{2\pi}\omega$ applies:

$$h(\theta) = \frac{1}{2\pi} \int_{-\infty}^{\infty} H(\omega) \; e^{j\omega\theta} \; d\omega = \int_{-\infty}^{\infty} H(f) \; e^{j2\pi f\theta} \; df \qquad . \tag{12.12}$$

Virtual time θ was introduced for this purpose. In a system that works in real time, the virtual time variable θ can sometimes also be imagined as the actual time t.

For the Fourier transform $h_d(\theta)$ of the sampled transfer function $H_d(\omega)$ in the virtual time domain θ or θ domain for short, we obtain with Eq. (12.1)

$$
\begin{aligned}
h_d(\theta) &= \int_{-\infty}^{\infty} H_d(f)\; e^{j2\pi f\theta}\; df \quad , \\
h_d(\theta) &= \sum_{n=-\infty}^{\infty} H(n\Delta f) \int_{-\infty}^{\infty} \delta(f-n\Delta f)\; e^{j2\pi f\theta}\; df \quad , \\
h_d(\theta) &= \sum_{n=-\infty}^{\infty} H(n\Delta f)\; e^{j2\pi n\Delta f\theta} \quad , \\
h_d(\theta) &= H(0) \;+\; 2\;\mathrm{Re}\left\{\sum_{n=1}^{\infty} H(n\Delta f)\; e^{j2\pi n\Delta f\theta}\right\} \quad . \qquad (12.13)
\end{aligned}
$$

The Eq. (12.13) is also a rule for a numerical calculation of $h_d(\theta)$. For a different representation of $h_d(\theta)$, we start with Eq. (12.5) and (12.9) and also apply the Fourier transform displacement theorem:

$$
\begin{aligned}
h_d(\theta) &= \int_{-\infty}^{\infty} H_d(f)\; e^{j2\pi f\theta}\; df \quad , \\
h_d(\theta) &= \int_{-\infty}^{\infty} H(f)\; P(f)\; e^{j2\pi f\theta}\; df \quad , \\
h_d(\theta) &= \frac{1}{\Delta\omega} \sum_{n=-\infty}^{\infty} \int_{-\infty}^{\infty} H(f)\; e^{j2\pi(\theta+nT)f}\; df \quad , \\
h_d(\theta) &= \frac{1}{\Delta\omega} \sum_{n=-\infty}^{\infty} h(\theta+nT) \quad . \qquad (12.14)
\end{aligned}
$$

We see that the frequency discretization leads to a virtual time function that repeats periodically in θ. The symbol $\mathrm{rep_T}\; h(t)$ can also be used for such a virtual time function that repeats periodically with the period T,

$$
h_d(\theta) = \frac{1}{\Delta\omega}\;\mathrm{rep_T}\; h(\theta) \;=\frac{1}{\Delta\omega} \sum_{n=-\infty}^{+\infty} h(\theta+nT) \quad . \qquad (12.15)
$$

Fig. 12.3 shows $h(\theta)$ and $\mathrm{rep_T}\; h(\theta)$ for a random progression of $h(\theta)$.

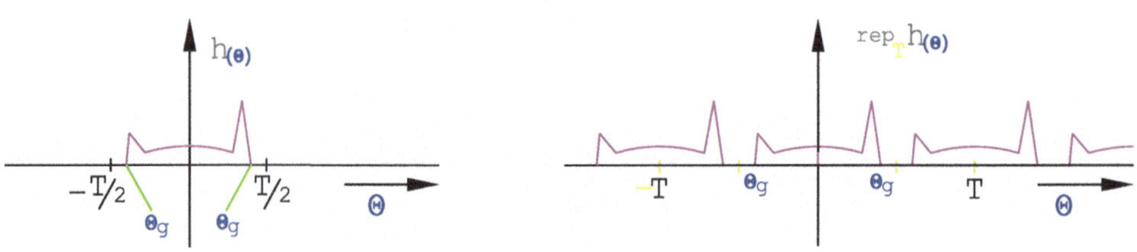

Fig. 12.3: Limited one-time and repeating pseudo-time function

If the system has a limited impulse response, as shown in Fig. 12.3, the following applies

$$h(\theta) = 0 \qquad \text{for} \qquad |\theta| > \theta_g \tag{12.16}$$

and will

$$T > 2\,\theta_g \tag{12.17}$$

elected, then the partial functions $h(\theta + nT)$ do not overlap and $h(\theta)$ can be obtained unambiguously from $\text{rep}_\text{T}\ h(\theta)$. You can also use the formalism with the rect function:

$$h(\theta) = \text{ rect}\left(\frac{2\theta}{T}\right) \text{ rep}_\text{T}\ h(\theta) \qquad . \tag{12.18}$$

The Fourier transform of the Eq. (12.18) provides the continuous spectrum of the transfer function from the sampled values:

$$H(\omega) = \sum_{n=-\infty}^{\infty} H(n\Delta\omega) \ \text{ si}\left(\pi\frac{\omega - n\Delta\omega}{\Delta\omega}\right) \qquad . \tag{12.19}$$

The result of the previous section can therefore be summarized as follows: Given the transfer function $H(\omega)$ with the corresponding limited impulse response $h(\theta)$, limited to $\pm\theta_g$.

If this transfer function is sampled in frequency steps Δf that are smaller than $\frac{1}{2\theta_g}$, both the continuous impulse response and the continuous transfer function can be reconstructed.

For accurate distance measurement, however, it is not necessary for the impulse response to be limited because the position of the maximum of the impulse response is generally taken as a measure of the distance. For band-limited transmission, it is also not possible to generate a limited impulse response.

Measuring devices with which the complex transfer function of components, lines and the like can be recorded very broadband and precisely are called network analyzers. In contrast to sampling in the time domain, an ideal sampling function $P(\omega)$ can be used here, as measurements are taken at discrete frequency points, whereas in the time domain a finite time window must always be taken into account.

To conclude this chapter, it should be pointed out that the impulse response calculated from the complex transfer function can be evaluated using similar methods as the received signal of an impulse radar and therefore plays a comparable role.

12.2.2 Frequency-Limited Sampling

The influence of frequency-limited sampling will be demonstrated using the example of an idealized interference point. For an `idealized interference point` at a distance proportional to half the transit time τ_1, the following applies to the impulse response

$$h_1(\theta) = A_1 \delta(\theta - \tau_1) \quad \circ\!\!-\!\!\bullet \quad H_1(\omega) = A_1 e^{-j\omega\tau_1} \quad . \tag{12.20}$$

Fig. 12.4 illustrates the spectrum and the impulse response under these ideal conditions.

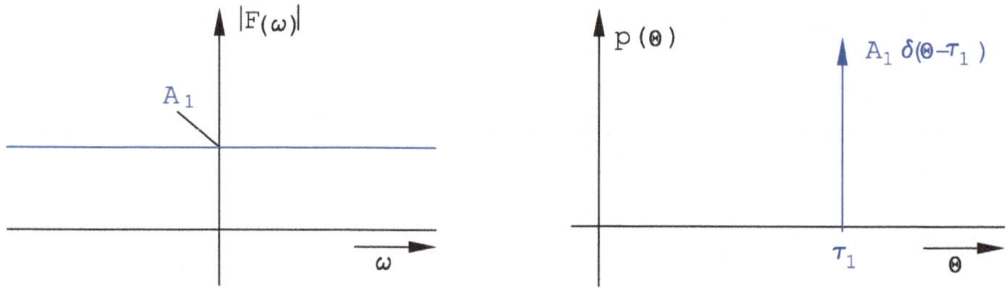

Fig. 12.4: Spectrum and impulse response of an idealized reflection point

The following section examines the impulse response that results when the transfer function of the idealized impurity $H_1(\omega)$ is measured with limited bandwidths.

Case 1: Low-pass limited spectrum

$$h_1'(\theta) = \int_{-\Omega_2}^{+\Omega_2} A_1 e^{-j\omega\tau_1} e^{j\omega\theta} df = \frac{\Omega_2}{\pi} A_1 \ \mathrm{si}\left(\Omega_2(\theta - \tau_1)\right) \tag{12.21}$$

with

$$\Omega_2 = 2\,\pi\, F_2 \tag{12.22}$$

The following figure shows the relationships graphically.

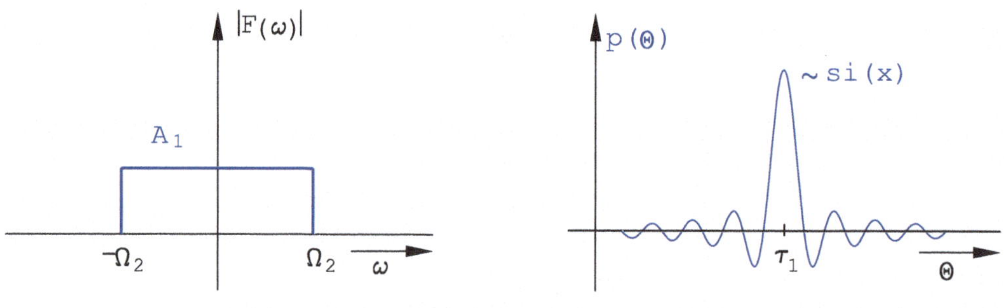

Fig. 12.5: Low-pass limited spectrum and associated impulse response of an idealized reflection point

Case 2: Band-pass limited spectrum

$$h_1'(\theta) = \int_{-\Omega_2}^{+\Omega_2} A_1 e^{j\omega(\theta-\tau_1)} df - \int_{-\Omega_1}^{+\Omega_1} A_1 e^{j\omega(\theta-\tau_1)} df \tag{12.23}$$

$$h_1'(\theta) = \frac{A_1}{\pi} \left\{ \Omega_2 \ \mathrm{si}\left(\Omega_2(\theta-\tau_1)\right) \ - \ \Omega_1 \ \mathrm{si}\left(\Omega_1(\theta-\tau_1)\right) \right\} \tag{12.24}$$

Fig. 12.6 shows the spectrum and the impulse response for this case.

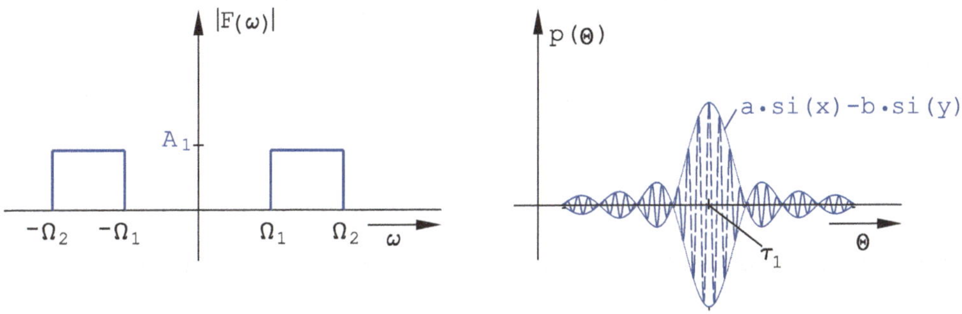

Fig. 12.6: Band-pass limited spectrum and associated impulse response of an idealized reflection point

Since the measuring device usually contains frequency-limiting components, e.g. the antenna, when measuring a transfer function of the free space, a band limitation is the typical case.

12.2.3 Sampling Method and Basic Hardware Structure

A special sampling method in the frequency range is the so-called FMCW method (**Fre**quency **M**odulated **C**ontinuous **W**ave).

It is characterized by the fact that a microwave signal with a frequency varying between two limits is evaluated rather than a mono-frequency signal (which would correspond to a CW method).

A very simple realization of an FMCW measuring system is shown in Fig. 12.7.

The microwave signal is generated in the generator and sent as a transmitted signal $s_t(t)$ on a transmission path.

With an FMCW distance meter, the transmission distance corresponds to the `free space distance`.

The reflected signal $s_r(t)$ is superimposed on the transmitted signal $s_t(t)$ in a mixer.

Two cases can be distinguished based on the type of continuous tuning (sweep) of the frequency. If you sweep the frequency range with a frequency that changes in finite steps, this is referred to as the `discrete` case and stepping.

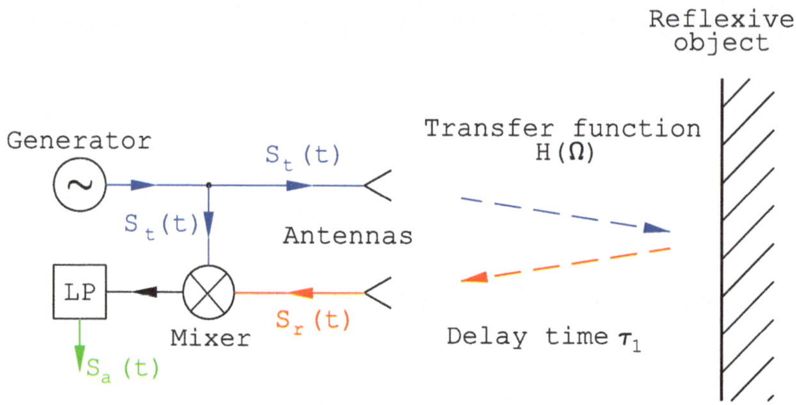

Fig. 12.7: Block diagram of an FMCW distance meter

If, on the other hand, the transmission signal is changed continuously, this is the **continuous** case and sweeping.

In the continuous case, which is purely with a voltage-controlled oscillator (VCO for **V**oltage **C**ontroled **O**scillator), it is necessary that the frequency ramp is traversed as evenly as possible, which does not work well in practice due to the non-linear characteristic curve of the varactor diode used for tuning.

The discrete frequency sweep requires the PLLs described in the previous chapter, which are now available as fast and highly integrated, inexpensive ICs.

First, the simpler discrete case will be explained.

12.2.4 The FMCW Method with Discrete Frequency Variancen

Fig. 12.8 shows how the output frequency of the generator is changed as a function of time. In general, the step sizes do not need to be equidistant, but only known.

Modern RF signal generators are connected to a quartz-stable reference oscillator via fixed and programmable dividers by means of a phase-locked loop. As a result, they only generate signals at discrete frequency points, but these are crystal-stable.

The following applies to the transmitted signal $s_t(t)$ in the time or frequency domain:

$$s_t(t) = \hat{U}_t \cos(\Omega_i t) \quad ○\!\!-\!\!\bullet \quad S_t(\omega) = \hat{U}_t \, \pi \left(\delta(\omega + \Omega_i) + \delta(\omega - \Omega_i)\right) \quad . \qquad (12.25)$$

The received signal $s_r(t)$ can be determined as follows.

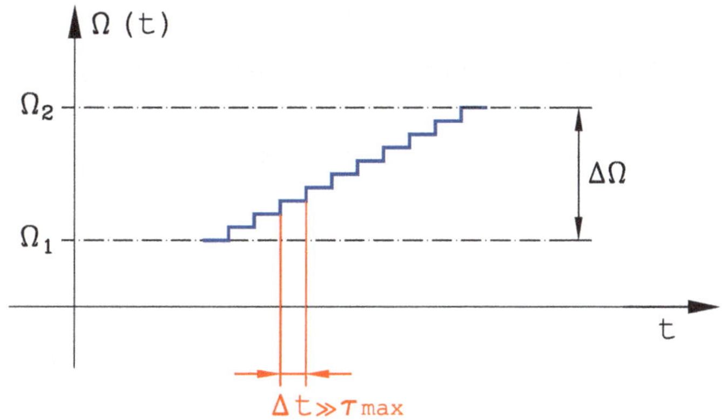

Fig. 12.8: Output frequency by stepping the generator with discrete sweep

$$S_r(\omega) = S_t(\omega)H(\omega) \quad \bullet\!\!-\!\!\circ \quad s_r(t) = \pi\,\hat{U}_t \int_{-\infty}^{\infty} H(\omega)(\delta(\omega+\Omega_i)+\delta(\omega-\Omega_i))\,e^{j\omega t}\,df \tag{12.26}$$

Utilizing the fade-out property of the delta function, the following relationship is obtained.

$$s_r(t) = \frac{\hat{U}_t}{2}\,\left(H(-\Omega_i)\,e^{-j\Omega_i t} + H(\Omega_i)\,e^{j\Omega_i t}\right) \tag{12.27}$$

$$s_r(t) = \frac{\hat{U}_t}{2}\,\left(H(\Omega_i)\,e^{j\Omega_i t} + (H(\Omega_i)\,e^{j\Omega_i t})^{\star}\right) \tag{12.28}$$

$$s_r(t) = \hat{U}_t \quad \mathrm{Re}\left\{H(\Omega_i)\,e^{j\Omega_i t}\right\} \tag{12.29}$$

This results in the following for the received signal

$$s_r(t) = \hat{U}_t\,\left|H(\Omega_i)\right|\,\cos\left(\Omega_i t + \arg\left(H(\Omega_i)\right)\right) \quad , \tag{12.30}$$

which is superimposed with the transmitted signal in the balanced mixer (with the conversion efficiency κ) and thus produces the output signal $s_a(t)$ after low-pass filtering.

The behavior of the mixer should correspond to that of an ideal analog multiplier. The output signal (DC voltage signal) has no time dependence but only a dependence on the instantaneous frequency Ω_i.

$$s_a(\theta) = \kappa\, \overline{s_t(\theta)\, s_r(\theta)} \tag{12.31}$$

$$s_a(i) = \frac{\kappa}{2}\, \hat{U}_t^2\ \mathrm{Re}\{H(\Omega_i)\} \tag{12.32}$$

The output signal has a typical cosine-shaped curve with a superimposed mixer output DC voltage. The basic cosine curve generally has an initial and final phase that depends on the frequency, the distance and the backscattering behavior.

12.2.5 The FMCW Method with Continuous Frequency Variation

If only a free-running oscillator is used as an RF signal generator, continuous frequency variation is ideally carried out, as shown in Fig. 12.9.

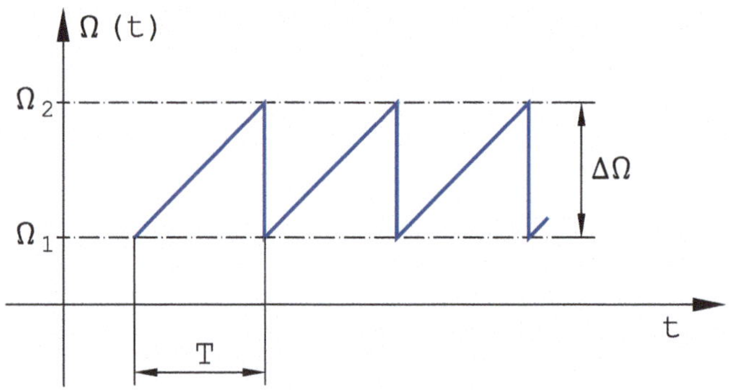

Fig. 12.9: Sawtooth-shaped continuous sweeping of the generator

As will be shown below, this problem can be solved exactly for an idealized reflection point and for a general transmission path in very good approximation.

A. Closed Solution for an Idealized Reflection Point

With this solution, it is appropriate to assume a linear sweep as shown in Fig. 12.9.

$$\Omega(t) = \Omega_1 + \Delta\Omega\, \frac{t}{T} \qquad \text{for} \qquad 0 \le t \le T \tag{12.33}$$

First, the phase response of the transmitted signal is to be examined more closely. The phase $\varphi(t)$ of the transmitted signal $s_t(t)$ can be determined by integrating the time-varying frequency $\Omega(t)$.

$$\varphi(t) = \int \left(\Omega_1 + \Delta\Omega \, \frac{t}{T} \right) \, dt \tag{12.34}$$

$$\varphi(t) = \Omega_1 \, t + \frac{\Delta\Omega}{2} \, \frac{t^2}{T} + \varphi_0 \tag{12.35}$$

This applies to the transmission signal:

$$s_t(t) = \hat{U}_t \cos\left(\Omega_1 \, t + \frac{\Delta\Omega}{2} \, \frac{t^2}{T} + \varphi_0 \right) \quad , \tag{12.36}$$

with the initial phase φ_0.

The received signal $s_r(t)$ for an idealized interference point is the transmitted signal delayed by the delay time τ_1 and attenuated by the factor A_1.

$$s_r(t) = \hat{U}_t A_1 \cos\left(\Omega_1 (t - \tau_1) + \frac{\Delta\Omega}{2} \, \frac{(t - \tau_1)^2}{T} + \varphi_0 \right) \tag{12.37}$$

After mixing the transmitted and received signals in an ideal multiplier and subsequent low-pass filtering, the alternating voltage signal $s_a(t)$ appears at the output.

$$s_a(t) = \frac{\kappa}{2} \, \hat{U}_t^2 A_1 \cos\left(\Omega_1 \, \tau_1 + \Delta\Omega \, \frac{t \, \tau_1}{T} - \frac{\Delta\Omega}{2} \, \frac{\tau_1^2}{T} \right) \tag{12.38}$$

B. Solution with Static Approach

In practice, the sweep time T for a sweep is much greater than the transit time τ_1 of the signal. Due to these differences, this static approach assumes that the frequency $\Omega(t)$, which changes slowly over time, does not change during the runtime τ_1.

Therefore, using the same derivation as for the discrete case, we obtain for the transmitted signal

$$s_t(t) = \hat{U}_t \cos\left(\Omega(t) \ t \right) \tag{12.39}$$

with a frequency $\Omega(t)$ that changes over time in any form the received signal

$$s_r(t) = \hat{U}_t \ \left| H(\Omega(t)) \right| \ \cos\left(\Omega(t) \ t + \arg\left(H(\Omega(t)) \right) \ \right) \quad . \tag{12.40}$$

The following time-varying signal is now present at the low-frequency output of the measuring system:

$$s_a(t) = \frac{\kappa}{2} \ \hat{U}_t^2 \ \mathrm{Re}\left\{ H(\Omega(t)) \right\} \quad . \tag{12.41}$$

Sampling linked to the sweep period eliminates the time dependency and produces a signal that is directly proportional to the real part of the transfer function, as in the discrete case.

If we again assume a linear sweep and an idealized interference point, we obtain the approximate result:

$$s_a(t) = \frac{\kappa}{2}\,\hat{U}_t^2\,A_1 \cos\left((\Omega_1 + \Delta\Omega\,\frac{t}{T})\,\tau_1\right) \quad . \tag{12.42}$$

Eq. (12.42) differs from the exact solution (12.38) only by one phase term, which is five orders of magnitude smaller than the other terms for realistic conditions, i.e. if the distance is not too great and the frequency change is not too fast.

If, as is usually the case, only the frequency component of the output signal is evaluated, then no errors occur due to the static approach.

12.2.6 Influence of the Mirror Impulses

It has been shown that an FMCW system can be used to measure the real part of a transmission path exactly in the case of a discrete frequency variation and to a very good approximation in the case of a continuous frequency variation.

For a single idealized reflection point, the real part of the transfer function must be specified as follows:

$$\mathrm{Re}\,\{H_1(\Omega)\} = \frac{A_1}{2}\,(e^{-j\Omega\,\tau_1} + e^{j\Omega\,\tau_1}) \quad . \tag{12.43}$$

If an inverse Fourier transformation is carried out with this output signal of the FMCW system, a result is obtained that is closely related to the impulse response of the transmission path. This result, known as `pseudo impulse response` $p(t)$, is strongly influenced by the bandwidths used in the measurement.

1. *Measurement with Infinitely Wide Frequency Sweep*

$$p_p(t) = \frac{1}{2\pi}\int_{-\infty}^{\infty} \mathrm{Re}\,\{H_1(\Omega)\}\,e^{j\Omega t}\,d\Omega = \frac{A_1}{2}\,(\delta(t-\tau_1) + \delta(t+\tau_1)) \tag{12.44}$$

The impulse response is obtained at time τ_1 and the mirrored impulse response (mirror impulse) at time $-\tau_1$. In this case, the distance can be determined exactly from the pseudo impulse response, as the pseudo impulse response corresponds to the impulse response of the transmission system for positive time points.

2. *Measurement with Low-Pass Limited Frequency Sweep up to Ω_2*

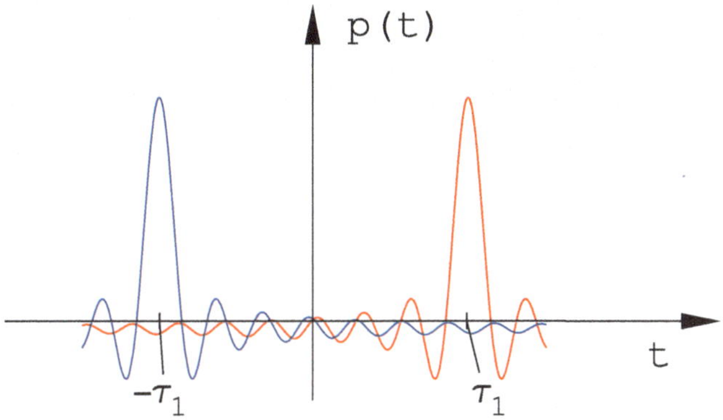

Fig. 12.10: Pseudo impulse response with low-pass limited frequency sweep

$$p_p(t) = \frac{A_1 \Omega_2}{2\pi} \left(\text{si}\left(\Omega_2(t - \tau_1)\right) + \text{si}\left(\Omega_2(t + \tau_1)\right)\right) \tag{12.45}$$

Fig. 12.10 illustrates that in this case the pseudo impulse response no longer corresponds to the impulse response, as parts of the mirror impulse overlap with the impulse response in the positive (real) time domain.

It is therefore no longer possible to exactly determine the transit time from the maximum of the pseudo impulse response, but the error is still small for sufficiently long transit times and correspondingly high frequency limits Ω_2.

3. Measurement with Band-Pass Limited Frequency Sweep in the Range from Ω_1 to Ω_2

$$p_p(t) = \frac{A_1}{2\pi} \left(\Omega_2 \left(\text{si}\left(\Omega_2(t - \tau_1)\right) + \text{si}\left(\Omega_2(t + \tau_1)\right)\right) - \Omega_1 \left(\text{si}\left(\Omega_1(t - \tau_1)\right) + \text{si}\left(\Omega_1(t + \tau_1)\right)\right)\right) \tag{12.46}$$

The same restrictions apply here as in the second case, as shown in Fig. 12.11.

So far, the main error influences due to the finite bandwidth of the measurement system and the fact that an FMCW system only measures the real part of the transfer function have been considered.

The fact that other errors occur in the evaluation of the output signal is discussed in the next section.

12.2.7 Determining the Distance of an Idealized Reflection Interference Point

In the case of an idealized reflection interference point, the output signal is obtained for a finite number of frequency support points with discrete frequency variation.

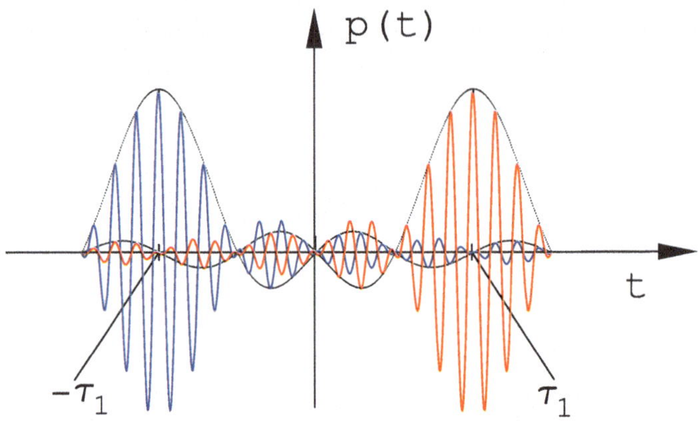

Fig. 12.11: Pseudo impulse response with band-pass limited frequency sweep

$$s_a(i) = \frac{\kappa}{2}\,\hat{U}_t^2\,A_1\,\cos\left(\Omega_i\,\tau_1\right) \qquad . \tag{12.47}$$

In contrast, the signal to be evaluated for the continuous frequency variation is the alternating component

$$s_a(t) = \frac{\kappa}{2}\,\hat{U}_t^2\,A_1\,\cos\left(\Omega_z\,t\right) \qquad \text{with} \qquad \Omega_z = \Delta\Omega\,\frac{\tau_1}{T} \qquad . \tag{12.48}$$

In the past, this signal was evaluated analogously using a spectrum analyzer. In modern rangefinders, the output signals for both methods are evaluated using an inverse Fourier transformation.

The fast Fourier transform (FFT) is used instead of a discrete Fourier transform (DFT) to achieve extreme savings in computing time. If, for example, 1024 frequency points are approached, an FFT requires 5120 multiplications, whereas a DFT requires 1048576 multiplications.

However, it turns out that the same problem occurs with an FFT as with analog spectrum analysis:

In both cases, the output signal is repeated periodically. With analog evaluation, the measurement process must be repeated several times (hence the sawtooth-shaped control of the generator) so that the spectrum analyzer is sufficiently sensitive.

The DFT calculates the spectrum of a unique signal, while an FFT calculates the spectrum of the same signal with infinite periodic repetition.

It is known that periodically repeated signals only produce a discrete spectrum. Thus, the evaluated signal with a low-pass limited frequency sweep looks as shown in Fig. 12.12.

The envelope curve in Fig. 12.12 results for a single signal. The discrete lines are multiples of the repetition frequency and are therefore generally not at the maximum of the envelope. In both cases, analog evaluation with sawtooth generator control and FFT, only two lines lie in the center of gravity of the envelope.

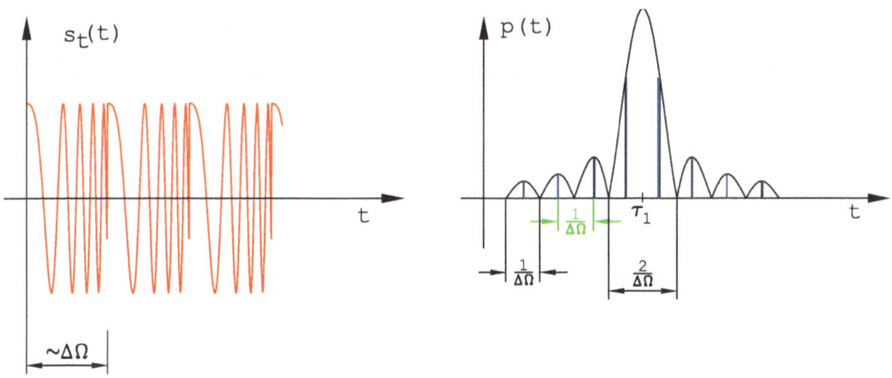

Fig. 12.12: Repeated transmission signal and associated magnitude of the pseudo impulse response

With experimental FMCW range finders, one must assume a band-pass limited frequency sweep. Consequently, a line located directly next to the main maximum can be very small. See Fig. 12.11 for an illustration.

In both cases, the number of lines in the maximum can be increased by introducing a dead time (Fig. 12.13). With analog counting, the dead time must be introduced between the ramps of the generator control, which has the disadvantage that the total energy of the received signal is reduced.

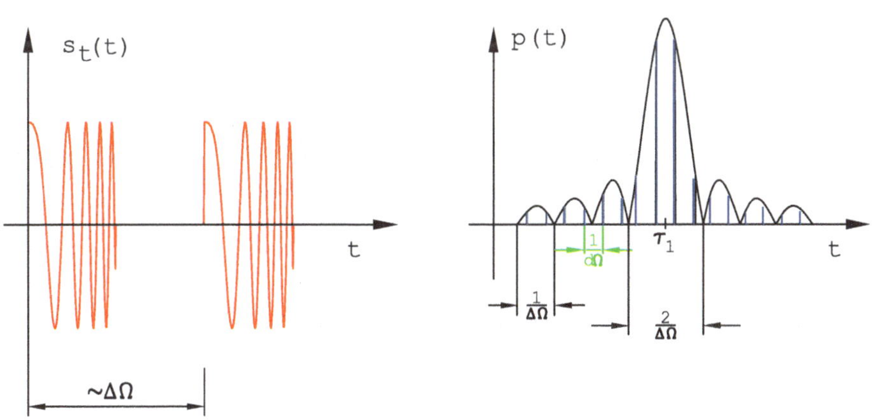

Fig. 12.13: Repeated transmission signal with dead times and associated magnitude of the pseudo impulse response

When evaluating using an FFT, the dead time must be added numerically to the measurement signal, so that the vector to be transformed and thus also the calculation time becomes longer.

The advantages of FMCW systems lie in their high sensitivity and ease of implementation. Such systems can be realized up to very high frequencies, provided that a sweep generator and a broadband mixer are available.

To avoid interference (DC drift, 1/f noise) with a discrete frequency variation, a simple amplitude modulation or 180° phase shift keying can be introduced on the transmitter side. This provides the measurement information in the amplitude of a narrow-band filterable and easily sampled IF signal with, for example, 50 kHz.

12.3 The SFCW Radar System

SFCW is the abbreviation for a stepped frequency continuous wave radar or measurement system. This is a radar that has a complete IQ receiver and therefore offers a complex reflection measurement (S_{11} measurement). The complex measuring IQ receiver eliminates the mirror frequency problems described above.

If you use a synthesizer as a transmitter, which is an easy way due to its wide availability, then in practice you carry out a measurement as with the network analyzer. This is a homodyne VNA measurement. The RF signal is used as a TX and LO signal. Their main disadvantage is the relatively slow measuring speed.

With a correspondingly fast computer technology and for shorter distances, these signal processing can also be used for distance measurements and thus as radar. This method of determining the distance from the phase values of the VNA measurement is shown in [50] as part of the mixed-frequency measurements for the sea rescue application, where this system was used as a radar even up to 6 km.

Alternatively, the VCO can be controlled using values that are read out via a memory and D/A converter, quickly. Then you work in FMCW mode with a continuous frequency sweep.

In addition to the real part measurement according to Eq. (12.38), you also obtain the result of the imaginary part measurement for an idealized reflection interference with

$$s_a(t) = \frac{\kappa}{2}\,\hat{U}_t^2\,A_1 \sin\left(\Omega_1\tau_1 - \frac{\Delta\Omega}{2}\frac{\tau_1^2}{T}\right) \quad . \tag{12.49}$$

The phase of the associated complex signal can be used to determine the transit time τ_1 at each measurement time and the values can be determined over a complete measurement sweep.

A block diagram of such a transceiver IC for a 24 GHz car radar application is shown in Fig. 12. The two measuring points can be used, for example, for two antennas at two separate reception locations for better spatial resolution.

The divided values of the set transmission frequency (Q1 and Q2) can be used either for PLL control or for linearization of the stored control values.

This IC can in principle be used as a unidirectional VNA for industrial S_{11} and S_{21} measurements at 24 GHz[2].

[2] Since it is an IC from a chip set, you may have to use other ICs from Infineon.

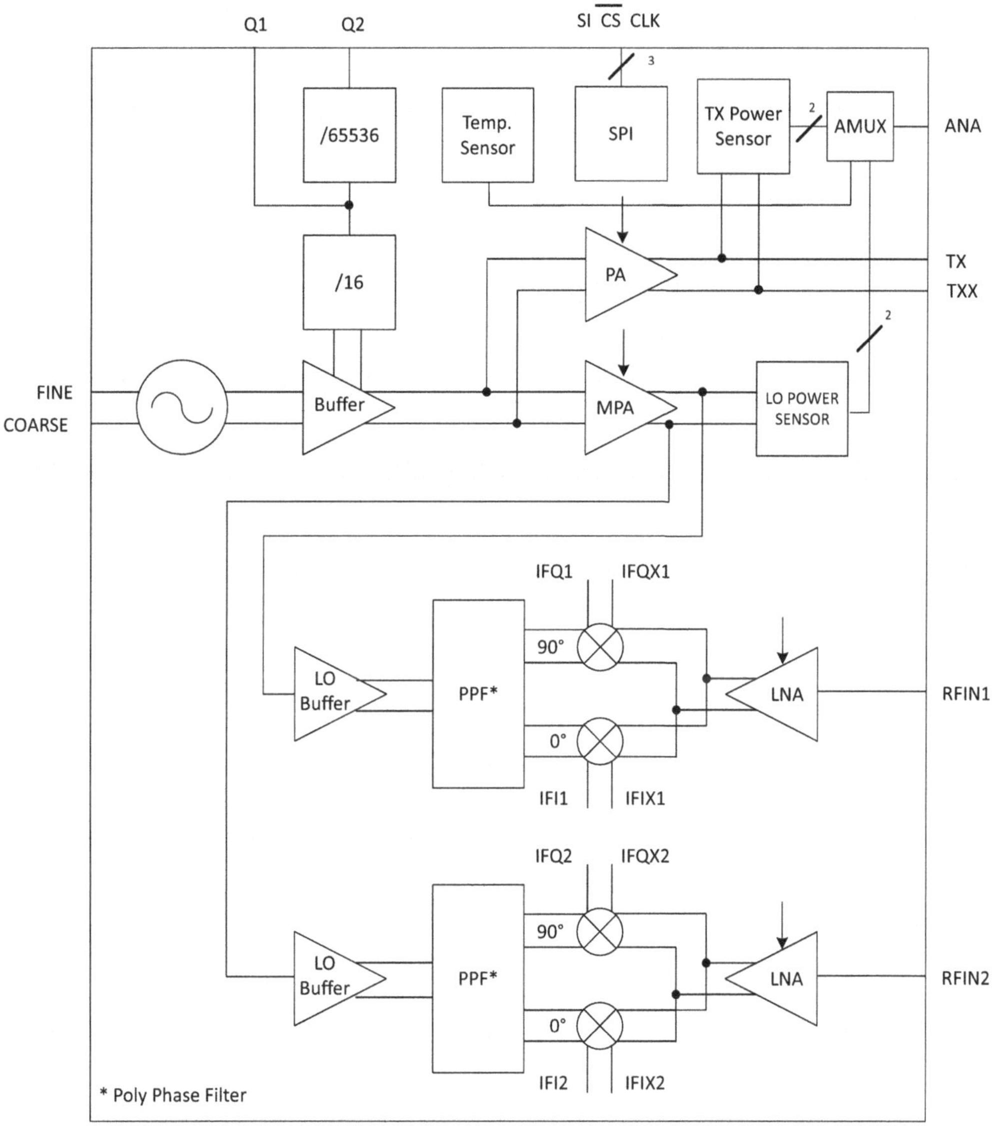

Fig. 12.14: Block diagram of an SFCW transceiver IC (Infineon Radar Sensor), which can be used for FMCW as well as VNA measurements

Appendix A

Appendix

A.1 Auxiliary Materials for Circuit Synthesis

H. Heuermann, *Radio Frequency Technologies*,
https://doi.org/10.1007/978-3-658-50170-9

S11 in dB = S11 (Return Loss pos.)	S21 in dB = S21 (Insertion Loss pos.)
-0.100 dB = 0.989	-16.428 dB = 0.151
-0.200 dB = 0.977	-13.467 dB = 0.212
-0.300 dB = 0.966	-11.756 dB = 0.258
-0.400 dB = 0.955	-10.556 dB = 0.297
-0.500 dB = 0.944	-9.636 dB = 0.330
-0.600 dB = 0.933	-8.893 dB = 0.359
-0.700 dB = 0.923	-8.272 dB = 0.386
-0.800 dB = 0.912	-7.741 dB = 0.410
-0.900 dB = 0.902	-7.278 dB = 0.433
-1.000 dB = 0.891	-6.868 dB = 0.454
-1.200 dB = 0.871	-6.172 dB = 0.491
-1.400 dB = 0.851	-5.598 dB = 0.525
-1.600 dB = 0.832	-5.112 dB = 0.555
-1.800 dB = 0.813	-4.694 dB = 0.583
-2.000 dB = 0.794	-4.329 dB = 0.607
-2.200 dB = 0.776	-4.007 dB = 0.630
-2.400 dB = 0.759	-3.721 dB = 0.652
-2.600 dB = 0.741	-3.463 dB = 0.671
-2.800 dB = 0.724	-3.231 dB = 0.689
-3.000 dB = 0.708	-3.021 dB = 0.706
-4.000 dB = 0.631	-2.205 dB = 0.776
-5.000 dB = 0.562	-1.651 dB = 0.827
-6.000 dB = 0.501	-1.256 dB = 0.865
-7.000 dB = 0.447	-0.967 dB = 0.895
-8.000 dB = 0.398	-0.749 dB = 0.917
-9.000 dB = 0.355	-0.584 dB = 0.935
-10.000 dB = 0.316	-0.458 dB = 0.949
-11.000 dB = 0.282	-0.359 dB = 0.959
-12.000 dB = 0.251	-0.283 dB = 0.968
-13.000 dB = 0.224	-0.223 dB = 0.975
-14.000 dB = 0.200	-0.176 dB = 0.980
-15.000 dB = 0.178	-0.140 dB = 0.984
-16.000 dB = 0.158	-0.110 dB = 0.987
-17.000 dB = 0.141	-0.088 dB = 0.990
-18.000 dB = 0.126	-0.069 dB = 0.992
-19.000 dB = 0.112	-0.055 dB = 0.994
-20.000 dB = 0.100	-0.044 dB = 0.995
-22.000 dB = 0.079	-0.027 dB = 0.997
-24.000 dB = 0.063	-0.017 dB = 0.998
-26.000 dB = 0.050	-0.011 dB = 0.999
-28.000 dB = 0.040	-0.007 dB = 0.999
-30.000 dB = 0.032	-0.004 dB = 0.999

Help Sheet A.1: Relationship between the reflection and transmission values of a lossless two-p (applies to any port impedances)

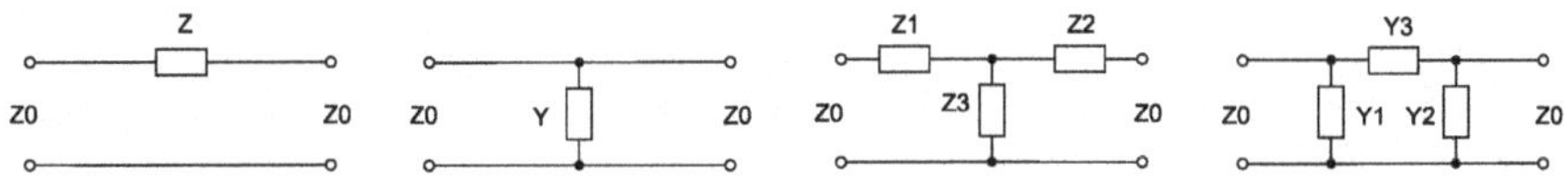

Fig. A.1: Series-Z Shunt-Y T-Circuit PI-Circuit

	Series Z	Shunt Y	T-Circuit	PI-Circuit
S_{11}	$\frac{z}{2+z}$	$\frac{-y}{2+y}$	$\frac{z_1+z_2-y_3(1-z_1+z_2-z_1z_2)}{2+z_1+z_2+y_3(1+z_1+z_2+z_1z_2)}$	$-\frac{y_1+y_2-z_3(1-y_1+y_2-y_1y_2)}{2+y_1+y_2+z_3(1+y_1+y_2+y_1y_2)}$
$S_{12}=S_{21}$	$\frac{2}{2+z}$	$\frac{2}{2+y}$	$\frac{2}{Denominator}$	$\frac{2}{Denominator}$
S_{22}	$\frac{z}{2+z}$	$\frac{-y}{2+y}$	$\frac{z_1+z_2-y_3(1+z_1-z_2-z_1z_2)}{Denominator}$	$-\frac{y_1+y_2-z_3(1+y_1-y_2-y_1y_2)}{Denominator}$
Z_{11}	/	Z	Z_1+Z_3	$\frac{Z_1(Z_2+Z_3)}{\Sigma Z}$
$Z_{12}=Z_{21}$	/	Z	Z_3	$\frac{Z_1 Z_2}{\Sigma Z}$
Z_{22}	/	Z	Z_2+Z_3	$\frac{Z_2(Z_1+Z_3)}{\Sigma Z}$
Y_{11}	Y	/	$\frac{Y_1(Y_2+Y_3)}{\Sigma Y}$	Y_1+Y_3
$Y_{12}=Y_{21}$	$-Y$	/	$\frac{-Y_1 Y_2}{\Sigma Y}$	$-Y_3$
Y_{22}	Y	/	$\frac{Y_2(Y_1+Y_3)}{\Sigma Y}$	Y_2+Y_3
A_{11}	1	1	$1+\frac{Z_1}{Z_3}$	$1+\frac{Y_2}{Y_3}$
A_{12}	Z	0	$Z_1+Z_2+\frac{Z_1 Z_2}{Z_3}$	$\frac{1}{Y_3}$
A_{21}	0	Y	$\frac{1}{Z_3}$	$Y_1+Y_2+\frac{Y_1 Y_2}{Y_3}$
A_{22}	1	1	$1+\frac{Z_2}{Z_3}$	$1+\frac{Y_1}{Y_3}$

Help sheet A.2: Network matrices of different circuits 1

With: $\Sigma Y = Y_1+Y_2+Y_3$, $\Sigma Z = Z_1+Z_2+Z_3$, $Y_i = \frac{1}{Z_i}$ $(i=1,2,3)$.

Standardization applies: $z_i = Z_i/Z_0$ resp. $y_i = Y_i\, Z_0$.

/ means that the values do not exist.

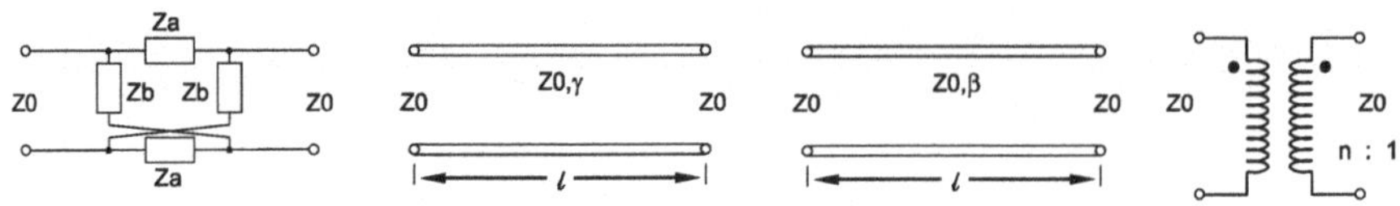

Fig. A.2: Cross-Circuit Line (losses) Line (ideal) Transformer

	Cross Circuit	Lossy Line	Ideal Line	Ideal Transformer
S_{11}	$\frac{z_a - y_b}{1+z_a+y_b+z_a y_b}$	0	0	$\frac{n^2-1}{n^2+1}$
$S_{12} = S_{21}$	$\frac{1-z_a y_b}{Denominator}$	$e^{-\gamma\ell}$	$e^{-j\beta\ell}$	$\frac{2\,n}{n^2+1}$
S_{22}	$\frac{z_a - y_b}{Denominator}$	0	0	$-\frac{n^2-1}{n^2+1}$
Z_{11}	$\frac{Z_a+Z_b}{2}$	$Z_0 \coth(\gamma\ell)$	$-jZ_0 \cot(\beta\ell)$	/
$Z_{12} = Z_{21}$	$\frac{Z_b-Z_a}{2}$	$\frac{Z_0}{\sinh(\gamma\ell)}$	$\frac{-j\,Z_0}{\sin(\beta\ell)}$	/
Z_{22}	$\frac{Z_a+Z_b}{2}$	$Z_0 \coth(\gamma\ell)$	$-jZ_0 \cot(\beta\ell)$	/
Y_{11}	$\frac{Y_a+Y_b}{2}$	$\frac{\coth(\gamma\ell)}{Z_0}$	$\frac{-j\cot(\beta\ell)}{Z_0}$	/
$Y_{12} = Y_{21}$	$\frac{Y_b-Y_a}{2}$	$\frac{-1}{Z_0\sinh(\gamma\ell)}$	$\frac{-1}{Z_0\sin(\beta\ell)}$	/
Y_{22}	$\frac{Y_a+Y_b}{2}$	$\frac{\coth(\gamma\ell)}{Z_0}$	$\frac{-j\cot(\beta\ell)}{Z_0}$	$Y_2 + Y_3$
A_{11}	$\frac{Z_a+Z_b}{Z_b-Z_a}$	$\cosh(\gamma\ell)$	$\cos(\beta\ell)$	n
A_{12}	$\frac{2\,Z_a\,Z_b}{Z_b-Z_a}$	$Z_0 \sinh(\gamma\ell)$	$j\,Z_0 \sin(\beta\ell)$	0
A_{21}	$\frac{2}{Z_b-Z_a}$	$\frac{\sinh(\gamma\ell)}{Z_0}$	$\frac{j\,\sin(\beta\ell)}{Z_0}$	0
A_{22}	$\frac{Z_a+Z_b}{Z_b-Z_a}$	$\cosh(\gamma\ell)$	$\cos(\beta\ell)$	$\frac{1}{n}$

Help sheet A.3: Network matrices of different circuits 2

With: $u_1 = n\,u_2$, $i_1 = -\frac{i_2}{n}$, $Y_x = \frac{1}{Z_x}$ $(x = a, b)$.
Both lines are ideally matched.

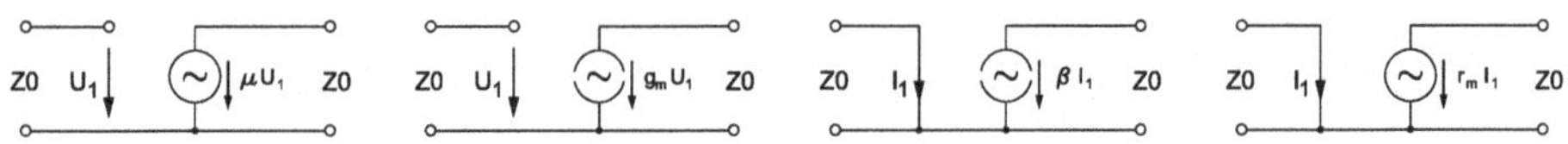

Fig. A.3: U=f(U) I=f(U) I=f(I) U=f(I)

	U-controlled U-source	U-controlled I-source	I-controlled I-source	I-controlled U-source
S_{11}	1	1	-1	-1
S_{12}	0	0	0	0
S_{21}	$2\,\mu$	$-2\,g_m$	$-2\,\beta$	$-2\,r_m$
S_{22}	-1	1	1	-1
Z_{11}	/	/	/	0
Z_{12}	/	/	/	0
Z_{21}	/	/	/	r_m
Z_{22}	/	/	/	0
Y_{11}	/	0	/	/
Y_{12}	/	0	/	/
Y_{21}	/	g_m	/	/
Y_{22}	/	0	/	/
A_{11}	$\frac{1}{\mu}$	0	0	0
A_{12}	0	$-\frac{1}{g_m}$	0	0
A_{21}	0	0	0	$\frac{1}{r_m}$
A_{22}	0	0	$-\frac{1}{\beta}$	0

Help sheet A.4: Network matrices of different circuits 3

With: U = Voltage; I = Current .

	[Z]	[Y]	[A]	[S]
Z_{11}	/	$Y_{22}/\Delta Y$	A_{11}/A_{21}	$z_{11} = \frac{(1+S_{11})(1-S_{22})+S_{12}S_{21}}{(1-S_{11})(1-S_{22})-S_{12}S_{21}}$
Z_{12}	/	$-Y_{12}/\Delta Y$	$\Delta A/A_{21}$	$z_{12} = \frac{2\ S_{12}}{Denominator}$
Z_{21}	/	$-Y_{21}/\Delta Y$	$1/A_{21}$	$z_{21} = \frac{2\ S_{21}}{Denominator}$
Z_{22}	/	$Y_{11}/\Delta Y$	A_{22}/A_{21}	$z_{22} = \frac{(1-S_{11})(1+S_{22})+S_{12}S_{21}}{Denominator}$
Y_{11}	$Z_{22}/\Delta Z$	/	A_{22}/A_{12}	$y_{11} = \frac{(1-S_{11})(1+S_{22})+S_{12}S_{21}}{(1+S_{11})(1+S_{22})-S_{12}S_{21}}$
Y_{12}	$-Z_{12}/\Delta Z$	/	$-\Delta A/A_{12}$	$y_{12} = \frac{-2\ S_{12}}{Denominator}$
Y_{21}	$-Z_{21}/\Delta Z$	/	$-1/A_{12}$	$y_{21} = \frac{-2\ S_{21}}{Denominator}$
Y_{22}	$Z_{11}/\Delta Z$	/	A_{11}/A_{12}	$y_{22} = \frac{(1+S_{11})(1-S_{22})+S_{12}S_{21}}{Denominator}$
A_{11}	Z_{11}/Z_{21}	$-Y_{22}/Y_{21}$	/	$a_{11} = \frac{(1+S_{11})(1-S_{22})+S_{12}S_{21}}{2\ S_{21}}$
A_{12}	$\Delta Z/Z_{21}$	$-1/Y_{21}$	/	$a_{12} = \frac{(1+S_{11})(1+S_{22})-S_{12}S_{21}}{2\ S_{21}}$
A_{21}	$1/Z_{21}$	$-\Delta Y/Y_{21}$	/	$a_{21} = \frac{(1-S_{11})(1-S_{22})-S_{12}S_{21}}{2\ S_{21}}$
A_{22}	Z_{22}/Z_{21}	$-Y_{11}/Y_{21}$	/	$a_{22} = \frac{(1-S_{11})(1+S_{22})+S_{12}S_{21}}{2\ S_{21}}$

Help sheet A.5: Conversions of various network matrices

	[Z]	[Y]	[A]
S_{11}	$\frac{(z_{11}-1)(z_{22}+1)-z_{12}z_{21}}{(z_{11}+1)(z_{22}+1)-z_{12}z_{21}}$	$\frac{(1-y_{11})(1+y_{22})+y_{12}y_{21}}{(y_{11}+1)(y_{22}+1)-y_{12}y_{21}}$	$\frac{a_{11}+a_{12}-a_{21}-a_{22}}{a_{11}+a_{12}+a_{21}+a_{22}}$
S_{12}	$\frac{2\ z_{12}}{Denominator}$	$-\frac{2\ y_{12}}{Denominator}$	$\frac{2\ \Delta a}{Denominator}$
S_{21}	$\frac{2\ z_{21}}{Denominator}$	$-\frac{2\ y_{21}}{Denominator}$	$\frac{2}{Denominator}$
S_{22}	$\frac{(z_{11}+1)(z_{22}-1)-z_{12}z_{21}}{Denominator}$	$\frac{(1+y_{11})(1-y_{22})+y_{12}y_{21}}{Denominator}$	$\frac{-a_{11}+a_{12}-a_{21}+a_{22}}{Denominator}$

Help sheet A.6: Conversion of various network matrices into S-parameters

The symmetrical counting arrow system applies.

The lower case related values are defined as follows:

$$z_{11} = Z_{11}/Z_0 \quad y_{11} = Y_{11}Z_0 \qquad a_{11} = A_{11} \ = A$$

$$z_{12} = Z_{12}/Z_0 \quad y_{12} = Y_{12}Z_0 \quad a_{12} = A_{12}/Z_0 \ = B$$

$$z_{21} = Z_{21}/Z_0 \quad y_{21} = Y_{21}Z_0 \quad a_{21} = A_{21} \cdot Z_0 = C$$

$$z_{22} = Z_{22}/Z_0 \quad y_{22} = Y_{22}Z_0 \qquad a_{22} = A_{22} \ = D$$

The following still applies:

$$\Delta X = X_{11}\, X_{22} - X_{12}\, X_{21} \qquad .$$

Components with Impedance Transformation

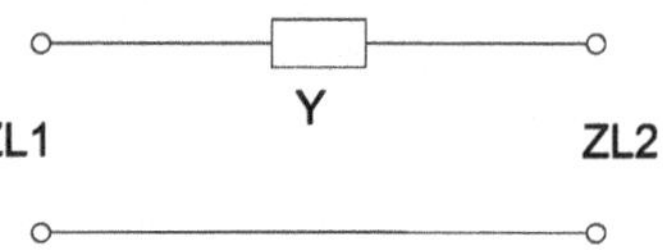

Fig. A.4: Series admittance

$$S = \begin{bmatrix} -\frac{-1-ZL2\,Y+ZL1\,Y}{1+ZL2\,Y+ZL1\,Y} & 2\,\frac{\sqrt{ZL1}Y\sqrt{ZL2}}{1+ZL2\,Y+ZL1\,Y} \\ 2\,\frac{\sqrt{ZL1}Y\sqrt{ZL2}}{1+ZL2\,Y+ZL1\,Y} & \frac{1+ZL1\,Y-ZL2\,Y}{1+ZL2\,Y+ZL1\,Y} \end{bmatrix}$$

$[Z]$ does not exist!

$$Y = \begin{bmatrix} ZL1\,Y & -\sqrt{ZL1}Y\sqrt{ZL2} \\ -\sqrt{ZL1}Y\sqrt{ZL2} & ZL2\,Y \end{bmatrix}$$

$$A = \begin{bmatrix} \frac{\sqrt{ZL2}}{\sqrt{ZL1}} & \frac{1}{\sqrt{ZL1}Y\sqrt{ZL2}} \\ 0 & \frac{\sqrt{ZL1}}{\sqrt{ZL2}} \end{bmatrix}$$

****

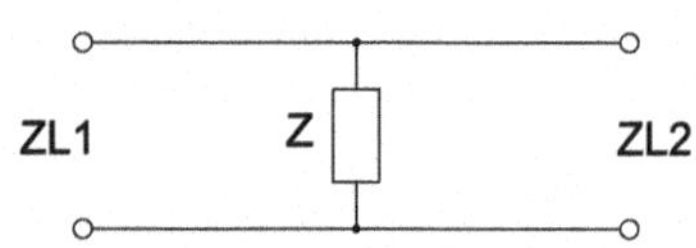

Fig. A.5: Shunt impedance

$$S = \begin{bmatrix} \frac{Z\,ZL2-ZL1\,Z-ZL1\,ZL2}{Z\,ZL2+ZL1\,Z+ZL1\,ZL2} & 2\,\frac{\sqrt{ZL1}Z\sqrt{ZL2}}{Z\,ZL2+ZL1\,Z+ZL1\,ZL2} \\ 2\,\frac{\sqrt{ZL1}Z\sqrt{ZL2}}{Z\,ZL2+ZL1\,Z+ZL1\,ZL2} & -\frac{Z\,ZL2-ZL1\,Z+ZL1\,ZL2}{Z\,ZL2+ZL1\,Z+ZL1\,ZL2} \end{bmatrix}$$

$$Z = \begin{bmatrix} \frac{Z}{ZL1} & \frac{Z}{\sqrt{ZL1}\sqrt{ZL2}} \\ \frac{Z}{\sqrt{ZL1}\sqrt{ZL2}} & \frac{Z}{ZL2} \end{bmatrix}$$

$[Y]$ does not exist!

$$A = \begin{bmatrix} \frac{\sqrt{ZL2}}{\sqrt{ZL1}} & 0 \\ \frac{\sqrt{ZL1}\sqrt{ZL2}}{Z} & \frac{\sqrt{ZL1}}{\sqrt{ZL2}} \end{bmatrix}$$

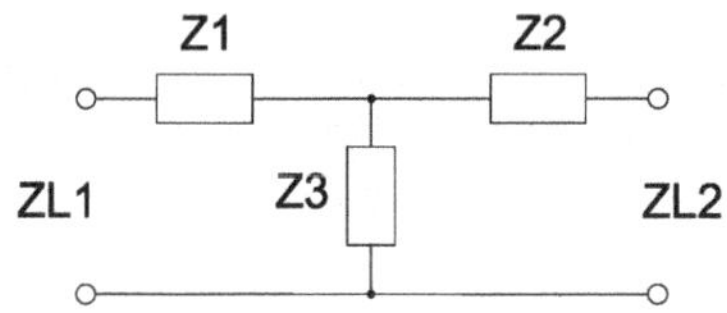

Fig. A.6: T-Circuit

$$S = \begin{bmatrix} 1 - \frac{2\,(Z2+Z3+Zl2)\,Zl1}{Z1\ Z2+Z1\ Z3+Z1\ Zl2+Z3\ Z2+Z3\ Zl2+Zl1\ Z2+Zl1\ Z3+Zl1\ Zl2} & 2\,\frac{\sqrt{Zl1}\,Z3\,\sqrt{Zl2}}{Denominator} \\ 2\,\frac{\sqrt{Zl1}\,Z3\,\sqrt{Zl2}}{Denominator} & 1 - \frac{2\,(Z1+Z3+Zl1)\,Zl2}{Denominator} \end{bmatrix}$$

$$Z = \begin{bmatrix} \frac{Z1+Z3}{Zl1} & \frac{Z3}{\sqrt{Zl1}\sqrt{Zl2}} \\ \frac{Z3}{\sqrt{Zl1}\sqrt{Zl2}} & \frac{Z2+Z3}{Zl2} \end{bmatrix} \qquad A = \begin{bmatrix} \frac{(Z1+Z3)\sqrt{Zl2}}{\sqrt{Zl1}\,Z3} & \frac{Z1\ Z2+Z1\ Z3+Z3\ Z2}{\sqrt{Zl1}\,Z3\,\sqrt{Zl2}} \\ \frac{\sqrt{Zl1}\sqrt{Zl2}}{Z3} & \frac{\sqrt{Zl1}\,(Z2+Z3)}{Z3\,\sqrt{Zl2}} \end{bmatrix}$$

$$Y = \begin{bmatrix} \frac{(Z2+Z3)\,Zl1}{Z1\ Z2+Z1\ Z3+Z3\ Z2} & -\frac{\sqrt{Zl1}\,Z3\,\sqrt{Zl2}}{Z1\ Z2+Z1\ Z3+Z3\ Z2} \\ -\frac{\sqrt{Zl1}\,Z3\,\sqrt{Zl2}}{Z1\ Z2+Z1\ Z3+Z3\ Z2} & \frac{(Z1+Z3)\,Zl2}{Z1\ Z2+Z1\ Z3+Z3\ Z2} \end{bmatrix}$$

****

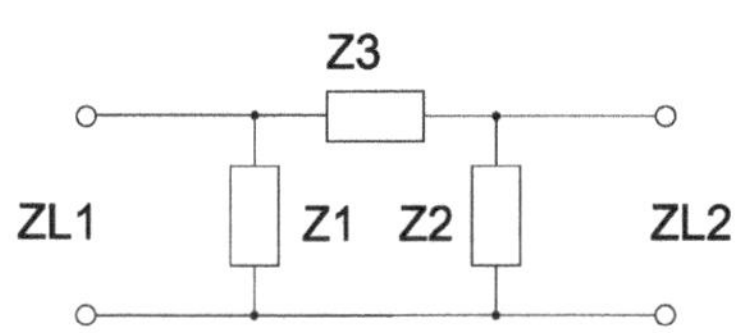

Fig. A.7: Π-Circuit

$$S = \begin{bmatrix} \frac{2\,(Z2\ Z3+Zl2\ Z3+Zl2\ Z2)\,Z1}{Denominator} - 1 & 2\,\frac{\sqrt{Zl1}\sqrt{Zl2}\,Z1\ Z2}{Denominator} \\ 2\,\frac{\sqrt{Zl1}\sqrt{Zl2}\,Z1\ Z2}{Denominator} & \frac{2\,(Z1\ Z3+Zl1\ Z3+Zl1\ Z1)\,Z2}{Denominator} - 1 \end{bmatrix}$$

$$\begin{aligned} Denominator \quad &= Z1\ Z3\ Z2 + Z1\ Z3\ Zl2 + Z1\ Zl2\ Z2 + Zl1\ Z3\ Z2 \\ &+ Zl1\ Z3\ Zl2 + Zl1\ Zl2\ Z2 + Zl1\ Z1\ Z2 + Zl1\ Z1\ Zl2 \end{aligned}$$

$$Z = \begin{bmatrix} \frac{(Z3+Z2)\,Z1}{(Z1+Z3+Z2)\,Zl1} & \frac{Z1\ Z2}{\sqrt{Zl1}\sqrt{Zl2}\,(Z1+Z3+Z2)} \\ \frac{Z1\ Z2}{\sqrt{Zl1}\sqrt{Zl2}\,(Z1+Z3+Z2)} & \frac{(Z3+Z1)\,Z2}{(Z1+Z3+Z2)\,Zl2} \end{bmatrix}$$

$$Y = \begin{bmatrix} \frac{(Z3+Z1)\,Zl1}{Z1\ Z3} & -\frac{\sqrt{Zl1}\sqrt{Zl2}}{Z3} \\ -\frac{\sqrt{Zl1}\sqrt{Zl2}}{Z3} & \frac{(Z3+Z2)\,Zl2}{Z2\ Z3} \end{bmatrix} \qquad A = \begin{bmatrix} \frac{(Z3+Z2)\sqrt{Zl2}}{\sqrt{Zl1}\,Z2} & \frac{Z3}{\sqrt{Zl1}\sqrt{Zl2}} \\ \frac{\sqrt{Zl1}\sqrt{Zl2}\,(Z1+Z3+Z2)}{Z1\ Z2} & \frac{\sqrt{Zl1}\,(Z3+Z1)}{\sqrt{Zl2}\,Z1} \end{bmatrix}$$

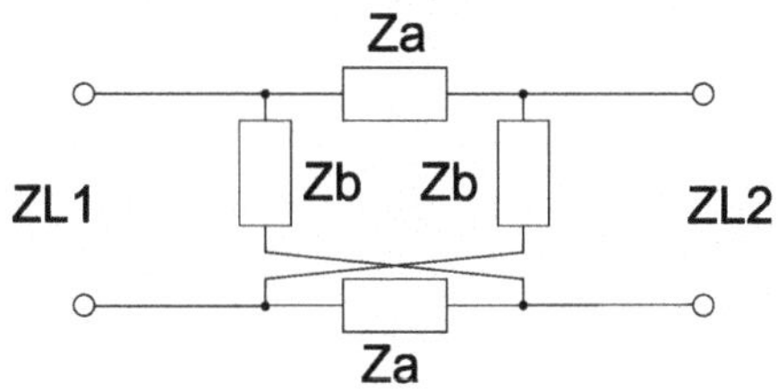

Fig. A.8: Cross-Circuit

$$S = \begin{bmatrix} 1-2\,\frac{(Za+Zb+2\,Zl2)\,Zl1}{Denominator} & 2\,\frac{\sqrt{Zl1}\,(Zb-Za)\sqrt{Zl2}}{Denominator} \\ 2\,\frac{\sqrt{Zl1}\,(Zb-Za)\sqrt{Zl2}}{Denominator} & 1-2\,\frac{(Za+Zb+2\,Zl1)\,Zl2}{Denominator} \end{bmatrix}$$

$$\begin{aligned} Denominator &= 2\,Za\,Zb + Za\,Zl2 + Zb\,Zl2 + \\ & \quad Zl1\,Za + Zl1\,Zb + 2\,Zl1\,Zl2 \end{aligned}$$

$$Z = \begin{bmatrix} 1/2\,\frac{Za+Zb}{Zl1} & 1/2\,\frac{Zb-Za}{\sqrt{Zl1}\sqrt{Zl2}} \\ 1/2\,\frac{Zb-Za}{\sqrt{Zl1}\sqrt{Zl2}} & 1/2\,\frac{Za+Zb}{Zl2} \end{bmatrix}$$

$$Y = \begin{bmatrix} 1/2\,\frac{(Za+Zb)\,Zl1}{Za\,Zb} & -1/2\,\frac{\sqrt{Zl1}\,(Zb-Za)\sqrt{Zl2}}{Za\,Zb} \\ -1/2\,\frac{\sqrt{Zl1}\,(Zb-Za)\sqrt{Zl2}}{Za\,Zb} & 1/2\,\frac{(Za+Zb)\,Zl2}{Za\,Zb} \end{bmatrix}$$

$$A = \begin{bmatrix} \frac{(Za+Zb)\sqrt{Zl2}}{\sqrt{Zl1}\,(Zb-Za)} & 2\,\frac{Za\,Zb}{\sqrt{Zl1}\,(Zb-Za)\sqrt{Zl2}} \\ 2\,\frac{\sqrt{Zl1}\sqrt{Zl2}}{Zb-Za} & \frac{(Za+Zb)\sqrt{Zl1}}{(Zb-Za)\sqrt{Zl2}} \end{bmatrix}$$

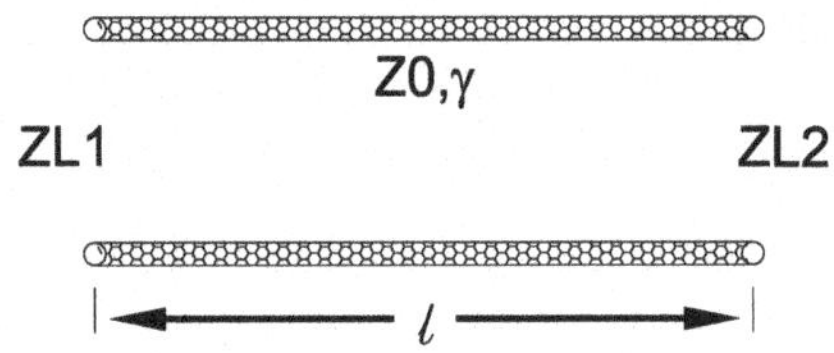

Fig. A.9: Lossless Line

$$S_{11} = \frac{Enumerator1}{Denominator} \qquad S_{12} = 2\,\frac{\sqrt{-1}\sqrt{ZL1}\,Z0\,\sin(\beta\,l)\sqrt{ZL2}}{Denominator}$$

$$S_{21} = 2\,\frac{\sqrt{-1}\sqrt{ZL1}\,Z0\,\sin(\beta\,l)\sqrt{ZL2}}{Denominator} \qquad S_{22} = -\frac{Enumerator2}{Denominator}$$

$$\begin{aligned} Denominator \;&= Z0^2\,(\cos(\beta\,l))^2 + \sqrt{-1}\sin(\beta\,l)\,Z0\,\cos(\beta\,l)\,ZL2 + \sqrt{-1}\sin(\beta\,l)\,ZL1\;Z0\,\cos(\beta\,l) \\ &\quad -ZL1\;ZL2 + ZL1\;ZL2\;(\cos(\beta\,l))^2 - Z0^2 \end{aligned}$$

$$\begin{aligned} Enumerator1 \;&= Z0^2\,(\cos(\beta\,l))^2 + \sqrt{-1}\sin(\beta\,l)\,Z0\,\cos(\beta\,l)\,ZL2 - \sqrt{-1}\sin(\beta\,l)\,ZL1\;Z0\,\cos(\beta\,l) \\ &\quad +ZL1\;ZL2 - ZL1\;ZL2\;(\cos(\beta\,l))^2 - Z0^2 \end{aligned}$$

$$\begin{aligned} Enumerator2 \;&= -Z0^2\,(\cos(\beta\,l))^2 + \sqrt{-1}\sin(\beta\,l)\,Z0\,\cos(\beta\,l)\,ZL2 - \sqrt{-1}\sin(\beta\,l)\,ZL1\;Z0\,\cos(\beta\,l) \\ &\quad -ZL1\;ZL2 + ZL1\;ZL2\;(\cos(\beta\,l))^2 + Z0^2 \end{aligned}$$

$$Z = \begin{bmatrix} \frac{\sqrt{-1}\cos(\beta\,l)\,Z0\,\sin(\beta\,l)}{ZL1\left((\cos(\beta\,l))^2-1\right)} & -\frac{\sqrt{-1}\,Z0}{\sqrt{ZL1}\sin(\beta\,l)\sqrt{ZL2}} \\ -\frac{\sqrt{-1}\,Z0}{\sqrt{ZL1}\sin(\beta\,l)\sqrt{ZL2}} & \frac{\sqrt{-1}\cos(\beta\,l)\,Z0\,\sin(\beta\,l)}{ZL2\left((\cos(\beta\,l))^2-1\right)} \end{bmatrix}$$

$$Y = \begin{bmatrix} \frac{\sqrt{-1}\cos(\beta\,l)\,ZL1\,\sin(\beta\,l)}{Z0\left((\cos(\beta\,l))^2-1\right)} & -\frac{\sqrt{-1}\sqrt{ZL1}\sin(\beta\,l)\sqrt{ZL2}}{Z0\left((\cos(\beta\,l))^2-1\right)} \\ -\frac{\sqrt{-1}\sqrt{ZL1}\sin(\beta\,l)\sqrt{ZL2}}{Z0\left((\cos(\beta\,l))^2-1\right)} & \frac{\sqrt{-1}\sin(\beta\,l)\cos(\beta\,l)\,ZL2}{Z0\left((\cos(\beta\,l))^2-1\right)} \end{bmatrix}$$

$$A = \begin{bmatrix} \frac{\cos(\beta\,l)\sqrt{ZL2}}{\sqrt{ZL1}} & \frac{\sqrt{-1}\,Z0\,\sin(\beta\,l)}{\sqrt{ZL1}\sqrt{ZL2}} \\ \frac{\sqrt{-1}\sqrt{ZL1}\sin(\beta\,l)\sqrt{ZL2}}{Z0} & \frac{\cos(\beta\,l)\sqrt{ZL1}}{\sqrt{ZL2}} \end{bmatrix}$$

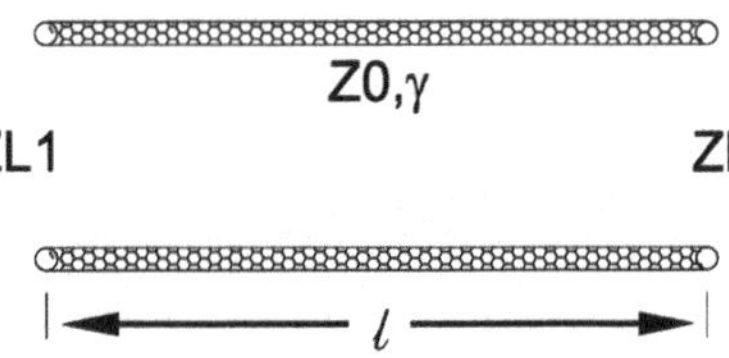

Fig. A.10: Lossy Line

$$S_{11} = -\frac{Enumerator1}{Denominator} \qquad S_{12} = 2\,\frac{\sqrt{ZL1}\,Z0\,\sinh(\gamma\,l)\sqrt{ZL2}}{Denominator}$$

$$S_{21} = 2\,\frac{\sqrt{ZL1}\,Z0\,\sinh(\gamma\,l)\sqrt{ZL2}}{Denominator} \qquad S_{22} = \frac{Enumerator2}{Denominator}$$

$$\begin{aligned} Denominator &= Z0^2\,(\cosh(\gamma\,l))^2 + \sinh(\gamma\,l) Z0\,\cosh(\gamma\,l) ZL2 + \sinh(\gamma\,l) ZL1\,Z0\,\cosh(\gamma\,l) \\ &\quad + ZL1\,ZL2\,(\cosh(\gamma\,l))^2 - ZL1\,ZL2 - Z0^2 \end{aligned}$$

$$\begin{aligned} Enumerator1 &= -Z0^2\,(\cosh(\gamma\,l))^2 - \sinh(\gamma\,l) Z0\,\cosh(\gamma\,l) ZL2 + \sinh(\gamma\,l) ZL1\,Z0\,\cosh(\gamma \\ &\quad + ZL1\,ZL2\,(\cosh(\gamma\,l))^2 - ZL1\,ZL2 + Z \end{aligned}$$

$$\begin{aligned} Enumerator2 &= Z0^2\,(\cosh(\gamma\,l))^2 - \sinh(\gamma\,l) Z0\,\cosh(\gamma\,l) ZL2 + \sinh(\gamma\,l) ZL1\,Z0\,\cosh(\gamma\,l) \\ &\quad - ZL1\,ZL2\,(\cosh(\gamma\,l))^2 + ZL1\,ZL2 - Z0^2 \end{aligned}$$

$$Z = \begin{bmatrix} \frac{\cosh(\gamma\,l) Z0}{ZL1\,\sinh(\gamma\,l)} & \frac{Z0}{\sqrt{ZL1}\,\sinh(\gamma\,l)\sqrt{ZL2}} \\ \frac{Z0}{\sqrt{ZL1}\,\sinh(\gamma\,l)\sqrt{ZL2}} & \frac{\cosh(\gamma\,l) Z0}{ZL2\,\sinh(\gamma\,l)} \end{bmatrix}$$

$$Y = \begin{bmatrix} \frac{\cosh(\gamma\,l)\sinh(\gamma\,l) ZL1}{Z0\left((\cosh(\gamma\,l))^2-1\right)} & -\frac{\sqrt{ZL1}\,\sinh(\gamma\,l)\sqrt{ZL2}}{Z0\left((\cosh(\gamma\,l))^2-1\right)} \\ -\frac{\sqrt{ZL1}\,\sinh(\gamma\,l)\sqrt{ZL2}}{Z0\left((\cosh(\gamma\,l))^2-1\right)} & \frac{\cosh(\gamma\,l) ZL2\,\sinh(\gamma\,l)}{Z0\left((\cosh(\gamma\,l))^2-1\right)} \end{bmatrix}$$

$$A = \begin{bmatrix} \frac{\cosh(\gamma\,l)\sqrt{ZL2}}{\sqrt{ZL1}} & \frac{Z0\,\sinh(\gamma\,l)}{\sqrt{ZL1}\sqrt{ZL2}} \\ \frac{\sqrt{ZL1}\,\sinh(\gamma\,l)\sqrt{ZL2}}{Z0} & \frac{\cosh(\gamma\,l)\sqrt{ZL1}}{\sqrt{ZL2}} \end{bmatrix}$$

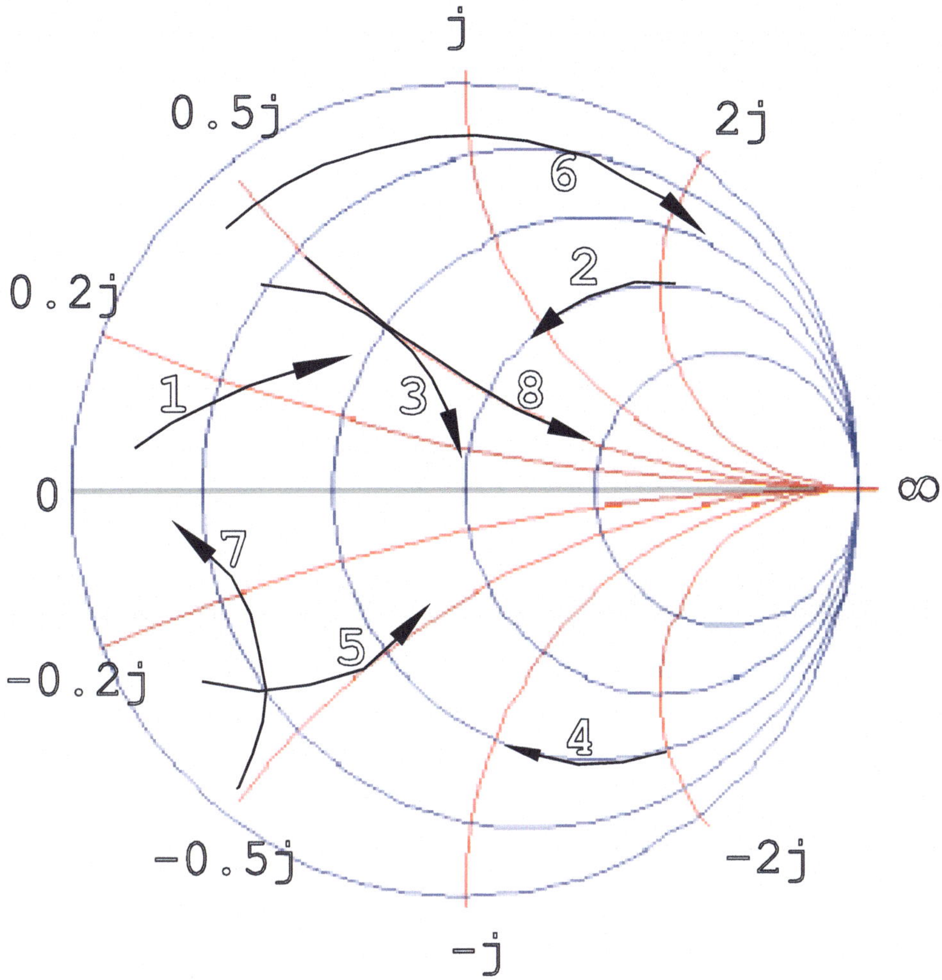

Fig. A.11: Transformation behavior of various components in the Smith chart

1. Transformer
2. Series-C
3. Shunt-C or open-ended stub-line
4. Series-L
5. Shunt-L or short-circuited stub-line
6. Series line
7. Shunt-R
8. Series-R

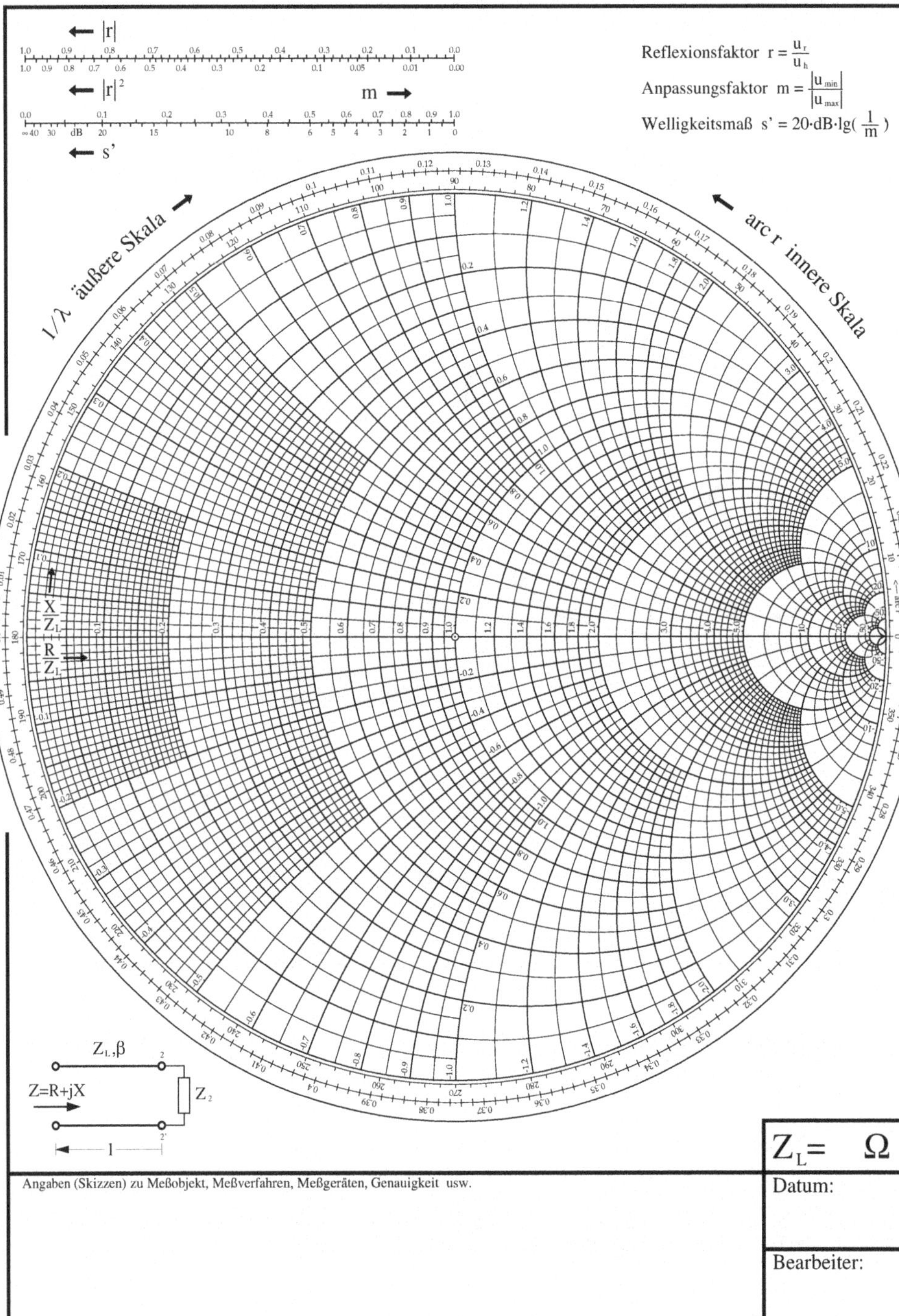

Fig. A.12: Smith-Chart 2. art, the old fashion work tool with high solution and German text

Components of Mixed-Mode Technology

Please note that, as an exception, Y and Z are normalized admittances or impedances.

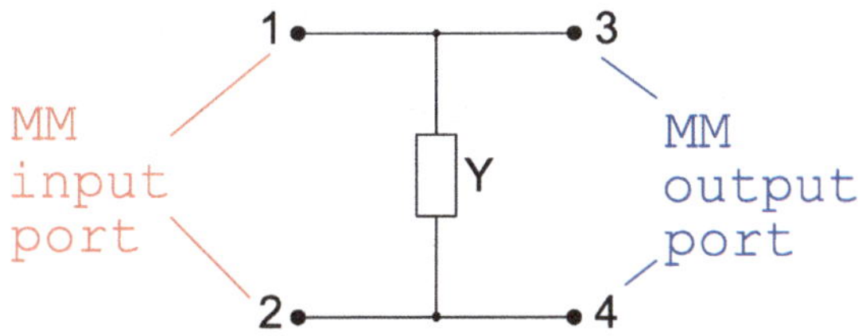

Fig. A.13: Cross admittance

$$M = \begin{bmatrix} -\frac{Y}{1+Y} & \frac{1}{1+Y} & 0 & 0 \\ \frac{1}{1+Y} & -\frac{Y}{1+Y} & 0 & 0 \\ 0 & 0 & 0 & 1 \\ 0 & 0 & 1 & 0 \end{bmatrix}$$

The following applies to all S-parameters: $S_{ij} = \frac{Y}{2+2Y}$.

****

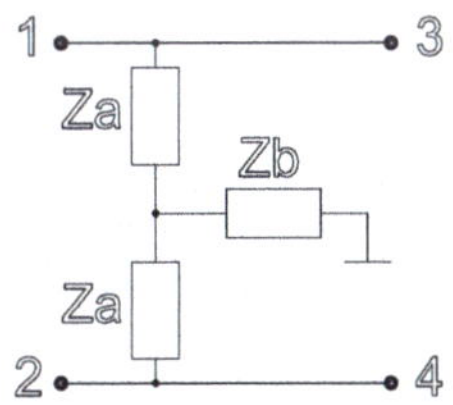

Fig. A.14: Cross-branch with ground connection

$$M = \begin{bmatrix} \frac{-1}{2Za+1} & \frac{2Za}{2Za+1} & 0 & 0 \\ \frac{2Za}{2Za+1} & \frac{-1}{2Za+1} & 0 & 0 \\ 0 & 0 & \frac{-1}{2(Za+2Zb)+1} & \frac{2(Za+2Zb)}{2(Za+2Zb)+1} \\ 0 & 0 & \frac{2(Za+2Zb)}{2(Za+2Zb)+1} & \frac{-1}{2(Za+2Zb)+1} \end{bmatrix}$$

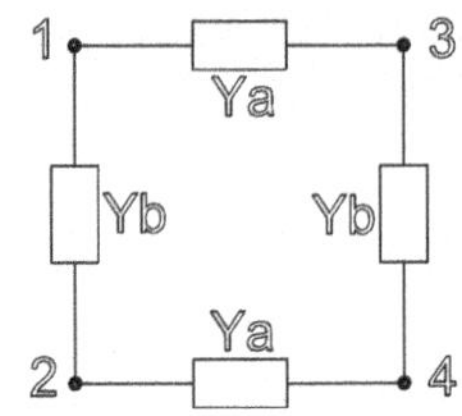

Fig. A.15: Symmetrical cross and series branches

$$M^- = \begin{bmatrix} -\frac{4\,Yb\,Ya-1+4\,Yb^2}{2\,Ya+4\,Yb\,Ya+1+4\,Yb+4\,Yb^2} & \frac{2\,Ya}{2\,Ya+4\,Yb\,Ya+1+4\,Yb+4\,Yb^2} \\ \frac{2\,Ya}{2\,Ya+4\,Yb\,Ya+1+4\,Yb+4\,Yb^2} & -\frac{4\,Yb\,Ya-1+4\,Yb^2}{2\,Ya+4\,Yb\,Ya+1+4\,Yb+4\,Yb^2} \end{bmatrix}$$

$$M^+ = \begin{bmatrix} (1+2\,Ya)^{-1} & 2\,\frac{Ya}{1+2\,Ya} \\ 2\,\frac{Ya}{1+2\,Ya} & (1+2\,Ya)^{-1} \end{bmatrix}$$

The M^{+-} and M^{-+} parameters are zero.

****

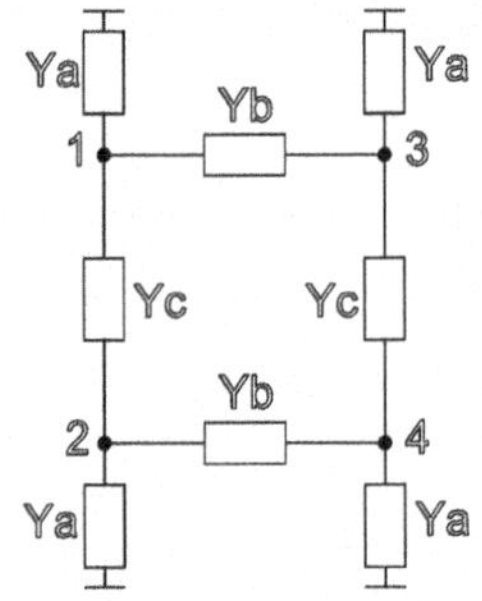

Fig. A.16: Cross and series branches and shunt elements

$$M_{11}^- = M_{22}^- = -\frac{2\,Ya\,Yb+4\,Yc\,Yb+Ya^2+4\,Ya\,Yc+4\,Yc^2-1}{2\,Ya\,Yb+2\,Yb+4\,Yc\,Yb+Ya^2+2\,Ya+4\,Ya\,Yc+4\,Yc+1+4\,Yc^2}$$

$$M_{12}^- = M_{21}^- = 2\,\frac{Yb}{2\,Ya\,Yb+2\,Yb+4\,Yc\,Yb+Ya^2+2\,Ya+4\,Ya\,Yc+4\,Yc+1+4\,Yc^2}$$

$$M^+ = \begin{bmatrix} -\frac{2\,Ya\,Yb+Ya^2-1}{1+2\,Yb+2\,Ya+2\,Ya\,Yb+Ya^2} & 2\,\frac{Yb}{1+2\,Yb+2\,Ya+2\,Ya\,Yb+Ya^2} \\ 2\,\frac{Yb}{1+2\,Yb+2\,Ya+2\,Ya\,Yb+Ya^2} & -\frac{2\,Ya\,Yb+Ya^2-1}{1+2\,Yb+2\,Ya+2\,Ya\,Yb+Ya^2} \end{bmatrix}$$

The M^{+-} and M^{-+} parameters are zero.

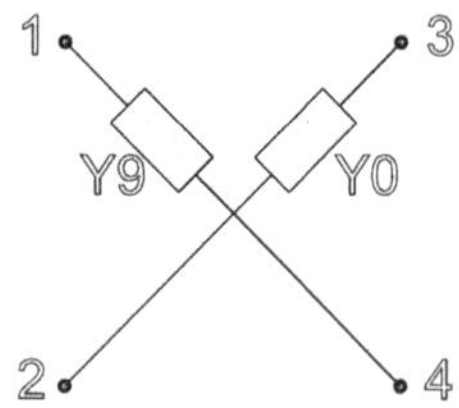

Fig. A.17: Crossing elements or branches

$$M = \begin{bmatrix} \frac{1+Y0+Y9}{(1+2\,Y0)(1+2\,Y9)} & -\frac{Y9+4\,Y0\,Y9+Y0}{(1+2\,Y0)(1+2\,Y9)} & -\frac{Y9-Y0}{(1+2\,Y0)(1+2\,Y9)} & \frac{Y9-Y0}{(1+2\,Y0)(1+2\,Y9)} \\ -\frac{Y9+4\,Y0\,Y9+Y0}{(1+2\,Y0)(1+2\,Y9)} & \frac{1+Y0+Y9}{(1+2\,Y0)(1+2\,Y9)} & -\frac{Y9-Y0}{(1+2\,Y0)(1+2\,Y9)} & \frac{Y9-Y0}{(1+2\,Y0)(1+2\,Y9)} \\ -\frac{Y9-Y0}{(1+2\,Y0)(1+2\,Y9)} & -\frac{Y9-Y0}{(1+2\,Y0)(1+2\,Y9)} & \frac{1+Y0+Y9}{(1+2\,Y0)(1+2\,Y9)} & \frac{Y9+4\,Y0\,Y9+Y0}{(1+2\,Y0)(1+2\,Y9)} \\ \frac{Y9-Y0}{(1+2\,Y0)(1+2\,Y9)} & \frac{Y9-Y0}{(1+2\,Y0)(1+2\,Y9)} & \frac{Y9+4\,Y0\,Y9+Y0}{(1+2\,Y0)(1+2\,Y9)} & \frac{1+Y0+Y9}{(1+2\,Y0)(1+2\,Y9)} \end{bmatrix}$$

****

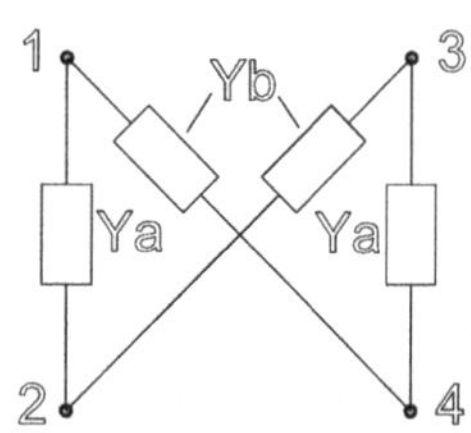

Fig. A.18: Symmetrical cross and intersection branches

$$M = \begin{bmatrix} -\frac{4\,Ya\,Yb-1+4\,Ya^2}{2\,Yb+4\,Ya\,Yb+1+4\,Ya+4\,Ya^2} & \frac{-2\,Yb}{2\,Yb+4\,Ya\,Yb+1+4\,Ya+4\,Ya^2} & 0 & 0 \\ \frac{-2\,Yb}{2\,Yb+4\,Ya\,Yb+1+4\,Ya+4\,Ya^2} & -\frac{4\,Ya\,Yb-1+4\,Ya^2}{2\,Yb+4\,Ya\,Yb+1+4\,Ya+4\,Ya^2} & 0 & 0 \\ 0 & 0 & \frac{1}{1+2\,Yb} & \frac{2\,Yb}{1+2\,Yb} \\ 0 & 0 & \frac{2\,Yb}{1+2\,Yb} & \frac{1}{1+2\,Yb} \end{bmatrix}$$

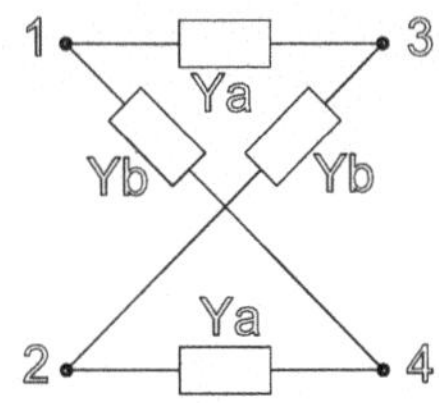

Fig. A.19: Symmetrical series and crossover branches

$$M = \begin{bmatrix} -\frac{4\,Ya\,Yb-1}{2\,Yb+4\,Ya\,Yb+2\,Ya+1} & \frac{-2\,(Yb-Ya)}{2\,Yb+4\,Ya\,Yb+2\,Ya+1} & 0 & 0 \\ \frac{-2\,(Yb-Ya)}{2\,Yb+4\,Ya\,Yb+2\,Ya+1} & -\frac{4\,Ya\,Yb-1}{2\,Yb+4\,Ya\,Yb+2\,Ya+1} & 0 & 0 \\ 0 & 0 & \frac{1}{2\,Yb+1+2\,Ya} & \frac{2\,(Ya+Yb)}{2\,Yb+1+2\,Ya} \\ 0 & 0 & \frac{2\,(Ya+Yb)}{2\,Yb+1+2\,Ya} & \frac{1}{2\,Yb+1+2\,Ya} \end{bmatrix}$$

****

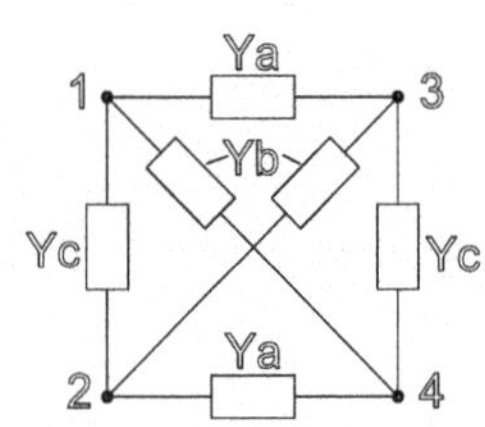

Fig. A.20: Cross and series branches with crossing

$$M_{11}^- = M_{22}^- = -\frac{4\,Ya\,Yb+4\,Yc\,Ya+4\,Yc^2-1+4\,Yc\,Yb}{2\,Ya+4\,Yc\,Ya+4\,Ya\,Yb+1+4\,Yc^2+2\,Yb+4\,Yc+4\,Yc\,Yb}$$
$$M_{12}^- = M_{21}^- = 2\,\frac{Ya-Yb}{2\,Ya+4\,Yc\,Ya+4\,Ya\,Yb+1+4\,Yc^2+2\,Yb+4\,Yc+4\,Yc\,Yb}$$

$$M^+ = \begin{bmatrix} (1+2\,Ya+2\,Yb)^{-1} & 2\,\frac{Ya+Yb}{1+2\,Ya+2\,Yb} \\ 2\,\frac{Ya+Yb}{1+2\,Ya+2\,Yb} & (1+2\,Ya+2\,Yb)^{-1} \end{bmatrix}$$

The M^{+-} and M^{-+} parameters are zero.

	M-Parameter
1	$\begin{matrix} 0 & 1 & 0 & 0 \\ 1 & 0 & 0 & 0 \\ 0 & 0 & 1 & 0 \\ 0 & 0 & 0 & 1 \end{matrix}$
2	$\begin{matrix} -1 & 0 & 0 & 0 \\ 0 & -1 & 0 & 0 \\ 0 & 0 & 0 & 1 \\ 0 & 0 & 1 & 0 \end{matrix}$
3	$\begin{matrix} \frac{1+Y8+Y6}{(1+2\,Y6)(1+2\,Y8)} & \frac{Y6+4\,Y6\,Y8+Y8}{(1+2\,Y6)(1+2\,Y8)} & -\frac{Y6-Y8}{(1+2\,Y6)(1+2\,Y8)} & \frac{Y6-Y8}{(1+2\,Y6)(1+2\,Y8)} \\ \frac{Y6+4\,Y6\,Y8+Y8}{(1+2\,Y6)(1+2\,Y8)} & \frac{1+Y8+Y6}{(1+2\,Y6)(1+2\,Y8)} & \frac{Y6-Y8}{(1+2\,Y6)(1+2\,Y8)} & -\frac{Y6-Y8}{(1+2\,Y6)(1+2\,Y8)} \\ -\frac{Y6-Y8}{(1+2\,Y6)(1+2\,Y8)} & \frac{Y6-Y8}{(1+2\,Y6)(1+2\,Y8)} & \frac{1+Y8+Y6}{(1+2\,Y6)(1+2\,Y8)} & \frac{Y6+4\,Y6\,Y8+Y8}{(1+2\,Y6)(1+2\,Y8)} \\ \frac{Y6-Y8}{(1+2\,Y6)(1+2\,Y8)} & -\frac{Y6-Y8}{(1+2\,Y6)(1+2\,Y8)} & \frac{Y6+4\,Y6\,Y8+Y8}{(1+2\,Y6)(1+2\,Y8)} & \frac{1+Y8+Y6}{(1+2\,Y6)(1+2\,Y8)} \end{matrix}$

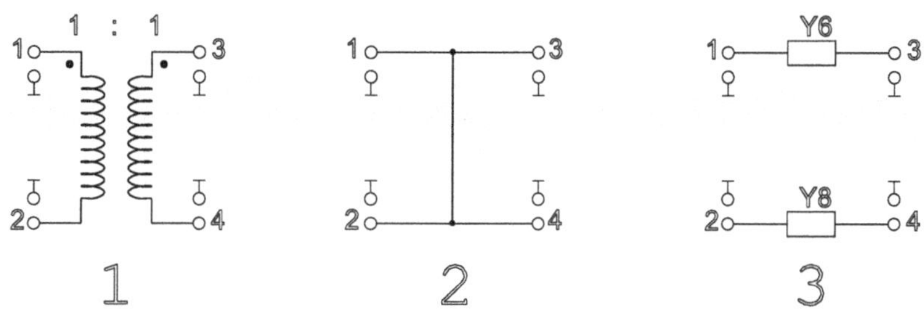

Fig. A.21: Various networks

Bibliography

[1] ADS, Agilent Inc., Westlake Village, CA, USA

[2] Agilent Technologies, Schottky Barrier Diode Video Detector, Application Note 923, 1999

[3] Amari, S., Le Coq, L., Bolomey, J. C., Influence of Discontinuities on Mode Conversion in Millimeter-Wave Microstrip Filters, IEEE Trans. on Microwave Theory and Tech., vol. 53, no. 6, pp. 2049-2055, Jun. 2005.

[4] Aneja, A., Li, X. J., Multiband JNAs for Software-Defined Radios, IEEE microwave magazine, Vol. 21, No. 7, July 2020

[5] Backes, A., Realisierung eines elektronisch durchstimmbaren High-Q-Kanalfilters mit Trägerunterdrückung für UHF-RFID-Frontends, Diplomarbeit, Lehrgebiet für Hoch- und Höchstfrequenztechnik, FH Aachen, 2007

[6] Bahl, I.J., Lumped Elements for RF and Microwave Circuits, Artech House Publishers, 2003

[7] Ballmann, R., Der HF-Netzwerkanalysator NWA S44, Mikrowellen & HF Telecommunications Magazin, Vol. 18, No. 2, 1992, pp. 100-103 (http://www.ballmann-gmbh.de)

[8] Bächthold, W., Mikrowellentechnik, Vieweg, Braunschweig/Wiesbaden, 1999

[9] Bächthold, W., Mikrowellenelektronik, Vieweg, Braunschweig/Wiesbaden, 2002

[10] Best, R., Theorie und Anwendungen des Phase-Locked Loops, 1982

[11] Bex, H.-J., Hoch- und Höchstfrequenztechnik, Teil 1, Vorlesungsskript, Aachen, 1992

[12] Blume, S., Theorie elektromagnetischer Felder, Hüthig, Heidelberg, 1988

[13] Bianco, B., Chiabrera, A., Granara, M., Ridella, S., Frequency Dependence of Microstrip Parameters, Alta Frequenza XLIII (1974), pp. 413-416

[14] Bronstein, I., Semendjajew, K., Taschenbuch der Mathematik, Harry Deutsch Verlag, Thun, Frankfurt/Main, 1987

[15] Cage, O. Electronic Measurements and Instrumentation, McGraw-Hill, Tokyo, 1971

[16] Caulton, M., Knight, S.P., Daly, D.A., Hybrid Integrated Lumped-Element Microwave Amplifiers, IEEE Trans. on Electron Devices, Vol. 15, Nr. 7, July 1968, pp 459-466

[17] Chen, Z. N., et al., Handbook of Antenna Technologies, 2016, Springer Nature

[18] Danesh, M., Long, J.R., Hadaway, R.A., Harame, D.L., A Q-Factor Enhancement Technique for MMIC Inductors, IEEE MTT-S Int. Microwave Symp. Dig., Vol. 1, June 1998, pp. 183-186

[19] Daywitt, W.C., Complex Admittance of a Lossy Coaxial Open Circuit with Hollow Center Conductor, Metrologia, vol. 24, 1987, pp. 1-10

H. Heuermann, *Radio Frequency Technologies*,
https://doi.org/10.1007/978-3-658-50170-9

[20] DENLINGER, E.J., A Frequency Dependent Solution for Microstrip Transmission Lines, IEEE Trans. on Microwave Theory and Techn., MTT-19, Jan. 1971, pp. 30-39

[21] DETLEFSEN, J., SIART, U., Grundlagen der Hochfrequenztechnik, Oldenbourg Verlag, München, 2003

[22] ENGEN, G.F., The Six-Port Reflectometer – An Alternative Network Analyzer, IEEE Trans. Microwave Theory Tech., MTT-25, Dec. 1977, pp. 1075-1080

[23] EISENSTADT, W.R., STENGEL, B., THOMPSON, B. M., Microwave Differntial Circuit Design, Artech House, Boston, ISBN 1-58053-933-5

[24] EPC-RFID: Electronic Product Code for Communications at 860 MHz - 960 MHz, http://www.gs1.org/standards/epc-rfid

[25] ENGEN, G.F., HOER, C.A., Thru-Reflect-Line: An Improved Technique for Calibrating the Dual Six Port Automatic Network Analyzer, IEEE Trans. Microwave Theory Tech., MTT-27, Dec. 1979, pp. 987-993

[26] ERKENS, H., HEUERMANN, H., Blocking Structures for Mixed-Mode-Systems, European Microwave Conf., Amsterdam, Oct. 2004, pp. 297-300

[27] ERKENS, H., HEUERMANN, H., Mixed-Mode Chain Parameters: Theory and Verification, IEEE Trans. Microwave Theory Tech., MTT-8, Aug. 2007, pp. 1704-1708

[28] ERKENS, H., HEUERMANN, H., Novel RF Switch Concepts for Differential Wireless Communication Frontends, IEEE Trans. Microwave Theory Tech., MTT-6, June 2006, pp. 2376-2482

[29] EUL, H.-J., Methoden zur Kalibrierung von heterodynen und homodynen Netzwerkanalysatoren, Dissertationsschrift, Institut für Hoch- und Höchstfrequenztechnik, Ruhr-Universität Bochum, 1990

[30] EUL, H.J., SCHIEK, B., A Generalized Theory and New Calibration Procedures for Network Analyzer Self-Calibration, IEEE Trans. Microwave Theory Tech., MTT-39, Apr. 1991, pp. 724-731

[31] EUL, H.J., SCHIEK, B., Thru-Match-Reflect: One Result of a Rigorous Theory for De-embedding and Network Analyzer Calibration, Proceedings of the 18th European Microwave Conference, Stockholm, 1988, pp. 909-914

[32] FELDTKELLER, R., Einführung in die Theorie der Hochfrequenz-Bandfilter, 5. Edition, Hirzel, Stuttgart, 1961

[33] FERRERO, A., PISANI, U., KERWIN, K.J., A New Implementation of a Multiport Automatic Network Analyzer, IEEE Trans. Microwave Theory Tech., vol. 40, Nov. 1992, pp. 2078-2085

[34] FINKENZELLER, K., RFID-Handbuch, 7. Auflage, Carl Hanser Verlag, München, 2015

[35] GÄRTNER, U., Ein homodynes Streuparametermessverfahren mit digitaler Phasenmodulation für den Mikrowellenbereich, Dissertationsschrift, Institut für Hoch- und Höchstfrequenztechnik, Ruhr-Universität Bochum, 1986

[36] GOLDBERG, B.-G., Digital Techniques in Frequency Synthesis, 1998

[37] GGB INDUSTRIES, INC., Picoprobe Model 67 A, Naples, FL 33941, 3/1993

[38] GRONAU, G., Rauschparameter- und Streuparameter-Messtechnik, Fortschritte der Hochfrequenztechnik, Band 4, Verlagsbuchhandlung Nellissen-Wolff, Aachen, 1992

[39] GRONAU, G., Höchstfrequenztechnik, Springer, Berlin, 2001

[40] GETSINGER, W.J., Microstrip Dispersion Model, IEEE Trans of Microwave Theory and Techn., MTT-21, 1973, pp. 34-39

[41] Getsinger, W.J., Measurement of the Characteristic Impedance of Microstrip over a Wide Frequency Range, IEEE MTT-S Digest, Dallas 1982, pp. 342-344

[42] Greenhouse, H., Design of Planar Rectangular Microelectronic Inductors, IEEE Transactions on Parts, Hybrids, and Packaging, Vol. 10, Nr. 2, June 1974, pp. 101-109

[43] Gupta, K.C., Microwave Integrated Circuits, Halsted Press, 1974.

[44] Gustrau, K.C., RF and Microwave Engineering, Wiley, 2012.

[45] Gupta, K.C., Garg, R., Bahl, I.J., Microstrip Lines and Slotlines, Artech House, Dedham, 1979

[46] Haase, M., Entwurf und Realisierung eines elektronisch optimierten Zirkulators mit integrierter Mikroprozessorsteuerung für das UHF-RFID-Band, Diplomarbeit, Lehrgebiet für Hoch- und Höchstfrequenztechnik, FH Aachen, 2007

[47] Hammerstad, E., Jensen, O., Accurate Models for Microstrip Computer-Aides Design, IEEE MTT-S Internat. Microwave Symp. Digest, 1980, pp. 407-409

[48] Hartnagel H. L., Quay R., Rohde, U. L., ... Funadamentals of RF and Microwave Techniques and Technologies, Vol. 1 + Vol. 2, Springer Nature, 2023

[49] Heuermann, H., Sichere Verfahren zur Kalibrierung von Netzwerkanalysatoren für koaxiale und planare Leitungssysteme, Dissertationsschrift, Institut für Hochfrequenztechnik, Ruhr-Universität Bochum, 1995, ISBN 3-8265-1495-5

[50] Heuermann, H., Microwave Technology, Springer Nature, ISBN 978-3-658-45685-6, 2024

[51] Heuermann, H., Schiek, B., Procedures for the Determination of the Scattering Parameters for Network Analyzer Calibration, IEEE Trans. Instrum. Meas., IM-2, Apr. 1993, pp. 528-531

[52] Heuermann, H., Schiek, B., Robust Algorithms for Txx Network Analyzer Procedures, IEEE Trans. Instrum. Meas., Feb. 1994, pp. 18-23

[53] Heuermann, H., Schiek, B., Results of Network Analyzer Measurements with Leakage Errors – Corrected with Direct Calibration Techniques, IEEE Trans. Instrum. Meas., IM-5, Oct. 1997, pp. 1120-1127

[54] Heuermann, H., Schiek, B., 15-Term Self-Calibration Methods for the Error-Correction of On-Wafer Measurements, IEEE Trans. Instrum. Meas., IM-5, Oct. 1997, pp. 1105-1110

[55] Heuermann, H., GSOLT: The Calibration Procedure for all Multi-Port Vector Network Analyzers, MTT-S International Microwave Symposium, Philadelphia, 2003

[56] Heuermann, H., Rumiantsev, A., Schott, St., Advanced On-Wafer Multiport Calibration Methods for Mono- and Mixed-Mode Device Characterisation, 63rd ARFTG Conf. Proc., Fort-Worth Texas, Jun. 2004

[57] Heuermann, H., A Synthesis Technique for Mono- and Mixed-Mode Symmetrical Filter, European Microwave Conf., Amsterdam, Oct. 2004, pp. 309-312

[58] Heuermann, H., Luenebach, M., A novel Dual-Mode-Design for Push-Push Oscillators with Improved Phase Noise Performance, European Microwave Conf., Paris, 2005

[59] Heuermann, H., Kalibrierverfahren zur Durchführung von Mehrtormessungen auf Halbleiterscheiben, Patentantrag P 103 14 463.3 vom 25.03.03

[60] Heuermann, H., Transmitter und Verfahren zur Erzeugung eines Signals mit digitaler Modulation, Hochschulpatent D102004049019.8, 12.10.2004

[61] Heuermann, H., Antennenarchitektur und Koppler, Hochschulpatent D102004054442.5, 16.11.2004

[62] Heuermann, H., Sadeghfam, A., Lünebach, M., Resonatorsystem und Verfahren zur Erhöhung der belasteten Güte eines Schwingkreises, Hochschulpatent D102004054443.3, 16.11.2004

[63] Heuermann, H., Aufbau von Dual-Mode-Funkübertragungsstrecken, University patent D102005061219.9 vom 20.12.2005

[64] Heuermann, H., Dual Mode Radio: A new Tranceiver Architecture for UWB- and 60 GHz-Applications, European Microwave Conf., Amsterdam, Oct. 2008

[65] Heuermann, H., A 14 W SPST Switch with four PIN-Diodes in 0402-SMD-Package, German Microwave Conf., Hamburg, Mar. 2008

[66] Heuermann, H., Breitbandige Schalter für symmetrische Schaltungen, Hochschulpatent, 30.06.2005

[67] Heuermann, H., Sadeghfam, A., Kompensationsschaltung für eine RFID-Readereinheit, Pepperl+Fuchs-Patent D102008013468 from 10.03.2008

[68] Heuermann, H., Sadeghfam, A., Einstellbare Schaltung und RFID-Readereinheit, Pepperl+Fuchs-Patent D102008013469 vom 10.03.2008

[69] Hirota, T., Tanaka, Y., Kobayashi, K., Dispersion Characteristics of High-Frequency Microstrip Lines in Millimeter-Wave Region, IEEE Trans. on Microwave Theory and Tech., vol. 47, no. 12, pp. 2431?2437, Dec. 1999

[70] Homepage: HF-Lehrgebiet der FH-Aachen, https://www.fh-aachen.de/fachbereiche/elektrotechnik-und-informationstechnik/labore-und-lehrgebiete/lehrgebiet-hoch-und-hoechstfrequenztechnik-prof-dr-ing-holger-heuermann/

[71] Homepage: Institut für Mikrowellen- und Plasmatechnik, IMP der FH-Aachen, https://www.fh-aachen.de/forschung/imp/

[72] Homepage, Heuermann HF-Technik GmbH, http://www.hhft.de

[73] Hoffmann, M.H.W., Hochfrequenztechnik, Springer-Verlag, Berlin-Heidelberg, 1997

[74] Ibrahim, I., Heuermann, H., Improvements in the Flicker Noise Reduction Technique for Oscillator Designs, European Microwave Conf., Rom, Oct. 2009

[75] IEEE microwave Magazine, Full-Duplex Technologies, Vol. 20, No. 2, Feb. 2019

[76] Infineon, Halbleiter, Publicis Corporate Publishing, 3. Auflage, Erlangen, 2004,

[77] Jansen, R.H., Kirschning, M., Arguments and an Accurate Mathematical Model for the Power-Current Formulation of Microstrip Characteristic Impedance, Arch. Elektronik u. übertragungstechn. 37, 1983

[78] Janssen, R., Streifenleiter und Hohlleiter, Hüthig-Verlag, Heidelberg, 1992

[79] Juroshek, J.R., Free, G.M., Measurements of the Characteristic Impedance of Coaxial Air Line Standards, IEEE Trans. Microwave Theory Tech., MTT-42, Feb. 1994, pp. 186-191

[80] Kark, K., Antennen und Strahlungsfelder, Vieweg-Verlag, Wiesbaden, 2004

[81] Kajfez, D., Guillon, P., Dielectric Resonators, Artech House, Dedham, 1986

[82] Kelly, D. u.a, The State-of-the-Art of Silicon-on-Sapphire CMOS RF Switches, IEEE Compound Semiconductor Integrated Circuit Symposium, CSIC 05, doi: 10.1109/CSICS.2005.1531812, 2005.

[83] Kern, C., Anwendung von RFID-Systemen, Springer Verlag, Berlin 2005

[84] Khannar, V. K., Integrated Nanoelectronics. Springer Nature, isbn: 978-81-322-3623-8, 2016

[85] Kirschning, M., Jansen, R.H., Accurate Model for Effective Dielectic Constant of Microstrip with Validity up to Millimetre-Wave Frequencies, Electron. Letters 18, 1982, pp. 272-273

[86] Kummer, M., Grundlagen der Mikrowellentechnik, VEB Verlag Technik, Berlin, 1989

[87] Kurokawa, Power Waves and the Scattering Matrix, IEEE Trans. Microwave Theory Tech., MTT, Mar. 1965, pp. 194-202

[88] Lee, T.H., The design of CMOS radio-frequency integrated circuits, second edition, Cambridge University Press, 2004

[89] Levy, R., Synthesis of General Asymmetric Singly- and Doubly-Terminated Cross-Coupled Filters, IEEE Trans. Microwave Theory Tech., Vol. 42, Dec. 1994, pp. 2468-2471

[90] Long, J.R., Copeland, M.A., The modeling, characterization, and design of monolithic inductors for silicon RF IC's, IEEE Journal of Solid-State Circuits, Vol. 32, Nr. 3, Mar. 1997, pp. 357-369

[91] Matthei, G., Young, L., Jones, E., Microwave Filters, Impedance-Matching Networks, and Coupling Structures, McGraw-Hill, New York, 1980

[92] Maas, S., A., Microwave Mixers, Artech House, London, 1993

[93] Menzel, E., Homodyne und heterodyne Netzwerkanalysatoren für breitbandige Anwendungen im Hochfrequenzbereich, Dissertationsschrift, Institut für Hochfrequenztechnik, Ruhr-Universität Bochum, 1992

[94] Miller, Technique Enhances the Performance of PLL Synthesizers, Microwave & RF, Jan. 1993

[95] Möller, M., Entwurf und Optimierung monolithisch integrierter Breitbandverstärker in Si-Bipolartechnologie für optische übertragungssysteme, Dissertationsschrift, Ruhr-Universität Bochum, 1999

[96] Mohan, S.S., del Mar Hershenson, M., Boyd, S.P., Lee, T.H., Simple Accurate Expressions for Planar Spiral Inductances, IEEE Journal of Solid-State Circuits, Vol. 34, Nr. 10, Oct. 1999, pp, 1419-1424

[97] Morgan, A.M., Boyd, T.A., Theoretical and Experimental Study of a New Class of Reflectionless Filter, IEEE Trans. on Microwave Theory and Tech., Vol. 59, Nr. 5, May 2011, pp. 1214-1220

[98] Morgan, A.M., Reflectionless Filter, Artech House, Boston 2017

[99] MAXIM, Datenblatt zum MAX4003, 0.1 - 2.5 GHz Leistungsdetektor, 19-2620; Rev 1, 03.2003

[100] Natarajan, R., Gupta, A., Shah, S., Design of Via-Shielded Microstrip-to-Stripline Transitions for Mode Purity at Millimeter-Wave Frequencies,? IEEE MTT?S International Microwave Symposium, 2016

[101] Niknejad, A.M. und Meyer, R.G., Analysis, Design and Optimization of Spiral Inductors and Transformers for Si RF IC's, IEEE Jour. of Solid-State Circuits, Vol. 33, Nr. 10, Oct. 1998, pp. 1470-1481

[102] LVDS-Grundlagen http://www.national.com/analog/interface/lvds_owners_manual

[103] Parisi, S. J., 180° Lumped Element Hybrid, IEEE Microw. Theory Tech. - Sym. Digest, 1989, pp. 1243-1246

[104] Peters, N., Schmitz, Th., Sadeghfam, A., Heuermann, H. Concept of Balanced Antennas with Load-Invariant Base Impedance Using a Two Element LC-Coupler, Proceedings of the European Microwave Association, 2005

[105] Pettenpaul, E., Kapusta, H., Weisberger, A., Mampe, H., Luginsland, J., Wolff, I., CAD Models of Lumped Elements on GaAs up to 18 GHz, IEEE Trans. on Microwave Theory and Tech., Vol. 36, Feb. 1988, pp. 294-304

[106] Philippow, E., Grundlagen der Elektrotechnik, VEB Verlag Technik, Berlin, 1988

[107] Pozar, D., Microwave Engineering, 4th Ed., Wiley, 2011

[108] Pucel, R.A., Design Considerations for Monolithic Microwave Circuits, IEEE Trans. on Microwave Theory and Tech., Vol. 29, Nr. 6, June 1981, pp. 513-534

[109] Raicu, D., Multiterminal Distributed Resistors as Microwave Attenuators, IEEE Trans. Microwave Theory Tech., vol. 42, No. 7, July 1994, pp. 1140-1148

[110] Rhodes, J.D., Theory of Electrical Filters, Wiley Verlag, New York, 1985

[111] Riley, Copeland and Kwasniewski, Digital PLL Frequency Synthesizers, 1983 Delta-Sigma Modulation in Fractional-N Frequency Synthesis, IEEE Journal of Solid State Circuits, Vol. 28, No. 5, May 1993

[112] Root, D.E., Verspecht, J., Horn, J., Marcu, M., X-Parameters, Cambridge University Press, Cambridge, 2013

[113] Rohde, U., Newkirk, D., RF/Microwave Circuit Design for Wireless Applications, Wiley Verlag, New York, 2000

[114] Rohde, U.L., Digital PLL Frequency Synthesizers: Theorie and Design, Prentice Hall, 1982

[115] Rohde, U.L., Microwave and Wireless Synthesizers, Wiley-Interscience, 1997

[116] Rohde & Schwarz, Applikationssoftware R&S ZVR-K11, (http://www.rohde-schwarz.com/)

[117] Rosenberger Hochfrequenztechnik, (http://www.rosenberger.de)

[118] Rosenberger Hochfrequenztechnik, Präzise Streuparametermessungen auf planaren Schaltungen, Applikationsschrift TI010, Tittmoning, 1996

[119] Rosenberger Hochfrequenztechnik, Streuparametermessungen in koaxialen Leitersystemen, Applikationsschrift TI020, Tittmoning, 1996

[120] Saal, R., Entenmann, W., Handbuch zum Filterentwurf, Hüthig-Verlag, Heidelberg, 1988

[121] Sadeghfam, A., Entwurfsverfahren für passive hochintegrierte Multimode-Schaltungen der Hochfrequenztechnik in mehrlagigen Herstellungsprozessen, Shaker Verlag, Aachen, 2008

[122] Sadeghfam, A., Heuermann, H., Novel Balanced Inductor for Compact Differential Systems, European Microwave Conf., Amsterdam, Oct. 2004, pp. 709-712

[123] Sadeghfam, A., Heuermann, H., Electrically tunable bandpass filter with integrated carrier suppression for UHF RFID systems, European Microwave Conf., Amsterdam, Oct. 2008

[124] Sadeghfam, A., Heuermann, H., On the Design of Multimode Integrated Circuits in Multilayered Processes, European Microwave Conf., Oct. 2007, pp. 516-519

[125] Schiek, B., Grundlagen der Hochfrequenz-Messtechnik, Springer Verlag, Berlin Heidelberg, 1999

[126] Schiek, B., Rolfes, I., Siweris, H.-J., Noise in High-frequency Circuits and Oszillators, J. Willey ans Sons, New Jersey, 2006

[127] Schiel, J.-Ch., Tatu, S. O., Wu,K., Bosisio R. G., Six-Port Direct Digital Receiver (SPDR) and Standard Direct Receiver (SDR) Results for QPSK Modulation at High Speeds, MTT-S International Microwave Symposium, Philadelphia, 2003, pp. 931-935

[128] Schmelz, Ch., Heuermann, H., Heide, P., Experimentelle Untersuchungen zur Anregung parasitärer Moden auf Koplanarleitungen bis 110 GHz, Kleinheubacher Berichte 1998, Bd. 41

[129] Schneider, M.V., Microstrip Lines for Microwave Integrated Circuits, Bell Syst. Tech. J. 48, 1969, pp. 1421-1444

[130] Schneider, J., Entwicklung eines homodynen Netzwerkanalysators für den Mikrowellenbereich 26,5-40 GHz, Dissertationsschrift, Institut für Hoch- und Höchstfrequenztechnik, Ruhr-Universität Bochum, 1987

[131] Simonyi, K., Theoretische Elektrotechnik, VEB Deutscher Verlag der Wissenschaften, Berlin, 1989

[132] Springer, A., Maurer, L., Weigel, R., RF System Concepts for Highly Integrated RFICs for W-CDMA Mobile Radio Terminals, IEEE Trans. on Microw. Theory and Tech., Vol. 50, No. 1, Jan. 2002

[133] Link zum Freeware-Download der RF-Synthese-Software SynRF:

`http://hhft.de/index.php?page=competences&subpage=syn_rf`

[134] Stengel, B., Thompson, B., Neutralized Differential Amplifiers using Mixed-Mode S-Parameters, IEEE MTT-S Digest, 2003

[135] Stolle, R., Heuermann, H., Schiek, B., Auswertemethoden zur Präzisions-Entfernungsmessung mit FMCW-Systemen und deren Anwendung im Mikrowellenbereich, tm – Technisches Messen, Heft 2/95, R. Oldenbourg Verlag, München, Feb. 1995, pp. 66-73

[136] SERENADE, Ansoft Inc., Paterson, N.J., USA

[137] Swanson, D.G., Hoefer, W., Microwave Circuit Modeling Using Electromagnetic Field Simulation, Artech House-Verlag, New-York, 2004

[138] Tietze, U., Schenk, Ch., Gamm, E., Halbleiter-Schaltungstechnik, 16. überarbeitete und erweiterte Auflage, Springer Verlag, ISBN: 978-3-662-48553-8, Heidelberg 2019

[139] Tombak, A. u.a., Design of High-Order Switches for Multimode Applications on a Silicon-on-Insulator Technology, IEEE Transactions on Microwave Theory and Techniques, doi: 10.1109/TMTT.2013.2277989, 2013, pp. 3639-3649

[140] Unger, H.-G., Harth, W., Hochfrequenz-Halbleiterelektronik, S. Hirzel Verlag, Stuttgart, 1972

[141] Unger, H.-G., Schultz, W., Elektronische Bauelemente und Netzwerke I, Vieweg Verlag, Braunschweig

[142] Unger, H.-G., Elektromagnetische Wellen auf Leitungen, Hüthig Verlag, Heidelberg

[143] van der Heijden, M. P., Spirito, M., de Vreede, L. C. N., van Straten, F., Burghartz, J. N., A 2 GHz High-Gain Differential InGaP HBT Driver Amplifier Matched for High IP3, IEEE MTT-S Digest, 2003

[144] Vincent, B.T., Microwave Transistor Amplifier Design, IEEE Microwave Symp. Dig., 1965, pp. 81-85

[145] Voges, E., Hochfrequenztechnik, Hüthig Verlag, Heidelberg, 2003

[146] Wadell, B. C., Transmission Line Design Handbook, Artech House, Boston, 1991

[147] Wang, D. u.a., High Performance SOI RF Switches for Wireless Applications, 210th IEEE International Conference on Solid-State and Integrated Circuit Technology, doi: 10.1109/ICSICT.2010.5667307.2010, pp. 611-614

[148] Wheeler, H. A., Transmission-Line Properties of Parallel Strips Separated by a Dielectric Sheet, IEEE Trans. Microwave Theory Tech., MTT-13, 1965, pp. 112-185

[149] Wheeler, H. A., Transmission-Line Properties of a Strip on a Dielectric Sheet on a Plane, IEEE Trans. Microwave Theory Tech., MTT-25, 1977, pp. 631-647

[150] Williams, A. B., *Electronic Filter Design Handbook,* New York, McGraw-Hill

[151] Wiltron, Microstrip Measurements with the Wiltron 360 Vector Network Analyzer, Application Note AN360-7, Jan. 1990

[152] Yakebe, T., Kinoshita, M., Yabe, H., Complete Calibration of a Six-Port Reflectometer with One Sliding Load and One Short, IEEE Trans. Microwave Theory Tech., vol. 42, Nov. 1994, pp. 2035-2039

[153] Zorzy, J., Skin-Effect Corrections in Immittance and Scattering Coefficient Standards Employing Precision Air-Dielectric Coaxial Lines, IEEE Trans. Instrum. Meas., IM-15, Dec. 1966, pp. 358–364

List of Frequently Used Formula Symbols and Abbreviations

Symbol	Meaning
A	Surface area
$[\mathbf{A}]$	A-chain matrix of two-port elements
A_c	Ripple in pass-band in dB
A_d	Insertion loss in dB
A_{ij}	Elements of the matrix $[\mathbf{A}]$
$[\mathbf{ABCD}]$	ABCD-Chain matrix of two-port elements
ADC	Analog-to-Digital Converter
AGC	Automatic Gain Control
AI	Artificial Intelligence
a_i, b_i	Complex wave sizes (mono-mode case)
a_i^+, b_i^+	Complex wave sizes of a common mode
a_i^-, b_i^-	Complex wave sizes of a push-pull mode
a_E	Insertion loss
a_T	Transducer loss
$\vec{a}$	Vector of complex wave sizes
B	Susceptance, reactive conductance
$[\mathbf{B}]$	Inverse chain matrix of two-port elements
B_{ij}	Elements of the matrix $[\mathbf{B}]$
BP	Abbreviation for Band-Pass filter
b_0	Complex generator wave
C	Capacitor, capacitance
C'	Capacitor, capacitance per unit length
$C2C$	Chip to Chip (connection)
CPW	Coplanar waveguide
CPWG	Coplanar waveguide with ground plane
CW	Continuous Wave
c_0	Speed of light
DM	Dual-Mode
DSB	Double Side Band
DSP	Digital Signal Processor
DUT	Device Under Test
dB	Decibel
dBc	Decibel in reference to a carrier
dBm	Decibel in reference to 1 mW
det ()	Determinant of a matrix
E	Electric field strength
EC	Equivalent Circuit
ECD	Equivalent Circuit Diagram
EPC	Electronic Product Code
ESD	ElectroStatic Discharge
e	Base of natural logarithm
F	Noise figure; farad
$F(j\omega)$	General function in the frequency range
FR4	Flame Resistant class 4
FET	Field Effect Transistor
f	Frequency
f_c	Corner frequency
f_d	Cutoff frequency
f_n	Standardized frequency
f_r	Resonant frequency
f_s	Stop-band edge frequency
$f(t)$	General function in the time domain
G	Conductance; active conductance; gain
G'	Conductance per unit length
GaAs	Gallium Arsenide
G_{av}	Available gain
GNSS	Global Navigation Satellite Systems
GPS	Global Positioning System
GSM	Global System for Mobile Communications (2G)
H	Magnetic field strength; henry
$H_{(\omega)}$	System function
HAS	High Accuracy Service
HF	High Frequency

H. Heuermann, *Radio Frequency Technologies*,
https://doi.org/10.1007/978-3-658-50170-9

HP	Abbreviation for High-Pass filter
h	Planck's constant; height
h_t	Impulse response
I	Complex current
$[\mathbf{I}]$	Unit matrix
IC	Integrated Circuit (semiconductor)
IF	Intermediate Frequency
Im()	Imaginary operator
IMP	Institute of Microwave and Plasma technology, FH Aachen
i	Complex current related to Z_0
i, j	Running indices for matrix and vector elements, ranging from n=1 to the respective dimension
$i(t)$	Current in the time domain
J	Current density
k	Wave number; stability factor (k-factor); coupling factor of resonators
k_d	Damping constant in the filter design
L	Inductance
L'	Inductance per unit length
LBT	Listen Before Talk
LINC	LInear amplification with Nonlinear Components
LNA	Low Noise Amplifier
LO	Local Oscillator
LP	Low-Pass filter
LTCC	Low Temperature Co-fired Ceramic
LTE	Long Term Evolution (4G)
LVDS	Low Voltage Differential Signaling
ℓ	Cable length
lg ()	Logarithm on the base 10, common logaritm
ln ()	Logarithm on the base e, natural logaritm
M	Match
M	Magnetization
M_{ij}	M-parameters resp. Mixed mode parameters
$[\mathbf{M}]$	M-parameter matrix resp. Mixed-mode parameter matrix
MAG	Maximum Available power Gain
MM	Mixed-Mode
MMIC	Microwave Monlithically Integrated Circuit
MW	MicroWave
m	Mass
m_i	Complex value measured value of a measuring point
N	Unknown two-port standard
n	Filter order
O	Open (calibration standard)
OP	OPerational amplifier
P	Electrical Power
PCI	Peripheral Component Interconnect
PD	Phase Discriminator
PFD	Phase-Frequency Discriminator
PI	Model of coil with 5 elements
PISK	Model of coil with 9 elements
P_w	Active electrical Power
PWM	Pulse width modulation
$[\mathbf{P}]$	Propagation matrix
p	Electrical length in degrees
Q	Quality of components and resonators, reactive power
Q_0	Unloaded quality factor of a resonator
Q_L	Loaded quality of a resonator
R	Effective resistance
R'	Effective resistance per unit length
Re()	Real part operator
RF	Radio Frequency
RFID	Radio Frequency IDentification

RX	Abbreviation for the receiving branch
Rg ()	Rank of a matrix
r_x	Reflection parameters of single-port and lumped elements
S	Short (calibration standard)
S_{ij}	Scattering parameter
S	Poynting vector, power density
$[\mathbf{S}]$	Scattering parameter matrix
$[\mathbf{S}]^{-1}$	Inverse scattering parameter matrix
$[\mathbf{S}]^{T}$	Transposed (scatter) matrix
$[\mathbf{S}]^{*}$	Conjugated complex (S) matrix
SAR	Search And Rescue
SC	SemiConductor
SFCW	Stepped-Frequency Continuous-Wave
Si	Silicon
$\mathrm{Si}_{(x)}$	Sine integral function
SMD	Surface Mounted Device
SNA	Scalar Network Analyzer
SOI	Silicon On Insulator
SPDT	Single Pole Double Thru
SPST	Single Pole Single Thru
SL	Stripline
SSB	Single Side Band
spur ()	Trace of a matrix
s	Second
T, T	Time interval; Through connection (Thru); Temperature
T_e	Effective noise temperature
TID	Tag IDentifier
T_n	Chebychev filter constants
TX	Abbreviation transmission branch
t	Time variable, transformation ratio of an impedance transformer thickness
UHF	Ultra High Frequency
U	Complex voltage
u	Complex voltage related to Z_0, normalized complex voltage
$u(t)$	Voltage in the time domain
ü	Transformation ratio of the currents and voltages of a transformer
UMTS	Universal Mobile Telecommunications System (3G)
V	Volume
VCO	Voltage Controlled Oscillator
VNA	Vectorial measuring Network Analyzer
v	Propagation speed
v_{gr}	Group speed
v_p	Power speed
v_{ph}	Phase velocity
$W_{e,m}$	Electrical or magnetic field energy
w	Width
X	Reactance
$[\mathbf{X}]$	Scattering parameter matrix for non-linear components
Y	Admittance, apparent conductance
$[\mathbf{Y}]$	Y-conductance matrix of two-port elements
y_x	Admittance parameters of single-port and lumped elements
Z	Impedance
Z_0	Characteristic impedance, reference impedance (usually $50\,\Omega$)
Z_E, Z_e	Input resistance or impedance
Z_{even}, Z_+	Common mode impedance (usually $25\,\Omega$)
Z_f	Field wave impedance of the free space ($377\,\Omega$)
Z_L	Characteristic impedance of a transmission line (usually $50\,\Omega$)
Z_{odd}, Z_-	Push-pull impedance (usually $100\,\Omega$)
$[\mathbf{Z}]$	Z-resistance matrix of two-port elements

z_x	Impedance parameters of single-port and lumped elements
α, β	Coefficients with alternating meaning
α	Attenuation constant of a transmission line
α_R	Attenuation due to radiative losses
α_G	Attenuation due to dielectric
β	Phase constant of a transmission line
γ	Complex propagation constant of a transmission line
$\gamma\ell$	Complex electrical length
$\Delta()$	Determinant of a matrix
δ	Penetration depth, skin depth; directivity of a coupler
$\delta(\omega)$	Dirac's delta function as a function of frequency
ϵ	Dielectric constant
ϵ_0, ϵ_r	Absolute and relative dielectric constant
η	Efficiency
κ	(Specific) conductivity; coupling transfer factor of a coupler
λ	Wavelength
λ_0	Wavelength in free space
μ	Permeability constant
μ_0	Permeability of free space
μ_r	Relative permeability
μ_n, μ_p	Mobility of electrons or holes
ν	Isolation of a coupler
ρ	Specific resistance; reflection value; self-calibration value
σ	Specific conductivity
τ	Runtime variable; pass-band factor of a coupler
φ	Phase
Ω	Circular frequency, frequency, normalized frequency
Ω_i	Fixed circular frequencies
ω	Circular frequency, frequency

Sachwortverzeichnis

H. Heuermann, *Radio Frequency Technologies*,
https://doi.org/10.1007/978-3-658-50170-9

The manufacturer's authorised representative in the EU is Springer Nature Customer Service Centre GmbH, Europaplatz 3, 69115 Heidelberg, Germany. If you have any concerns regarding our products, please contact ProductSafety@springernature.com

Printed and bound by CPI Group (UK) Ltd, Croydon, CR0 4YY
03/08/2026
02184507-0001